INTEGRATED INTELLIGENT SYSTEMS FOR ENGINEERING DESIGN

Frontiers in Artificial Intelligence and Applications

Volume 149

Published in the subseries
Knowledge-Based Intelligent Engineering Systems
Editors: L.C. Jain and R.J. Howlett

Recently published in KBIES:

Vol. 132. K. Nakamatsu and J.M. Abe (Eds.), Advances in Logic Based Intelligent Systems – Selected Papers of LAPTEC 2005
Vol. 115. G.E. Phillips-Wren and L.C. Jain (Eds.), Intelligent Decision Support Systems in Agent-Mediated Environments
Vol. 104. A. Abraham, M. Köppen and K. Franke (Eds.), Design and Application of Hybrid Intelligent Systems
Vol. 102. C. Turchetti, Stochastic Models of Neural Networks
Vol. 87. A. Abraham et al. (Eds.), Soft Computing Systems – Design, Management and Applications
Vol. 86. R.S.T. Lee and J.H.K. Liu, Invariant Object Recognition based on Elastic Graph Matching – Theory and Applications
Vol. 83. V. Loia (Ed.), Soft Computing Agents – A New Perspective for Dynamic Information Systems
Vol. 82. E. Damiani et al. (Eds.), Knowledge-Based Intelligent Information Engineering Systems and Allied Technologies – KES 2002
Vol. 79. H. Motoda (Ed.), Active Mining – New Directions of Data Mining
Vol. 72. A. Namatame et al. (Eds.), Agent-Based Approaches in Economic and Social Complex Systems
Vol. 69. N. Baba et al. (Eds.), Knowledge-Based Intelligent Information Engineering Systems and Allied Technologies – KES'2001

Recently published in FAIA:

Vol. 148. K. Kersting, An Inductive Logic Programming Approach to Statistical Relational Learning
Vol. 147. H. Fujita and M. Mejri (Eds.), New Trends in Software Methodologies, Tools and Techniques – Proceedings of the fifth SoMeT_06
Vol. 146. M. Polit et al. (Eds.), Artificial Intelligence Research and Development
Vol. 145. A.J. Knobbe, Multi-Relational Data Mining

ISSN 0922-6389

Integrated Intelligent Systems for Engineering Design

Edited by

Xuan F. Zha

University of Maryland and NIST, USA

and

R.J. Howlett

University of Brighton, UK

IOS Press

Amsterdam • Berlin • Oxford • Tokyo • Washington, DC

ISBN 1-58603-675-0
Library of Congress Control Number: 2006933008

Publisher
IOS Press
Nieuwe Hemweg 6B
1013 BG Amsterdam
Netherlands
fax: +31 20 687 0019
e-mail: order@iospress.nl

Distributor in the UK and Ireland
Gazelle Books Services Ltd.
White Cross Mills
Hightown
Lancaster LA1 4XS
United Kingdom
fax: +44 1524 63232
e-mail: sales@gazellebooks.co.uk

Distributor in the USA and Canada
IOS Press, Inc.
4502 Rachael Manor Drive
Fairfax, VA 22032
USA
fax: +1 703 323 3668
e-mail: iosbooks@iospress.com

PRINTED IN THE NETHERLANDS

Integrated Intelligent Systems for Engineering Design
X.F. Zha and R.J. Howlett (Eds.)
IOS Press, 2006

Preface

Background

With the ever increasing complexity of products and customer demands, companies are adopting new strategies to meet the changing technological requirements, shorter product life cycles, and globalization of manufacturing operations. Product design requires more sophisticated procedures and processes and requires designers and engineers possessing different expertise, knowledge and experience to work together. To address these challenges, techniques based on artificial intelligence (AI) are increasingly being used to improve effectiveness and efficiency in the product-design life cycle. Intelligent systems can be beneficially applied to many aspects of design and also design-related tasks at different stages; for example, identifying customer demands and requirements, design and planning, production, delivery, marketing and customer services, etc. Individual intelligent paradigms (such as fuzzy logic, neural network, genetic algorithm, case-based reasoning, and especially expert systems) have been applied to specific stages of the design process (product planning, conceptual design, detailed design). However, increasingly, hybrid solutions that integrate multiple individual intelligent techniques are required to solve complex design problems. The integrated intelligent environment can provide various types of information and knowledge for supporting rapid and intelligent decision-making throughout the entire design process. This is in line with the evolutionary trends of the product design process, from the traditional CAD systems into the knowledge-based engineering and integrated intelligent design systems through a combination of concurrent engineering, collaborative engineering and integrated intelligent techniques.

In recent years, with the advancement of artificial intelligence and information science and technology, there has been a resurgence of work in combining individual intelligent paradigms (knowledge-based systems, fuzzy logic, neural networks, genetic algorithms, case-based reasoning, machine learning and knowledge discovery, data mining algorithms, intelligent agents, soft computing, user intelligent interfaces, etc.) into integrated intelligent systems to solve complex problems. Hybridization of different intelligent systems is an innovative approach to constructing computationally intelligent systems consisting of artificial neural networks, fuzzy inference systems, approximate reasoning and derivative-free optimization methods such as evolutionary computation and so on. The integration of different learning and adaptation techniques, to overcome individual limitations and achieve synergetic effects through hybridization or fusion of these techniques, has contributed to a large number of new intelligent system designs. Hybrid intelligent systems are becoming a very important problem-solving methodology affecting researchers and practitioners in areas ranging from science, technology, business and commerce. Integrated intelligent systems are gaining better acceptance in engineering design. The driving force behind this is that integrated intelligence and distributed 3C (collaboration, cooperation, and coordination) allows the capture of human knowledge and the application of it so as to achieve high-quality designs/products. Further motivation arises from steady advances in individual and hybrid intelligent-systems techniques, and the widespread availability of computing resources and communications capability through intranets and the web.

There is a need for an edited collection of articles to reflect emerging integrated intelligent technologies and their applications in engineering design. The great breadth and expanding significance of AI and integrated intelligent systems (IIS) fields on the international scene requires a major reference work for an adequately substantive treatment of the subject. It is intended that this work will fulfill this need.

The Objective of the Book

This book aims to describe recent findings and emerging techniques that use intelligent systems (particularly integrated and hybrid paradigms) in engineering design, and examples of applications. The goal is to take a snapshot of progress relating to research into systems for supporting design and to disseminate the way in which recent developments in integrated, knowledge-intensive, and computational AI techniques can improve and enhance such support. The selected articles provide an integrated, holistic perspective on this complex set of challenges and provide rigorous research results. The focus of this book is on the integrated intelligent methodologies, frameworks and systems for supporting engineering design activities. The subject pushes the boundaries of the traditional topic of engineering design into new areas.

The Target Audience of the Book

We intend this book to be of interest to researchers, graduate students and practicing engineers involved in engineering design and applications using integrated intelligent techniques. In addition, managers and others can use it to obtain an overview of the subject, and gain a view about the applicability of this technology to their business. As AI and intelligent systems technologies are fast evolving, we certainly hope that this book can serve as a useful insight to the readers on the state-of-the-art applications and developments of such techniques at the time of compilation.

The Organization of the Book

The chapters provide an integrated, holistic perspective on the complex set of challenges, combined with practical experiences of leading experts in industry. Some of the chapters provide rigorous research results, while others are in-depth reports from the field. All chapters have been rigorously reviewed and carefully edited. There is a logical flow through this book, starting with intelligence foundations, emerging intelligent techniques, frameworks, systems and tools then continuing integrated and hybrid intelligent systems followed by their applications for engineering design. The treatment of the subject in the book can be described as:

1) Examines emerging technologies and recent research results on AI and integrated intelligent systems (IIS) in engineering design, including integrated intelligent systems, soft computing, distributed artificial intelligence (DAI), computer-integrated information systems (CIIS), etc.
2) Introduces new knowledge-intensive problem-solving strategies and their implementations based on AI and integrated intelligent systems techniques.

3) Presents theoretical fundamental principles and implementation technologies as well as engineering design applications and case studies, including, for example, electro-mechanical assemblies and systems, process control system, embedded and mechatronic systems design.

This book consists of 20 chapters organized into three thematic sections. An overview of each section and a brief description of the component chapters are presented here.

Part I: *Intelligence Foundations for Engineering Design.* This section, consisting of Chapters 1 to 5, provides the theoretical foundations of specific AI and IIS-based technologies for engineering design, including the principles of directed mutations for evolutionary algorithms, fuzzy logic and many valued logic, swarm intelligence, constraint satisfaction problem, fuzzy set and logic, fuzzy linear programming, Bayesian model, decision tree, uncertainty, etc.

Chapter 1, by Stefan Berlik and Bernd Reusch, introduces directed mutation as well as different operators in one single place. Their characteristics such as a multivariate skew distribution as a mutation operator in a covariance matrix adaptation algorithm are presented. An application scenario and experimental results solving a real world optimization task in this scenario are presented to show how evolutionary algorithms and directed mutation can be applied in engineering design.

Chapter 2, by Kalle Saastamoinen, studies the properties and usability of basic many valued-structures known as t-norms, t-conorms, implications and equivalences in comparison tasks. It shows how these measures can be aggregated with generalized mean and what kind of measures for comparison can be achieved from this procedure.

Chapter 3, by Arun Khosla, Shakti Kumar, K. Aggarwal, and Jagatpreet Singh, reports a swarm intelligence (SI) technique, Particle Swarm Optimization (PSO), which is a robust stochastic evolutionary computation engine. This is emerging as an innovative and powerful computational metaphor for solving complex problems in design, optimization, control, management, business and finance. The focus of this chapter is to present the use of the PSO algorithm for building optimal fuzzy models from the available data in design of the rapid Nickel-Cadmium (Ni-Cd) battery charger.

Chapter 4, by Arijit Bhattacharya and Pandian Vasant, outlines an intelligent fuzzy linear programming (FLP) method that uses a flexible logistic membership function (MF) to determine fuzziness patterns at disparate level of satisfaction for theory of constraints (TOC) based product-mix design problems. The fuzzy-sensitivity of the decision has been focused for a bottle-neck-free, optimal product-mix solution of TOC problem.

Chapter 5, by Vitaly Schetinin, Jonathan Fieldsend, Derek Partridge, Wojtek, Krzanowski, Richard Everson, Trevor Bailey, and Adolfo Hernandez, proposes a new approach to decision trees (DTs) for the Bayesian Markov Chain Monte Carlo technique to estimate uncertainty of decisions in safety-critical engineering applications. It also proposes a new procedure of selecting a single DT and describes an application scenario.

Part II: *Techniques, Frameworks, Tools and Standards*. This section, containing Chapters 6 to 11, explores techniques, models and frameworks, both current and emerging, and potential architectures for intelligent integrated engineering design.

Chapter 6, by Xiang Li, Junhong Zhou, and WenFeng Lu, presents a set of customer requirement discovery methodologies to achieve broad and complex market studies for new products. The proposed approach uses data mining and text mining technologies to discover customer multi-preference and corresponding customer motivation. A prototype system that allows for on-line customer feedback collection, digitization of the language feedbacks, numerical descriptions of customer preferences and customer motivation of a product is developed to demonstrate the feasibility of the proposed methodologies. It is shown that the proposed work could significantly shorten the survey and analysis time for customer preference and is thus expected to help companies to reduce design cycle time for new product design.

Chapter 7, by Paulo Gomes, Nuno Seco, Francisco Pereira, Paulo Paiva, Paulo Carreiro, José Ferreira and Carlos Bento, introduces an approach to reusing the knowledge gathered in the design phase of software development. An intelligent CASE tool using case-based reasoning (CBR) techniques and WordNet is developed to support software design and provide a framework for storage and reuse of design knowledge. This Chapter presents the approach to exploiting a knowledge base and several reasoning mechanisms that reuse the stored knowledge.

Chapter 8, by W.D. Li, S.K. Ong, A.Y.C., Nee, L. Ding, and C.A. Mcmahon, proposes and develops three intelligent optimization methods, i.e., Genetic Algorithm (GA), Simulated Annealing (SA) and Tabu Search (TS). These are applied to the solution of intractable decision-making issues in process planning with complex machining constraints. These algorithms can determine the optimal or near-optimal allocation of machining resources and sequence of machining operations for a process plan simultaneously, and a fuzzy logic-based Analytical Hierarchical Process technique is applied to evaluate the satisfaction degree of the machining constraints for the process plan.

Chapter 9, by Andrew Feller, Teresa Wu, and Dan Shunk, reviews existing research and industry-based practices relating to collaborative product design (CPD) information systems. An information framework is proposed called the 'Virtual Environment for Product Development' (VE4PD) that is based on the integration of Web services and agent technologies to manage the CPD process. The VE4PD architecture is designed to support CPD functions such as design synchronization, timely notification, distributed control, role based security, support for distributed intelligent agents, and varying design rule standards. An implementation system including intelligent agents for design negotiation is also described that validates the application approach.

Chapter 10, by Zhu Fan, Mogens Andreasen, Jiachuan Wang, Erik Goodman, and Lars Hein, proposes an integrated evolutionary engineering design framework that integrates the chromosome model in the domain theory, the evolutionary design, and human interaction. The evolvable chromosome model can help the designer to improve creativity in the design process, suggesting to them unconventional design concepts, and preventing them from looking for solutions only in a reduced solution space. The systematic analytical process to obtain a chromosome model followed by evolutionary design algorithms also helps the designer to have a complete view of design requirements and intentions. Human interaction is integrated to the framework due to the complex and dynamic nature of engineering design. It also helps the designer to accumulate design knowledge and form a design knowledge base. An example of the design of a vibration absorber for a typewriter demonstrates the feasibility of the technique.

Chapter 11, by Xuan F. Zha, proposes an integrated intelligent approach and a multi-agent framework for the evaluation of the assemblability and assembly sequence of electro-mechanical assemblies (EMAs). The proposed approach integrates the STEP (STandard for the Exchange of Product model data, officially ISO 10303) based assembly model and XML schema with a fuzzy analytic hierarchy process. Through integration with the STEP-based product modeling agent system, a CAD agent system and assembly planning agent system, the developed assembly evaluation agent system can effectively incorporate, exchange, and share concurrent engineering knowledge into the preliminary design process so as to provide users with suggestions for improving a design and also helping obtain better design ideas.

Part III: *Applications*. Chapters 12 to 20 in this section address the important issue of the ways that integrated intelligent systems are applied in engineering design. Case studies examine a wide variety of application areas including benchmarking and comparative analysis. The basic question is how accumulated data and expertise from engineering and business operations can be abstracted into useful knowledge, and how such knowledge can be applied to support engineering design. In this part of the book, chapters report case studies of innovative actual IIS-ED applications deploying specific AI-based technologies, such as logic rule-based systems, neural networks, fuzzy logic, cased-based reasoning, genetic algorithms, data mining algorithms, intelligent agents, and user intelligent interfaces, among others, and the integrations of these paradigms.

Chapter 12, by Sarawut Sujitjorn, Thanatchai Kulworawanichpong, Deacha Puangdownrecong and Kongpan Areerak, presents detailed step-by-step description of an intelligent search algorithm known as 'Adaptive Tabu Search' (ATS). The proof of its convergence and its performance evaluation are illustrated. The chapter demonstrates the effectiveness and usefulness of the ATS through various engineering applications and designs in the following fields: power system, identification, and control.

Chapter 13, by Shi-Shang Jang, David Shun-Hill Wong and Junghui Chen, addresses a technique known as 'experimental design' describes The design of new processes in modern competitive markets is mainly empirical because the short life-cycle does not allow the development of first-principle models. A systematic methodology know as 'experimental design', based on statistical data analysis and decision making, is used to optimise the number of experiments and direct process development. However, such methods are unsatisfactory when the number of design variables becomes very large and there are non-linearities in the input-output relationship. The new approach described in this chapter uses artificial neural networks as a meta-model, and a combination of random-search, fuzzy classification, and information theory as the design tool. An information free energy index is developed which balances the needs for resolving the uncertainty of the model and the relevance to finding the optimal design. The procedure involves iterative steps of meta-model construction, designing new experiments using meta-model and actual execution of designed experiments. The effectiveness of this approach is benchmarked using a simple optimization problem. Three industrial examples are presented to illustrate the applicability of the method to a variety of design problem.

Chapter 14, by Miki Fukunari and Charles J. Malmborg, proposes computationally efficient cycle time models for Autonomous Vehicle Storage and Retrieval System that use scalable computational procedures for large-scale design conceptualization. Simulation based validation studies suggest that the models produce high accuracy. The procedure is demonstrated for over 4,000 scenarios corresponding to enumeration of the design spaces for a range of sample problems.

Chapter 15, by Hirotaka Nakayama, Koichi Inoue and Yukihiro Yoshimori, discusses approximate optimization methods developed using computational intelligence, in which optimization is performed in parallel with the prediction of the form of the objective function. In this chapter, radial basis function networks (RBFN) are employed in predicting the form of objective function, and genetic algorithms (GA) used in searching for the optimal value of the predicted objective function. The effectiveness of the suggested method is shown through some numerical examples along with an application to seismic design in reinforcement of cable-stayed bridges.

Chapter 16, by Glenn Semmel, Steven R. Davis, Kurt W. Leucht, Daniel A. Rowe, Kevin E. Smith, Ladislau Bölöni, discusses a rule-based telemetry agent used for Space Shuttle ground processing. It presents the problem domain along with design and development considerations such as information modeling, knowledge capture, and the deployment of the product. It also presents ongoing work with other condition monitoring agents.

Chapter 17, by Mitun Bhattacharyya, Ashok Kumar, Magdy Bayoumi, proposes two techniques in two different sub areas of Wireless Sensor Networks (WSN) to reduce energy using learning methods. In the first technique, a watchdog/blackboard mechanism is introduced to reduce query transmissions, and a learning approach is used to determine the query pattern from the cluster head. Once the pattern is learnt, data are automatically sent back even in the absence of queries from the cluster head. In the second technique a learning agent method of profiling the residual energies of sensors within a cluster is proposed.

Chapter 18, Juan Vidal, Manuel Lama, and Alberto Bugarin, describes a knowledge-based system approach that combines problem-solving methods, workflow and machine learning technologies for dealing with the furniture estimate task. The system integrates product design in a workflow-oriented solution, and is built over a workflow management system that delegates the execution of activities to a problem-solving layer. An accurate estimation of the manufacturing cost of a custom furniture client order allows competitive prices, better profits adjustment, and increments the client portfolio too.

Chapter 19, by Martin Böhner, Hans Holm Frühauf and Gabriella Kókai, discusses the suitability of ant colony optimization (ACO) to an employment with blind adaptation of the directional characteristic of antenna array systems. In order to fulfill the hard real time constraints for beam forming in ranges of few milliseconds a very efficient hardware implementation for a highly parallel distributed logic is proposed in this chapter. The application requirements are given because of the high mobility of wireless subscribers in modern telecommunication networks. Such a dynamic alignment of the directional characteristic of a base-station antenna can be achieved with the help of a hardware-based Ant Colony Optimization methodology, by controlling the steering antenna array system parameters as digital phase shifts and amplitude adjustment. By means of extensive simulations it was confirmed that the suggested ACO fulfills the requirements regarding the highly dynamic changes of the environment. Based on these results a concept is presented to integrate the optimizing procedure as high-parallel digital circuit structure in a customized integrated circuit of a reconfigurable gate array.

Chapter 20, by Ankur Agarwal, Ravi Shankar, A. S. Pandya, presents the application of genetic algorithms to system level design flow to provide best effort solutions for two specific tasks, viz., performance tradeoff and task partitioning. Multiprocessor system on chip (MpSoC) platform has set a new innovative trend for the system-on-chip (SoC) design. Demanding Quality of Service (QOS) and performance metrics are leading to the adoption of a new design methodology for MpSoC. These will have to be built around highly scalable and reusable architectures that yield high speed at low cost and high energy efficiency for a variety of demanding applications. Designing such a system, in the presence of such aggressive QOS and Performance requirements, is an NP-complete problem.

Summary

There are over 48 coauthors of this notable work and they come from 19 countries. The chapters are clearly written, self-contained, readable and comprehensive with helpful guides including introduction, summary, extensive figures and examples and future trends in the domain with comprehensive reference lists. The discussions in these parts and chapters provide a wealth of practical ideas intended to foster innovation in thought and consequently, in the further development of technology. Together, they comprise a significant and uniquely comprehensive reference source for research workers, practitioners, computer scientists, academics, students, and others on the international scene for years to come.

Acknowledgment

The original intention of this edited book came from the discussion on the proposition to publish a Special Issue on Integrated and Hybrid Intelligent Systems in Product Design and Development in the International Journal of Knowledge-based and Intelligent Engineering Systems (KES) (Bob is Chief Editor of KES). The special issue was out in June 2005. We thought that based on the Special Issue in KES the selected quality papers could be extended into chapters, supplementing with additional chapters and forming the whole into a book that would fit well into the knowledge-based intelligent engineering systems collection of IOS Press. We then started working together in this direction. We were quite excited about this movement and immediately contacted the contributors and spread call for papers. The response from all the contributors was very positive and the proposal for a book was submitted to IOS for evaluation. The good news that the IOS Editorial Committee had approved the publishing of the book was conveyed to us in August 2005. We were overjoyed that the call-for-papers of the Special Issue and chapters had attracted favorable responses from many top researchers in this field. As the original intention was a peer reviewed Special Issue, and all the papers were either in the process of being reviewed or had already gone through the reviewing process, we informed the contributors that the quality of each paper, now each chapter, had followed the same standard of a rigorously peer-reviewed international journal.

We are most grateful to the kind cooperation of all the contributors who had promptly responded to all the questions and had followed our requests for additional information. We would also like to thank IOS for giving us this opportunity of publishing this book.

Xuan F. (William) Zha
Gaithersburg, Maryland

Robert J. (Bob) Howlett
Brighton, UK

Contents

Section I

Intelligence Foundations

Integrated Intelligent Systems for Engineering Design
X.F. Zha and R.J. Howlett (Eds.)
IOS Press, 2006

Foundations of Directed Mutation

Stefan Berlik and Bernd Reusch
Dortmund University, Dep. of Computer Science, 44221 Dortmund, Germany

Abstract. Directed mutation abandons the so-called *random mutation hypothesis* postulating mutations to occur at random, regardless of fitness consequences to the resulting offspring. By introducing skewness into the mutation operators, bigger portions of offspring can be created in the area of higher fitness with respect to the elder and thus promising directions of the evolution path can be favored. The aim of this work is to present the foundations of directed mutation as well as different operators in one single place. Their characteristics will be presented and their advantages and disadvantages are discussed. Furthermore, an application scenario will be presented that shows how evolutionary algorithm and directed mutation can be applied in engineering design. In addition, some experimental results solving a real world optimization task in this scenario are provided. Finally some first, preliminary results of a multivariate skew distribution as mutation operator in a covariance matrix adaptation algorithm will be presented.

Keywords. Directed mutation, mutation operator, Evolutionary Algorithm, Evolution Strategy, CMA-ES, skew-normal distribution, multivariate skew-normal distribution.

Introduction

Evolutionary Algorithms (EAs hereafter) are a set of stochastic optimization algorithms with the common feature of being inspired by biology, especially by those processes that allow populations of organisms to adapt to their surrounding environment. Namely these are genetic inheritance and survival of the fittest. Among the class of EAs, Evolution Strategies (ESs) are the most popular algorithms for solving continuous optimization problems, i.e. for optimizing a real-valued function f, called objective function, defined on a subspace $S \subseteq \mathbb{R}^d$ for some dimension d [1]. EAs work with sets of points of the search space in an iterative way: first some points of the search space are sampled randomly, then their fitness according to the objective function is evaluated, and last some of them are selected as basis for the next iteration. This loop is repeated until some stopping criterion is met. In ESs, one major component in the generation of new points is the to the so-called mutation operator. Given a point $\boldsymbol{x} \in \mathbb{R}^d$, also termed as parent, a new point or offspring is created by adding a random vector to this point. Usually a normally distributed random vector is used and the mutation thus reads

$$\boldsymbol{x} \mapsto \boldsymbol{x} + N(\boldsymbol{0}, \boldsymbol{C}) \qquad (1)$$

where $\boldsymbol{C}$ denotes the covariance matrix of the distribution.

Since the early work of Rechenberg [2] and Schwefel [3] the design of mutation operators turned out to be one of the most critical points in ESs. These early works relied on just one single mutation strength, i.e. step-size, for all problem dimensions (isotropic mutation) and were concerned mainly with determining the optimal step-size for a faster search. To put it in a more general light, the covariance matrix was considered to be the identity matrix. Soon Schwefel extended this approach and proposed to self-adapt one step-size per variable, i.e. to use a diagonal covariance matrix with positive terms. Consequently, as the most general case, he later suggested self-adapting of all parameters of the covariance matrix (correlated mutation). A more detailed review of the field's history is given e.g. by Bäck et al. [4].

However, all of the previously presented methods rely on normally distributed mutations, and relatively little effort has been put into examining different distributions as mutation operators. One such example is the so-called Fast Evolution Strategy by Yao [5], where a Cauchy distribution is proposed as mutation operator. Nevertheless, Rudolph [6] later proofed that the order of local convergence is identical to that of normal mutations. Just to exchange the mutation distribution seems in general to be a questionable idea. Our intention on the other hand is to introduce a different *mutation principle*, called *directed mutation*. We will abandon the random mutation hypothesis – a fundamental tenet postulating that mutations occur at random, regardless of fitness consequences to the resulting offspring. This seems to be justified by the fact that the ES knows its optimization history and is thus able to extrapolate the evolution path to some extent. Under the assumption of a locally similar objective function it is obviously reasonable to generate a bigger portion of offspring along the successful path.

We argue in this contribution that such a strategy can in fact improve the performance of ESs. In section 1 we therefore present directed mutation in detail, recapitulate several univariate skew distributions that meet the demands, and discuss their advantages and disadvantages. An application scenario showing how to apply evolutionary algorithm and directed mutation in engineering design and some experimental results on a real-world optimization are provided in section 2. We then present, in section 3, some recent results of a multivariate skew distribution within a covariance matrix adaptation framework. Finally, section 4 gives a conclusion and future research opportunities in the domain of directed mutation.

1. Directed Mutation Operators

To be able to do directed mutation we need to generate skew distributed random numbers whereby expected values unequal to zero are introduced. Obviously this means that the mutation operator is not compliant to standard ES any longer postulating an expected value of zero, i.e. remaining on average in the current position in search space. However, it has to be ensured that the expected value is convergent, thus forcing the mutation operator to continue mutating near by the current position. It should already be pointed out that convergence of the expectation in the symmetrical case only is not enough. Convergence has rather to be guaranteed for all degrees of skewness.

The first directed mutation operators proposed by Hildebrand [7] violate this demand. Based on his Ξ–distribution, his operators suffer from diverging expected values caused by increasing skewness, which can result in wide jumps in the search space. Of even greater interest is the variance. It can be seen as a measure of the mutation strength and is a strategy parameter in most ESs. Because of this it should not be modified by the skewness parameter. In the ideal case the variance is an invariant, independent of the skewness. At least convergence is necessary and a small spread between minimal and maximal value is desired to limit the impact of the skewness on the mutation strength. Again, the Ξ–distribution violates this demand. To overcome this, several alternative directed mutation operators have been developed, all with convergent moments. The first one was the naïve skew-normal mutation that is strongly related to the asymmetric mutation. A completely different approach follows the class of skew-normal mutation operators that will be presented afterwards.

1.1. Asymmetric Mutation

For his *asymmetric mutation* Hildebrand has chosen an additive approach [7]. The density function is defined in sections and made up of two parts, one function for the negative domain and another function for the positive domain. One of these functions always equals the standard normal density while the other one is an expanded normal density. To transform the whole function into a density again, its integral is then normalized to one.

1.1.1. Probability Density Function

The complete definition of the density of the Ξ_c–distribution splits into four cases and reads

$$\xi_c(x) = \begin{cases} \dfrac{\sqrt{2}}{\sqrt{\pi}\left(1+\sqrt{1-c}\right)} e^{-\frac{x^2}{2(1-c)}} & \text{for } c<0, x<0 \\ \dfrac{\sqrt{2}}{\sqrt{\pi}\left(1+\sqrt{1-c}\right)} e^{-\frac{1}{2}x^2} & \text{for } c<0, x\geq 0 \\ \dfrac{\sqrt{2}}{\sqrt{\pi}\left(1+\sqrt{1+c}\right)} e^{-\frac{1}{2}x^2} & \text{for } c\geq 0, x<0 \\ \dfrac{\sqrt{2}}{\sqrt{\pi}\left(1+\sqrt{1+c}\right)} e^{-\frac{x^2}{2(1+c)}} & \text{for } c\geq 0, x\geq 0. \end{cases} \tag{2}$$

For some values of c, ξ_c density functions are plotted in Figure 1. Note the fat tails which lead to diverging expectation and variance.

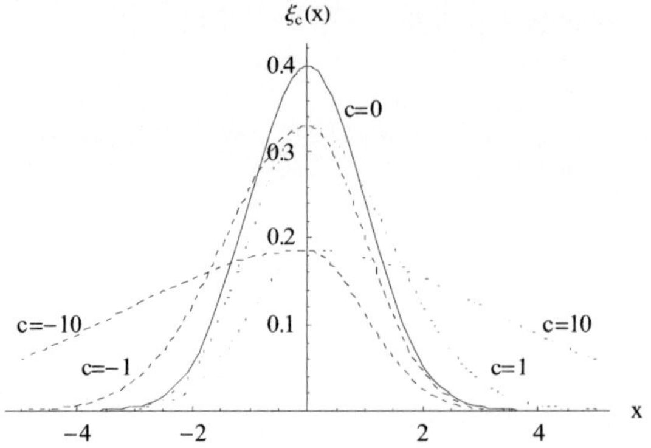

Figure 1. Density functions ξ_{-10}, ξ_{-1}, ξ_0, ξ_1, and ξ_{10}

1.1.2. Moments

Given below are the formulae of the expected value and the variance of a Ξ_c distributed random variable X.

$$\mathrm{E}(X) = \sqrt{\frac{2}{\pi}} \frac{c}{1+\sqrt{1+|c|}} \tag{3}$$

$$\mathrm{V}(X) = \left(2 - \sqrt{1+|c|} + |c| - \frac{2c^2}{\pi\left(1+\sqrt{1+|c|}\right)^2} \right) \tag{4}$$

As mentioned before, the expected value as well as the variance diverge, see Eqs. (5). Their graphs are depicted in Figure 2.

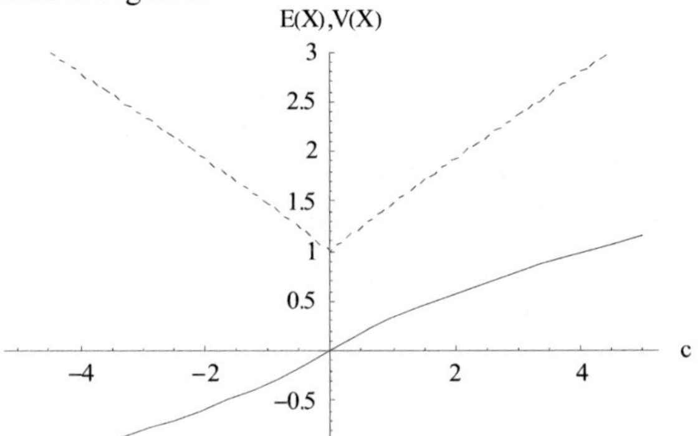

Figure 2. Expectation (solid) and variance (dashed) of a Ξ_c distributed random variable

$$\lim_{c\to-\infty}(\mathrm{E}(X)) = -\infty \quad \lim_{c\to\infty}(\mathrm{E}(X)) = \infty$$
$$\lim_{c\to-\infty}(\mathrm{V}(X)) = \infty \quad \lim_{c\to\infty}(\mathrm{V}(X)) = \infty \tag{5}$$

1.1.3. Random Variate Generation

Generating Ξ_c distributed random numbers is demanding and cumbersome. It is done using the inverse function of the Ξ_c–distribution. Random numbers thus are created by multiplying uniform distributed random numbers with the inverse distribution function. The latter is defined as:

$$\Xi_c(y) = \begin{cases} \sqrt{2(1-c)}\,\mathrm{inverf}\left(y\left(1+\frac{1}{\sqrt{1-c}}\right)-1\right) & \text{for } c<0,\ y<\frac{\sqrt{1-c}}{1+\sqrt{1-c}} \\ \sqrt{2}\,\mathrm{inverf}\left(y\left(1+\sqrt{1-c}\right)-\sqrt{1-c}\right) & \text{for } c<0,\ y\geq\frac{\sqrt{1-c}}{1+\sqrt{1-c}} \\ \sqrt{2}\,\mathrm{inverf}\left(y\left(1+\sqrt{1+c}\right)-1\right) & \text{for } c\geq 0,\ y<\frac{1}{1+\sqrt{1+c}} \\ \sqrt{2(1+c)}\,\mathrm{inverf}\left(y+\frac{y-1}{\sqrt{1+c}}\right) & \text{for } c\geq 0,\ y\geq\frac{1}{1+\sqrt{1+c}} \end{cases} \tag{6}$$

Note that there are two case differentiations, one calculation of the transcendent inverse error function, and several arithmetic operations necessary to generate a Ξ_c distributed random number.

1.2. Naïve Skew-Normal Mutation

The naïve skew-normal (NSN) distribution is built in a similar manner to the Ξ_c-distribution [8]. The main difference is to compress on half of the normal density instead of expanding it. This avoids fat tails (see Figure 3) and guarantees convergent expectation and variance, Eqs. (10).

1.2.1. Probability Density Function

A random variable Z is said to be naïve skew-normal with parameter λ, written Z ~ NSN(λ), if its probability density function is

$$f_{NSN}(z;\lambda) = \begin{cases} \sqrt{\frac{2}{\pi}} \frac{\sqrt{1-\lambda}}{\left(1+\sqrt{1-\lambda}\right)} e^{-\frac{1}{2}z^2} & \text{for } \lambda \le 0, z \le 0 \\ \sqrt{\frac{2}{\pi}} \frac{\sqrt{1-\lambda}}{\left(1+\sqrt{1-\lambda}\right)} e^{-\frac{(1-\lambda)z^2}{2}} & \text{for } \lambda \le 0, z > 0 \\ \sqrt{\frac{2}{\pi}} \frac{\sqrt{1+\lambda}}{\left(1+\sqrt{1+\lambda}\right)} e^{-\frac{(1+\lambda)z^2}{2}} & \text{for } \lambda > 0, z \le 0 \\ \sqrt{\frac{2}{\pi}} \frac{\sqrt{1+\lambda}}{\left(1+\sqrt{1+\lambda}\right)} e^{-\frac{1}{2}z^2} & \text{for } \lambda > 0, z > 0. \end{cases} \tag{7}$$

Graphs for several degrees of skewness of the NSN density are shown in Figure 3.

1.2.2. Moments

The formulae for the expected value and the variance of a NSN distributed random variable Z take the following form:

$$E(Z) = \sqrt{\frac{2}{\pi}} \frac{\lambda}{1+|\lambda|+\sqrt{1+|\lambda|}} \tag{8}$$

$$V(X) = \frac{4\left(\sqrt{1+|\lambda|}-1\right)+|\lambda|(\pi-2)+\pi\left(2-\sqrt{1+|\lambda|}\right)}{\pi\left(1+|\lambda|\right)}. \tag{9}$$

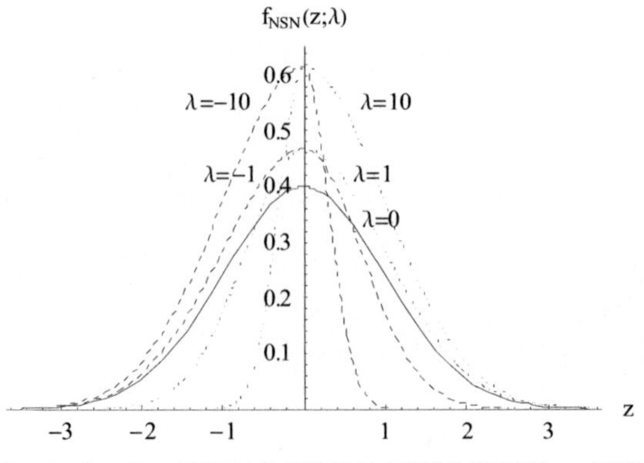

Figure 3. Density functions NSN(-10), NSN(-1), NSN(0),NSN(1), and NSN(10)

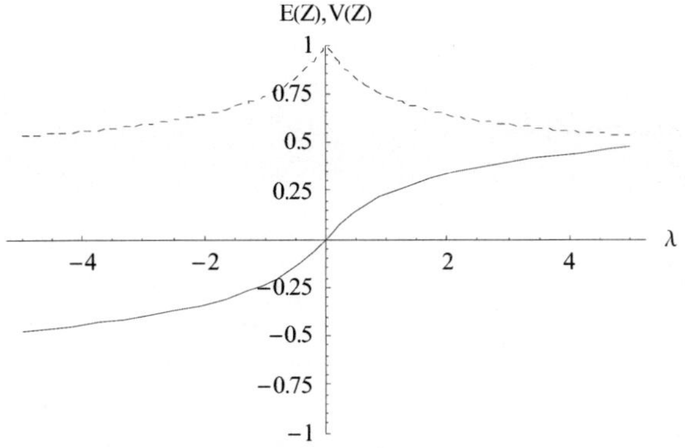

Figure 4. Expectation (solid) and variance (dashed) of a NSN(λ) distributed random variable

The limits are:

$$\lim_{\lambda\to-\infty}(\mathrm{E}(Z)) = -\sqrt{\frac{2}{\pi}},\quad \lim_{\lambda\to\infty}(\mathrm{E}(Z)) = \sqrt{\frac{2}{\pi}}$$
$$\lim_{\lambda\to-\infty}(\mathrm{V}(Z)) = \frac{(\pi-2)}{\pi},\quad \lim_{\lambda\to\infty}(\mathrm{V}(Z)) = \frac{(\pi-2)}{\pi} \tag{10}$$

Their graphs are depicted in Figure 4. One can see that the variance is convergent, but still spreads about 0.64. To make the variance an invariant, a linear transformation has to be applied to the NSN distributed random variable leading to the standardized NSN distribution.

1.2.3. Random Variate Generation

NSN distributed random variables can be generated using the method described in 1.1.3 with the appropriate inverse distribution given below.

$$\overline{\mathrm{F}}_{\mathrm{NSN}}(y;\lambda) = \begin{cases} \sqrt{2}\,\mathrm{inverf}\left(y\left(1+\frac{1}{\sqrt{1-\lambda}}\right)-1\right) & \text{for } \lambda \le 0,\ y \le \frac{\sqrt{1-\lambda}}{1+\sqrt{1-\lambda}} \\ \frac{\sqrt{2}}{\sqrt{1-\lambda}}\,\mathrm{inverf}\left(y\left(1+\sqrt{1-\lambda}\right)-\sqrt{1-\lambda}\right) & \text{for } \lambda \le 0,\ y > \frac{\sqrt{1-\lambda}}{1+\sqrt{1-\lambda}} \\ \frac{\sqrt{2}}{\sqrt{1+\lambda}}\,\mathrm{inverf}\left(y\left(1+\sqrt{1+\lambda}\right)-1\right) & \text{for } \lambda > 0,\ y \le \frac{1}{1+\sqrt{1+\lambda}} \\ \sqrt{2}\,\mathrm{inverf}\left(y\left(1+\frac{1}{\sqrt{1+\lambda}}\right)-\frac{1}{\sqrt{1+\lambda}}\right) & \text{for } \lambda > 0,\ y > \frac{1}{1+\sqrt{1+\lambda}} \end{cases} \tag{11}$$

1.3. Standardized Naïve Skew-Normal Mutation

Obviously the NSN distribution can be transformed into a version with invariant variance. The standardization term that has to be applied is

$$\sigma_{\mathrm{Std}}(\lambda) = \sqrt{\frac{\pi\left(1+|\lambda|\right)}{4\left(\sqrt{1+|\lambda|}-1\right)+|\lambda|(\pi-2)+\pi\left(2-\sqrt{1+|\lambda|}\right)}}\ . \tag{12}$$

For further details on the standardized naïve skew-normal mutation see [8].

1.4. Skew-Normal Mutation

The class of distributions used to build the following directed mutation operator is called skew-normal (SN) distribution and was introduced by Azzalini [9]. A detailed presentation of the SN distribution, some extensions, and a small historical review are given by Arnold and Beaver [10]. Some advantages of the skew-normal distribution are in short:

- Natural extension of the normal distribution to accommodate asymmetry
- Strict inclusion of the normal distribution
- Salient mathematical tractability
- Moment generating function has closed form
- Straightforward random variate generation
- Properties of the distribution have been studied extensively.

1.4.1. Probability Density Function

The SN density function is defined by

$$\mathrm{f}_{\mathrm{SN}}(z;\lambda) = 2\phi(z)\Phi(\lambda z) \tag{13}$$

where ϕ and Φ represents the probability density function and the cumulative distribution function of the standard normal density, respectively. λ is a real parameter that controls the skewness, where positive (negative) values indicate positive (negative) skewness. In the case $\lambda = 0$ the SN density gets back to the normal density (see Figure 5). With $Z \sim \mathrm{SN}(\lambda)$ we denote a random variable that has density (13).

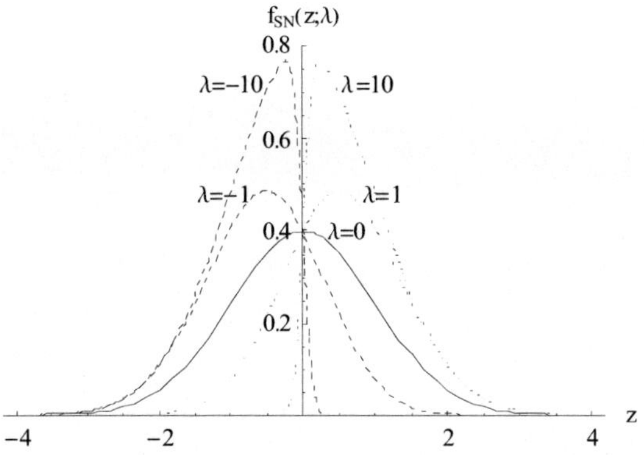

Figure 5. The density functions SN(-10), SN(-1), SN(0), SN(1), and SN(10)

1.4.2. Moments

The first four Moments are given by

$$\begin{aligned}
\mathrm{E}(Z) &= b\delta \\
\mathrm{V}(Z) &= 1-(b\delta)^2 \\
\gamma_1(Z) &= \frac{1}{2}(4-\pi)\,\mathrm{sign}(\lambda)\left(\frac{(E(Z))^2}{V(Z)}\right)^{3/2} \\
\gamma_2(Z) &= 2(\pi-3)\,\mathrm{sign}(\lambda)\left(\frac{(E(Z))^2}{V(Z)}\right)^{2},
\end{aligned} \tag{14}$$

where

$$b=\sqrt{\frac{2}{\pi}} \quad \text{and} \quad \delta=\frac{\lambda}{\sqrt{1+\lambda^2}}. \tag{15}$$

$\gamma_1(Z)$ and $\gamma_2(Z)$ denote the skewness and kurtosis. As desired, both expectation and variance converge. The limits are

$$\begin{aligned}
\lim_{\lambda\to\pm\infty}\left(\mathrm{E}(Z)\right) &= \mathrm{sign}(\lambda)\sqrt{\frac{2}{\pi}} \\
\lim_{\lambda\to\pm\infty}\left(\mathrm{V}(Z)\right) &= 1-\frac{2}{\pi}.
\end{aligned} \tag{16}$$

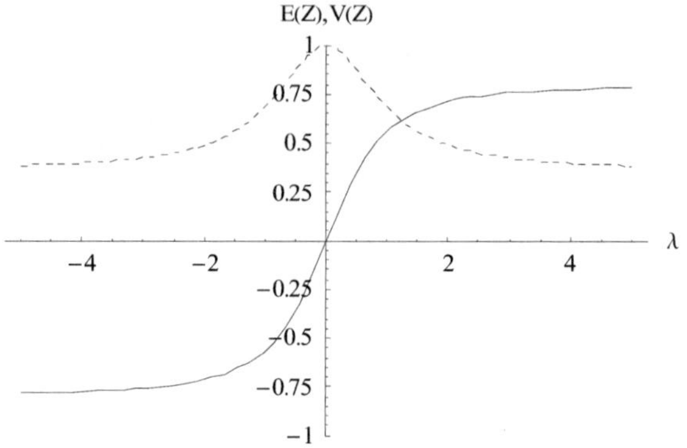

Figure 6. Expectation (solid) and variance (dashed) of a SN(λ) distributed random variable

Their graphs are depicted in Figure 6. One can see that the variance is convergent but still spreads. Like in the case of the variance of a NSN distributed random variable, the spread is about 0.64. Consequently, using a linear transformation the variance can again be made invariant, leading to the standardized SN distribution.

1.4.3. Random Variate Generation

Generation of SN distributed random numbers is straightforward and fast. A random variable Z with density (13) can be generated by virtue of its stochastic representation, Eq. (17). Therefore sample Y and W from ϕ and Φ', respectively. Then Z is defined to be equal to Y or $-$Y, conditionally on the event $\{W < \lambda Y\}$:

$$Z = \begin{cases} Y & \text{if } W < \lambda Y \\ -Y & \text{otherwise.} \end{cases} \tag{17}$$

Thus simply two standard normal random variables are needed to generate one SN distributed random variable.

1.5. Standardized Skew-Normal Mutation

As already mentioned, using a linear transformation the SN distribution can be changed to a version where the skewness does not influence the variance any longer. The variance then becomes an invariant [12]. This is achieved using the transformed random variable sZ with

$$s = \frac{1}{\sqrt{V(Z)}} = \frac{1}{\sqrt{1 - (b\delta)^2}} = \sqrt{\frac{\pi(1+\lambda^2)}{\pi + (\pi - 2)\lambda^2}}\,. \tag{18}$$

1.6. Comparison

The characteristics of the presented operators are summed up in Table 1. One sees that SN and SSN perform considerably better than the other mutation operators and that they are the sole variants given in closed form. While all but asymmetric mutation have convergent expectations and variances, these two are the only that also provide acceptable random variate generation procedures. Taking into account that during an optimization process a vast amount of random numbers has to be generated, this issue becomes very important – which is also reflected in the point *Usefulness*. Asymmetric mutation is unusable because of its diverging moments, whereas the head start of the SN and SSN mutations compared to the naïve versions originates from the random number generation and mathematical tractability. If an invariant variance is desired the SSN variant should be used, causing only a slight overhead compared to the SN mutation.

Table 1. Comparison of the mutation operators

	Asymmetric	Naïve skew-normal	Std. naïve skew-normal	Skew-normal	Std. skew-normal
Convergent expectation	-	+	+	+	+
Convergent variance	-	+	+	+	+
Invariant variance	-	-	+	-	+
Mathematical tractability	o	o	o	+	+
Given in closed form	-	-	-	+	+
Random variate generation	o	o	o	+	+
Usefulness	-	o	o	+	+

2. Application Scenario and Experimental Results

With the different mutation operators presented in the previous section, seven simple and well-known test functions have been examined in [12]. The essence of this investigation in short is that directed mutations by means of the SN distribution perform considerably better than the other variants and the classical mutation operators. We omit the presentation of these results here for space reasons and would rather like to focus on the more relevant case of solving a real world optimization problem in an engineering scenario [19].

The task of the optimization is to enhance the rotor geometry of a screw-type machine. The most common form of screw-type machines are rotary compressors, especially the helical twin screw-type. Meshing male and female screw-rotors rotate inside a housing in opposite directions and thereby trap air, reducing the volume of the air along the rotors to the air discharge point. Rotary screw-type compressors have low initial cost, compact size, low weight, and are easy to maintain.

A special topic during the construction of screw machines is to be seen in the design of the rotor geometry of an individual stage. One can differentiate here three-dimensional characteristics such as rotor length and wrap angle as well as two-dimensional characteristics such as rotor diameters, numbers of lobes and the lobe profile.

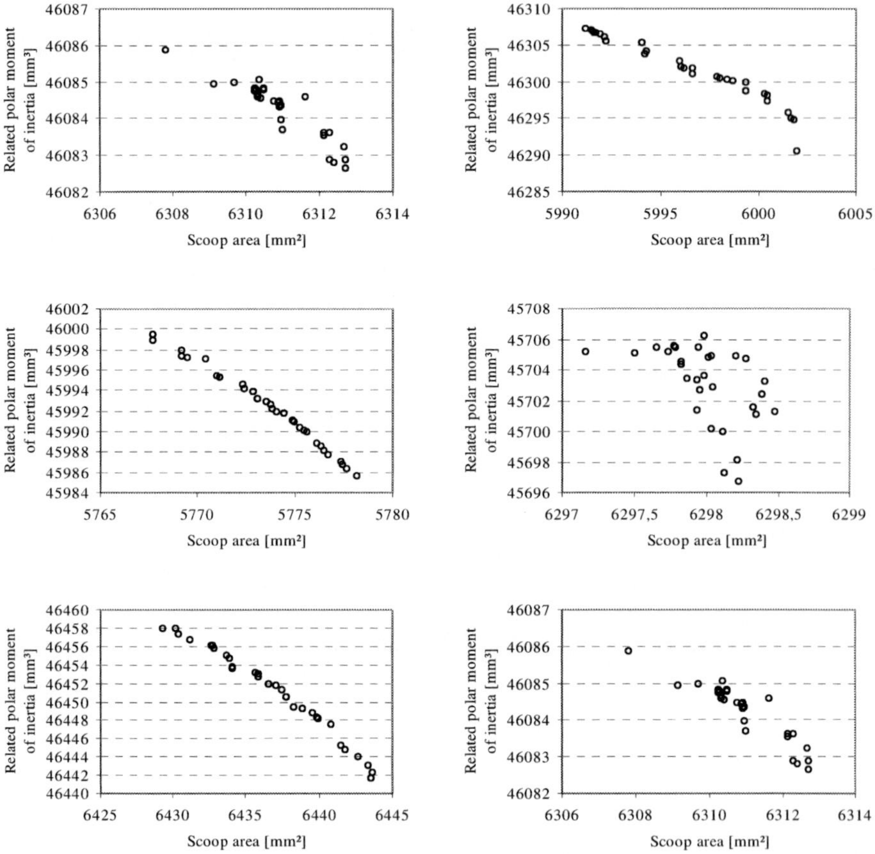

Figure 7. Final populations of the optimizations using directed mutation

At the beginning of the design process, there always stands the draft of a suitable two-dimensional front section, since this already significantly exerts influence on the operational behavior, the thermodynamic and mechanical characteristics, and the manufacturing. The meaning of the front sections results among other things from the operating behavior of the screw machine. A change of the lobe profile or the number of lobes affects on the one hand the contact line and by this the form and position of working chamber limiting clearances, on the other hand the size of the work space itself and the utilization of the construction volume. Both affect the quality of the thermodynamic processing.

The development of front sections was based in the past mainly on the combination of few geometrically simple curve sections by hand. Frequently used curve types are e.g. straight lines, circular arcs, involutes, cycloids or equidistant ones to cycloids. Here, splines were chosen because of their great flexibility and thus having the convenience to operate

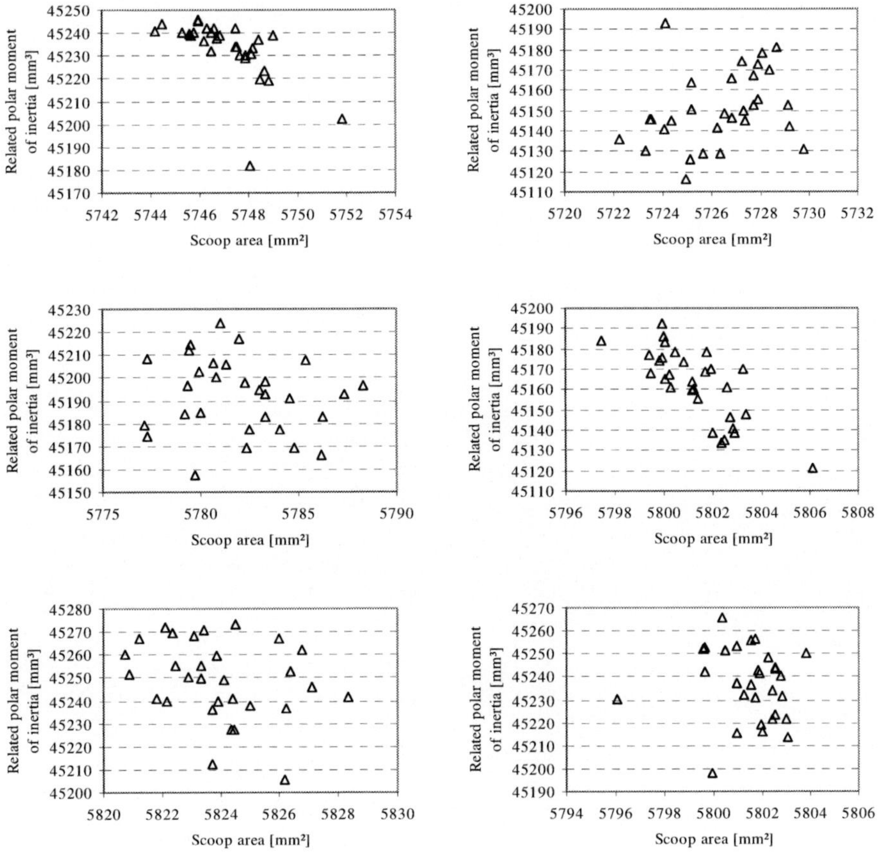

Figure 8. Final populations of the optimizations using simple-n mutation

with one single curve type only. In this example eight splines are used to describe a single lobe of the male profile, leading to a 32-dimensional optimization problem. The whole male rotor consists of four of these lobes; the female rotor with six lobes is calculated to fit the male rotor. Several constraints have to be fulfilled by every generated profile pair to be valid; in this case ten constraints are considered. While in a real-world optimization also several objectives are to be treated, for the sake of clarity in this example their number is limited. Just two important objectives are treated, the scoop area and the related polar moment of inertia: the first is a measure of the volume flow through the machine; the second can be seen as a simplified measure of the stiffness of the rotors. These are conflicting goals that obviously both have to be maximized. For technical details on screw-type machines and their optimization in general see [13]–[15]. More details on multi-objective optimization of screw-type machines are provided in [16] and an extensive

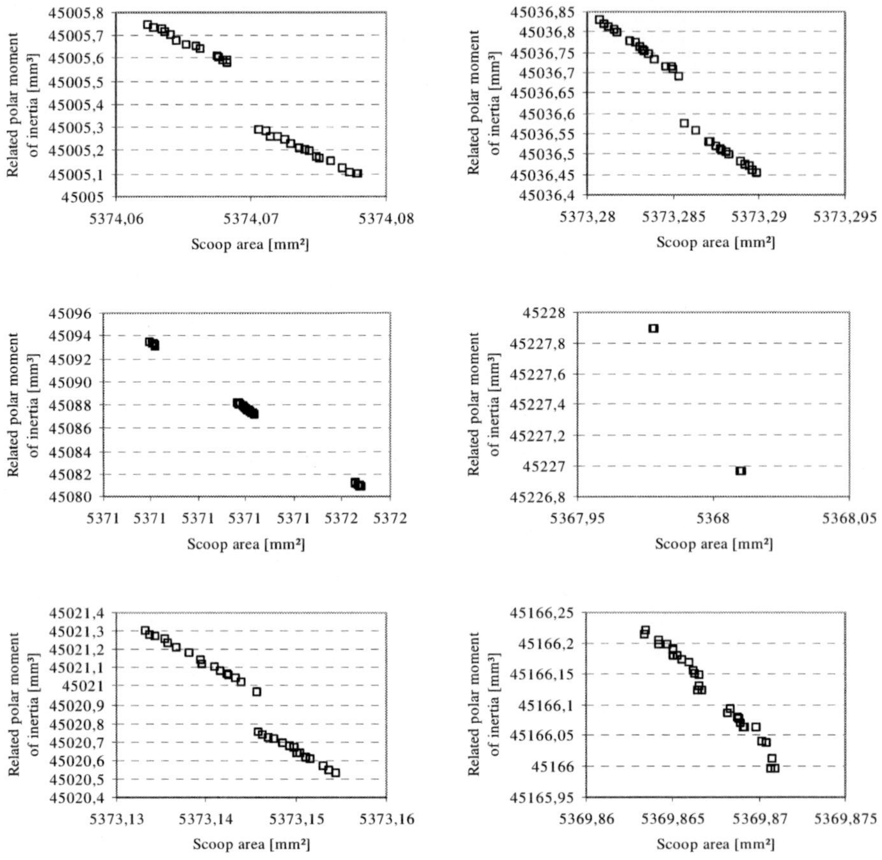

Figure 9. Final populations of the optimizations using simple-1 mutation

presentation of multi-objective optimization is given by Deb [17].

The experiments have been done using a NSGA-II like evolution strategy [18] with self-adaptive standard deviations. The (30,100)-ESs run 5000 generations. The same initial profile was used, initial mutation strength has been set to 10^{-5}, and skewness parameters have initial values of zero. Because of the complexity of the optimization task, each experiment has been carried out only six times. Three mutation operators have been investigated:

- Directed mutation, by virtue of the SN distribution
- Simple-n mutation, with a separate mutation strength for every dimension
- Simple-1 mutation, with only one mutation strength for all dimensions together.

The results of the runs with directed mutation are given in Figure 7. One can see that the populations form relatively sharp Pareto fronts and cover a good spectrum of the search space with respect to both criteria (cf. also the average of the standard deviations of the single runs for the both criteria given in [19]).

Figure 8 shows the results using simple-n mutation. Here the populations do not form Pareto fronts as sharp as it was the case with directed mutation. This means that there is a distinct fraction of inefficient individuals in the population. Also, the two criteria are not distributed equally (The standard deviations of the populations with respect to the second criterion are about 10 times larger.).

At last Figure 9 depicts the results for the simple-1 mutation. Note that there are two runs with collapsing populations. At the end of the other runs relatively sharp Pareto fronts emerged. It holds for all of them that they cover only a very small region of the search space.

To compare the results of the different mutation operators they have been compiled in Figure 10. It is apparent that directed mutation clearly outperforms the other two mutation strategies. All runs but one dominate all runs of the other strategies, i.e. they are better in both criteria. The directed mutation also shows the greatest diversity under the different runs and within them. For a more detailed analysis of these results and the optimization task itself see [19].

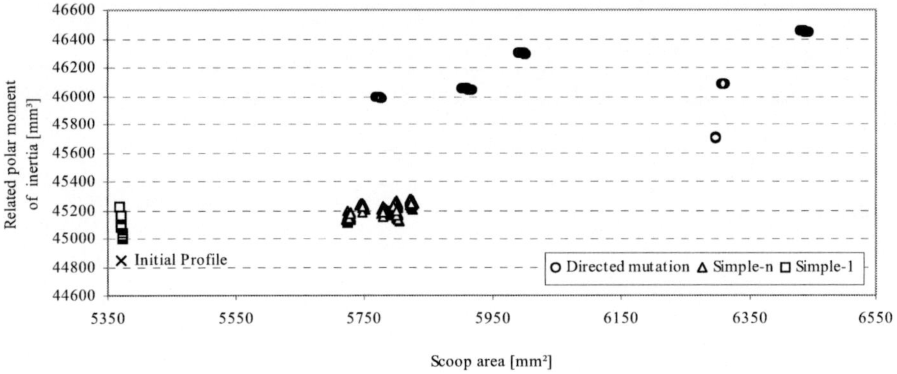

Figure 10. Comparison of the results

3. Covariance Matrix Adaptation

So far we were concerned with uncorrelated mutation models only. However, as already mentioned in the introduction, there are several ES approaches that rely on the flexibility of correlated mutations. The performance of the EA depends obviously highly on the choice of the covariance matrix $\boldsymbol{C}$, which has to be adjusted not only to the problem at hand, but also to the current state of the evolution process. Several methods have been proposed, from the self-adaptation of the mutation parameters in ES (SA-ES) [20] to the Covariance Matrix Adaptation ES (CMA-ES) [21]. While the first removes the need to manually adjust the covariance matrix, the latter takes into account the history of evolution and deterministically adapts the covariance matrix from the last moves of the algorithm, thereby directing the search to use the most recent descent direction. In [22] an advanced version of the CMA-ES is presented, that is computationally more efficient. Again, the present approaches use symmetric normally distributed random numbers. The aim of the sequel is therefore to accommodate the CMA-ES with a multivariate skew-normal distribution. Recent studies have shown remarkable results. However, much further research is necessary and the results are in that sense preliminary.

The rest of this section is organized as follows: first we present a multivariate version of the skew-normal distribution and then give a hint how to generate corresponding random vectors. Afterwards a possible way to integrate it into the CMA-ES framework is proposed. Finally, some first experimental data is provided.

3.1. The Multivariate Skew-Normal Distribution

An extension of the skew-normal distribution to the multivariate setting was proposed by Azzalini and Dalla Valle [11]. An n-dimensional random vector X is said to have a multivariate skew-normal distribution, denoted by $\mathrm{SN}_n(\boldsymbol{\mu},\boldsymbol{\Omega},\boldsymbol{\alpha})$, if it is continuous with probability density function

$$\mathrm{f}_{\mathrm{SN}_n}(z;\boldsymbol{\mu},\boldsymbol{\Omega},\boldsymbol{\alpha}) = 2\phi_n(z;\boldsymbol{\mu},\boldsymbol{\Omega})\,\Phi(\boldsymbol{\alpha}^T(z-\boldsymbol{\mu})), \qquad (19)$$

where $\phi_n(z;\boldsymbol{\mu},\boldsymbol{\Omega})$ is the n-dimensional probability density function with mean $\boldsymbol{\mu}$ and correlation matrix $\boldsymbol{\Omega}$, Φ is the standard normal distribution function N(0,1), and $\boldsymbol{\alpha}$ is an n-dimensional shape vector.

To generate SN_n distributed random vectors again their stochastic representation is used. Let Y have the probability density function $\phi_n(z;\mathbf{0},\boldsymbol{\Omega})$ and W be a N(0,1) distributed random variable. If

$$\mathrm{Z} = \begin{cases} \mathrm{Y}+\boldsymbol{\mu} & \text{if } \mathrm{W} < \boldsymbol{\alpha}^T\mathrm{Y} \\ -\mathrm{Y}+\boldsymbol{\mu} & \text{otherwise,} \end{cases} \qquad (20)$$

then $\mathrm{Z} \sim \mathrm{SN}_n(\boldsymbol{\mu},\boldsymbol{\Omega},\boldsymbol{\alpha})$, see e.g. [23].

3.2. Skew-Normal Covariance Matrix Adaptation-ES

As already mentioned, integration of the multivariate skew-normal distribution into the CMA-ES framework is at a very early stage. As an ad hoc implementation we use the mechanics of step-size adaptation to adjust the shape vector. Shape control then reads

$$\boldsymbol{p}_{\alpha}^{(g+1)} = (1-c_{\alpha})\boldsymbol{p}_{\alpha}^{(g)} + \sqrt{c_{\alpha}(2-c_{\alpha})\mu_{eff}}\,\boldsymbol{C}^{(g)^{-1/2}} \frac{\boldsymbol{m}^{(g+1)} - \boldsymbol{m}^{(g)}}{\sigma^{(g)}} \tag{21}$$

with learning rate

$$c_{\alpha} = \frac{\mu_{eff} + 2}{n + \mu_{eff} + 3} \tag{22}$$

and all other constants as given Hansen and Kern [24]. Neither the shape control nor the learning rate has been adapted to the special demands of shape vector control till now. However, even with this crude treatment of the shape vector some very promising results have been obtained.

3.3. Experimental Results

Two different CMA-ES are experimentally investigated: the classical variant, as described in [24], using $N(\mathbf{0},\boldsymbol{C})$ distributed random vectors and the SN_n-CMA-ES, using instead $SN_n(\mathbf{0},\boldsymbol{C},\boldsymbol{\alpha})$ distributed random vectors. Both have been used to solve four well known test problems: the sphere function, the ellipsoid function, the cigar function, and the tablet function. The problem dimension has always been set to ten, as stopping criterion fitness better then 10^{-10} has been chosen. For each combination 20 runs were carried out. Depicted in Figure 11 are the best, median, and worse run in each case.

Table 2. Test functions

Function	n	S	f^{stop}	f^{min}
$f_{\text{sphere}}(x) = \sum_{i=1}^{n} x_i^2$	10	$[-1,1]^n$	10^{-10}	0
$f_{\text{cigar}}(x) = x_1^2 + \sum_{i=2}^{n}(1000x_i)^2$	10	$[-1,1]^n$	10^{-10}	0
$f_{\text{tablet}}(x) = (1000x_1)^2 + \sum_{i=2}^{n} x_i^2$	10	$[-1,1]^n$	10^{-10}	0
$f_{\text{elli}}(x) = \sum_{i=1}^{n}\left(1000^{\frac{i-1}{n-i}} y_i\right)$	10	$[-1,1]^n$	10^{-10}	0

As can be seen in Figure 11, the SN_n-CMA-ES performs under the mentioned conditions surprisingly well. Even at this early stage of development it clearly outperforms the CMA-ES on all but the sphere function. For this test problem it performs only slightly worse than the CMA-ES. Another thing is striking: strong fluctuations occur during the SN_n-CMA-ES optimizations. They might indicate that the shape vector adaptation is still inadequate, because the local losses significantly reduce the overall performance. Also, the learning rate has to be investigated and optimized.

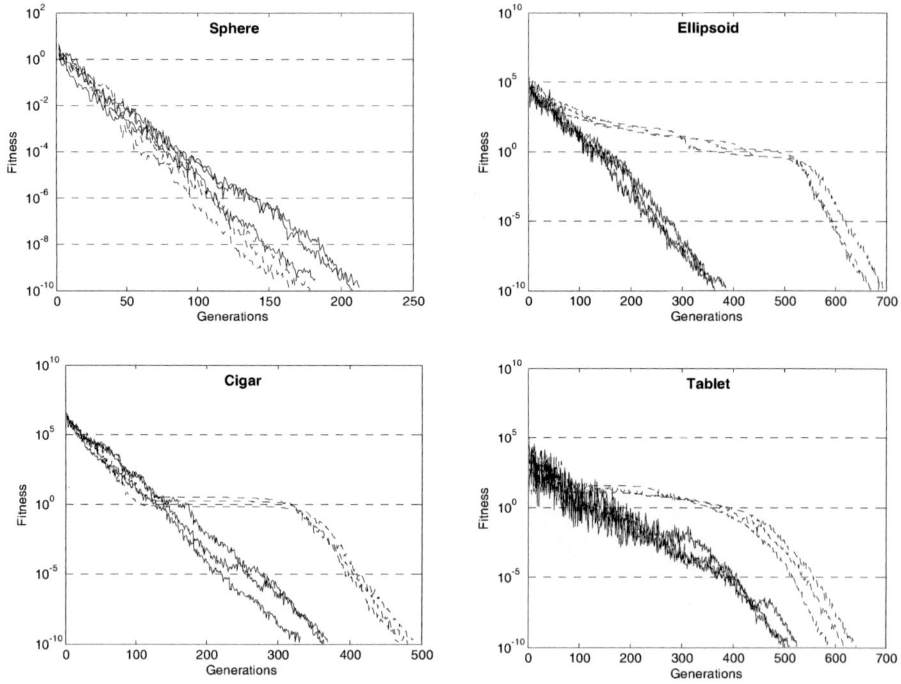

Figure 11. Experimental results of the SN_n-CMA-ES (solid) and CMA-ES (dashed) solving different problems.

4. Conclusions

With the directed mutation a promising new mutation principle for EAs has been presented. We pointed out some drawbacks of the first variant of this kind, the asymmetric mutation and then compared different univariate distributions. It was then argued that several advantages arise when the skew-normal distribution as basis for the mutation operator is chosen. First, it is a natural extension of the normal distribution with strict inclusion of the latter. Also, it is of salient mathematical tractability and the properties of the distribution have been studied extensively. Last but not least it is of stunningly mathematical beauty.

Another advantage is the straightforward and fast random number generator. By means of this distribution a mutation operator is given that clearly outperforms the classical mutation operators. This was shown here for a high dimensional, constraint multi-objective real-world optimization problem in a mechanical engineering scenario. First results with the multivariate skew-normal distribution in a CMA-ES context left us optimistic about the potential of this approach. Nevertheless, much work is left to be done. Adaptation of the shape vector is still inadequate and the learning rate has to be further investigated. Finally, comprehensive experimental studies are undone, which should also include multi-objective problems.

References

[1] T. Bäck and H.-P. Schwefel. An Overview of Evolutionary Algorithms for Parameter Optimization. Evolutionary Computation, 1(1):1–23, 1993.

[2] I. Rechenberg. Evolutionsstrategie: Optimierung technischer Systeme nach Prinzipien der biologischen Evolution. Frommann-Holzboog Verlag, Stuttgart, 1973.

[3] H.-P. Schwefel. Adaptive Mechanismen in der biologischen Evolution und ihr Einfluß auf die Evolutionsgeschwindigkeit. Technical report, Technical University of Berlin, 1974.

[4] T. Bäck, F. Hoffmeister, and H.-P. Schwefel. A survey of Evolution Strategies. In R. K. Belew, L. B. Booker, editors, ICGA'91, pages 2–9. Morgan Kaufmann, 1991.

[5] X. Yao and Y. Liu. Fast evolution strategies. Control and Cybernetics, 26(3):467–496, 1997.

[6] G. Rudolph. Local convergence rates of simple evolutionary algorithms with Cauchy mutations. IEEE Transactions on Evolutionary Computation, 1(4):249–258, 1998.

[7] L. Hildebrand. Asymmetrische Evolutionsstrategien. PhD thesis, Universität Dortmund, 2001.

[8] S. Berlik. A directed mutation framework for evolutionary algorithms. In R. Matoušek and P. Ošmera, editors, Proc. of the 10th Int. Conf. on Soft Computing, MENDEL, pages 45–50, 2004.

[9] A. Azzalini. A class of distributions which includes the normal ones. Scand. J. Statist., 12:171–178, 1985.

[10] B.C. Arnold and R.J. Beaver. Skewed multivariate models related to hidden truncation and/or selective reporting. Test. Sociedad de Estadística e Investigación Operativa, Vol. 11-1, 7–54, 2002.

[11] A. Azzalini and A. Dalla Valle. The multivariate skew-normal distribution. Biometrika, 83(4): 715–726, 1996.

[12] S. Berlik. Directed mutation by means of the skew-normal distribution. In Proc. of the Int. Conf. on Computational Intelligence, FUZZY DAYS, Advances in Soft Computing. Springer-Verlag Berlin Heidelberg, 2005.

[13] K. Kauder, B. Reusch, M. Helpertz, S. Berlik. Automated Geometry-Optimisation of Rotors of Twin-Screw Compressors, Part 1. Schraubenmaschinen Vol. 9, pp. 27–47, 2001.

[14] K. Kauder, B. Reusch, M. Helpertz, S. Berlik. Automated Geometry-Optimisation of Rotors of Twin-Screw Compressors, Part 2. Schraubenmaschinen Vol. 10, pp. 17–34, 2002.

[15] K. Kauder, B. Reusch, M. Helpertz, S. Berlik. Automated Geometry-Optimisation of Rotors of Twin-Screw Compressors, Part 3. Schraubenmaschinen Vol. 11, pp. 15–29, 2003.

[16] K. Kauder, B. Reusch, M. Helpertz, S. Berlik. Optimisation Methods for Rotors of Twin-Screw Compressors. VDI Berichte 1715, VDI-Verlag, 2002.

[17] K. Deb, Multi-Objective Optimization using Evolutionary Algorithms, John Wiley & Sons: Chichester, UK , 2001.

[18] K. Deb, S. Agarwal, A. Pratap, T. Meyarivan. A Fast Elitist Non-Dominated Sorting Genetic Algorithm for Multi-Objective Optimization: NSGA-II. KanGAL report 200001, Indian Institute of Technology, Kanpur, India, 2000.

[19] S. Berlik and M. Fathi. Multi-Objective Optimization using Directed Mutation. In: Proceedings of the International MultiConference in Computer Science & Computer Engineering, International Conference on Artificial Intelligence - ICAI 2005, Las Vegas, USA, 2005.

[20] H.-P. Schwefel. Evolution and Optimum Seeking. John Wiley & Sons, New York, 1995.

[21] N. Hansen and A. Ostermeier. Adapting arbitrary normal mutation distributions in evolution strategies: The covariance matrix adaptation. In IEEE International Conference on Evolutionary Computation, pages 312–317, 1996.
[22] N. Hansen, S. Müller, and P. Koumoutsakos. Reducing the Time Complexity of the Derandomized Evolution Strategy with Covariance Matrix Adaptation (CMA-ES). Evolutionary Computation, 11(1):1–18, 2003.
[23] J. Wang, J. Boyer, and M. G. Genton. A skew-symmetric representation of multivariate distributions. Statistica Sinica, 14(4):1259–1270, 2004.
[24] N. Hansen and S. Kern. Evaluating the CMA evolution strategy on multimodal test functions. In X. Yao et al, editors, Parallel Problem Solving from Nature, 282–291, Springer-Verlag Berlin Heidelberg, 2004.

Integrated Intelligent Systems for Engineering Design
X.F. Zha and R.J. Howlett (Eds.)
IOS Press, 2006

Many Valued Algebraic Structures as the Measures for Comparison[1]

Kalle O. Saastamoinen [2]
Lappeenranta University of Technology
kalle.saastamoinen@lut.fi

Abstract. In this chapter we will study properties and usability of basic many valued structures called t-norms, t-conorms, implications and equivalences in comparison tasks. We will show how these measures can be aggregated with generalized mean and what kind of measures for comparison can be achieved from this procedure. New classes for comparison measures are suggested, which are combination measure based on the use of t-norms and t-conorms and pseudo equivalence measures based on S-type implications.

In experimental part of this chapter we will show how some of the comparison measures presented here work in comparison task. For comparison task we use classification. We show by comparison to results that can be achieved through some known public domain classifier results that our classification results are highly competitive.

Keywords. Fuzzy logic, Comparison measures, Aggregation, Adaptive weighting, Classification

1. Introduction

William James has written the following about the sense of sameness [1]: "This sense of sameness is the very keel and backbone of our thinking." The fields of problem solving, categorization, data mining, classification, memory retrieval, inductive reasoning, and generally cognitive processes require that we understand how to assess the sameness. Sometimes comparison measures used to measure sameness in practice in the field of soft computing are based on a very intuitive understanding of the theoretical backgrounds of mathematics or a naive idea of coupling the measure of sameness to the Euclidean distance. Takin measures intuitively can be crucial mistake and can lead to so called black-box systems which can work but no one is able to say how and why are they working. An example of this kind of systems are neural networks [2]. In fact one of the first things to consider in systems should be measures used. In this chapter we will present two types of measures namely combination of many valued intersection and union and many valued equivalence, which both can be used in various different applications and have many good properties for further analysis. We use t-norms and t-conorms as our many valued

[1]This work was partially supported by EU Asia Link Project (Contract Reference no.: ASI/B7-301/98/679-023).

[2]Correspondence to: Kalle Saastamoinen, Huuhtakatu 3, 53850 Lappeenranta, Finland. Tel.: +358 40 7606489; Fax: +358 5 6212898; E-mail: kalle.saastamoinen@lut.fi.

intersections and unions and we use S-type of equivalences as our equivalence structure. Generalized mean we use to combine various membership degrees into one numerical value. The definitions one can find from the next section of this chapter. The benefits what one can get by applying basic connectives from many valued logic with generalized mean in comparison tasks come apparent when these connectives are used to model the systems which are somewhat graded in their nature. Examples of these kind of systems are expert systems and decision analysis, where management of uncertainty often plays an important part.

In soft computing there are variety of measures what have been applied for comparison. Some of them are set-theoretical, distance based, logical or heuristical. Measures called similarities are many times used as synonyms for the measures which are used for comparison. The well-known definition for similarity in the field of fuzzy logic is the one created by Lotfi Zadeh [3], which is a fuzzy logical generalization of the equivalence relation, basically it is, however, just inverse ultra-metrics. The more general approach uses any suitable t-norm in the definition of transitivity and is called fuzzy equivalence relation [4], indistinguishability operator [5], fuzzy equality [6] or proximity relation [7], depending taste of the author. If we use the previous definitions we assume that similarity used is reflexive, symmetric and t-transitive. Sometimes demands of reflexivity, symmetricity or transitivity can be unsuitable for the problems or data sets in hand [8,9,10,11]. It seems that similarity measure, if considered as the measure for sameness, is over simplified if it is defined correspondingly to be ultra-metrics. The meaning of the measure for sameness is wider than just this definition of metric based similarity. For example t-norms can be seen as the measures of sameness but they almost never are reflexive.

Much of Fuzzy Set Theory original inspiration and further developments are originated from the problems from pattern classification and cluster analysis. Essentially this is the reason why we have chosen to use here classification as our test bench for our comparison measures. When we do classification, the question is not whether a given object is or not a member of a class, but the degree in which the object belongs to the class; which amounts to saying that most classes in real situations are fuzzy in their nature [12]. This fuzzy nature of real world classification problems may throw some light on the general problem of decision making [13].

This chapter is organized for two parts. At the first part we show how different many valued comparison measures can be created from simple logical rules. At the second part we represent examples how these comparison measures can be used in classification. In order to make our measures more flexible we make them adaptive by using weights and parameterizations. We show that measures created in this chapter give very good results when they are adapted for classification.

2. Mathematical background

In this section we give short mathematical background for the terms what we use in this chapter.

2.1. Generalized mean as the compensative operator

A highly compelling feature of Fuzzy Set Theory is to provide categories for the sets of measures called aggregation connectives [14,15]. In the papers [16,17,18,19] we have used generalized mean as our aggregative operator in order to compensate and combine vectors. Vectors compensated in the articles mentioned are coming from t-norms, t-conorms, uninorms and S-type of many valued implications. This subsection will explain why we have chosen to use generalized mean as our aggregative operator.

Generalized mean [20] or quasilinear mean [14] is an aggregation operator that belong to the class called averaging or mean operators, well-known averaging operators are also OWA operators [21]. The name averaging operator originates from the fact that they combine arguments by giving them some kind of compensation value. Averaging operators can also be considered as extending the space of universal quantifier $\forall$ (for all) and $\exists$ (at least one) from the pair $\forall, \exists$ to the interval $[\forall, \exists]$. More recently generalized mean has been implemented into OWA operators in [22] to reach generalized version of these operators. This GOWA operator seems to be an important special case of the use of the generalized mean. We can give the following form for aggregation operators in general [23].

Definition 1 *For the aggregation of a number n of arguments, it holds that*

$$A(a_1, \ldots, a_n) = f^{-1}\left(\sum_{i=1}^{n} f(a_i)\right), \tag{1}$$

where a_i denotes arguments and f is a generator function.

Now if we choose generator function $f(a_i) = a_i^m$, which is invertible as long as $m \neq 0$ we get the the following form for our aggregation of number n of arguments a_i. Obviously this is the same thing as taking of p-norm from arguments.

$$A(a_1, \ldots, a_n) = \left(\sum_{i=1}^{n} a_i^m\right)^{\frac{1}{m}}, \tag{2}$$

this can be transformed into the weighted form of the generalized mean.

Definition 2 *Weighted form of the generalized mean operator of dimension n and $m \neq 0$ is in the form*

$$A(a_1, \ldots, a_n) = \left(\sum_{i=1}^{n} w_i a_i^m\right)^{\frac{1}{m}}, \tag{3}$$

where a_i denotes arguments, w_i denotes weights and m denotes the corresponding parameter.

The formula above corresponds the formulation, which can be found in [20].

Remark 1 *If we set* $w_i = \frac{1}{n}, \forall i \in N$ *we get the normal formulation of the generalized mean.*

From the articles of [20,22] we find that weighted form of the generalized mean defined in 3 has the properties that it is i) commutative ii) monotonic and iii) bounded, which means that it is averaging operator. It has also several other good properties listed in [20].

Definition 3 *Grade of compensation. Since the generalized mean is monotonically increasing with respect to* m*, grade of compensation is achieved by any strictly monotone increasing transformation* γ *of the compensation parameter* m *from* $[-\infty, \infty]$ *onto* $[0, 1]$.

Example 1 *This can be done for instance by*

$$\gamma = 0.5\left(1 + \frac{m}{1 + |m|}\right)$$

or

$$\gamma = 0.5\left(1 + \frac{2}{\pi}\arctan(m)\right).$$

Both yield:

$$\begin{array}{ll}
\gamma = 0.00 \textit{ for minimum} & m \to -\infty \\
\gamma = 0.25 \textit{ for harmonic mean} & m = -1 \\
\gamma = 0.50 \textit{ for geometric mean} & m = 0 \\
\gamma = 0.75 \textit{ for arithmetic mean} & m = 1 \\
\gamma = 1.00 \textit{ for maximum} & m \to \infty,
\end{array}$$

where $\gamma = 0$ *and* $\gamma = 1$ *characterize* min *and* max *type of operators [24], which are only* t*-norms and* t*-conorms that are idempotent. From this we see that* t*-norms and* t*-conorms provide lower and upper bound for averaging operators.*

Compensative property clearly means that generalized mean is a valuable tool, when we combine measures. It also provides more freedom than the use of well-known arithmetic, geometric or harmonic means. In the becoming sections we show how generalized mean has been used for combining measures like t-norms, t-conorms, S-type of implications and many valued equivalences.

2.2. Combined measure from t-norms and t-conorms

A standard approach is to create mathematical models with a kind of logic where every axiom, sentence, connective etc. is based on the classical, Aristotelian, bivalent logic, which is all about interpretations of two values 0 and 1. The real world, however, is not this black and white, so sometimes models offered by bivalent logic are inaccurate from the beginning, and inaccuracy originates from the basic nature of this logic. In the type of a many-valued logic named Fuzzy Logic every interpretation is in the closed interval of 0 and 1. It is this characteristic of offering an infinite number of possible interpretations, which makes this logic generally more suitable. Connectives play an important role when we try to model reality by equations. For example when we use connectives such as AND or OR we often do not need or mean the exact AND or OR, but these connectives in some degree. Here the connectives of Fuzzy Logic provide tools for both theory and practically orientated research, for example many-valued intersections and unions called t-norms and t-conorms. These are used in the fields of statistical metric spaces as the tool for generalizing the classical triangular inequality [25,26], successfully used in expert knowledge systems since 1976 [27] and many other fields.

Definition 4 *The t-norm is a binary operation T on the unit interval that satisfies at least the following for all $\alpha, \beta, \gamma \in I$:*

a1) $T(\alpha, 1) = \alpha$ (boundary condition),
b) $T(\alpha, \beta) = T(\beta, \alpha)$ (commutativity),
c) $\beta \leq \gamma$ implies $T(\alpha, \beta) \leq T(\alpha, \gamma)$ (monotonicity),
d) $T(\alpha, T(\beta, \gamma)) = T(T(\alpha, \beta), \gamma)$ (associativity).

Only difference between t-norms and t-conorms is the choice of the identity element in the boundary condition. For t-conorms this is defined by the following.

a2) $T_{co}(\alpha, 0) = \alpha$ (boundary condition).

The most important additional requirements that restrict the class of t-norms and t-conorms are continuity, monotonicity, subidempotency for t-norms and superidempotency for t-conorms. These can be expressed as follows:

e) T and T_{co} are continuous functions (continuity).
f1) $T(\alpha, \alpha) < \alpha$ (subidempotency).
g1) $\alpha_1 < \alpha_2$ and $\beta_1 < \beta_2$ implies $T(\alpha_1, \beta_1) < T(\alpha_2, \beta_2)$ (strict monotonicity).
f2) $T_{co}(\alpha, \alpha) > \alpha$ (superidempotency)
g2) $\alpha_1 < \alpha_2$ and $\beta_1 < \beta_2$ implies $T_{co}(\alpha_1, \beta_1) < T_{co}(\alpha_2, \beta_2)$ (strict monotonicity)

Definition 5 *A continuous t-norm (correspondingly for t-conorm) that satisfies subidempotency (for t-conorm superidempotency) is called an Archimedean t-norm (correspondingly Archimedean t-conorm).*

Definition 6 *An Archimedean t-norm that satisfies strict monotonicity is called a strict Archimedean t-norm (correspondingly strict Archimedean t-conorm).*

The t-norm gives a minimum type of compensation, while the t-conorm gives a maximum type of compensation. This means that t-norms tend to give more value to the small values, while t-conorms give more value to the higher values in the intervals they are used. In practice, neither of these connectives fit the collected data appropriately. There is still a great deal of information that is left between them. An important issue when dealing with t-norms and t-conorms is the question of how to combine them in a meaningful way since neither of these connectives alone gives general compensation for the values where they are adapted. For this reason we should use a measure that somewhat compensates this gap of values between them.

In the paper [16] we have defined measures based on the use of the generalized mean, weights, t-norms and t-conorms. Here we will add to these results the definition of a combination measure of a t-norm and a t-conorm. The first ones to try the compensation of t-norms and t-conorms have been Zimmermann and Zysno in [28]. They used the weighted geometric mean in order to compensate for the gap between many-valued intersections and unions. In equation 4 we have given a more general definition where many-valued unions and intersections are combined with a generalized mean and weights. Archimedean t-norms and t-conorms are a good choice for combination since they are continuous and monotonic [29].

The motivation for the equations 4, 5 and 6 is that when we use the generalized mean for aggregation it is possible to go through all the possible variations of the magnitude of the combined values of t-norms and t-conorms.

Definition 7 *Combined measure based on t-norm and t-conorm with generalized mean:*

$$F_p(x_1, x_2) = \left(\sum_{i=1}^{n} \left(w_i (\omega_{ci} c_i (x_1, x_2)) + (1 - w_i) (\omega_{di} d_i (x_1, x_2)) \right)^m \right)^{\frac{1}{m}} \tag{4}$$

, where $i = 1, \ldots, n$, p is a parameter combined to the corresponding class of weighted t-norm c_i and t-conorm d_i, ω_{ci}, ω_{di} and w_i are weights.

We see that if we set $w_i = 1, \forall i \in \mathbb{Z}$ we get the following equation:

Definition 8 *Generalized measure based on t-norm with generalized mean:*

$$C_p(x_1, x_2) = \left(\sum_{i=1}^{n} \left(\omega_{ci} c_i (x_1, x_2) \right)^m \right)^{\frac{1}{m}} \tag{5}$$

, where p is a parameter combined to the corresponding class of weighted t-norms and $i = 1, \ldots, n$ and ω_{ci} is the weight belonging to the t-norm.

We see that if we set $w_i = 0, \forall i \in \mathbb{Z}$ we get the following equation:

Definition 9 *Generalized measure based on t-conorm with generalized mean:*

$$D_p(x_1, x_2) = \left(\sum_{i=1}^{n} \left(\omega_{di} d_i (x_1, x_2) \right)^m \right)^{\frac{1}{m}} \tag{6}$$

, where p is a parameter combined to the corresponding class of weighted t-conorms and $i = 1, \ldots, n$ and ω_{di} is the weight belonging to the t-conorm.

From the formulation of 4, 5 and 6 can be seen that combined measure of t-norms and t-conorms have inside all the valuations what weighted t-norms and t-conorms can give and in this way will always give at least as good results as any t-norm or t-conorm can give alone as long as weights are chosen in the sensible way. In the experimental part of this chapter we will present how this measures worked in classification tasks.

2.3. S-Type Equivalences and Implications as the Measures for Comparison

Equivalence is logically a sentence, which states that something exists if and only if something else also exists. For this reason it is naturally suitable for the comparison of different objects. Implication means that if something exist then something else will also exist. This means implications are suitable for decision-making. Rule based classifiers are quite popular in classification processes [30]. They are normally used as counterparts for fuzzy control systems. Here we will first present comparison measures, which rise from the Fuzzy Set Theoretic class of implications called S-implications. Secondly we will present comparison measures based on functional form of implications presented first time by Smetz and Magrez, 1987 in [31].

In the articles [32,33] we studied the use of Łukasiewicz type of equivalence, with means and weights. In the article [18] we take this study further by the use of generalized mean and weights. In the equations presented in [18] we have parameterized the implications by replacing the variables in the formulas by their exponential forms. When we use these formulations we come to the conclusion that they hold all the properties of the corresponding implications since the exponent is a monotonic operator. We will demonstrate in application part of this chapter that these new measures give competitive results when they were tested in classification tasks.

One way of extending the implication is first to use the classical logic formula $x \to y \equiv \neg x \vee y$ for all $x, y \in \{0,1\}$. This is done by interpreting the disjunction and negation as a t-conorm and the use of standard fuzzy complement ($\neg x \equiv 1 - x$). This results in defining implication by the formula $a \to b \equiv d(\neg a, b)$ for all $a, b \in [0,1]$, which gives rise to the family of many valued implications called S-implications. We use equivalence of the form $a \leftrightarrow b \equiv c(a \to b, b \to a)$.

We can give the following procedure for defining logical equivalences from implications.

1. Take an implication for example of the type $x \to y \equiv \neg x \vee y$ here disjunction symbol $\vee$ refers to the t-conorm, if we choose to use ordinary Łukasiewicz implication, which is obtained by the use of bounded sum as a t-conorm we get $I_Ł(x, y) = \min(1, 1 - x + y)$.
2. Next we will form parameterized form from the implication chosen by setting variables into exponential forms. Taking an exponent is the monotonic operation this does not change any properties concerning the equation that it is used to. For example by setting $x = x^p$ and $y = y^p$ to the ordinary Łukasiewicz implication, leads to the following formula $I_Ł(x, y) = \min(1, 1 - x^p + y^p)$.
3. Thirdly we will use the equation $x \leftrightarrow y \equiv \wedge((x \to y), (y \to x))$, where $\wedge$ refers to the t-norm and implication $\to$ used is also many valued. For example in the case of Łukasiewicz choose to use bounded product as conjunction which leads to the equivalence equation $E_Ł(x, y) = 1 - |x^p - y^p|$.
4. On the fourth step we generalize this measure more by weights w_i, $i \in N$ and generalized mean m into it. In our example case this leads to the equation
$$E_Ł(x_i, y_i) = \left(\sum_{i=1}^{n} w_i (1 - |x_i^p - y_i^p|)^m\right)^{\frac{1}{m}}.$$

We easily see that in general this kind of measures does not hold the definition of equivalence or the usual definition of similarity that is that they would be reflexive, symmetric and transitive. It is common belief that measures which are used as the measures for comparison should hold these properties. This belief is originating from the blinkered view that comparison of objects should always have something to do with distance. In practise it seems that these properties have little or no affect at all for the results that can be achieved from the use of different comparison measure. This becomes empirically clear when one looks the test result from the end of this chapter. We suggest that comparison measure could be any logically rightly formulated equation, which holds the following properties:

1. Comparison measure used has clear logical structure e.g. it is nilpotent t-norm or t-conorm (like Frank 9, 14) or S-equivalence (19, 26, 23).
2. Comparison measure is monotonically increasing or decreasing.
3. Comparison measure is continuous.

Next approach is a sort of combination of syntactic approach originating from mathematics and logical approach.

1. First we use the equation $I(x, y) = f^{-1}(f(1) - f(x) + f(y))$ that is a mapping $I : [0,1]^2 \to [0,1]$, where $f : [0,1] \to [0,\infty[$ is a strict increasing continuous function such that $f(0) = 0$, for all $x, y \in [0,1]$ [31]. The implications that follows from the usage of previous equation has many important properties, next we are going to mention some. It is monotonic in first and second argument, which means that the truth value of many valued implications increase if truth value of antecedent decrease or truth value of the consequent increases. It is continuous, which is the property that ensures that small changes in the truth values of the antecedent or the consequent do not produce large (discontinuous) changes in the truth values of many valued implications. It is bounded, this means that many valued implications are true if and only if the consequent is at least as true the antecedent. For example we can select $f(x) = x^p$, which leads to the implication of the form $I(x, y) = \min\left\{1, (1 - x^p + y^p)^{\frac{1}{p}}\right\}$, also known as pseudo-Łukasiewicz type 2 implication [34] and if we allow $p \in [-\infty, \infty]$ this can also be seen Schweizer and Sklar type 1 implication [35,34].
2. Next phase we combine two many valued implications by the use of proper many valued conjunction into the equivalence of the form $x \leftrightarrow y \equiv \wedge((x \to y), (y \to x))$. In our example case we can choose fuzzy conjunction to be of the form $(\max\{x^p + y^p - 1, 0\})^{\frac{1}{p}}$, where $p \in [-\infty, \infty]$. This leads to equivalence equation of the form $E(x, y) = (1 - |x^p - y^p|)^{\frac{1}{p}}$, where $p \in [-\infty, \infty]$.
3. On the third step we generalize previous step more by using weights w_i, $i \in N$ and generalized mean m. In example case this leads to the equation $E(x_i, y_i) = \left(\sum_{i=1}^{n} w_i (1 - |x_i^p - y_i^p|)^{\frac{m}{p}}\right)^{\frac{1}{m}}$.

Application section of this chapter shows that there is no significant difference between the measures achieved by semantic manipulations of equations versus to the measures achieved at first place by using functions mathematical properties.

2.4. Created comparison measures

Next we will present measures what we have created using techniques presented in previous sections.

In the following comparison measures all relations are of the form $R : ([0,1]^n)^2 \to [0,1]$. Tested vectors $f_i \in [0,1], \forall i \in N$, mean value $m \in R$ and $m \neq 0$, all the weights (w_i, ω_{ci} and ω_{di}) holds that $w_i \geq 0$ with $\sum_{i=1}^{n} w_i = 1$ and p is a parameter value, which corresponds to the comparison measure class.

2.4.1. T-norms and t-conorms

Definition 10 *Combined measure based on t-norm and t-conorm with generalized mean and weights:*

$$F_p \langle f_1(i), f_2(i) \rangle = \left(\sum_{i=1}^{n} \left(w_i C_i^p \langle f_1(i), f_2(i) \rangle + (1 - w_i) \left(D_i^p \langle f_1(i), f_2(i) \rangle \right) \right)^m \right)^{\frac{1}{m}} \tag{7}$$

, where $i = 1, \ldots, n$, p is a parameter combined to the corresponding class of weighted t-norms C_i and t-conorms D_i and w_i are weights.

The measure 7 has been used by combining into it the following t-norms and t-conorms. Tested t-norms and t-conorms have been the following ones.

Definition 11 *Measure based on Dombi (1982), [23] class of t-norm with generalized mean and weights:*

$$C_D \langle f_1(i), f_2(i) \rangle = \left(\sum_{i=1}^{n} \omega_{ci} \left(1 + \left[\left(\frac{1}{f_1(i)} - 1 \right)^p + \left(\frac{1}{f_2(i)} - 1 \right)^p \right]^{\frac{1}{p}} \right)^{-m} \right)^{\frac{1}{m}} \tag{8}$$

, where $p > 0$ and $i = 1, \ldots, n$.

Definition 12 *Measure based on Frank (1979) [36] class of t-norm with generalized mean and weights:*

$$C_F \langle f_1(i), f_2(i) \rangle = \left(\sum_{i=1}^{n} \omega_{ci} \left(\log_p \left[1 + \frac{\left(p^{f_1(i)} - 1\right)\left(p^{f_2(i)} - 1\right)}{p - 1} \right] \right)^m \right)^{\frac{1}{m}} \tag{9}$$

, where $p > 0$, $p \neq 1$ and $i = 1, \ldots, n$.

Definition 13 *Measure based on Schweizer & Sklar 1 (1963) [37] class of t-norm with generalized mean and weights:*

$$C_{SS} \langle f_1(i), f_2(i) \rangle = \left(\sum_{i=1}^{n} \omega_{ci} \left(\max\left\{0, (f_1(i))^p + (f_2(i))^p - 1\right\} \right)^{\frac{m}{p}} \right)^{\frac{1}{m}} \tag{10}$$

, where $p \neq 0$ and $i = 1, \ldots, n$.

Definition 14 *Measure based on Yager (1980) [38] class of t-norm with generalized mean and weights:*

$$C_Y \langle f_1(i), f_2(i) \rangle = \left(\sum_{i=1}^{n} \omega_{ci} \left(1 - \min \left\{ 1, [(1 - f_1(i))^p + (1 - f_2(i))^p]^{\frac{1}{p}} \right\} \right)^m \right)^{\frac{1}{m}} \tag{11}$$

, where $p > 0$ and $i = 1, \ldots, n$.

Definition 15 *Measure based on Yu (1985) [39] class of t-norm with generalized mean and weights:*

$$C_{Yu} \langle f_1(i), f_2(i) \rangle = \left(\sum_{i=1}^{n} \omega_{ci} \left(\max \{0, (1 + p)(f_1(i) + f_2(i) - 1) - p \cdot f_1(i) f_2(i)\} \right)^m \right)^{\frac{1}{m}} \tag{12}$$

, where $p > -1$ and $i = 1, \ldots, n$.

Definition 16 *Measure based on Dombi (1982) [23] class of t-conorm with generalized mean and weights:*

$$D_D \langle f_1(i), f_2(i) \rangle = \left(\sum_{i=1}^{n} \omega_{di} \left(1 + \left[\left(\frac{1}{f_1(i)} - 1 \right)^{-p} + \left(\frac{1}{f_2(i)} - 1 \right)^{-p} \right]^{-\frac{1}{p}} \right)^{-m} \right)^{\frac{1}{m}} \tag{13}$$

, where $p > 0$ and $i = 1, \ldots, n$.

Definition 17 *Measure based on Frank (1979) [36] class of t-conorm with generalized mean and weights:*

$$D_F \langle f_1(i), f_2(i) \rangle = \left(\sum_{i=1}^{n} \omega_{di} \left(1 - \log_p \left[1 + \frac{(p^{1-f_1(i)} - 1)(p^{1-f_2(i)} - 1)}{p - 1} \right] \right)^m \right)^{\frac{1}{m}} \tag{14}$$

, where $p > 0$, $p \neq 1$ and $i = 1, \ldots, n$.

Definition 18 *Measure based on Schweizer & Sklar 1 (1963) [37] class of t-conorm with generalized mean and weights:*

$$D_{SS} \langle f_1(i), f_2(i) \rangle = \left(\sum_{i=1}^{n} \omega_{di} \left(1 - (\max \{0, (f_1(i))^p + (f_2(i))^p - 1\})^{\frac{1}{p}} \right)^m \right)^{\frac{1}{m}} \tag{15}$$

, where $p \neq 0$ and $i = 1, \ldots, n$.

Definition 19 *Measure based on Yager (1980) [38] class of t-conorm with generalized mean and weights:*

$$D_Y \langle f_1(i), f_2(i) \rangle = \left(\sum_{i=1}^{n} \omega_{di} \left(\min \left\{ 1, [(f_1(i))^p + (f_2(i))^p]^{\frac{1}{p}} \right\} \right)^m \right)^{\frac{1}{m}} \quad (16)$$

, where $p > 0$ *and* $i = 1, \ldots, n$.

Definition 20 *Measure based on Yu (1985) [39] class of t-conorm with generalized mean and weights:*

$$D_{Yu} \langle f_1(i), f_2(i) \rangle = \left(\sum_{i=1}^{n} \omega_{di} \left(\min \{ 1, f_1(i) + f_2(i) + p \cdot f_1(i) f_2(i) \} \right)^m \right)^{\frac{1}{m}} \quad (17)$$

, where $p > -1$ *and* $i = 1, \ldots, n$.

2.4.2. Equivalences, Pseudo Equivalences and Implications

1. **Kleene-Dienes Measures**

Kleene-Dienes implication [34] is obtained by using standard fuzzy disjunction as a t-conorm:

$$I_{K-D}(x, y) = \max(1 - x, y)$$

We can create parameterized form of Kleene-Dienes by setting $x = x^p$ and $y = y^p$, which leads to the equation

$$I_{K-D}(x, y) = \max(1 - x^p, y^p) \quad (18)$$

If we now use standard fuzzy conjunction to combine two implications $I_{K-D}(x, y) = \max(1 - x^p, y^p)$ and $I_{K-D}(y, x) = \max(1 - y^p, x^p)$ we will reach

Definition 21 *The equivalence based on Kleene-Dienes implication:*

$$E_{K-D}(x, y) = \min(\max(1 - x^p, y^p), \max(1 - y^p, x^p)) \quad (19)$$

Since for example $E_{K-D}(x, x) \neq 1$ in general so this measure can not hold reflexivity we see that this form is not the similarity as defined in [3]. For 19 we can apply generalized mean and weights, which we can use to obtain more values for evaluation. This approach has been proven to be practically effective in many previous researches [33,17,19,22].

Definition 22 *Generalized semantic weighted equivalence based on Kleene-Dienes implication:*

$$E_{K-D}(f_1(i), f_2(i)) = \left(\sum_{i=1}^{n} w_i \left(E_{K-D}(f_1(i), f_2(i)) \right)^m \right)^{\frac{1}{m}} \quad (20)$$

2. **Reichenbach Measures**

Reichenbach implication [34] is obtained by using the algebraic sum as a t-conorm:

$$I_R(x, y) = 1 - x + xy$$

We can set parameterized form of Kleene-Dienes by setting $x = x^p$ and $y = y^p$, which leads to the equation

$$I_R(x, y) = 1 - x^p + x^p y^p \tag{21}$$

If we now use algebraic product as conjunction to combine two implications $I_R(x, y) = 1 - x^p + x^p y^p$ and $I_R(y, x) = 1 - y^p + x^p y^p$ we will reach

Definition 23 *The equivalence based on Reichenbach implication:*

$$E_R(x, y) = (1 - x^p + x^p y^p)(1 - y^p + x^p y^p) \tag{22}$$

The formula above can not hold the reflexivity so it can not be considered as the similarity in the way defined by Zadeh [3]. To this form we can also apply generalized mean and weights in order to get an extra parameter, which we can use to obtain more values for evaluation.

Definition 24 *Generalized semantic weighted equivalence based on Reichenbach implication:*

$$E_R(f_1(i), f_2(i)) = \left(\sum_{i=1}^{n} w_i \left(E_R(f_1(i), f_2(i)) \right)^m \right)^{\frac{1}{m}} \tag{23}$$

3. **Łukasiewicz measure**

Łukasiewicz implication [34] is obtained by using bounded sum $\min(1, x + y)$ as a t-conorm:

$$I_L(x, y) = \min(1, 1 - x + y)$$

We can set parameterized form of Łukasiewicz by setting $x = x^p$ and $y = y^p$, which leads to the equation

$$I_L(x, y) = \min(1, 1 - x^p + y^p) \tag{24}$$

If we now use algebraic bounded product as conjunction to combine two implications $I_L(x, y) = \min(1, 1 - x^p + y^p)$ and $I_L(y, x) = \min(1, 1 - y^p + x^p)$ we will reach

Definition 25 *The equivalence based on Łukasiewicz implication:*

$$E_L(x, y) = 1 - |x^p - y^p| \tag{25}$$

Equation above is reflexive, symmetric and transitive like normal pseudo type Łukasiewicz structures are. When implemented to the generalized mean we get the following form of equation.

Definition 26 *Generalized logical weighted equivalence based on Łukasiewicz implication:*

$$E_L(f_1(i), f_2(i)) = \left(\sum_{i=1}^{n} w_i (E_L(f_1(i), f_2(i)))^m\right)^{\frac{1}{m}} \tag{26}$$

4. **Combined Lukasiewicz and Shweizer & Sklar Based Measure**

Here we are using the measure which rise from the functional definition for the implications given in [31]. We note that pseudo Łukasiewicz type 2 [34] and Shweizer and Sklar type 1 [35,34] implications form almost the same equivalence measure when these equivalences are formed by using fuzzy conjunction to combine corresponding implications. Łukasiewicz equivalence is included to Shweizer & Sklar by taking the parameter values which go from negative side to the positive, so $p \in]-\infty, \infty[$.

Definition 27 *Generalized logical-functional weighted equivalence based on Shweizer & Sklar - Łukasiewicz:*

$$E_{SSL}(f_1(i), f_2(i)) = \left(\sum_{i=1}^{n} w_i (1 - |f_1^p(i) - f_2^p(i)|)^{\frac{m}{p}}\right)^{\frac{1}{m}} \tag{27}$$

3. Classification

Many time there are given a set of data which are already grouped into classes and the problem is then to predict which class each new data belongs to this is normally referred to as classification problem. First set of data is normally referred to as training set, while this new set of data is referred to as test set [40]. Classification is here seen as comparison between training set and test set.

In this section we show comparison of how well some of the comparison measures presented in this chapter manage from the classification tasks compared to results reported in [41].

3.1. Data sets

We tested our measures with five different data sets which are available from the [42] Data sets chosen for the test were: Ionosphere, Post-operative, Iris, Pima Indians diabetes and Wine. These are derived from the variety of sources: medical, biological and engineering domains. These learning sets differ greatly in the magnitude of instances and number of predictive attribute values. We tested the stability of classification results with our measures for different parameter values and optimized weight values.

- **Post Operative, row 1**: Task of this database is to determine where patients in a postoperative recovery area should be sent to next. The attributes correspond roughly to body temperature measurements. The number of Instances is 90. The number of attributes is 9 including the decision (class attribute). Attribute 8 has 3 missing values. Missing values has in this study been replaced by the average of attribute 8 values.

- **Ionosphere, row 2**: This is radar data, where the targets were free electrons in the ionosphere. Here are two classes: "Good" and "Bad". "Good" radar returns are those showing evidence of some type of structure in the ionosphere. "Bad" returns are those that do not; their signals pass through the ionosphere. The number of instances is 351. The number of attributes is 34 plus the class attribute.
- **Wine, row 3**: The data is the result of a chemical analysis of wines grown in the same region in Italy but derived from three different cultivars. The analysis determined the quantities of 13 constituents found in each of the three types of wines. The number and deviation of instances: class 1 59, class 2 71, class 3 48.
- **Pima, row 4**: The diagnostic, binary-valued variable investigated is whether the patient shows signs of diabetes. All instances here are females at least 21 years old of Pima Indian heritage. Number of Instances is 768. Number of Attributes is 8 plus class. Class 1 (negative for diabetes) 500, Class 2 (positive for diabetes) 268.
- **Iris, row 5**: Perhaps the best-known database to be found in the pattern recognition literature. The number of attributes is 4 and the class. The data set contains 3 classes of 50 instances each, where each class refers to a type of iris plant.

3.2. Description of the classifiers

We have done classification in this study by using comparison measure based classifier described in the following section. We have compared results achieved to the results that has been achieved in the report [41] using classifiers called C4.5, C4.5 rules, ITI, LMDT, CN2, LVQ, OC1, Nevprop, K5, Q*, RBF, SNNS.

3.2.1. Description of the comparison measure based classifier

We want to classify objects, each characterized by one feature vector in $[0,1]^n$ to different classes. Assumption that vectors belongs to $[0,1]^n$ is not restrictive since appropriate shift and normalization can be done for any space $[a,b]^n$. Comparison measure equations above can be used to compare objects to classes.

In the algorithm below we denote feature vectors by u_m, $m = 1,\ldots,M$ i.e. we have M objects to be classified. We also denote the number of different classes by L. General classification procedure that we have used can be described now as follows:

Step 1: Choose range of m and p used with Comparison measure marked here as CM.

Step 2: Choose ideal vectors f_l that presents classes as well as possible. These ideal vectors can be either given by expert knowledge or calculated in some way from training set. We have calculated one ideal vector f_l for each class l by using generalized mean i.e.

$$f_l(i) = \left(\frac{1}{n_l} \sum_{j=1}^{n_l} (v_{j,l}(i))^m \right)^{\frac{1}{m}} \quad \forall i \in \{1,\ldots,n\}, \tag{28}$$

, where vectors $v_{j,l}$ are known to belong to class l and n_l is number of those vectors.

Step 3: Choose values for weights w_i. For example evolutionary algorithms can be used when training data is available.

Step 4: Compare each feature vector u_i to each ideal vector f_l, i.e. calculate $CM\,(u_m, f_l; m; p; w)$ for all $m \in \{1, \ldots, M\}$ and $l = 1, \ldots, L$.

Step 5: Make decision that feature vector u_m belongs to that class k for which $CM\,(u_m, f_k; m; p; w) = \max\{CM\,(u_m, f_l; m; p; w) \mid l = 1, \ldots, L\}$

We see that respect to L, the number of classes, classification time is $\mathcal{O}(L)$ after parameters have been fixed and ideal vectors calculated. However, finding good parameters could be hard but it seemed that all measures presented above are relatively stable with respect to parameters, which is demonstrated in examples.

Classifier used in this comparison tasks is presented more precisely in [43]. The advantage of using this classifier here is that it results mainly depends on which comparison measure we choose to use.

3.2.2. *Short description of other classifiers*

Followings are very short descriptions of classifiers that we use for comparison of our classification results and much better descriptions can be found from [41].

- C4.5 and C4.5 rules are popular decision tree classifiers.
- ITI is described as being incremental decision tree induction classifier.
- LMDT is described as being linear machine decision tree classifier.
- CN2 is a rule-based inductive learning system classifier.
- LVQ is learning vector quantization based classifier.
- OC1 is called Oblique Classifier and it is a system for induction of oblique decision trees.
- Nevprop in other words Nevada backpropagation classifier is a feed-forward backpropagation multi-layer perceptron simulator.
- K5 is a k-nearest neighbor classifier with k$= 5$
- Q* is similar classifier as LVQ.
- RBF can be described radial basis function network classifier.
- SNNS is described as Stuttgart neural network simulator.

In the following we are going to show table of average classification results some figures and shortly discuss about the results.

3.3. *Classification results*

In the table 1 we have compared average classification results vs. to the results available in the report [41]. Results are represented as percents of correctly classified data. Best average results are marked bold.

Data sets are by rows as follows $1 =$ Post Operative, $2 =$Ionosphere, $3 =$Wine, $4 =$Pima Indian, $5 =$Iris. **Sixth row** shows average result from classification averages. If the corresponding classifier has been unable to give classification results it result has been considered as zero.

In column marked as CM we have used comparison measure based classification and used comparison measures have been combined measure 7 from Dombi t-norm 8 and t-conorm 13 (row 1, no optimization), Frank combo 7 from Frank t-norm 9 and t-conorm 14 (row 2), Łukasiewicz equivalence 26 (row 3, no optimization), Frank combo 7 from Frank t-norm 9 and t-conorm 14 (row 4), Łukasiewicz equivalence 26 (row 5, no

Table 1. Average classification results

#	CM	C4.5	C4.5 rules	ITI	LMDT	CN2	LVQ	OC1	Nevprop	K5	Q*	RBF	SNNS
1	**78.04**	62.57	60.05	59.48	66.88	57.91	x	x	x	x	x	x	x
2	90.02	91.56	91.82	**93.65**	86.89	90.98	88.58	88.29	83.80	85.91	89.70	87.60	x
3	**99.24**	91.09	91.90	91.09	95.40	91.09	68.90	87.31	95.41	69.49	74.35	67.87	x
4	**78.99**	71.02	71.55	73.16	73.51	72.19	71.28	50.00	68.52	71.37	68.50	70.57	x
5	**100**	91.60	91.58	91.25	95.45	91.92	91.44	93.89	90.34	91.94	92.10	85.64	93.55
6	**89.26**	81.57	81.38	81.73	83.63	80.82	64.04	63.90	67.61	63.74	64.93	62.34	18.71

optimization). Words *no optimization* refers that we did not managed to get better result with optimized weights.

Following figures show some main features of the comparison measures that we want to underline.

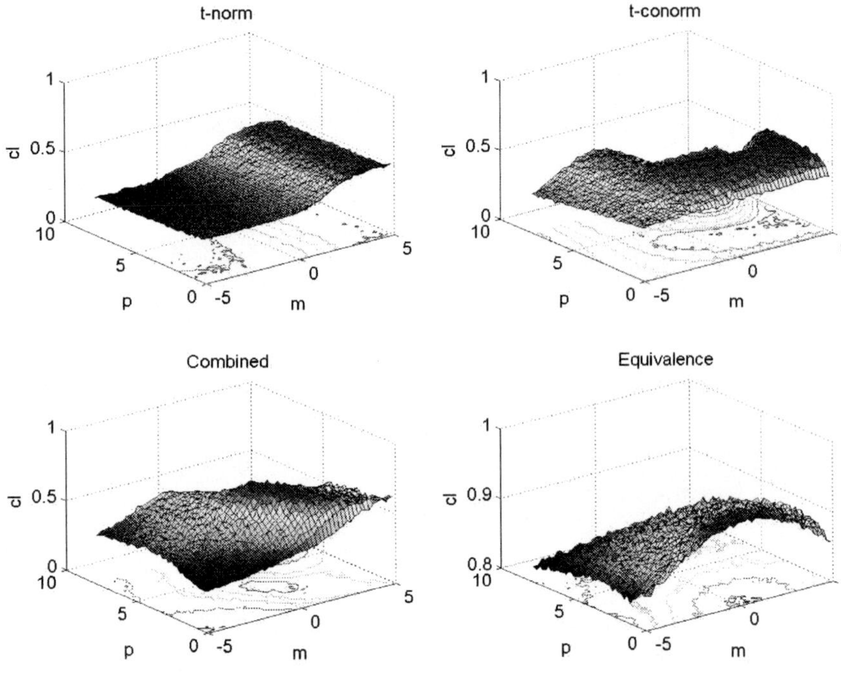

Figure 1. Average classification results from Wine data without weight optimization

Figure 1 represents the mean classification results from the Wine data set. Figure 1 clearly shows how the GWNO measure 7 (title: Combined) compensates the classification results achieved by the t-norm measure 9 or the t-conorm measure 14 alone. We also see that results from the combined measure are significantly different from the results achieved from the use of equivalence 2.

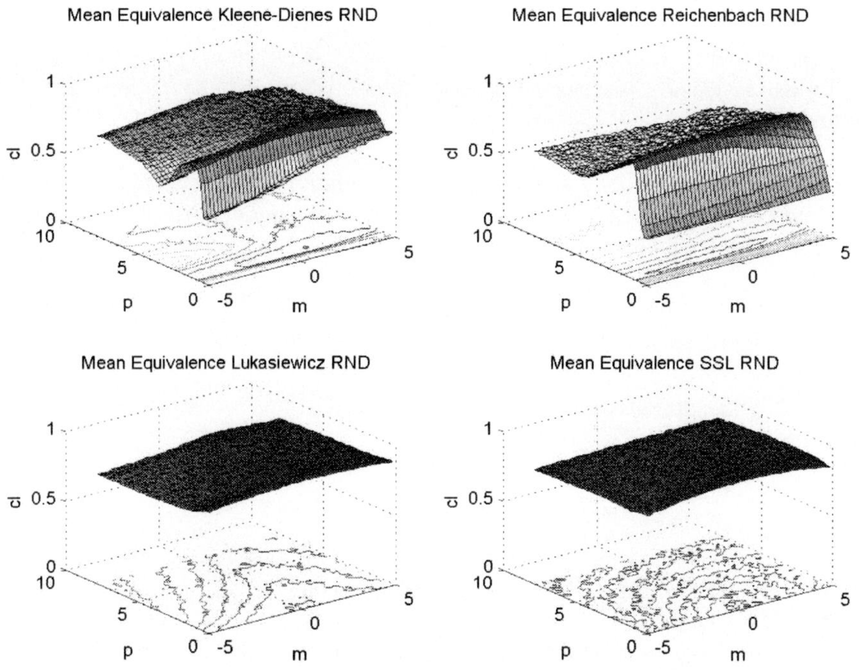

Figure 2. Average classification results using equivalences without weight optimization for wine data

Figure 2 shows the typical classification results for equivalences from which we see that the generalized Kleene-Dienes pseudo equivalence measure 20 and the generalized Reichenbach pseudo equivalence measure 23 give results which have very similar topology. Same phenomena can be seen between the generalized semantical Łukasiewics equivalence 26 and the generalized Schweizer & Sklar - Łukasiewicz equivalence 27. This picture also shows that the results achieved by the use of 26 and 27 are highly stable, and from the contours it seems that 26 more stable than 27.

4. Conclusions and future

In classification and the development of the expert systems one often faces the problem of choosing the right functions for comparison. Usually simplest operators are selected, which are not normally the optimal choice. Basically every area of humans involved some kind of measures for comparison are needed. In soft computing these areas are for example classification, pattern recognition, clustering, expert systems, medical diagnosis systems, decision support systems, fuzzy control etc. Recently new areas are for example in web-search engines, where information retrieval is of high importance.

In this chapter we offered two different approaches to create comparison measures, which are on theoretically sound basis. We also formulated several comparison measures what can be created form these approaches.

In practical part we showed that mainly results achieved by using these simple comparison measures see table 1 were better in classification tasks chosen to an example of comparison than any of the public domain classifier results found from [41]. We also see from pictures 1 and 2 that normally these classification results stay quite stable in any parameter values.

We used differential evolution for optimization of weights and parametrization in order to give high degree of adaptivity for our comparison measures. The another interesting approach is to create measures and aggregation operators straight from the data in hand like Gleb Beliakov has done in [44], [45], [46]

Acknowledgements

This work was supported by EU Asia Link Project (Contract Reference no.: ASI/B7-301/98/679-023).

References

[1] W. James, *The Principles of Psychology* , Dover: New York, Vol. 1, Chapter 7, 1890/1950.

[2] S. Haykin *Neural Networks: A Comprehensive Foundation, 2nd Edition*, Prentice Hall, 1998.

[3] L.A. Zadeh, *Similarity relations and fuzzy orderings*, Inform. Sci. 3, 177-200, 1971.

[4] F. Klawonn, J.L. Castro, *Similarity in Fuzzy Reasoning*. Mathware and Soft Computing, 3(2): 197-228, 1995.

[5] E. Trillas, L. Valverde, *An Inquiry into Indistinguishability Operators*, in: 'Aspects of Vaguenes', Skala, H.J., Termini, S., Trillas, E., eds., Reidel, Dordrecht, 231–256, 1984.

[6] U. Höhle, L.N. Stout, *Foundations of Fuzzy Sets*, Fuzzy Sets and Systems, 40(2), 257-296, 1991.

[7] D. Dubois, H. Prade, *Similarity-Based Approximate Reasoning*, in: 'Computational Intelligence Imitating Life.' Zurada, J.M., Marks II, R.J., Robinson, C.J., eds., IEEE Press, New York, 69–80, 1994.

[8] A. Tversky, D.H. Krantz, *The Dimensional Representation and the Metric Structure of Similarity Data*, Journal of Mathematical Psychology, 572-596, 1970.

[9] S. Santini , R. Jain, *Similarity Measures*, IEEE Transactions on Pattern Analysis and Machine Intelligence, 21(9), 871-883, 1999.

[10] M. De Cock, E. E. Kerre *Why Fuzzy T-Equivalence Relations do Not Resolve the Poincaré Paradox, and Related Issues*, Fuzzy Sets and Systems 133(2), 181-192, 2003.

[11] R. Jain, S. N. Murthy, L. Tran, S. Chatterjee, *Similarity Measures for Image Databases*, SPIE Proceedings, Storage and Retrieval for Image and Video Databases, 58-65 1995.

[12] L.A. Zadeh, *Fuzzy Sets and Their Application to Pattern Classification And Clustering Analysis*, in J. Van Ryzin (Ed.): Classification and Clustering, Academic Press, 251-299, 1977.

[13] S.K. Pal S.K. and D.K. Dutta-Majumder, *Fuzzy Mathematical Approach to Pattern Recognition* John Wiley & Sons (Halsted), N. Y. 1986.

[14] D. Dubois and H. Prade, *A review of fuzzy set aggregation connectives*, Information Sciences, 36, 85-121, 1985.

[15] J. Fodor and R. R. Yager, *Fuzzy set-theoretic operators and quantifiers*, In: 'Fundamentals of Fuzzy Sets', edited by Dubois, D. and Prade, H., Kluwer Academic Publishers: Norwell, Ma, 125-193, 2000.

[16] K. Saastamoinen, *On the Use of Generalized Mean with T-norms and T-conorms*, proceedings of the IEEE 2004 Conference on Cybernetic and Intelligent Systems, Singapore.

[17] K. Saastamoinen and J. Sampo *On General Class of Parameterized $3\pi-$ Uninorm Based Comparison*, WSEAS TRANSACTIONS on MATHEMATICS, 3(3), 482-486, 2004.

[18] K. Saastamoinen, *Semantic Study of the Use of Parameterized S Implications and Equivalences in Comparison*, proceedings of the IEEE 2004 Conference on Cybernetic and Intelligent Systems, Singapore.

[19] K. Saastamoinen, J. Ketola, *Defining Athlete's Anaerobic and Aerobic Thresholds by Using Similarity Measures and Differential Evolution*, proceedings of the IEEE SMC 2004 conference, Hague, Netherlands.

[20] H. Dyckhoff, W. Pedrycz, *Generalized Means as Model of Compensative Connectives*, Fuzzy Sets and Systems, 14, pp. 143-154, 1984.

[21] R. R. Yager, *On ordered weighted averaging aggregation operators in multi-criteria decision making* IEEE Transactions on Systems, Man and Cybernetics, 18, 183-190, 1988.

[22] R. R. Yager, *Generalized OWA aggregation operators*, Fuzzy Optimization and Decision Making, 3, 93-107, 2004.

[23] J. Dombi, : *Basic Concepts for a theory of evaluation: The aggregative operator*, European Journal of Operational Research 10, 282-293, 1982.

[24] L.A. Zadeh, *Fuzzy Sets*, Information and Control, 8, 338-353, 1965.

[25] K. Menger, Statistical Metrics. CProc. Nat. Acad. Sci. U.S.A. (1942) 37:535-537.

[26] B. Schweizer and A. Sklar, Statistical Metric Spaces. Pacific J. Math. (1960) 10:313-334.

[27] E. H. Shortliffe, Computer-Based Medical Consultation: MYCIN. Elsevier, New York (1976).

[28] H.-J. Zimmermann and P. Zysno, Latent connectives in human decision making. Fuzzy Sets and Systems (1980) 4:37-51.

[29] T. Bilgiç and I.B. Türkşen, Measurement-theoretic Justification of Connectives in Fuzzy Set Theory. Fuzzy Sets and Systems (1995) 76(3):289-308.

[30] L.I. Kuncheva, How good are fuzzy if-then classifiers? IEEE Transactions on Systems, Man, and Cybernetics, Part B: Cybernetics, 30 (4), pp. 501-509, 2000.

[31] P. Smets, P. Magrez, *Implication in fuzzy logic*, Int. J. Approx. Reasoning 1(4): 327-347, 1987.

[32] K. Saastamoinen, V. Könönen and P. Luukka, *A Classifier Based on the Fuzzy Similarity in the Łukasiewicz-Structure with Different Metrics*, proceedings of the FUZZ-IEEE 2002 Conference, Hawaii, USA.

[33] K. Saastamoinen and P. Luukka, *Testing Continuous t-norm called Łukasiewicz Algebra with Different Means in Classification*, proceedings of the FUZZ-IEEE 2003 Conference, St Louis, USA.

[34] G.J. Klir, B. Yuan, *Fuzzy Sets and Fuzzy Logic: Theory and Applications*, Prentice Hall, PTR, Upper Saddle River, NJ: pp. 304-321, 1995.

[35] T. Whalen, *Parameterized R-implications*, Fuzzy Sets and Systems 134(2): pp. 231-281, 2003.

[36] M.J. Frank, *On the simultaneous associativity of $F(x, y)$ and $x + y - F(x, y)$*, Aeqationes Math. (19), pp. 194-226, 1979.

[37] B. Schweizer and A. Sklar, *Associative functions and abstract semigroups*, Publ. Math. Debrecen 10, pp. 69-81, 1963.

[38] R. R. Yager, *On a General Class of Fuzzy Connectives*, Fuzzy Sets and Systems 4, pp. 235-242, 1980.

[39] Y. Yu, *Triangular norms and TNF-sigma algebras*, Fuzzy Sets and Systems 16, pp. 251-264 1985.

[40] T. Hastie and R. Tibshirani *The Elements of Statistical Learning: Data Mining, Inference, and Prediction*, Springer Series in Statistics, Springer, New York, 2001.

[41] P.W. Eklund, *A Performance Survey of Public Domain Supervised Machine Learning Algorithms*, KVO Technical Report 2002, The University of Queensland, submitted, 2002.

[42] UCI ML Rep., various authors. UCI Repository of Machine Learning Databases network document. http://www.ics.uci.edu/~mlearn/MLRepository.html, Accessed March 14, 2004.

[43] P. Luukka, emphSimilarity measure based classification PhD thesis, Lappeenranta University of Technology, 2005.

[44] G. Beliakov, *Definition of general aggregation operators through similarity relations*, Fuzzy Sets and Systems 114 (3), pp. 437 - 453, 2000.

[45] G. Beliakov, *How to build aggregation operators from data*, Int. J. Intell. Syst. 18(8), pp. 903-923, 2003.

[46] G. Beliakov, R. Mesiar, L. Valaskova, *Fitting Generated Aggregation Operators To Empirical Data*, International Journal of Uncertainty, Fuzziness and Knowledge-Based Systems 12(2), pp. 219-236, 2004.

Integrated Intelligent Systems for Engineering Design
X.F. Zha and R.J. Howlett (Eds.)
IOS Press, 2006

Design of Fuzzy Models through Particle Swarm Optimization

Arun KHOSLA [a,1], Shakti KUMAR [b], K.K. AGGARWAL [c] and Jagatpreet SINGH [d]
[a] *National Institute of Technology, Jalandhar – 144011. India*
[b] *Haryana Engineering College, Jagadhari – 135003. India*
[c] *GGS Indraprastha University, Delhi – 110006. India*
[d] *Infosys Technologies Limited, Chennai – 600019. India*

Abstract. Particle Swarm Optimization (PSO), which is a robust stochastic evolutionary computation engine, belongs to the broad category of swarm intelligence (SI) techniques. SI paradigm has been inspired by the social behavior of ants, bees, wasps, birds, fishes and other biological creatures and is emerging as an innovative and powerful computational metaphor for solving complex problems in design, optimization, control, management, business and finance. SI may be defined as any attempt to design distributed problem-solving algorithms that emerges from the social interaction. The objective of this chapter is to present the use of PSO algorithm for building optimal fuzzy models from the available data. The fuzzy model identification procedure using PSO as an optimization engine has been implemented as a Matlab toolbox and is also presented in this chapter. For the purpose of illustration and validation of the approach, the data from the rapid Nickel-Cadmium (Ni-Cd) battery charger developed by the authors has been used.

Keywords. Fuzzy models, particle swarm optimization, strategy parameters, fitness function, search-space.

1. Introduction

Developing models of complex real-systems is an important topic in many disciplines of engineering. Models are generally used for simulation, identifying the system's behavior and design of controllers etc. Last few years have witnessed a drastic growth of sub-disciplines in science and engineering that have adopted the concepts of fuzzy set theory. This development can be attributed to successful applications in consumer electronics, robotics, signal processing, image processing, finance, management etc.

---

[1] Corresponding Author: Arun Khosla, Department of Electronics and Communication Engineering, National Institute of Technology, Jalandhar. India. Email: khoslaak@nitj.ac.in, arun.khosla@gmail.com

Design of fuzzy model or fuzzy model identification is the task of finding the parameters of fuzzy model so as to get the desired behavior. Two principally different approaches are used for the design of fuzzy models: heuristic-based design and model-based design. In the first approach, the design is constructed from the knowledge acquired from the expert, while in the second, the input-output data is used for building model. It is also possible to integrate both the approaches. In this chapter, we have presented the use of PSO algorithm for the identification of fuzzy models from the available data.

This chapter is organized as follows. In Section 2, a brief introduction to PSO algorithm is presented. Overview of fuzzy models alongwith various issues about fuzzy model identification problem are described in Section 3. A methodology for fuzzy model identification using PSO algorithm is presented in Section 4. This methodology has been implemented as a Matlab toolbox and is described in Section 5. Simulation results generated from this toolbox are presented in Section 6. The concluding remarks are made in Section 7 and some of the future trends are discussed in Section 8.

2. PSO Algorithm

The origin of PSO is best described as sociologically inspired, since it was initially developed as a tool by Reynolds [1][2] for simulating the flight patterns of birds, which was mainly governed by three major concerns: collision avoidance, velocity matching and flock centering. On the other hand, the reasons presented for the flocking behaviour observed in nature are: protection from predator and gaining from a large effective search with respect to food. The last reason assumes a great importance, when the food is unevenly distributed over a large region. It was realized by Kennedy and Eberhart that the bird flocking behavior can be adapted to be used as an optimizer and resulted in the first simple version of PSO [3] that has been recognized as one of the computational intelligence technique intimately related to evolutionary algorithms. Like evolutionary computation techniques, it uses a population of potential solutions called particles that are flown through the hyperspace/search-space. In PSO, the particles have an adaptable velocity that determines their movement in the search-space. Each particle also has a memory and hence it is capable of remembering the best position in the search-space ever visited by it. The position corresponding to the best fitness is known as *pbest* and the overall best out of all the particles in the population is called *gbest*.

Consider that the search-space is d-dimensional and i-th particle in the swarm can be represented by $X_i = (x_{i1}, x_{i2}, \ldots, x_{id})$ and its velocity can be represented by another d-dimensional vector $V_i = (v_{i1}, v_{i2}, \ldots, v_{id})$. Let the best previously visited position of this particle be denoted by $P_i = (p_{i1}, p_{i2}, \ldots, p_{id})$. If g-th particle is the best particle and the iteration number is denoted by the superscript, then the swarm is modified according to the Eqs. (1) and (2) suggested by Shi & Eberhart [4].

$$v_{id}^{n+1} = wv_{id}^{n} + c_1 r_1^n (p_{id}^n - x_{id}^n) + c_2 r_2^n (p_{gd}^n - x_{id}^n) \quad (1)$$

$$x_{id}^{n+1} = x_{id}^{n} + v_{id}^{n+1} \quad (2)$$

where,
w – inertia weight
c_1 – cognitive acceleration
c_2 – social acceleration
r_1, r_2 – random numbers uniformly distributed in the range (0,1).

These parameters viz. inertia weight (w), cognitive acceleration (c_1), social acceleration (c_2), alongwith V_{max} [4] are known as the strategy/operating parameters of PSO algorithm. These parameters are defined by the user before the PSO run. The parameter V_{max} is the maximum velocity along any dimension, which implies that, if the velocity along any dimension exceeds V_{max}, it shall be clamped to this value. The inertia weight governs how much of the velocity should be retained from the previous time step. Generally the inertia weight is not kept fixed and is varied as the algorithm progresses so as to improve performance [4][5]. This setting allows the PSO to explore a large area at the start of simulation run and to refine the search later by a smaller inertia weight. The parameters c_1 and c_2 determine the relative pull of *pbest* and *gbest.* Random numbers r_1 and r_2 help in stochastically varying these pulls, that also account for slight unpredictable natural swarm behavior. Figure 1 depicts the position update of a particle for a two-dimensional parameter space. Infact, this update is carried out as per Eqs. (1) and (2) for each particle of swarm for each of the M dimensions in an M-dimensional optimization.

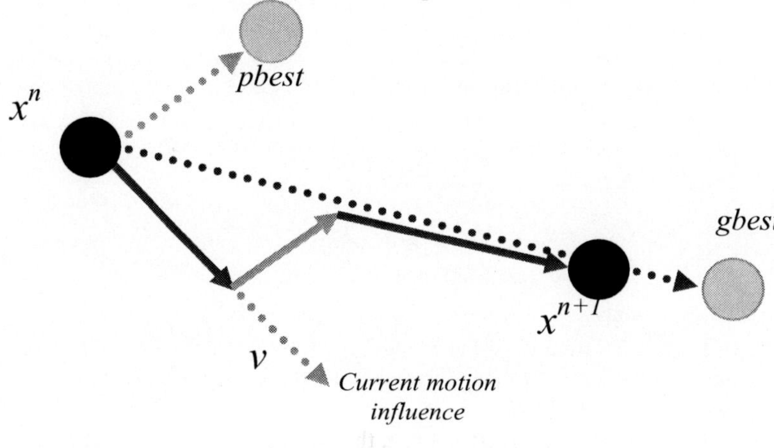

Figure 1. Depiction of position updates in particle swarm optimization for 2-D parameter space

3. Fuzzy Models

This section reviews the fuzzy model structures and the various issues associated with the fuzzy model identification. Basic knowledge about the fuzzy sets, fuzzy logic and fuzzy inference system is assumed.

3.1. Overview of Fuzzy Models

Three commonly used types of fuzzy models are [6]:

- Mamdani fuzzy models
- Takagi-Sugeno fuzzy models
- Singleton fuzzy models

In Mamdani models, each fuzzy rule is of the form:
$\boldsymbol{R_i}$: **If** x_1 is A_{i1} **and… and** x_n is A_{in} **then** y is B
In Takagi-Sugeno models, each fuzzy rule is of the form:

$\boldsymbol{R_i}$: **If** x_1 is A_{i1} **and… and** x_n is A_{in} **then** y is $\sum_{i=1}^{n} a_i x_i + C$

whereas, in Singleton models, each fuzzy rule is of the form:
$\boldsymbol{R_i}$: **If** x_1 is A_{i1} **and… and** x_n is A_{in} **then** y is C

where,
$x_1,\ldots, x_n$ are the input variables and y is the output variable, $A_{i1},\ldots, A_{in}$, B are the linguistic values of the input and output variables in the i-th fuzzy rule and a_i and C are constants. Infact, Singleton fuzzy model can be seen as a special case of Takagi-Sugeno model when a_i=0. The input and output variables take their values in their respective universes of discourse or domains. Identification of Mamdani and Singleton fuzzy models has been considered in this chapter.

3.2. Fuzzy Model Identification Problem

Generally, the problem of fuzzy model identification includes the following issues [6][7]:

- Selecting the type of fuzzy model
- Selecting the input and output variables for the model
- Identifying the structure of the fuzzy model, which includes determination of the number and types of membership functions for the input and output variables and the number of fuzzy rules
- Identifying the parameters of antecedent and consequent membership functions
- Identifying the consequent parameters of the fuzzy rulebase

Some commonly used techniques for creating fuzzy models from the available input-output data are Genetic Algorithms [8][9][10][11], Fuzzy C-Means (FCM) clustering algorithm [12][13], Neural Networks [6] and Adaptive Neuro Fuzzy Inference System model (ANFIS) [14][15].

4. A Methodology for Fuzzy Models Identification through PSO

Fuzzy model identification can be formulated as a search and optimization problem in high-dimensional space, where each point corresponds to a fuzzy system i.e. represents membership functions, rule-base and hence the corresponding system behaviour. Given some objective/fitness function, the system performance forms a hypersurface and designing the optimal fuzzy system is equivalent to finding the optimal location on this hypersurface. The hypersurface is generally found to be infinitely large, nondifferentiable, complex, noisy, multimodal and deceptive [16], which make evolutionary algorithms very suitable for searching the hypersurface than the traditional gradient-based methods. PSO algorithms like GAs have the capability to find optimal or near optimal solution in a given complex search-space and can be used to modify/learn the parameters of fuzzy model. The methodology to identify the optimal fuzzy models using PSO as an optimization engine is shown in Figure 2.

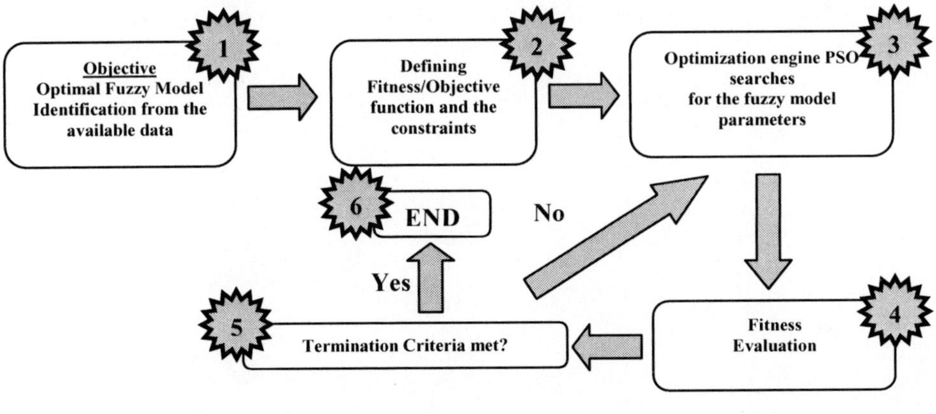

Figure 2. Optimal fuzzy model identification using PSO as an optimization engine

An optimization problem can be represented as a tuple of three components as represented in Figure 3 and explained below:

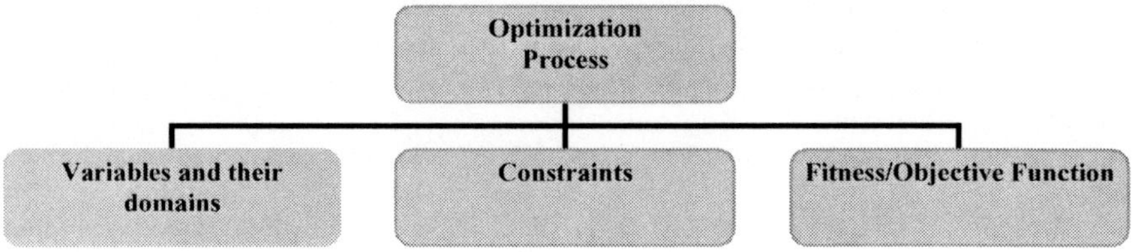

Figure 3. Representation of optimization process

- **Solution Space** – The first step in the optimization step is to pick up the variables to be optimized and define the domain/range in which to search for the optimal solution.
- **Constraints** – It is required to define a set of constraints which must be followed by the solutions.
- **Fitness/Objective Function** – The fitness/objective function represents the quality of each solution and also provides a link between the optimization algorithm and the problem under consideration.

The objective of optimization problem is to look for the values of the variables being optimized, that satisfy the defined constraints, which maximizes or minimizes the fitness function. Hence, it is required to define the solution space, constraints and the fitness function when using PSO for the identification of optimized fuzzy models.

In this chapter, we have used Mean Square Error (MSE) defined in Eq. (3) as fitness/objective function for rating the fuzzy model.

$$\mathrm{MSE} = \frac{1}{Z}\sum_{k=1}^{Z}[\mathrm{y(k)} - \tilde{\mathrm{y}}\mathrm{(k)}]^2 \qquad (3)$$

where,
y(k) – desired output
$\tilde{y}$ (k) – actual output of the model
Z – number of data points taken for model validation

A very important consideration is to completely represent a fuzzy system by a particle, and for this, all the needed information about the rule-base and membership functions is required to be specified through some encoding mechanism. It is also suggested to modify the membership functions and rule-base simultaneously, since they are codependent in a fuzzy system [16].

For the purpose of fuzzy model encoding, consider a multi-input single-output (MISO) system with n number of inputs. The number of fuzzy sets for the inputs are m_1, m_2, m_3,……, m_n respectively. Following assumptions have been made for encoding:

i) Fixed numbers of triangular membership functions were used for both input and output variables with their centres fixed and placed symmetrically over corresponding universes of discourse.

ii) First and last membership functions of each input and output variable were represented with left- and right-skewed triangles respectively.

iii) Complete rule-base was considered, where all possible combinations of input membership functions of all the input variables were considered for rule formulation.

iv) Overlapping between the adjacent membership functions for all the variables was ensured through some defined constraints.

In this chapter, we have considered only MISO fuzzy models, as multi-input multi-output (MIMO) models can be constructed by the parallel connection of several MISO models.

4.1. Encoding Mechanism (Membership functions)

Consider a triangular membership function and let parameters x_k^l, x_k^c and x_k^r represent the coordinates of the left anchor, cortex and right anchor of the kth linguistic variable as shown in Figure 4.

A straightforward way to characterize this membership function is by means of 3-tuple (x_k^l, x_k^c, x_k^r). Therefore, a particle carrying details about the parameters of the membership functions of all the input and output variables can be represented as follows:

$$(x_1^l, x_1^c, x_1^r, x_2^l, x_2^c, x_2^r, \ldots, x_n^l, x_n^c, x_n^r, x_{n+1}^l, x_{n+1}^c, x_{n+1}^r)$$

The index $n+1$ is associated with the membership functions of the output variable.

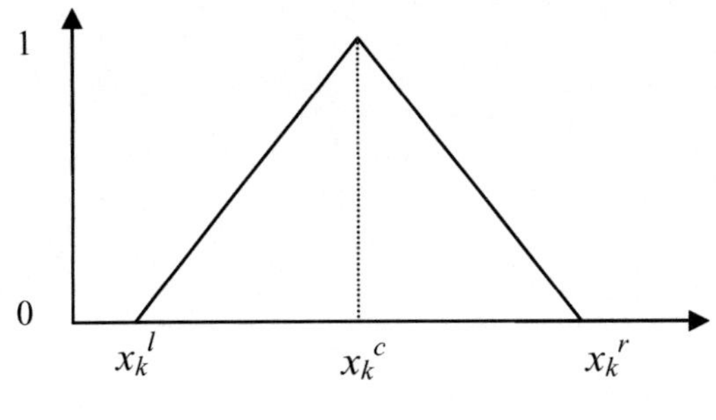

Figure 4.. Representation of a triangular membership function

It was ensured that following constraints are followed by every membership function of input and output variables.

$$x_k^l < x_k^c < x_k^r$$

At the same time, the overlapping between the adjacent membership functions was also ensured by defining some additional constraints. Let's assume that a variable is represented by three fuzzy sets as in Figure 4, then those additional constraints to ensure overlapping can be represented by the following inequality.

$$x_{min} \leq x_2^l < x_1^r < x_3^l < x_2^r \leq x_{max}$$

where, x_{min} and x_{max} are the minimum and maximum values of the variable respectively.

The dimensions of the particle representing Mamdani fuzzy model can be worked out from Figure 5, which represents the membership functions for any one of the input/output variables with three membership functions. Thus, four dimensions are required for each variable, which are to be modified during PSO run.

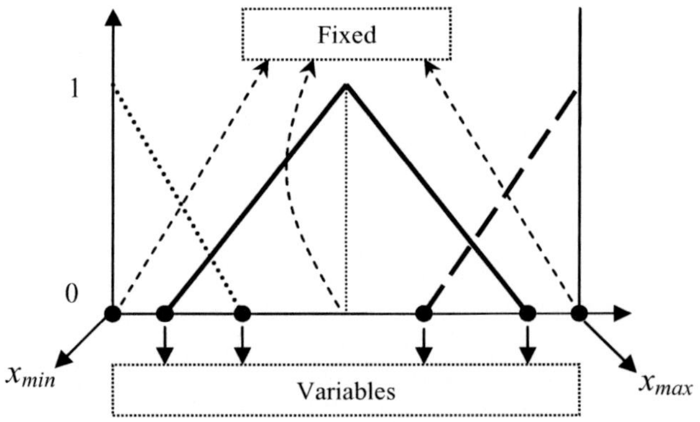

Figure 5.. Representation of a variable with 3 membership functions with centre of each membership function fixed with overlapping between the adjacent membership functions

The representation can be generalized to Eq. (4).

$$\text{Particle Size} = 2m_i - 2 \tag{4}$$

Thus the particle size for representing the membership functions of input and output variables for a Mamdani model is given by Eq. (5).

$$\text{Particle Size (for membership functions)} = \sum_{i=1}^{n+1}(2m_i - 2) \tag{5}$$

where,
n – number of input variables
m_i – number of fuzzy sets for i-th input and the index $n+1$ corresponds to the membership functions of the output variable.

4.2. Encoding Mechanism (Fuzzy rules)

Considering the complete rule base, the particle size required for its representation is given by Eq. (6).

$$\text{Particle Size (for rule base)} = \prod_{i=1}^{n} m_i \tag{6}$$

Thus, the particle size required for representing the complete Mamdani fuzzy model can be calculated through Eq. (7), obtained by adding Eqs. (5) and (6).

$$\text{Particle Size (Mamdani model)} = \sum_{i=1}^{n+1}(2m_i - 2) + \prod_{i=1}^{n} m_i \tag{7}$$

If Singleton fuzzy model is considered with possible t number of consequent singleton values, then the particle dimensions required for representing this model can be obtained from Eq. (7) after a little modification, which is represented by Eq. (8).

$$\text{Particle Size (Singleton model)} = \sum_{i=1}^{n}(2m_i - 2) + t + \prod_{i=1}^{n} m_i \tag{8}$$

A particle representing a fuzzy model whose membership function parameters of input/output variables and rule consequents can be modified through PSO algorithm is shown in Figure 6.

The suggested methodology can be extended to increase the flexibility of search by incorporating additional parameters so as to execute the search for optimal solutions in terms of types of membership functions for each variable and the number of rules also. Particle representing fuzzy model and implementing this approach is shown in Figure 7.

For such an implementation, the expression for particle size for encoding Mamdani fuzzy model would be as in Eq. (9).

$$\text{Particle Size (Mamdani model)} = 3\sum_{i=1}^{n+1} m_i + 2\prod_{i=1}^{n} m_i \quad (9)$$

The corresponding expression for the particle size to encode Singleton fuzzy model would be as given in the Eq. (10).

$$\text{Particle Size (Singleton model)} = 3\sum_{i=1}^{n} m_i + t + 2\prod_{i=1}^{n} m_i \quad (10)$$

In Figure 7, each membership function is represented by three dimensions representing the start value, end value and the type of membership function like sigmodial, triangular etc. Two dimensions have been reserved for each rule, one representing the consequent value and other a flag. If the rule flag is '1', the rule is included, and for '0', it won't be part of the rule-base.

Let's consider a system with two-inputs and single output. If we further consider that each input and output variable for this system is represented by three fuzzy sets, and five possible consequent values for the Singleton model, then the particle size for two different models implementing the two frameworks are represented in the Table 1.

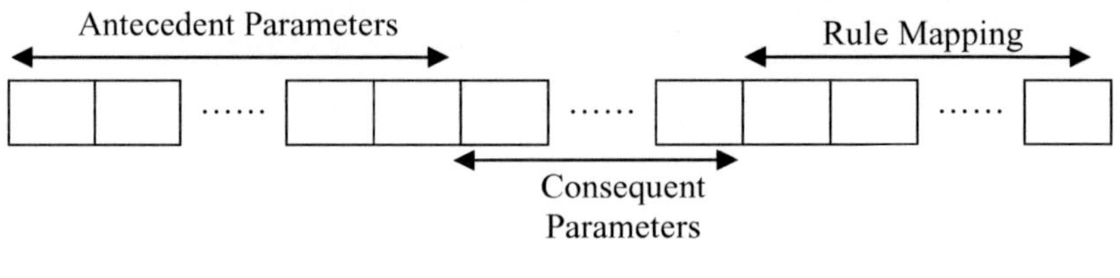

Figure 6. Representation of a fuzzy model by a particle

Table 1. Particle Size for different fuzzy models

Fuzzy model parameters modified/not modified through PSO	Mamdani	Singleton
Parameters modified: MFs parameters, rule consequents Parameters not modified: Types of MFs, Centres of MFs, Number of MFs, rule-set (complete rule base)	21 (Eq. (7))	22 (Eq. (8))
Parameters modified: MF parameters, MFs type, rule consequents, rule-set Parameters not modified: Number of MFs	45 (Eq. (9))	41 (Eq. (10))

MF - *membership function*

5. Fuzzy Model Identification through PSO: A Matlab Implementation

Matlab is a high-level technical computing language and environment for computation, visualization and programming [17] and is equally popular in academia and industry. Matlab toolbox, which is a collection of Matlab functions, helps in extending its capabilities to solve problems related to some specific domain. The methodology presented in the previous section for the identification of optimized fuzzy models has been implemented as a Matlab toolbox viz. *PSO Fuzzy Modeler for Matlab*. All functions for this toolbox have been developed using Matlab with the Fuzzy Logic toolbox and are listed in Table 2. The toolbox is hosted on SourceForge.net [18], which is world's largest development and download repository of open source code and application. The role of each of the implemented function is explained below in the context of identification of fuzzy models using PSO algorithm, which can be conveniently understood with the help of Figure 8.

Table 2. List of Matlab functions

i)	*RandomParticle*
ii)	*limitSwarm*
iii)	*limitParticle*
iv)	*limitMembershipFunctions*
v)	*limitRules*
vi)	*GetFIS*
vii)	*calculateMSE*

i) *RandomParticle* – To begin searching for the optimal solution in the search-space, each particle begins from some random location with a velocity that is random both in magnitude and direction. The role of this function is to generate such random particles representing the fuzzy models in the search-space. The particle dimensions equal to the search-space dimensions and the number of particles is as defined by the swarm size.

ii) *limitSwarm* – This function calls another function *limitParticle* for each particle.

iii) *limitParticle* – It is important to always ensure that the particles are confined to the search-space and represent feasible solutions. There are possibilities that during the movement of the swarm, some particles may move out of the bounds defined by the system constraints. It is therefore necessary to constrain the exploration to remain inside the valid search-space. Thus, all the particles in the swarm are scrutinized after every iteration to ensure that they represent only valid solutions. To illustrate this, consider that the search-space is three-dimensional represented by a cube as shown in Figure 9(a). During exploration, some particle may move out of the search-space as shown in Figure 9(b). All such particles are required to be brought back to the valid search-space by applying some limiting mechanism shown in Figure 9(c). The function *limitParticle* is further made up of two functions viz. *limitMembershipFunctions* and *limitRules.*

iv) *limitMembershipFunctions* (implemented as a separate function) – The role of this function is to ensure that membership function parameters for every input and output variable are confined within the respective universe of discourse and at the same time satisfy the constraint represented in Figure 5.

v) *limitRules* (implemented within *limitParticle* function) – A fuzzy rule consequent can only refer to one of the membership functions of the output variable. In other words, it can have possible values equal to the number of membership functions of output variable. This limiting can be achieved by using the modulus operator. For example, if there are three (3) membership functions for the output, mod-3 of the consequent values for each fuzzy rule is calculated.

vi) *GetFIS* – Every particle in the search-space is basically representing a fuzzy model and after every iteration the performance of each fuzzy model is to be worked out to determine the movement of all the particles in the swarm. The role of this function is to generate fuzzy inference system (FIS) from each particle. The Fuzzy Logic Toolbox for Matlab has a structure that can be easily modified. This flexibility has been used for modifying the parameters of fuzzy models through PSO. The FIS structure is the Matlab object that contains all the information about the fuzzy inference system viz. variables names, membership function definitions, rule base etc. [19]. The structure is basically a hierarchy of structures as shown in Figure 10, which can be easily modified by editing its .fis text file. The parameters of fuzzy model that are being modified by PSO are represented by the shaded blocks in Figure 10.

vii) *calculateMSE* – As discussed earlier in Section 4, it is imperative to define the fitness/objective function to rate the quality of solutions during the optimization process. This function calculates after every iteration the MSE for each of the fuzzy model represented by each particle of swarm.

Another toolbox developed by the authors is PSO toolbox for Matlab, which is also hosted on SourceForge.net [20]. This toolbox is also a part of PSO Fuzzy Modeler for Matlab and implements the PSO loop. A graphical user interface has also been designed for the user convenience and is shown in Figure 11 and the organization of various modules of PSO Fuzzy Modeler is shown in Figure 12.

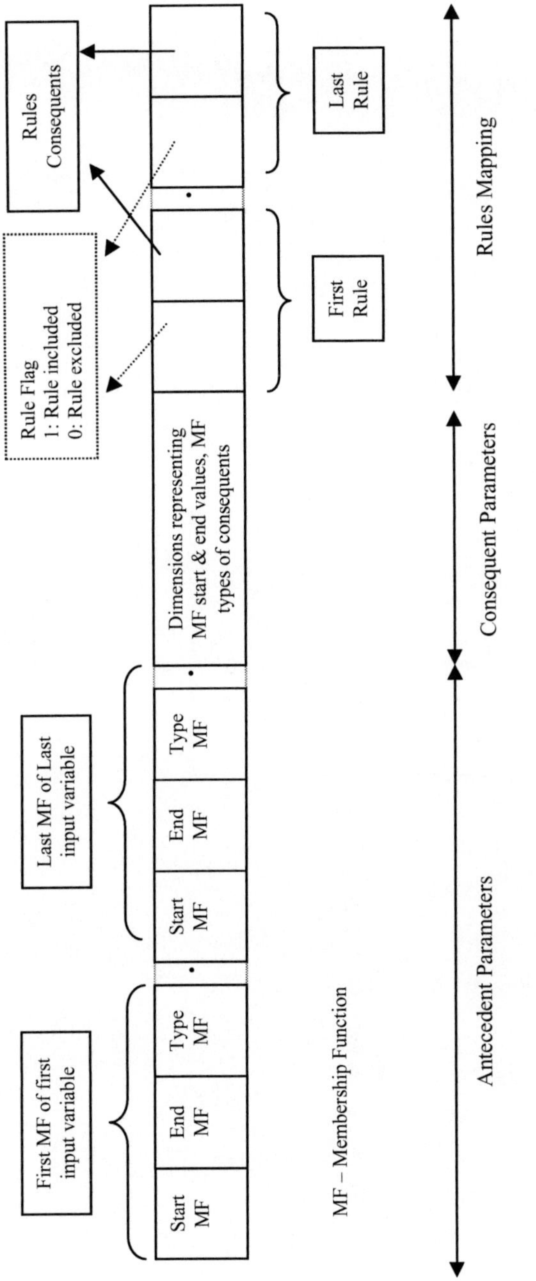

Figure 7. Particle representing Mamdani fuzzy model, where membership function parameters, types and ruleset can be modified through PSO Algorithm

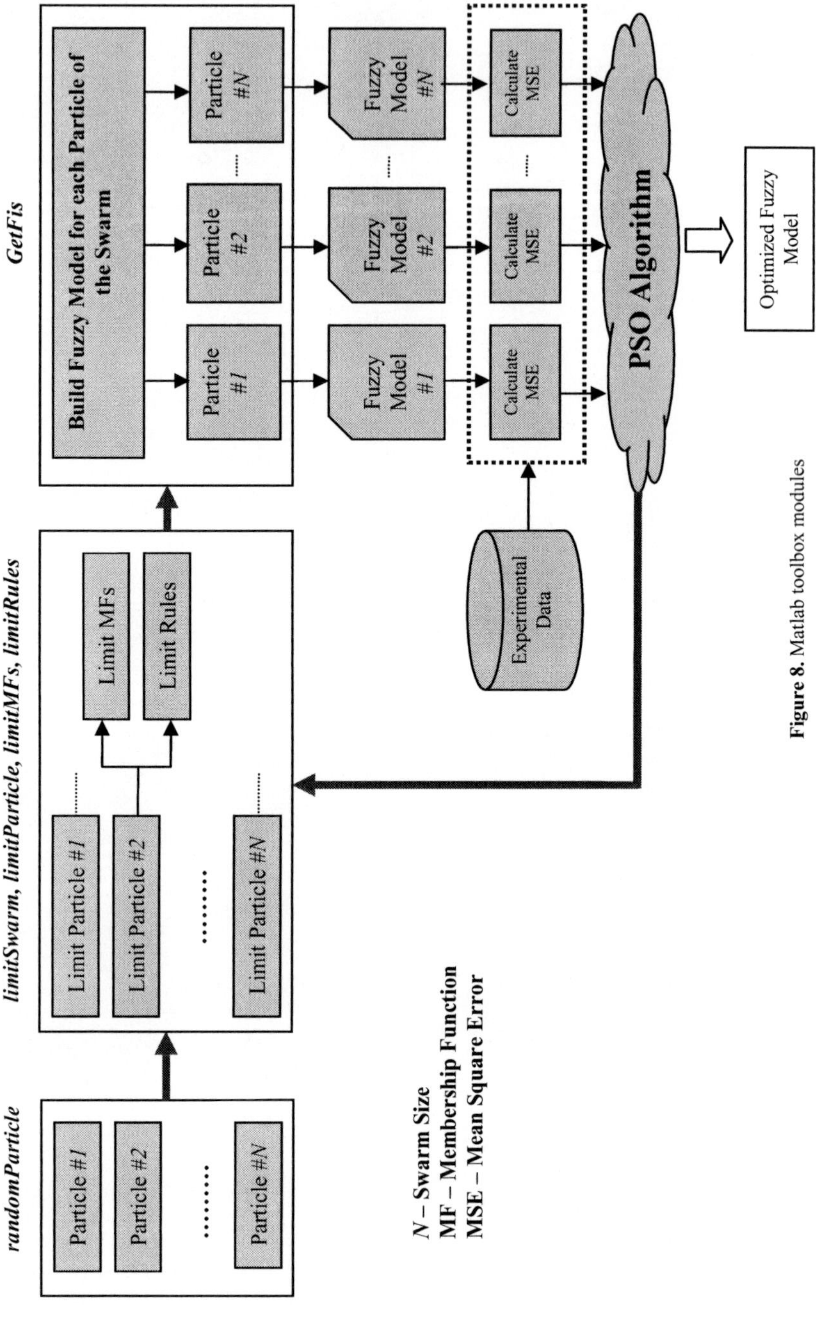

Figure 8. Matlab toolbox modules

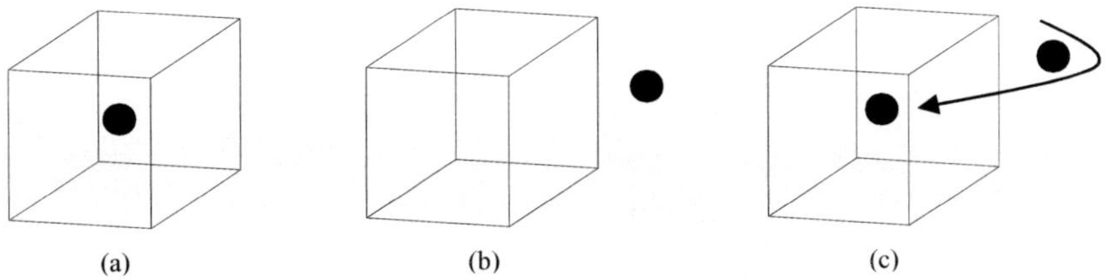

Figure 9. Limiting mechanism representation

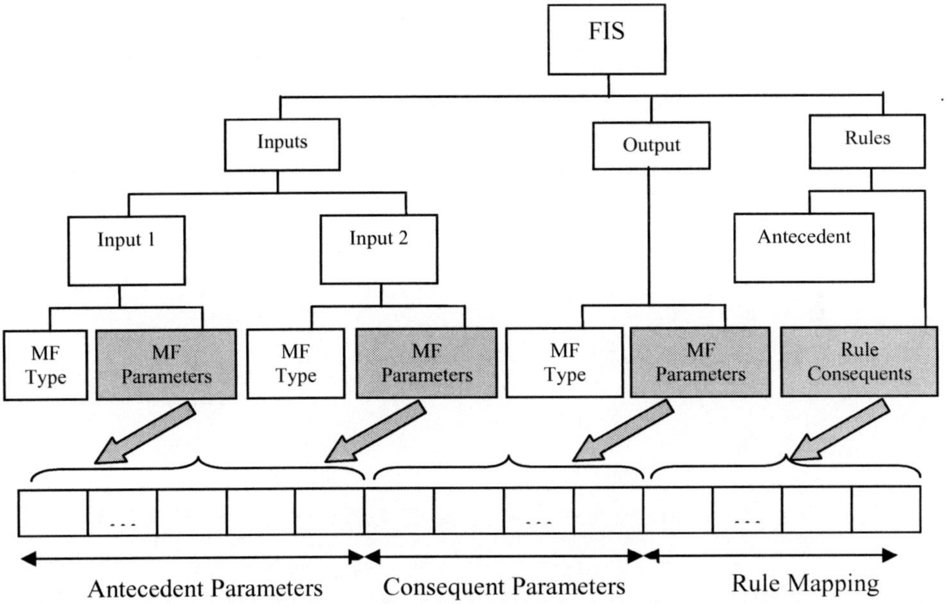

Figure 10. The FIS Structure

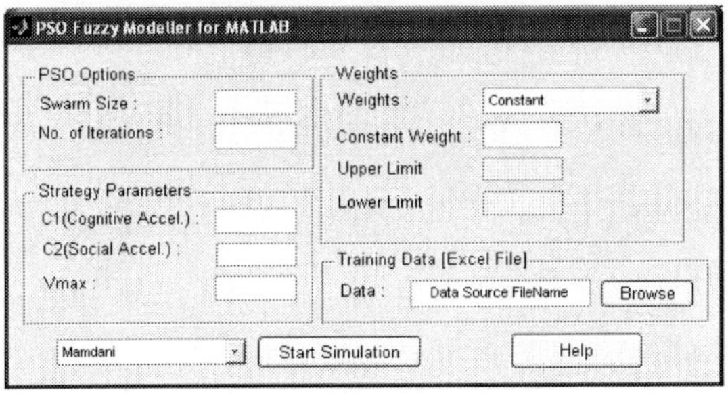

Figure 11. Toolbox Graphical User Interface

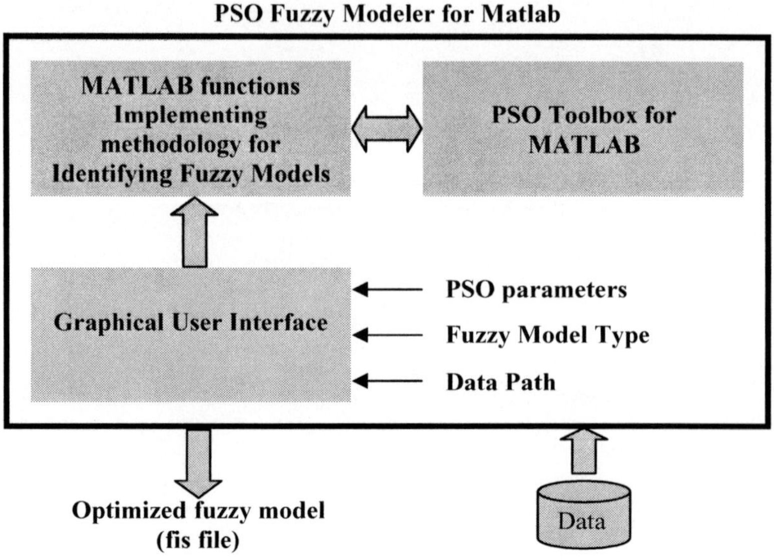

Figure 12. Organization of toolbox modules

6. Simulation Results

The proposed methodology has been applied for identification of fuzzy model for the rapid Ni-Cd battery charger, developed by the authors [21]. Based on the rigorous experimentation with the Ni-Cd batteries, it was observed that the two input variables used to control the charging rate (Ct) are absolute temperature of the batteries (T) and its temperature gradient (dT/dt). From the experiments performed, input-output data was tabulated and that data set consisting of 561 points is available at http://research.4t.com. The input and output variables identified for Ni-Cd batteries along with their universes of discourse are listed in Table 3.

Table 3. Input and Output Variables(s) alongwith their Universes of Discourse

Input Variables	**Universe of Discourse**
Temperature (T)	0-50°C
Temperature Gradient (dT/dt)	0-1 (°C/sec)
Output Variable	
Charging Rate (Ct)	0-8C*

**Charging rates are expressed as multiple of rated capacity of the battery, e.g. C/10 charging rate for a battery of C=500 mAh is 50 mA* [22].

As mentioned earlier, the Matlab toolbox presented in the previous section has been used for the identification of Mamdani and Singleton fuzzy models from the data. The strategy parameters of PSO algorithm selected for the identification of both the models are listed in Table 4 and the simulation results obtained are presented in Table 5. Centre of Gravity and Weighted Average defuzzification techniques [7] were selected for Mamdani and Singleton fuzzy models respectively.

Table 4. Strategy parameters used for Identification of fuzzy models

Swarm Size	30
Iterations	2500
c_1	2
c_2	2
w_{start} (Inertia weight at the start of PSO run)	0.9
w_{end} (Inertia weight at the end of PSO run)	0.3
Vmax	75

Table 5. Simulation Results

Model	MSE of Fuzzy Model corresponding to Swarm's *gbest*		Simulation time
	After 1st Iteration	After 2500 Iterations	
Mamdani	12.10	0.0488	19.424 hours
Singleton	46.95	0.1118	16.633 hours

Graphical representation of the input data and the fuzzy models identified through PSO is given in Figures 13(a) to 13(c). The numerical values presented in Table 5 and the surface plots shown in Figure 13 clearly depict the effectiveness of the proposed methodology and its implementation, as considerable improvement in the performance of fuzzy models was achieved after the complete run of PSO algorithm. More simulation time for Mamdani fuzzy model is due to more complicated defuzzification process.

7. Conclusions

In this chapter, the use of PSO algorithm for identification of optimized fuzzy model from the available input-output data is proposed. The suggested approach has been implemented as a Matlab toolbox viz. *PSO Fuzzy Modeler for MATLAB* and is also presented in this chapter. This toolbox, which is hosted on SourceForge.net as an open source initiative, has the capabilities to generate Mamdani and Singleton fuzzy model from the available data and is going to help the designers build fuzzy systems from their data in a short time. The data from the rapid Ni-Cd battery charger developed by the authors was used for the presentation and validation of the approach. Simulation results presented in this chapter have been generated through this toolbox and give a clear indication about the ability of PSO algorithm for fuzzy model identification. The proposed technique is of universal nature and there are no limitations in its usage.

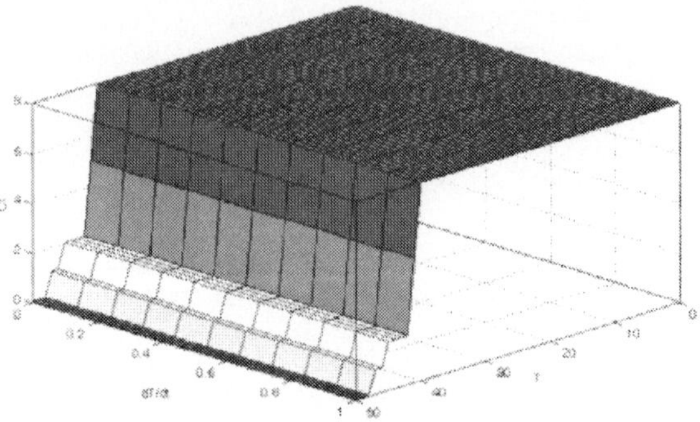

(a) Graphical representation for the input-output data

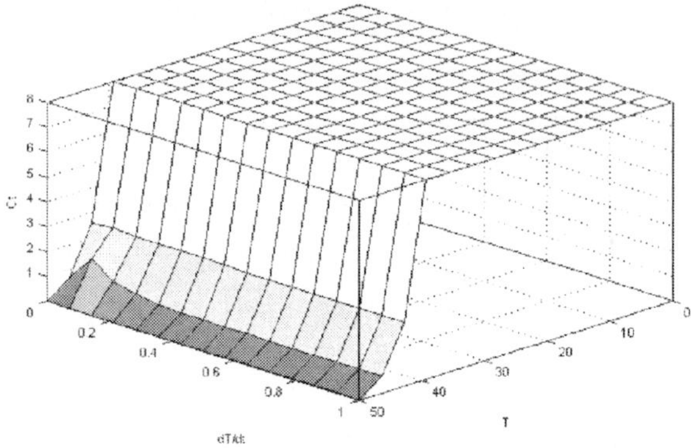

(b) Surface plot for the identified Mamdani fuzzy model

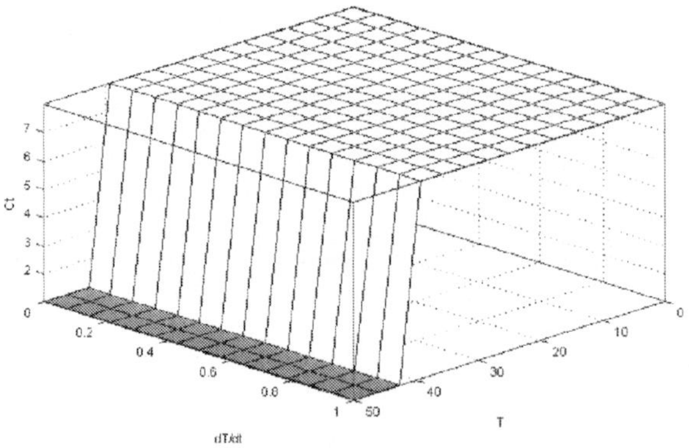

(c) Surface plot for the identified Singleton fuzzy model

Figure 13. Graphical representation

Two broad variants of PSO algorithm were developed: one with a global neighborhood called *gbest* model and the other with local neighborhood known as *lbest* model. The *gbest* model maintains only a single best solution and each particle moves towards its previous best position and towards the best particle in the whole swarm. The best particle acts as an attractor, pulling all the particles towards it. In the *lbest* model, each particle moves towards its previous best position and also towards the best particle in its restricted neighborhood and thus maintains multiple attractors. Although the *gbest* model is most commonly used, it is vulnerable to premature convergence. In this chapter, we have implemented only the *gbest* model for fuzzy model identification.

8. Future Trends

It is a well recognized fact that the performance of evolutionary algorithms largely depends on the choice of appropriate operating/strategy parameters [23][24]. Many users adjust the strategy parameters manually and this decision is usually taken either in terms of most common values given in the literature or by means of trial and error, which is unsystematic and requires unnecessary experimentation. Thus, one of the important directions for the future work is to address the issue of parameters selection in PSO algorithm.

Many variants of PSO algorithm have been suggested by different researchers. Another possible direction could be to use these PSO variants for identifying fuzzy systems with an objective to improve their performance further.

The parallel nature of evolutionary algorithms requires lot of computational efforts, which is evident from the simulation time reported in Table 5 for the given data. The computer time is directly proportional to the complexity of the problem under consideration and for a practical system, the simulation time may run into many days or even months. This can be reduced considerably through cluster computing. A cluster is a group of independent computers working as a single, integrated computing resource. The cluster computing has become the paradigm of choice for executing large-scale science, engineering and commercial applications. Thus, in an attempt to reduce the processing time, the toolbox presented in this chapter can be modified so as to run on cluster.

The toolbox in true sense can be called an open-source only if both the toolbox and the platform on which the toolbox runs should be free. Since the Matlab environment is commercial, this may become a hindrance in exchanging ideas and further improvements in the toolbox design from people who doesn't use Matlab. One of the important tasks for future would be to develop such tools/applications in Java or other high level languages so as to make them platform independent for wider usage, exchange and improvements.

Another direction for the future work could be applying this methodology for other fields and applications.

References

[1] Reynolds, C. W., "*Flocks, herds and schools: A distributed behavioral model*", Computer Graphics, 1987, pp. 25-34.

[2] J. Kennedy and R. Eberhart, *Swarm Intelligence*, Morgan Kaufmann, 2001.

[3] J. Kennedy and R. Eberhart, *Particle Swarm Optimization*, Proceedings of IEEE Conference on Neural Networks, vol. IV, Perth, Australia, 1995, pp. 1942-1948.

[4] Eberhart, R.C and Shi, Y., *Particle Swarm Optimization: Developments, Applications and Resources*, Proceedings of the Congress on Evolutionary Computation, Seoul, Korea, 2001, pp. 81-86

[5] K.E. Parsopoulos and M.N. Vrahatis, *Recent Approaches to Global Optimization Problems through Particle Swarm Optimization*, Natural Computing, Kluwer Academic Publishers, 2002, pp.235-306.

[6] H.Hellendoorn and D. Driankov (Eds.), *Fuzzy Model Identification - Selected Approaches*, Springer-Verlag, 1997.

[7] John Yen and Reza Langari, *Fuzzy Logic - Intelligence, Control and Information*, Pearson Education, Delhi, First Indian Reprint, 2003.

[8] A. Bastian, *A Genetic Algorithm for Tuning Membership Functions*, Fourth European Congress on Fuzzy and Intelligent Technologies EUFIT(96), Aachen, Germany, vol.1, 1996, pp. 494-498.

[9] B. Carse, T.C. Fogarty and A. Munro, *Evolving Fuzzy Rule-based Controllers using GA*, Fuzzy Sets and Systems, 1996, pp.273-294.

[10] O. Nelles, *FUREGA--Fuzzy Rule Extraction by GA*, Fourth European Congress on Fuzzy and Intelligent Technologies EUFIT(96), Aachen, Germany, vol. 1, 1996, pp. 489-493.

[11] K. Nozaki, T. Morisawa, H. Ishibuchi, *Adjusting Membership Functions in Fuzzy Rule-based Classification Systems*, Third European Congress on Fuzzy and Intelligent Technologies, EUFIT(95), Aachen, Germany, vol. 1, 1995, pp. 615-619.

[12] M. Setnes, J.A. Roubos, *Transparent Fuzzy Modelling using Clustering and GAs*, North American Fuzzy Information Processing Society (NAFIPS) Conference, June 10-12, New York, USA, 1999, pp.198-202.

[13] Arun Khosla, Shakti Kumar and K.K. Aggarwal, *Identification of Fuzzy Controller for Rapid Nickel-Cadmium Batteries Charger through Fuzzy c-means Clustering Algorithm*, Proceedings of North American Fuzzy Information Processing Society (NAFIPS) Conference, Chicago, July 24-26, 2003, pp. 536-539.

[14] Patricia Melin and Oscar Castillo, *Intelligent Control of a Stepping Motor Drive using an Adaptive Neuro-Fuzzy Inference System*, Information Sciences, Volume 170, Issues 2-4,February 2005, pp 133-151.

[15] Arun Khosla, Shakti Kumar and K.K. Aggarwal, *Fuzzy Controller for Rapid Nickel-Cadmium Batteries Charger through Adaptive Neuro-Fuzzy Inference System (ANFIS) Architecture*, Proceedings of North American Fuzzy Information Processing Society (NAFIPS) Conference, Chicago, July 24-26, 2003, pp. 540-544.

[16] Y.Shi, R. Eberhart and Y.Chen, *Implementation of Evolutionary Fuzzy Systems*, IEEE Transactions on Fuzzy Systems, April 1999, vol. 7, pp. 109-119.

[17] Matlab Documentation

[18] http://sourceforge.net/projects/fuzzymodeler

[19] J.S. Roger Jang, Ned Gulley, "*Fuzzy Logic Toolbox User's Guide*", The Mathworks Inc., USA, 1995.

[20] http://sourceforge.net/projects/psotoolbox

[21] Arun Khosla, Shakti Kumar and K.K. Aggarwal, *Design and Development of RFC-10: A Fuzzy Logic Based Rapid Battery Charger for Nickel-Cadmium Batteries*, HiPC (High Performance Computing) 2002 Workshop on Soft Computing, Bangalore, pp. 9-14, 2002.

[22] David Linden (Editor-in-Chief), *Handbook of Batteries*, McGraw Hill Inc., 1995.

[23] Odetayo M. O., "*Empirical Studies of Interdependencies of Genetic Algorithm Parameters*", Proceedings of the 23rd EUROMICRO 97 Conference 'New Frontiers of Information Technology', 1997, pp. 639-643.

[24] David E Goldberg, "*Genetic Algorithms in Search, Optimization and Machine Learning*", Pearson Education Asia, New Delhi, 2001.

Integrated Intelligent Systems for Engineering Design
X.F. Zha and R.J. Howlett (Eds.)
IOS Press, 2006

Product-mix Design Decision under TOC by Soft-sensing of Level of Satisfaction using Modified Fuzzy-LP

Arijit BHATTACHARYA[a] and Pandian VASANT[b, 1]
[a] *Examiner of Patents & Designs, The Patent Office, India*
[b] *Research Lecturer, Universiti Teknologi Petronas, Malaysia*

Abstract. This chapter outlines an intelligent fuzzy linear programming (FLP) having a flexible logistic membership function (MF) in finding out fuzziness patterns at disparate level of satisfaction for theory of constraints-based (TOC) product-mix design problems. One objective of the present work is to find out degree of fuzziness of product-mix decisions having disparate level of satisfaction of decision-maker (DM). Another objective is to provide a robust, quantified monitor of the level of satisfaction of DMs and to calibrate these levels of satisfaction against DMs' expectations. Fuzzy-sensitivity of the decision has been focused for a bottle-neck-free, optimal product-mix solution of TOC problem.

Keywords. Engineering design, Theory-of-Constraints, Degree of fuzziness, Level of satisfaction, Fuzzified Linear Programming, Intelligent decision.

1. Introduction

Design is usually considered a reflective and ill-structured process [12]. Variables in the design evaluation stage are mostly fuzzy. It has been widely accepted that design is an ill-structured task [36], which requires a significant effort to understand the 'structure' of the problem. Thus, tools, such as fuzzy set theory, should be used to reduce the degree of imprecision and to view the designers' level of satisfaction. Vanegas and Labib [42] present several fuzzy approaches to design evaluation. Weights of criteria and performance levels are captured by fuzzy numbers, and the overall performance of an alternative is calculated through the new fuzzy weighted average. A way of comparing the overall performance of a design-candidate by drawing an aggregate profile of performance is proposed by Vanegas and Labib [42]. It is concluded that fuzzy set theory is a powerful and flexible tool for dealing with the imprecision in different types of problems in design evaluation.

An effective engineering design process should balance many different factors, such as customer requirements, performance, cost, safety, system integration, manufacturability, operability, reliability, and maintainability. Designers involved in engineering design use different knowledge sources to tackle design problems [11].

---

[1] EEE Program, Research Lecturer, Universiti Teknologi Petronas, 31750 Tronoh, BSI, Perak DR, Malaysia, E-mail: pvasant@gmail.com.

Two identified different types of knowledge sources are: (i) well structured explicit knowledge and (ii) tactic, implicit and experienced-based knowledge. Each of these plays a particular role in engineering design. These are known as knowledge-based design support systems (KBDSS) [11].

Knowledge-based design support system (KBDSS) is a decision support system (DSS) enabling designers to explore the structure of design problems and their solutions by combining human expertise with domain and design knowledge stored in a computational tool [38]. This definition is particularly well emphasising the 'hybridness' of KBDSS involving designers and computer-based tools. Computational tools simplify the exploration and navigation in usually vast design spaces.

Designers often begin with conceptual solutions before developing or combining them further [35, 8]. 'Concepts' deliver a desired function, encompass basic knowledge of principles governing its behaviour, but neglect the detailed structure.

A computational model of design processes that can be used for the construction of product-mix KBDSS is presented in this chapter. The key feature of the model is the understanding of design as a problem of requirements explication through simultaneous development of partial design solutions and reflection on them. The reflection from the space of solutions to the space of requirements is supported by an intelligent computer-based tool.

The organization of the chapter is as follows. The chapter first provides a brief introduction on the correlation between design and decision-support. Summary of the prior art is outlined thereafter to frame the objectives and scopes of this work. Next section highlights the concept of interactive design decisions by suitably designing a membership function as well as incorporating intelligence in the product-mix design decisions. Intelligent product-mix design decision under TOC is illustrated comparing with that of the traditional procedure and subsequent validation of the proposed model is carried out with a known product-mix design problem. Computational results are tabulated and interpretations are made thereafter. Last section concludes the meaningful significance and usability of the proposed intelligent model comparing the obtained solution with that of the prior models.

2. Prior Art

Design of product-mix is one of the major types of manufacturing flexibility, referring to the ability to produce a broad range of products or variants with presumed low changeover costs [4]. The value of such a capability is important to establish for an industrial firm in order to ensure that the flexibility provided will be at the right level and used profitably rather than in excess of market requirements and consequently costly [4]. Product-mix acts through capacity management decisions to reduce performance from the level implied by direct effects alone [1].

Goldratt [15, 16] demonstrates a technique known as theory of constraints (TOC). One of the features of the TOC is to calculate/design the product-mix for a bottleneck-free manufacturing process. In the 1993 Goldratt [16] improved the concept of TOC by the management philosophy on improvement based on identifying the constraints to increasing profits. It was shown that product-mix design under TOC could be mathematically tackled as a linear programming (LP) model. Luebbe and Finch [27] compare the TOC and LP using the five-step improvement process in TOC. It is stated that the algorithm could optimize the product-mix as ILP [27]. Further, it is revealed that the algorithm is inefficient in handling two types of problems. The first type includes problems associated with adding new product alternatives to an existing production line [25]. The second type includes problems concerning more than one bottleneck in which the algorithm could not reach the feasible optimum solution. Later on the concept of the dominant bottleneck is proposed as a remedy for finding our feasible optimum solution. Lee and Plenert [25] illustrate two examples of product-mix decision problem and conclude that TOC solution is inferior to the optimum solution and has the possibility of being infeasible when multiple constrained resources in a plant exists. Hsu and Chung [19] present a dominance rule-based algorithm that classifies non-critically constrained resources into three levels for solving the TOC product-mix problem when multiple constrained resources exists. Fredendall and Lea [13] revise the TOC product-mix heuristic to identify the optimal product-mix under conditions where the original TOC heuristic failed. Methods to design a product-mix that maximizes profit have been studied extensively. One method, known as integer linear programming (ILP), is often used to optimize the product mix. But it requires a high level of expertise to formulate and may take hours to solve. Researches reveal that TOC heuristic is simpler to use than an ILP [27]. But some researchers identify conditions under which TOC could create a non-optimal product-mix [25, 32]. Extensive studies have been carried out to design product-mix that maximizes profit by Buxey [7].

Lea and Fredendall [24] examine how three types of management accounting systems and two methods to determine product-mix interact in both the short term and the long term to affect the manufacturing performance of two shops – one with a flat and the other with a deep product structure – in a highly automated industry that has a significantly high overhead content. Through a large-scale computer simulation, Lea and Fredendall's [24] study provides insights into the product-mix decision considering fluctuations caused by environmental uncertainty, using an integrated information system that integrates a manufacturing system and a management accounting system, considering the decision-outcome dynamic over time, the choice of cost content, and using both financial and non-financial performance measures.

Letmathe and Balakrishnan [26] present two mathematical models that can be used by firms to determine their optimal product-mix and production quantities in the presence of several different types of environmental constraints, in addition to typical production constraints. The first model, which assumes that each product has just one operating procedure, is a linear program while the second model, which assumes that the firm has the option of producing each product using more than one operating procedure, is a mixed integer linear program [26]. Chung *et al.* [9] propose an application of the analytic network process (ANP) [34] for product-mix design for efficient manufacturing in a semiconductor fabricator.

Product-mix and the acquisition of the assets needed for their production are interdependent decisions [20]. Kee [21] modifies activity-based costing (ABC) to reflect separate flexible and committed cost driver rates for an activity enabling the ABC to reflect the difference in the behaviour of an activity's flexible and committed costs needed for operational planning decisions.

Several works are reported till date in the field of product-mix design by Haka *et al.* [18], Kee and Schmidt [22], Vasant [39], Vasant and Barsoum [40], Vasant *et al.* [41], Balakrishnan and Cheng [3] and Coman and Ronen [10]. Köksal [23] proposes an improvement of a TOC-based algorithm by incorporating quality loss with it. Aryanezhad and Komijan [2] propose an "improved algorithm", which can reach optimum solution in determining product-mix under TOC.

Many researchers propose variations of Goldratt's [15, 16] product-mix problem. Lee and Plenert [25] demonstrate that TOC is inefficient when new product was introduced. Their observation is that the solution from TOC during introduction of new product produces a non-optimal product-mix. Plenert [32] discuss an example having multiple constrained resources to show that the TOC heuristic doesn't provide an optimal feasible solution. Patterson [31] and Goldratt [16] test the TOC heuristic using problems where the solution fully utilizes the bottleneck. On the contrary, both Lee and Plenert [25] and Plenert [32] test the TOC heuristic where the optimal solution leaves idle time on bottleneck. They use an ILP formulation that identifies a product-mix. The product-mix fully utilizes the bottleneck. Their conclusion is that ILP solution is more efficient than the TOC heuristic. Mayday [28] and Posnack [33] criticize Lee and Plenert [25] and Plenert [32]. They [28, 33] argue that without having integer solution, TOC heuristic's solution will be optimal. Thus, still there is a need still to make the TOC heuristic more efficient using a new approach.

Onwubolu [29] compares the performance of the *Tabu search*-based approach to both the original TOC heuristic, the ILP solution and the revised-TOC algorithm. Further, large-scale problems that are difficult, if not impossible for other methods to address, are generated randomly and solved [29]. The research work of Boyd and Cox [6] is focused to compare the TOC solution to the product-mix problem with an optimal solution given by LP or ILP. Bhattacharya *et al.* [5] propose *De-novo programming* approach as an alternative to LP approach where multiple constraint resources exist.

Souren *et al.* [37] discuss some premises, in the form of a checklist, for generating optimal product-mix decisions using a TOC-base approach. The checklist for TOC-based product-mix solution appears to be a hypothetical one when it leads to "optimal solution". In reality product-mix constraints do not appear to be single; multiple constraint resources exist. So, the assumption, which leads Souren *et al.* [37] to conclude in getting an ideal solution in the checklist, is not valid. The authors are not aware of literature where intelligent FLP approach has been used in making product-mix design decisions under TOC heuristic more explicit.

After an extensive survey on existing literatures on product-mix design decisions, the following criticisms of the existing TOC product-mix decision literatures are made:

(i) TOC heuristic is implicit and infeasible, for multiple constrained resource product-mix design decision problems;

(ii) Crisp data contain certain degree of imprecision (fuzziness) in TOC product-mix decisions;

(iii) DMs are not aware of the level of satisfaction of design decisions while performing product-mix decisions through TOC;

(iv) Fuzzy-sensitivity should be carried out in the product-mix design decisions under TOC to achieve a certain pre-specified level of satisfaction of the DM; and

(v) An intelligent tripartite relationship among DM, analyst and implementer for TOC product-mix decision is essential.

In order to obviate the aforesaid criticisms, the next sections are outlined.

3. Concept of Interactive Design

Earlier studies on product-mix decision problems were considered on the basis of bipartite relationship of the DM and analyst. This notion is now outdated. Now tripartite relationship is to be considered, as shown on Figure 1. In tripartite relationship, the DM, analyst and implementer will interact in finding fuzzy satisfactory solution.

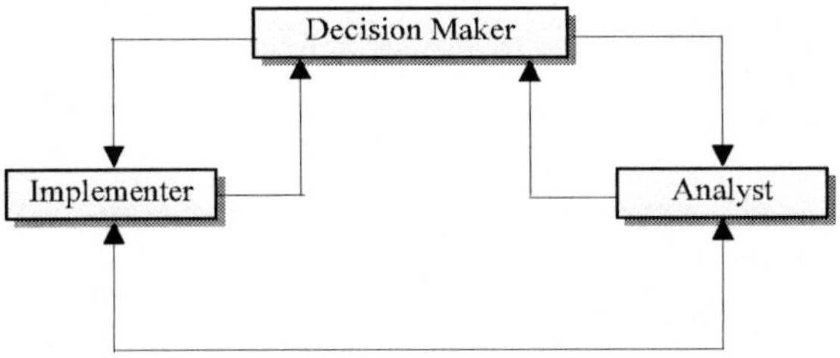

Figure 1. Tripartite FLP for solving TOC product-mix decision

In case of tripartite system the DM communicates and describes the problem to an analyst. Based on the data that are provided by the DM, the analyst designs MFs, solves the problems and provides the solution with feedback to the DM. The DM provides the design solutions with a trade-off to the implementer for implementation. Implementer interacts with the DM to obtain an efficient and high productive design solution. A tripartite relationship, e.g., DM-analyst-implementer interaction, is essential to solve any TOC-based product-mix decisions. This is because any change in the values of three criteria for TOC, viz., throughput, inventory and operating expenses [14] results in change in the bottom-line financial measurements.

3.1. Design of Membership Function

In this paper, we modify and employ a logistic non-linear MF as given in equation (1), to fit into the product-mix design decision problem under TOC heuristic.

$$f(x) = \frac{B}{1 + Ce^{\gamma x}}, \tag{1}$$

In equation (1) B and C are scalar constants and γ, $0 < \gamma < \infty$ is a fuzzy parameter for measuring degree of imprecision, wherein $\gamma = 0$ indicates crisp.

The logistic function is a monotonically non-increasing function. The MF is non-increasing as

$$\frac{df}{dx} = \frac{BC\alpha e^{\gamma x}}{(1 + Ce^{\gamma x})^2} \tag{2}$$

A MF is flexible when it has vertical tangency, inflexion point and asymptotes. Since B, C, γ and x are all greater than zero, $\frac{df}{dx} \le 0$. Furthermore it can be shown that equation (1) has asymptotes at f(x) = 0 and f(x) = 1 at appropriate values of *B* and *C*. It can also be shown that the logistic function equation has a vertical tangent at $x = x_0$, x_0 is the point where $f(x_0) = 0.5$.

The said logistic function has a point of inflexion at $x = x_0$, such that $f''(x_0) = \infty$, $f''(x)$ being the second derivative of f(x) with respect to x. A MF of *S*-curve nature, in contrast to linear function, exhibits the real life problem. Therefore, the generalized logistic MF is defined as:

$$f(x) = \begin{cases} 1 & x < x_L \\ \dfrac{B}{1 + Ce^{\gamma x}} & x_L < x < x_U \\ 0 & x > x_U \end{cases} \tag{3}$$

S-curve MF is a particular case of the logistic MF defined in equation (3). The *S*-curve MF has got specific values for *B*, *C* and γ. The logistic function as defined in equation (1) was indicated as *S*-curve MF by Goguen [17] and Zadeh [44, 45, 46].

Equation (3) is modified and re-defined in order to fit into TOC product-mix design decision problem:

$$\mu(x) = \begin{cases} 1 & x < x^a \\ 0.999 & x = x^a \\ \dfrac{B}{1 + Ce^{\gamma x}} & x^a < x < x^b \\ 0.001 & x = x^b \\ 0 & x > x^b \end{cases} \tag{4}$$

Figure 2 shows the nature of the *S*-curve. In equation (1) the MF is re-defined as $0.001 \le \mu(x) \le 0.999$. Design has to be made within this range because in real-world TOC heuristic, the physical capacity requirement cannot be 100% of the total plant requirement. Similarly, the capacity requirement cannot be 0%. Therefore there is a range between x_0 and x_1 with $0.001 \le \mu(x) \le 0.999$.

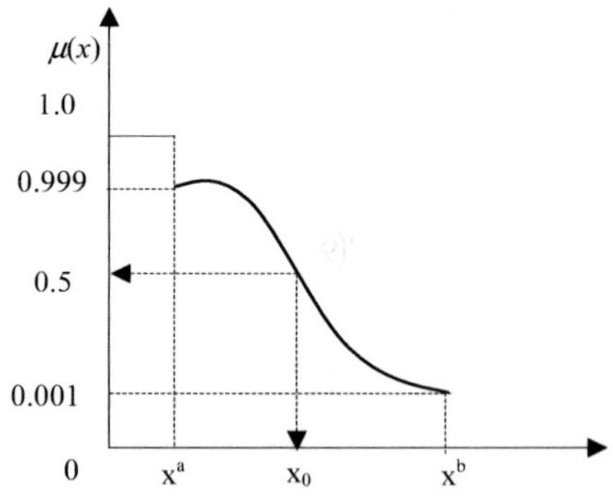

Figure 2. *S*-shaped membership function used in the proposed FLP approach

We rescale the x-axis as $x^a = 0$ and $x^b = 1$ in order to find the values of B, C and γ. The values of B, C and γ are obtained from equation (4) as

$$B = 0.999\,(1 + C) \tag{5}$$

$$\frac{B}{1+Ce^{\gamma}} = 0.001 \tag{6}$$

By substituting equation (5) into equation (6) one gets

$$\frac{0.999(1+C)}{1+Ce^{\gamma}} = 0.001 \tag{7}$$

Rearranging equation (7) one gets

$$\gamma \;=\; \ln\frac{1}{0.001}\left(\frac{0.998}{C}+0.999\right) \tag{8}$$

Since, B and γ depend on C, we require one more condition to get the values for B, C and γ.

Let, when $x_0 = \dfrac{x^a + x^b}{2}$, $\mu(x_0) = 0.5$. Therefore,

$$\frac{B}{1+Ce^{\frac{\gamma}{2}}} = 0.5 \text{'} \tag{9}$$

and hence

$$\gamma = 2\ln\left(\frac{2B-1}{C}\right) \tag{10}$$

Substituting equation (8) and equation (9) in to equation (10), we obtain

$$2\ln\left(\frac{2(0.999)(1+C)-1}{C}\right) = \ln\frac{1}{0.001}\left(\frac{0.998}{C}+0.999\right) \tag{11}$$

which in turn yields $(0.998 + 1.998C)^2 = C(998 + 999C)$ (12)

Equation (12) is solved and it is found that

$$C = \frac{-994.011992 \pm \sqrt{988059.8402 + 3964.127776}}{1990.015992} \tag{13}$$

Since C has to be positive, equation (13) gives $C = 0.001001001$ and from equations (5) and (10) one gets $B = 1$ and $\gamma = 13.81350956$.

The proposed S-shaped MF is used in this design due to the followings:

(i) $\mu(x)$ is continuous and strictly monotonously non-increasing;

(ii) $\mu(x)$ has lower and upper asymptotes at $\mu(x) = 0$ and $\mu(x) = 1$ as $x \to \infty$ and $x \to 0$, respectively;

(iii) $\mu(x)$ has inflection point at $x_0 = \frac{1}{\gamma}\ln(2 + \frac{1}{C})$ *with* $A = 1 + C$;

3.2. Intelligent Product-Mix Design Decision

Intelligent computing research aims at to bring intelligence, reasoning, perception, information gathering and analysis. According to Watada [43], if the obtained membership value of the solution is appropriate and proper, i.e., it is included in [0, 1], regardless of the shape of a MF, both solutions are not different so much. Nevertheless, it is possible that non-linear MF changes its shape according to the vague parameters (γ) values. Therefore a DM, analyst and implementer are able to apply their strategy to in designing optimal outcome using these parameters.

The present work uses an intelligent design decision rule that generates the coefficients of the fuzzy constraints in the decision variables. The rule declares a function C_j and assigns the constants in the MF. The aim is to produce a rule that works well on previously unseen data, i.e., the decision rule should "generalize" well. An example is appended below:

```
function [cj]=mpgen(cj0,cj1,gamma,mucj)
B=(0.998/((0.001*exp(gamma))-0.999));
A=0.999*(1+B);
cj=cj0+((cj1-cj0)/gamma)*(log((1/B)*((A/mucj)-1)));
```

The MATLAB® function *"linprog"* is called in the following way while using the designed MF:

```
[X,Z]=linprog(f,A,b,[ ],[ ],0,inf);
cj=cj0+((cj1-cj0)/gamma)*(log((1/B)*((A/mucj)-1)));
```

The rule supports this design by allowing the call to the function *linprog* to contain a variable which is automatically set to different values as one may request. The way in which the logic acts as an agent in the designed intelligent system includes many *if – else* rules.

4. Designing the Product-mix Decision under TOC

4.1. Conventional TOC Approach

A product-mix problem reported by Hsu and Chung [19] (refer to Fig. 3) is considered to show the effectiveness of the proposed model. With an objective to maximize the throughput when multiple CCR exist, the problem can be designed as a dual simplex LP problem. Four different types of products, viz., R, S, T & U, are produced. There are seven different resources, A to G. Each resource centre has a capacity of 2400 minutes (refer to Table 1).

Onwubolu and Mutingi [30] report that CCR will be there with resources A and D, as these two resources exceeds the available maximum capacity of 2400 minutes. Thus, it appears that TOC solution is infeasible when multiple CCR exist.

Table 1. Load requirements for producing products

Products	Weekly market potential (units)	Unit selling price ($ / unit)	Processing time per unit (min.)							Raw material cost per unit ($ / unit)	Throughput per unit ($ /unit)
			A	B	C	D	E	F	G		
R	70	90	20	5	10	--	5	5	20	10	80
S	60	80	10	10	5	30	5	5	5	20	60
T	50	70	10	5	10	15	20	5	10	20	50
U	150	50	5	15	10	5	5	15	--	20	30

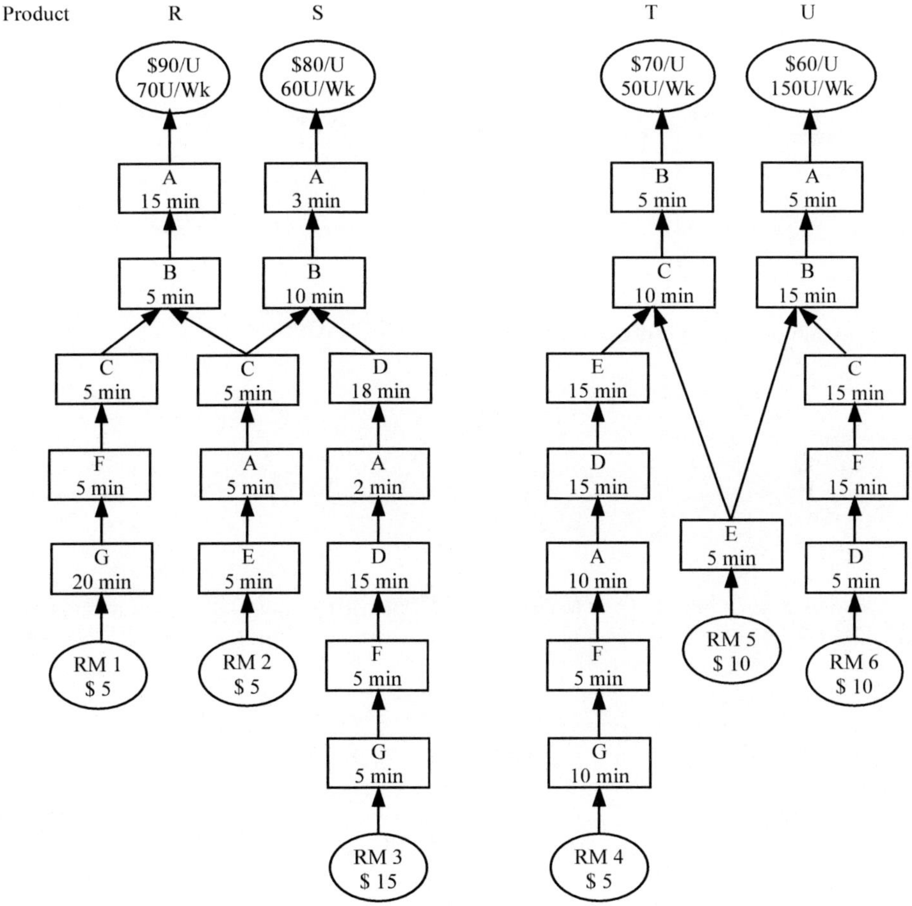

Figure 3. Product mix problem from Hsu and Chung [19]

4.2. Intelligent Product-mix Design under TOC

From the preceding sections it is evident that TOC heuristic seems to be implicit for designing product-mix decision problem when multiple CCR exist. Moreover, it is known that TOC-based product-mix design decision can never be better than a correctly formulated LP approach [37]. The following computational with the proposed design model will make the TOC heuristic more explicit.

The dual simplex LP formulation for the product-mix decision of Figure 3 is as follows:

$$\text{Maximize } Z = 80R + 60S + 50T + 30U \tag{14}$$

subject to the following technological constraints:

$$20R + 10S + 10T + 5U \leq 2400 \tag{15}$$

$$5R + 10S + 5T + 15U \leq 2400 \tag{16}$$

$$10R + 5S + 10T + 10U \leq 2400 \tag{17}$$

$$0R + 30S + 15T + 5U \leq 2400 \tag{18}$$

$$5R + 5S + 20T + 5U \leq 2400 \tag{19}$$

$$5R + 5S + 5T + 15U \leq 2400 \tag{20}$$

$$20R + 5S + 10T + 0U \leq 2400 \tag{21}$$

and subject to the following market constraints:

$$0 \leq R \leq 70 \tag{22}$$

$$0 \leq S \leq 60 \tag{23}$$

$$0 \leq T \leq 50 \tag{24}$$

$$0 \leq U \leq 150 \tag{25}$$

The above equations (14) to (25) are solved using the rule-based logics coded in MATLAB® M-files. The optimal feasible product-mix solution are found to be as R = 50.667; S = 38.167; T = 50.000 and U = 101.000. The corresponding throughput is US $11873. These values correspond to the values found by Hsu and Chung [19].

A critical look at the Table 2 reveals that the present approach is far better while designing optimal product-mix maximising throughput. The *genetic algorithm* (GA) model presented by Onwubolu and Mutingi [30] fails to maximise the throughput. It doesn't sense the decision-maker's level of satisfaction of the design decision. Even a cumbersome *dominance rule* [19] is not well suited compared to the present approach. The product-mix design decision, maximizing total throughput is made at a bottleneck free state. The fusion of DM's level of satisfaction as well as vagueness of decision into a single model makes the proposed fuzzified intelligent approach robust and more efficient.

Table 2. Throughput comparison

Problem	No. of Resources	TOC solution	LP solution	Dominance rule solution	GA solution of Onwubolu and Mutingi [30]	FLP solution
Hsu and Chung [19]	7	14100	----	11873	11860	11873

5. Computing Level of Satisfaction and Degree of Fuzziness

MATLAB® computation using the M-files finds relationship among the level of satisfaction, the degree of vagueness and the Z-value. The results are summarised in Table 3.

Table 3. Profit for disparate fuzziness and level of satisfaction

Z^{new}	*Vagueness* γ								
ϕ %	3	5	7	9	11	13.8	19	27	35
0.10	10690	10690	10690	10690	10690	10690	10690	10690	10690
5.09	11704	11434	11146	10988	10898	10816	10728	10699	10694
10.08	11786	11594	11309	11081	10973	10876	10771	10703	10697
15.07	11816	11672	11406	11164	11020	10913	10798	10708	10699
20.06	11832	11719	11474	11226	11055	10941	10819	10722	10701
25.05	11842	11752	11528	11277	11089	10965	10836	10734	10702
30.04	11849	11775	11572	11322	11127	10985	10850	10744	10703
35.03	11854	11793	11610	11361	11161	11003	10864	10754	10704
40.02	11857	11807	11643	11398	11193	11021	10876	10763	10705
45.01	11860	11818	11672	11432	11223	11037	10888	10771	10708
50.00	11863	11827	11698	11466	11253	11054	10900	10780	10714
54.99	11864	11835	11721	11498	11283	11070	10912	10788	10720
59.98	11866	11842	11743	11531	11313	11092	10924	10796	10727
64.97	11867	11848	11763	11563	11344	11118	10937	10805	10734
69.96	11868	11853	11782	11597	11377	11146	10950	10815	10741
74.95	11869	11857	11799	11632	11414	11176	10965	10825	10749
79.94	11870	11861	11816	11669	11455	11210	10982	10837	10758
84.93	11871	11864	11831	11709	11503	11252	11002	10851	10770
89.92	11872	11868	11846	11755	11566	11307	11030	10871	10784
94.91	11872	11870	11860	11807	11658	11396	11074	10902	10808
99.99	11873	11873	11873	11873	11873	11873	11873	11873	11873

A minute observation at Table 3 reveals that the data behaves as a monotonically increasing function. Figure 4 illustrates characteristics of Z-values with respect to the level of satisfaction at disparate fussiness. It is to be noted that the higher the fuzziness values (γ), lesser will be the degree of vagueness inherent in the decision. Therefore, it is understood that higher level of outcome of decision variable for a particular level of satisfaction point, results in a lesser degree of fuzziness inherent in the said decision variable.

The proposed methodology calculates product-mix at disparate degree of fuzziness (γ) and level of satisfaction (φ) (refer to Table 4). Figure 5 depicts relationships between the number of products and the level of satisfaction φ at disparate degree of fuzziness (γ). It is to be noted that $\mu = (1 - \varphi)$ for all the cases of Figure 5, μ being the degree of possibility. From the Figure 5 it is understood that the decision variables allow the proposed model to achieve a higher level of satisfaction with a lesser degree of fuzziness.

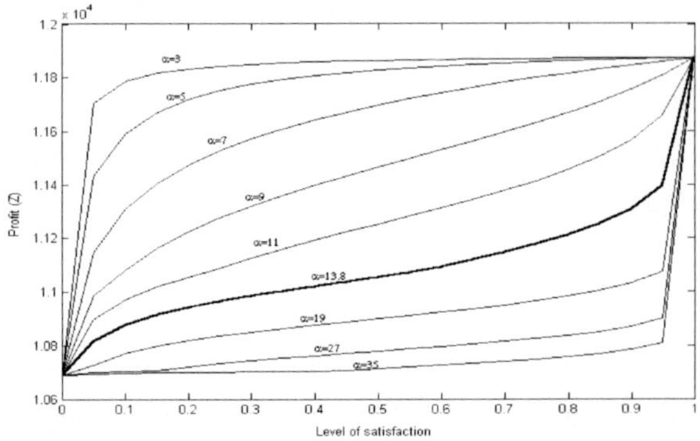

Figure 4. Behaviour of Z-values with respect to the level of satisfaction at disparate fuzziness

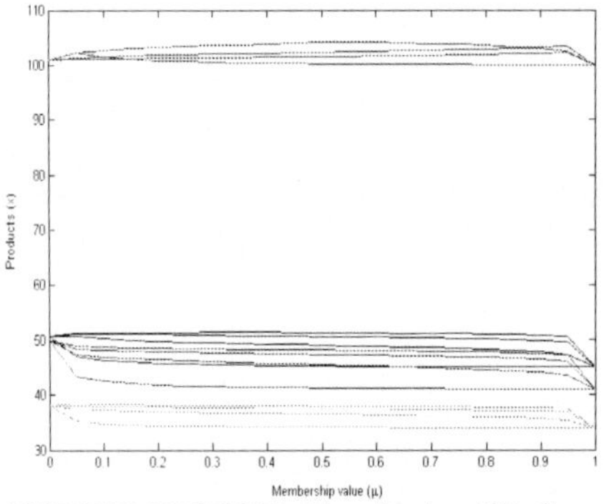

[COLOUR LEGENDS: Blue: Product 'R', Green: Product 'S', Red: Product 'T', Pink: Product 'U']

Figure 5. Relationships between the number of products and the level of satisfaction at disparate degree of fuzziness

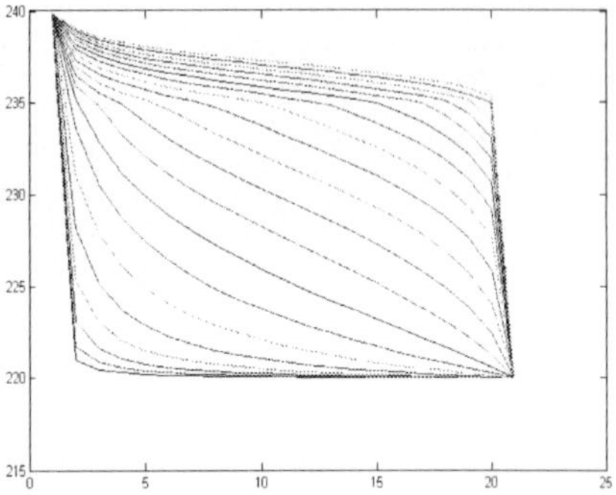

Figure 6. Fuzziness, γ, versus total number of products, P

Table 4. Product-mix at disparate fuzziness (γ) and level of satisfaction (ϕ)

Products	*Vagueness* γ															
	5				13.8				23				33			
ϕ %	R	S	T	U	R	S	T	U	R	S	T	U	R	S	T	U
0.10	50.67	38.17	50.00	101.00	50.67	38.17	50.00	101.00	50.67	38.17	50.00	101.00	50.67	38.17	50.00	101.00
5.09	47.12	35.32	43.38	102.38	50.69	37.56	47.40	102.26	51.39	38.14	48.44	101.30	51.08	38.39	48.91	101.07
10.08	46.35	34.84	42.51	101.51	50.26	37.29	46.92	102.74	51.36	37.97	48.15	101.51	51.21	38.29	48.71	101.17
15.07	45.97	34.61	42.09	101.09	49.99	37.12	46.62	103.04	51.19	37.87	47.97	101.69	51.29	38.22	48.59	101.23
20.06	45.74	34.46	41.83	100.83	49.79	36.99	46.39	103.27	51.08	37.79	47.83	101.83	51.35	38.16	48.49	101.28
25.05	45.58	34.37	41.66	100.66	49.63	36.89	46.21	103.46	50.98	37.74	47.73	101.94	51.40	38.12	48.42	101.31
30.04	45.47	34.29	41.53	100.53	49.49	36.80	46.05	103.62	50.89	37.68	47.63	102.04	51.45	38.08	48.35	101.35
35.03	45.39	34.24	41.44	100.44	49.35	36.72	45.89	103.77	50.81	37.63	47.54	102.13	51.48	38.05	48.28	101.38
40.02	45.32	34.19	41.36	100.36	49.23	36.64	45.76	103.91	50.74	37.59	47.46	102.21	51.42	38.01	48.23	101.44
45.01	45.27	34.17	41.29	100.29	49.11	36.57	45.62	104.04	50.67	37.54	47.38	102.29	51.37	37.98	48.17	101.49
50.00	45.22	34.14	41.25	100.25	48.99	36.49	45.49	104.17	50.59	37.49	47.29	102.37	51.33	37.95	48.12	101.55
54.99	45.18	34.11	41.21	100.21	48.88	36.43	45.37	104.30	50.53	37.46	47.22	102.45	51.28	37.92	48.06	101.61
59.98	45.15	34.09	41.17	100.17	48.76	36.35	45.23	104.23	50.46	37.41	47.14	102.53	51.23	37.89	48.01	101.66
64.97	45.12	34.07	41.14	100.14	48.64	36.27	45.09	104.09	50.38	37.36	47.05	102.61	51.18	37.86	47.95	101.72

Table 4. *(Continued.)*

69.96	45.09	34.06	41.11	100.11	48.51	36.19	44.94	103.94	50.30	37.31	46.97	102.70	51.12	37.83	47.89	101.78
74.95	45.08	34.05	41.08	100.08	48.36	36.10	44.78	103.78	50.22	37.26	46.87	102.79	51.06	37.79	47.82	101.85
79.94	45.06	34.04	41.07	100.07	48.19	35.99	44.59	103.59	50.12	37.19	46.75	102.91	50.99	37.74	47.74	101.93
84.93	45.04	34.03	41.05	100.05	47.99	35.87	44.37	103.37	49.99	37.12	46.62	103.05	50.91	37.69	47.64	102.02
89.92	45.03	34.02	41.03	100.03	47.73	35.70	44.07	103.07	49.83	37.02	46.44	103.23	50.79	37.62	47.52	102.15
94.91	45.01	34.01	41.01	100.02	47.30	35.44	43.59	102.59	49.57	36.86	46.15	103.52	50.61	37.51	47.31	102.35
99.99	45.00	34.00	41.00	100.00	45.00	34.00	41.00	100.00	45.00	34.00	41.00	100.00	45.00	34.00	41.00	100.00

Figure 6 shows a set of indifference curves at upwardly increasing level of satisfaction for fuzziness versus total number of products. Figure 6 reveals that at higher degree of fuzziness lesser total number of product-mix is obtained. It is also clear from the Figure 6 that even at higher level of satisfaction, the optimal product-mix cannot be obtained when the degree of fuzziness is very high.

Figure 7 illustrates another set of indifference curves at upwardly decreasing degree of fuzziness for degree of possibility versus total number of products. Keeping $\mu = (1 - \varphi)$ in mind, it is observed that at higher level of satisfaction of the DM, total number of product is highest. But, during this choice the DM should be aware of the amount of fuzziness hidden in the said decision. Therefore, a suitable curve should be selected during product-mix decision under TOC.

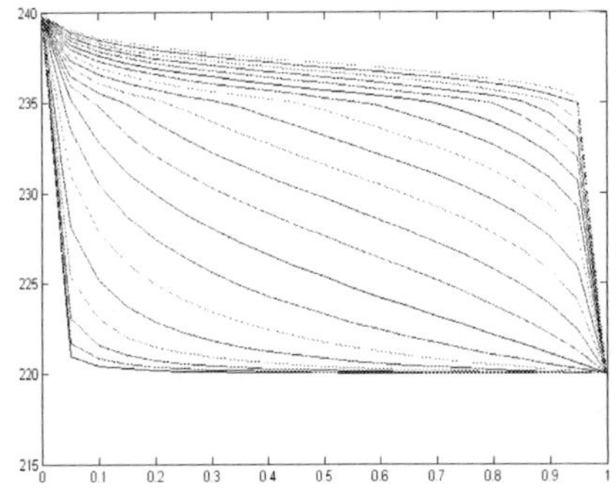

Figure 7. Degree of possibility, μ, versus total number of products, P

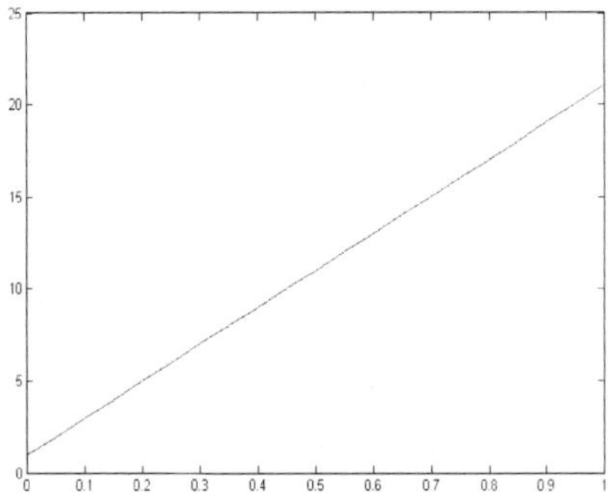

Figure 8. Fuzziness, γ, versus degree of possibility, μ

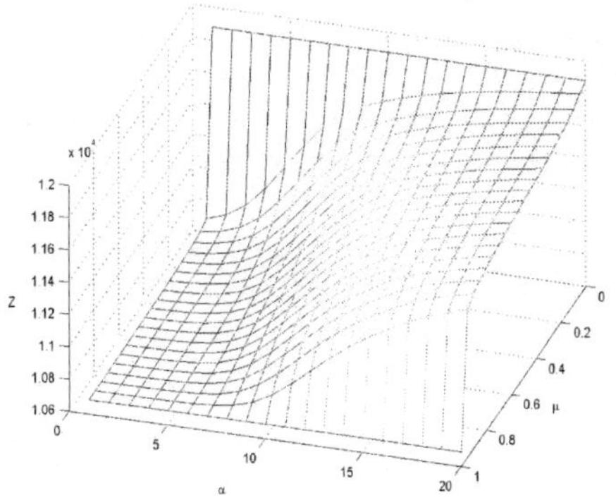

Figure 9. Relationships among three design parameters focusing the degree of fuzziness

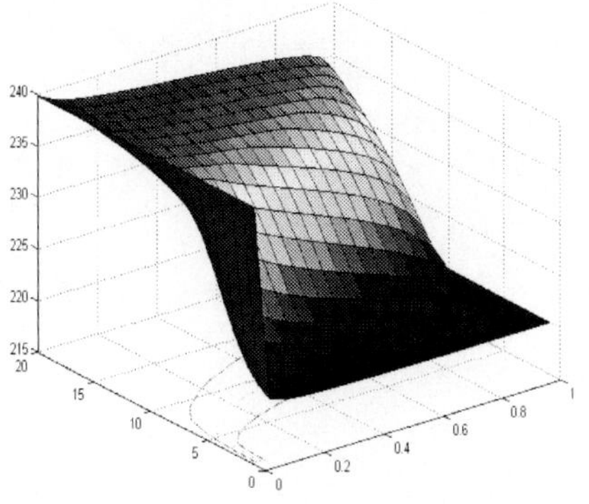

Figure 10. Surface and contour plot illustrating relationships among total number of products (P), vagueness (γ) and level of satisfaction (φ)

Figure 8 teaches that at lower level of satisfaction (φ) the chances of getting involved higher degree of fuzziness (γ) increase. Therefore, a DM's level of satisfaction should lie at least at a moderate level in order to avoid higher degree of fuzziness.

Combining all the aspects of Table 3, we get Figure 9. Figure 9 elucidates a plot combining fuzziness and level of satisfaction, and illustrates relationships among three prime parameters of TOC product-mix decision focusing the degree of fuzziness and fuzzy patterns.

Another check is needed to validate and confirm the intelligent product-mix decision-making methodology. Figure 10 depicts another surface-cum-contour plot illustrating relationships among total number of products (P), vagueness (γ) and level of satisfaction (μ) of intelligent product-mix decision under TOC focusing the degree of fuzziness. The DM can adopt the intelligent decision ranging the said maximum and minimum values at his/her desired level of satisfaction.

6. Discussions and General Conclusions

There may be many optimal design solution for different level of satisfaction (φ) at a particular α. As the chief aim of the LP model is to maximize the throughput at optimal product mix, we consider the maximum throughput for a particular level of satisfaction (φ) as the optimal throughput so that a bottleneck-free decision can be made. Thus, the optimal throughput for any of the stated level of satisfaction (φ) is found to be US $ 11,873. This value is identical to the optimal throughput found by Hsu and Chung [19] using their *dominance rule* methodology. Question arises that when there exists a LP solution suitable to tackle multi-capacity constraint resources, what is the utility of using *dominance rule* and *GA-based approach* of Onwubolu and Mutingi [30]? Perhaps the mottos inherent in employing those approaches were laid to show TOC product-mix heuristic implicit. Hsu and Chung's [19] solution using *dominance rule* technique yields the optimal throughput as US $ 11,873 that is identical to the solution found by proposed FLP methodology. But what is lacking in the said methodologies [19, 30] for product-mix decision is that the DMs are unable to view their level of satisfaction of the decision made.

Another focusing point that DMs might want to view is the associated degree of imprecision incurred during decision-making process. Crisp data sometime contain information that are partially vague in nature. The present approach, in a nutshell, fuses two most important measures for DMs as well as implementers, into the existing TOC-LP product-mix heuristic. This fusion strengthens the existing TOC product-mix decision and simultaneously worthwhile in removing bottlenecks. The dual-simplex LP with bounded variables, as such, is not capable in taking into account the DMs' preferences of level of satisfaction. The newer version of the intelligent TOC product-mix solution presented in this research work enables to deal with DMs' as well as implementers' preferences.

During computation with the proposed fuzzified intelligent approach an interesting phenomenon has been observed. The approach doesn't yield any result at the high vagueness and highest level-of-satisfaction condition. At $\gamma = 45$ and $\varphi = 99.9\%$ DM contradicts his/her decision. In reality, at highest level-of-satisfaction the vagueness is drastically reduced. Therefore, the proposed intelligent approach takes into account natural as well as human mind decisions prevailing in nature.

There is a possibility to design a computationally interactive intelligent man-machine intelligent system for the product-mix design decision using a neuro-fuzzy hybrid model in order to find more realistic solution. The risk involving capital investment on the decisions made can also be tackled suitably when other relevant criteria are considered in combination with the presented fuzzified approach. In future, involvement of more design criteria can be tackled in the TOC product-mix problem devising a new rule-based hybrid compromise multi-objective linear programming (HCMOLP).

ACKNOWLEDGEMENT

We thankfully acknowledge the anonymous referees for their valuable suggestions that improve the quality of this chapter.

References

[1] S.W. Anderson, Direct and indirect effects of product mix characteristics on capacity management decisions and operating performance, *International Journal of Flexible Manufacturing Systems* **13**(3) (2001), 241–265.

[2] M.B. Aryanezhad and A.R. Komijan, An improved algorithm for optimizing product mix under the theory of constraints, *International Journal of Production Research* **42**(20) (2004), 4221–4233.

[3] J. Balakrishnan and C.H. Cheng, Theory of constraints and linear programming: a re-examination, *International Journal of Production Research* **38**(6) (2000), 1459–1463.

[4] J. Bengtsson and J. Olhager, The impact of the product mix on the value of flexibility, *Omega* **30**(4) (2002), 265–273.

[5] A. Bhattacharya, B. Sarkar and S.K. Mukherjee, Use of De Novo programming in theory of constraints (TOC), *Industrial Engineering Journal* **XXXIII**(7) (2004), 6–11.

[6] L.H. Boyd and J.F. Cox, Optimal decision making using cost accounting information, *International Journal of Production Research* **40**(8) (2002), 1879–1898.

[7] G. Buxey, Production scheduling: Practice and theory, *European Journal of Operational Research* **39**(1) (1989), 17–31.

[8] L. Candy and E. Edmonds, Creative design of the Lotus bicycle: implications for knowledge support systems research, *Design Studies*, **17**(1996), 71-90.

[9] S.-H. Chung, A.H.I. Lee and W.L. Pearn, Analytic network process (ANP) approach for product mix planning in semiconductor fabricator, *International Journal of Production Economics* **96**(1) (2005), 15–36.

[10] A. Coman and B. Ronen, Production outsourcing: a linear programming model for the Theory-Of-Constraints, *International Journal of Production Research* **38**(7) (2000), 1631–1639.

[11] M. Dzbor, Explication of design requirements through reflection on solutions, *4th IEEE Conf. on Knowledge-based Intelligent Engineering Systems & Allied Technologies*, Brighton, UK, August 2000, IEEE Press.

[12] M. Dzbor and Z. Zdrahal, Design as interactions of problem framing and problem solving, *15th European Conference on Artificial Intelligence (ECAI)*, Lyon, France, July 2002, Ohmsha / IOS Press.

[13] L.D. Fredendall and B.R. Lea, Improving the product mix heuristic in the theory of constraints, *International Journal of Production Research* **35**(6) (1997), 1535–1544.

[14] E.M. Goldratt and J. Cox, *The Goal*, North River Press, New York, 1984.

[15] E.M. Goldratt, *What is This Thing called Theory of Constraints*, North River Press, New York, 1990.
[16] E.M. Goldratt, What is the theory of constraints?, *APICS–The Performance Advantage* **3**(6) (1993), 18–20.
[17] J.A. Goguen, The logic of inexact concepts, *Syntheses* **19**(1969), 325–373.
[18] S. Haka, F. Jacobs and R. Marshall, Fixed cost allocation and the constrained product mix decision, *Advances in Accounting* **19**(2002), 71–88.
[19] T.–C. Hsu and S.–H. Chung, The TOC-based algorithm for solving product mix problems, *Production Planning and Control* **9**(1998), 36–46.
[20] R. Kee, Evaluating product mix and capital budgeting decisions with an activity-based costing system, *Advances in Management Accounting* **13**(2004), 77–98.
[21] R.C. Kee, Operational planning and control with an activity-based costing system, *Advances in Management Accounting* **11**(2003), 59–84.
[22] R. Kee, and C. Schmidt, A comparative analysis of utilizing activity-based costing and the theory of constraints for making product-mix decisions, *International Journal of Production Economics*, **63**(1) (2000), 1–17.
[23] G. Köksal, Selecting quality improvement projects and product mix together in manufacturing: an improvement of a theory of constraints-based approach by incorporating quality loss, *International Journal of Production Research* **42**(23) (2004), 5009–5029.
[24] B.-R. Lea and L.D. Fredendall, The impact of management accounting, product structure, product mix algorithm, and planning horizon on manufacturing performance, *International Journal of Production Economics* **79**(3) (2002), 279–299.
[25] T.N. Lee and G. Plenert, Optimizing theory of constraints when new product alternatives exist, *Production and Inventory Management Journal* **34**(3) (1993), 51–57.
[26] P. Letmathe and N. Balakrishnan, Environmental considerations on the optimal product mix, *European Journal of Operational Research* **167**(2) (2005), 398–412.
[27] R. Luebbe and B. Finch, Theory of constraints and linear programming: a comparison, *International Journal of Production Research* **30**(6) (1992), 1471–1478.
[28] C.J. Maday, Proper use of constraint management, *Production and Inventory Management Journal* **35**(1) (1994), 84.
[29] G.C. Onwubolu, Tabu search-based algorithm for the TOC product mix decision, *International Journal of Production Research* **39**(10) (2001), 2065–2076.
[30] G.C. Onwubolu and M. Mutingi, A genetic algorithm approach to the theory of constraints product mix problems, *Production Planning and Control* **12**(1) (2001), 21–27.
[31] M.C. Patterson, The product mix decision: a comparison of theory of constraints and labor-based management accounting, *Production and Inventory Management Journal* **33**(1992), 80–85.
[32] G. Plenert, Optimizing theory of constraints when multiple constrained resources exist, *European Journal of Operational Research* **70**(1) (1993), 126–133.
[33] A.J. Posnack, Theory of constraints: improper applications yield improper conclusions. *Production and Inventory Management Journal* **35**(1) (1994), 85–86.
[34] T.L. Saaty, *The Analytic Network Process*, RWS Publications: Pittsburgh, 1996.
[35] D.A. Schön, *Reflective Practitioner – How professionals think in action*, Basic Books, Inc, USA, 1983.
[36] H.A. Simon, The structure of ill-structured problems, *Artificial Intelligence*, **4** (1973), 181-201.
[37] R. Souren, H. Ahn and C. Schmitz, Optimal product mix decisions based on the Theory of Constraints? Exposing rarely emphasized premises of throughput accounting, *International Journal of Production Research* **43**(2) (2005), 361–374.
[38] M. Tang, A knowledge-based architecture for intelligent design support, *Knowledge Engineering Review*, **12**(4) (1997): 387-406.
[39] P. Vasant, Application of fuzzy linear programming in production planning, *Fuzzy Optimization and Decision Making* **2**(3) (2003), 229–241.
[40] P. Vasant and N.N. Barsoum, Fuzzy optimization of units products in mix-product selection problem using fuzzy linear programming approach, *Soft Computing – A Fusion of Foundations, Methodologies and Applications* (Published online 7 April 2005), In Press.
[41] P. Vasant, R. Nagarajan and S. Yaacob, Fuzzy linear programming with vague objective coefficients in an uncertain environment, *Journal of the Operational Research Society* **56**(5) (2005), 597–603.
[42] LV. Vanegas and A.W. Labib, Fuzzy approaches to evaluation in engineering design, *Journal of Mechanical Design*, **127**(1) (2005), 24–33.
[43] J. Watada, Fuzzy portfolio selection and its applications to decision making, *Tatra Mountains Mathematics Publication* **13**(1997), 219–248.

[44] L.A. Zadeh, The concept of a linguistic variable and its application to approximate reasoning I, *Information Sciences* **8**(1975), 199–251.
[45] L.A. Zadeh, The concept of a linguistic variable and its application to approximate reasoning II, *Information Sciences* **8**(1975), 301–357.
[46] L.A. Zadeh, The concept of a linguistic variable and its application to approximate reasoning III, *Information Sciences* **9**(1975), 43–80.

Integrated Intelligent Systems for Engineering Design
X.F. Zha and R.J. Howlett (Eds.)
IOS Press, 2006

A Bayesian Methodology for Estimating Uncertainty of Decisions in Safety-Critical Systems

Vitaly SCHETININ[a,1], Jonathan E. FIELDSEND[b], Derek PARTRIDGE[b], Wojtek J. KRZANOWSKI[b], Richard M. EVERSON[b], Trevor C. BAILEY[b] and Adolfo HERNANDEZ[b]

[a]*Department of Computing and Information Systems, University of Luton, LU1 3JU, UK*

[b]*School of Engineering, Computer Science and Mathematics, University of Exeter, EX4 4QF,* UK

Abstract. Uncertainty of decisions in safety-critical engineering applications can be estimated on the basis of the Bayesian Markov Chain Monte Carlo (MCMC) technique of averaging over decision models. The use of decision tree (DT) models assists experts to interpret causal relations and find factors of the uncertainty. Bayesian averaging also allows experts to estimate the uncertainty accurately when *a priori* information on the favored structure of DTs is available. Then an expert can select a single DT model, typically the Maximum a Posteriori model, for interpretation purposes. Unfortunately, *a priori* information on favored structure of DTs is not always available. For this reason, we suggest a new prior on DTs for the Bayesian MCMC technique. We also suggest a new procedure of selecting a single DT and describe an application scenario. In our experiments on real data our technique outperforms the existing Bayesian techniques in predictive accuracy of the selected single DTs.

Keywords. Uncertainty, decision tree, Bayesian averaging, MCMC.

Introduction

The assessment of uncertainty of decisions is of crucial importance for many safety-critical engineering applications [1], e.g., in air-traffic control [2]. For such applications Bayesian model averaging provides reliable estimates of the uncertainty [3, 4, 5]. In theory, uncertainty of decisions can be accurately estimated using Markov Chain Monte Carlo (MCMC) techniques to average over the ensemble of diverse decision models. The use of decision trees (DT) for Bayesian model averaging is attractive for experts who want to interpret causal relations and find factors to account for the uncertainty [3, 4, 5].

Bayesian averaging over DT models allows the uncertainty of decisions to be estimated accurately when *a priori* information on favored structure of DTs is available as described in [6]. Then for interpretation purposes, an expert can select a single DT

[1]Corresponding Author: Vitaly Schetinin, Department of Computing and Information Systems, University of Luton, Luton, LU1 3JU, The UK; E-mail: vitaly.schetinin@luton.ac.uk.

model which provides the Maximum a Posteriori (MAP) performance [7]. Unfortunately, in most practical cases, *a priori* information on the favored structure of DTs is not available. For this reason, we suggest a new prior on DT models within a sweeping strategy that we described in [8].

We also suggest a new procedure for selecting a single DT, described in Section 3. This procedure is based on the estimates obtained within the Uncertainty Envelope technique that we described in [9]. An application scenario, which can be implemented within the proposed Bayesian technique, is described in Section 5.

In this Chapter we aim to compare the predictive accuracy of decisions obtained with the suggested Bayesian DT technique and the standard Bayesian DT techniques. The comparison is run on air-traffic control data made available by the National Air Traffic Services (NATS) in the UK. In our experiments, the suggested technique outperforms the existing Bayesian techniques in terms of predictive accuracy.

1. Bayesian Averaging over Decision Tree Models

In general, a DT is a hierarchical system consisting of splitting and terminal nodes. DTs are binary if the splitting nodes ask a specific question and then divide the data points into two disjoint subsets [3]. The terminal node assigns all data points falling in that node to the class whose points are prevalent. Within a Bayesian framework, the class posterior distribution is calculated for each terminal node, which makes the Bayesian integration computationally expensive [4].

To make the Bayesian averaging DTs a feasible approach, Denison *et al.* [5] have suggested the use of the MCMC technique, taking a stochastic sample from the posterior distribution. During sampling, the parameters $\boldsymbol{\theta}$ of candidate-models are drawn from the given proposal distributions. The candidate is accepted or rejected accordingly to Bayes rule calculated on the given data **D**. Thus, for the m-dimensional input vector **x**, data **D** and parameters $\boldsymbol{\theta}$, the class posterior distribution $p(y \mid \mathbf{x}, \mathbf{D})$ is

$$p(y \mid \mathbf{x}, \mathbf{D}) = \int p(y \mid \mathbf{x}, \boldsymbol{\theta}, \mathbf{D}) p(\boldsymbol{\theta} \mid \mathbf{D}) d\boldsymbol{\theta} \approx \frac{1}{N} \sum_{i=1}^{N} p(y \mid \mathbf{x}, \boldsymbol{\theta}^{(i)}, \mathbf{D}),$$

where $p(\boldsymbol{\theta} \mid \mathbf{D})$ is the posterior distribution of parameters $\boldsymbol{\theta}$ conditioned on data **D**, and N is the number of samples taken from the posterior distribution.

Sampling across DT models of variable dimensionality, the above technique exploits a Reversible Jump (RJ) extension suggested by Green [10]. When *priori* information is not distorted and the number of samples is reasonably large, the RJ MCMC technique, making birth, death, change-question, and change-rule moves, explores the posterior distribution and as a result provides accurate estimates of the posterior.

To grow large DTs from real-world data, Denison *et al.* [5] and Chipman *et al.* [6] suggested exploring the posterior probability by using the following types of moves:

Birth. Randomly split the data points falling in one of the terminal nodes by a new splitting node with the variable and rule drawn from the corresponding priors.

Death. Randomly pick a splitting node with two terminal nodes and assign it to be a single terminal with the united data points.

Change-split. Randomly pick a splitting node and assign it a new splitting variable and rule drawn from the corresponding priors.

Change-rule. Randomly pick a splitting node and assign it a new rule drawn from a given prior.

The first two moves, *birth* and *death*, are reversible and change the dimensionality of $\boldsymbol{\theta}$ as described in [10]. The remaining moves provide jumps within the current dimensionality of $\boldsymbol{\theta}$. Note that the *change-split* move is included to make "large" jumps which potentially increase the chance of sampling from a maximal posterior whilst the *change-rule* move does "local" jumps.

For the birth moves, the proposal ratio R is written

$$R = \frac{q(\boldsymbol{\theta} \mid \boldsymbol{\theta}')p(\boldsymbol{\theta}')}{q(\boldsymbol{\theta}' \mid \boldsymbol{\theta})p(\boldsymbol{\theta})},$$

where $q(\boldsymbol{\theta} \mid \boldsymbol{\theta}')$ and $q(\boldsymbol{\theta}' \mid \boldsymbol{\theta})$ are the proposed distributions, $\boldsymbol{\theta}'$ and $\boldsymbol{\theta}$ are $(k + 1)$ and k-dimensional vectors of DT parameters, respectively, and $p(\boldsymbol{\theta})$ and $p(\boldsymbol{\theta}')$ are the probabilities of the DT with parameters $\boldsymbol{\theta}$ and $\boldsymbol{\theta}'$:

$$p(\boldsymbol{\theta}) = \left\{ \prod_{i=1}^{k-1} \frac{1}{N(s_i^{\text{var}})} \frac{1}{m} \right\} \frac{k}{S_k} \frac{1}{K},$$

where $N(s_i^{\text{var}})$ is the number of possible values of s_i^{var} which can be assigned as a new splitting rule, S_k is the number of ways of constructing a DT with k terminal nodes, and K is the maximal number of terminal nodes, $K = n - 1$.

The proposal distributions are as follows

$$q(\boldsymbol{\theta} \mid \boldsymbol{\theta}') = \frac{d_{k+1}}{D_{Q'}},$$

where $D_{Q1} = D_Q + 1$ is the number of splitting nodes whose branches are both terminal nodes.

Then the proposal ratio for a *birth* is given by

$$R = \frac{d_{k+1}}{b_k} \frac{k}{D_{Q1}} \frac{S_k}{S_{k+1}}.$$

The number D_{Q1} is dependent on the DT structure and it is clear that $D_{Q1} < k \; \forall \; k = 1, \ldots, K$. Analyzing the above equation, we can also assume $d_{k+1} = b_k$. Then letting the DTs grow, i.e., $k \rightarrow K$, and considering $S_{k+1} > S_k$, we can see that the value of $R \rightarrow c$, where c is a constant lying between 0 and 1.

Alternatively, for the death moves the proposal ratio is written as

$$R = \frac{b_k}{d_{k-1}} \frac{D_Q}{(k-1)} \frac{S_k}{S_{k-1}}.$$

However, in practice the lack of *a priori* information brings bias to the posterior estimates, and as a result the evaluation of classification uncertainty may be incorrect [11].

Within the RJ MCMC technique, the prior on the number of splitting nodes should be given properly. Otherwise, most samples may be taken from the posterior calculated for DTs that are located far away from a region containing the desired DT models. Likewise, when the prior on the number of splits is assigned as uniform, the minimal number of data points, p_{min}, allowed to be at nodes may be set inappropriately small. In this case, the DTs will grow excessively and most of the samples will be taken from the posterior distribution calculated for over-fitted DTs. As a result, the use of inappropriately assigned priors leads to poor results [5, 6].

For the special cases when there is knowledge of the favored DT structure, Chipman *et al.* [6] suggested the prior probability, with which a terminal node should be split further. This probability is dependent on how many splits have been made above it. For the given constants $\gamma > 0$ and $\delta \geq 0$, the probability P_s of splitting the ith node is

$$P_s(i) = \gamma(1 + d_i)^{-\sigma},$$

where d_i is the number of splits made above node i. Here the additional parameters γ and δ serving as hyperpriors should be given properly.

2. A Sweeping Strategy

Clearly, the lack of *a priori* knowledge on the favored DT structure, which often happens in practice, increases the uncertainty in results of the Bayesian averaged DTs. To decrease the uncertainty of decisions, a new Bayesian strategy of sampling DT models has been suggested [8]. The main idea behind this strategy is to assign *a priori* probability of further splitting DT nodes dependent on the range of values within which the number of data points will be not less than p_{min}. This prior is explicit because at the current partition the range of such values is unknown.

Within the above prior, the new splitting value q_j' for variable j is drawn from a uniform distribution:

$$q_j' \sim U(x_{\min}^{1,j}, x_{\max}^{1,j}),$$

and from a Gaussian with a given variance δ_j:

$$q_j' \sim N(q_j, \delta_j),$$

for the birth and change moves, respectively.

Because of the hierarchical structure, new moves applied to the first partition levels can heavily change the shape of the DT and, as a result, at its bottom partitions the terminal nodes can contain fewer data points than p_{min}. However, if the DT contains one such node, we can sweep it and then make the death move. Less likely, after birth or change moves the DT will contain more than one node containing fewer than p_{min} data points. In such cases we have to resample the DT.

3. Selection of a Single DT

In this Section we describe our method of interpreting Bayesian DT ensembles. This method is based on the estimates of confidence in the outcomes of the DT ensemble which can be quantitatively estimated on the training data within the Uncertainty Envelope technique.

3.1. Selection Techniques

There are two approaches to interpreting DT ensembles. The first approach is based on searching a DT of MAP [11]. The second approach is based on the idea of clustering DTs in the two-dimensional space of DT size and DT fitness [12].

Our approach is based on the quantitative estimates of classification confidence, which can be made within the Uncertainty Envelope technique described in [9]. The idea behind our method of interpreting the Bayesian DT ensemble is to find a single DT which covers most of the training examples classified as confident and correct. For multiple classification systems the confidence of classification outputs can be easily estimated by counting the consistency of the classification outcomes.

Indeed, within a given classification scheme the outputs of the multiple classifier system depend on how well the classifiers were trained and how representative were the training data. For a given data sample, the consistency of classification outcomes depends on how close this sample is to the class boundaries. So for the *i*th class, the confidence in the set of classification models can be estimated as a ratio γ_i between the number of classifier outcomes of the *i*th class, N_i, and the total number of classifiers N: $\gamma_i = N_i / N$, $i = 1, \ldots, C$, where C is the number of classes.

Clearly the classification confidence is maximal, equal to 1.0, if all the classifiers assign a given input to the same class, otherwise the confidence is less than 1.0. The minimal value of confidence is equal to $1/C$ if the classifiers assign the input datum to the C classes in equal proportions. So for a given input the classification confidence in the set of classifiers can be properly estimated by the ratio γ.

Within the above framework in real-world applications, we can define a given level of the classification confidence, γ_0: $1/C \leq \gamma_0 \leq 1$, for which cost of misclassification is small enough to be accepted. Then for the given input, the outcome of the set of classifiers is said to be *confident* if the ratio $\gamma \geq \gamma_0$. Clearly, on the labeled data we can distinguish between *confident and correct* outcomes and *confident but incorrect* outcomes. The latter outcomes may appear in a multiple classifier system due to noise or overlapping classes in the data.

3.2. A Selection Procedure

In practice, the number of DTs in the ensemble as well as the number of the training examples can be large. Nevertheless, counting the number of confident and correct outcomes as described above, we can find a desired DT which can be used for interpreting the confident classification. The performance of such a DT can be slightly worse than that of the Bayesian DT ensemble. Within the Chapter we provide the experimental comparison of their performances. The main steps of the selection procedure are next.

All that we need is to find a set of DTs which cover the maximal number of the training samples classified as confident and correct while the number of misclassifications on the remaining examples is kept minimal. To find such a DT set, we can remove the conflicting examples from the training data and then select the DTs with a maximal cover of the training samples classified by the DT ensemble as confident and correct.

Thus the main steps of the selection procedure are as follows:

1. Amongst a given Bayesian DT ensemble find a set of DTs, S1, which cover a maximal number of the training samples classified as confident and correct with a given confidence level γ_0.
2. Find the training samples which were misclassified by the Bayesian DT ensemble and then remove them from the training data. Denote the remaining training samples as D1.
3. Amongst the set S1 of DTs find those which provide a minimal misclassification rate on the data D1. Denote the found set of such DTs as S2.
4. Amongst the set S2 of DTs select those whose size is minimal. Denote a set of such DTs as S3. The set S3 contains the desired DTs.

The above procedure finds one or more DTs and puts them in the set S3. These DTs cover a maximal number of the training samples classified as confident and correct with a given confident level γ_0. The size of these DTs is minimal and any of them can be finally selected for interpreting the confident classification.

4. Experimental Results

In this Section first we describe the data used in our experiments. Then we show how the suggested Bayesian technique runs on these data. The resultant Bayesian averaging over DT models gives us a feature importance diagram. The suggested selection procedure gives us the single DTs for each run and finally we compare the predictive accuracies obtained with the existing procedures.

4.1. The Experimental Data

The data used in our experiments are related to the Short-Term Conflict Alert (STCA) problem which emerges when the distance between two aircraft, landing or taking off, might be critically short. Table 1 lists 12 features selected for predicting STCA.

In this table Δ_x, Δ_y, and Δ_z are the distances between pairs of aircraft on the X, Y, and height Z axes, respectively. Feature $x_4 = \sqrt{\Delta_x^2 + \Delta_y^2 + \Delta_z^2}$ is the distance between pairs of aircraft in 3-dimensional space. $V_{x,1}$ is the velocity of craft 1 on axis X, …, $V_{z,2}$ is the velocity of craft 2 on height Z. T_1 and T_2 are the times since the last correlated plot in the lateral plane for aircraft 1 and aircraft 2, respectively.

In our experiments we used 2500 examples of radar cycles taken each 6 seconds. From these examples 984 cycles were labeled as alerts. The number of cycles relating to one pair is dependent on the velocity and, on average, is around 40. All the examples of radar cycles were split into halves for training and test data sets within 5 fold cross-validation.

Table 1. The features selected for predicting the STCA.

#	Name	Feature
1	x_1	Δ_x
2	x_2	Δ_y
3	x_3	Δ_z
4	x_4	$\sqrt{\Delta_x^2 + \Delta_y^2 + \Delta_z^2}$
5	x_5	$V_{x,1}$
6	x_6	$V_{y,1}$
7	x_7	$V_{z,1}$
8	x_8	$V_{x,2}$
9	x_9	$V_{y,2}$
10	x_{10}	$V_{z,2}$
11	x_{11}	T_1
12	x_{12}	T_2

Fig. 1 shows two pairs of aircraft flying with different velocities: the distance between the aircraft of a pair is shown here by x_4 versus the radar cycles. The alert cycles are shown as stars, and the cycles, recognized by experts as normal, are shown as circles.

From Fig. 1, we can see that the left hand trace seems more complex for predicting the STCA than the right hand trace. First the series of alert cycles on the left hand trace is disrupted by 2 normal cycles, and second the aircraft having passed a critical 30th cycle remain in the alert zone. In contrast, the right hand trace seems straight and predictable.

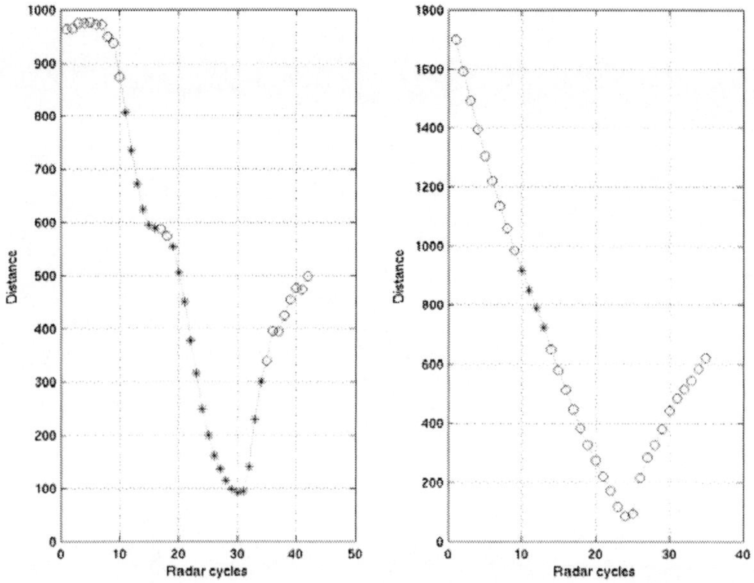

Figure 1. Two examples of alert cycles denoted here by the stars.

4.2. Performance of the Bayesian DT averaging technique

We ran the Bayesian DT technique without *a priori* information on the preferable DT shape and size. The minimal number of data points allowed in the splits, p_{min}, was set equal to 15 or 1.2% of the 1250 training examples. The proposal probabilities for the death, birth, change-split and change-rules were set to 0.1, 0.1, 0.2, and 0.6, respectively. The numbers of burn-in and post burn-in samples were set equal to 100k and 10k, respectively. The sampling rate was set equal to 7, and the proposal variance was set at 0.3 in order to achieve the rational rate of acceptance rate around 0.25, which was recommended in [5].

5 fold cross-validation was used to estimate the variability of the resultant DTs. The performances of all the 5 runs were nearly the same, and for the first run Fig. 2 depicts samples of log likelihood and numbers of DT nodes as well as the densities of DT nodes for burn-in and post burn-in phases.

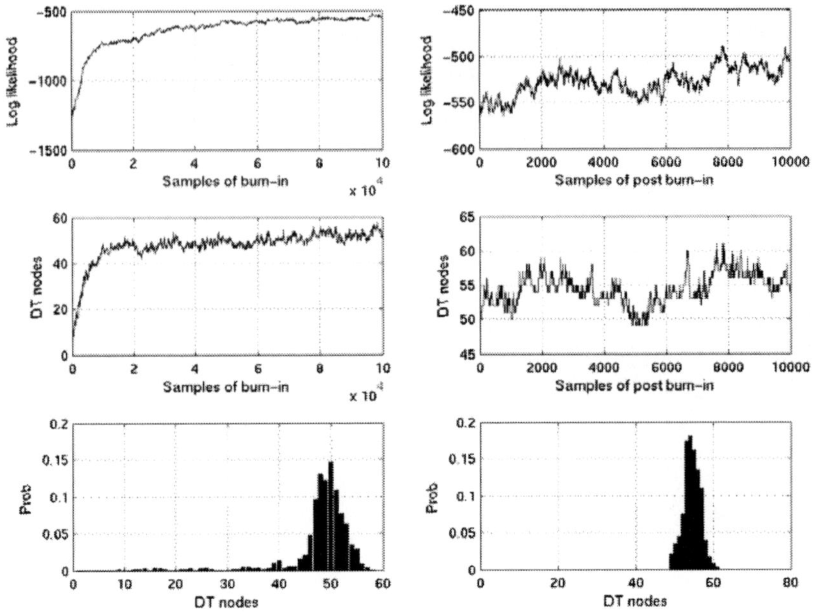

Figure 2. Samples of log likelihood and DT size during burn-in (the left side) and post burn-in (the right side). The bottom plots are the distributions of DT sizes.

From the top left plot we can see that the Markov chain converges to the stationary value of log likelihood near to –500 after starting around –1200. During the post burn-in phase the values of log likelihood slightly oscillate between –550 and –500.

The acceptance rates were 0.24 for the burn-in and 0.22 for the post burn-in phases. The average number of DT nodes and its variance were equal to 54.4 and 2.2, respectively.

On the first run, the Bayesian DT averaging technique misclassified 14.3% of the test examples. The rate of the confident and correct outcomes was 62.77%.

4.3. Feature Importance

Table 2 lists the average posterior weights of all the 12 features sorted by value. The bigger the posterior weight of a feature, the greater is its contribution to the outcome. On this basis, Table 2 provides ranks for all the 12 features.

Fig. 3 shows us the error bars calculated for the contributions of the 12 features to the outcome averaged over the 5 fold cross-validation. From this figure we can see that such features as x_8, x_1, and x_9 are used in the Bayesian DTs, on average, more frequently than the others. In contrast, feature x_{12} is used with a less frequency. Additionally, the widths of the error bars in Fig. 3 give us the estimates of variance of the contributions.

Table 2. Posterior weights of the features sorted on their contribution to the outcome.

Feature	Posterior weight	Rank
x_8	0.168	1
x_1	0.137	2
x_9	0.120	3
x_6	0.110	4
x_4	0.095	5
x_5	0.090	6
x_3	0.078	7
x_2	0.061	8
x_{10}	0.050	9
x_7	0.042	10
x_{11}	0.001	11
x_{12}	0.008	12

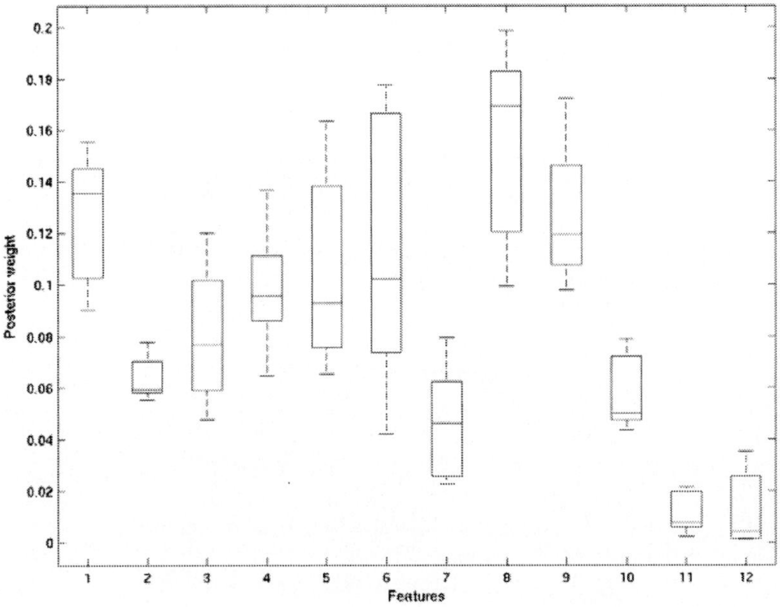

Figure 3. Feature importance averaged over 5 fold cross-validation.

4.4. A Resultant DT

The resultant DT selected by the SC procedure is presented as a machine diagram in Fig. 4. Each splitting node of the DT provides a specific question that has a yes/no answer, and two branches. The terminal nodes provide the predictive probabilities of alert, whose values range between 0.0 and 1.0.

node01 X04 < 1847.05, then node03, otherwise node45
node03 X04 < 1459.91, then node06, otherwise node28
node06 X05 < -281.95, then node15, otherwise node07
node07 X03 < 1713.61, then node08, otherwise node12

```
node08  X01 <   -1.64, then node02, otherwise node04
node02  X07 <    6.19, then node10, otherwise node14
node04  X04 <  324.47, then node18, otherwise node11
node10  X08 <   69.80, then node20, otherwise alert(0.99)
node11  X10 <  -13.43, then node19, otherwise node17
node12  X08 < -105.10, then alert(1.00), otherwise node09
node14  X08 <  150.30, then node23, otherwise alert(0.45)
node17  X05 <   82.84, then node25, otherwise node13
node18  X06 < -235.08, then alert(0.09), otherwise node31
node19  X08 <  -45.98, then node43, otherwise node05
node20  X04 <  415.29, then alert(0.13), otherwise node21
node21  X09 <   31.87, then alert(0.89), otherwise node39
node15  X09 <   81.89, then node34, otherwise node29
node25  X09 < -138.94, then node27, otherwise node41
node13  X01 <    2.31, then node44, otherwise node22
node22  X06 < -275.55, then alert(0.50), otherwise alert(0.99)
node28  X08 <  -28.49, then node16, otherwise node30
node29  X08 <  -46.49, then alert(0.06), otherwise alert(1.00)
node27  X05 <   11.31, then node42, otherwise alert(0.00)
node34  X01 <   -1.24, then alert(0.96), otherwise node32
node05  X11 <    0.00, then alert(0.07), otherwise node33
node31  X09 < -317.08, then alert(0.68), otherwise node37
node30  X03 < 4075.28, then alert(0.86), otherwise alert(0.00)
node39  X05 <  142.61, then alert(0.98), otherwise alert(0.27)
node37  X05 < -212.06, then node26, otherwise node40
node42  X02 <   -1.28, then node38, otherwise node51
node43  X01 <   -0.02, then alert(0.38), otherwise alert(1.00)
node16  X07 <   29.42, then node24, otherwise alert(0.25)
node41  X08 <  314.03, then alert(0.97), otherwise alert(0.56)
node33  X08 <   76.40, then alert(0.16), otherwise alert(0.86)
node40  X09 <  174.57, then node35, otherwise alert(0.12)
node38  X12 <    0.00, then alert(0.08), otherwise alert(1.00)
node35  X02 <    0.28, then alert(0.27), otherwise alert(0.79)
node51  X06 <   23.34, then alert(0.65), otherwise alert(1.00)
node44  X05 <  216.38, then alert(1.00), otherwise alert(0.62)
node45  X01 <  -15.93, then alert(0.96), otherwise alert(0.89)
node32  X01 <   13.80, then alert(0.00), otherwise alert(0.61)
node09  X10 <    0.64, then alert(0.89), otherwise node49
node23  X09 <   26.46, then alert(0.17), otherwise alert(0.07)
node49  X01 <    9.91, then alert(0.24), otherwise alert(0.57)
node24  X09 <  -99.68, then alert(1.00), otherwise alert(0.83)
node26  X05 < -239.50, then alert(0.06), otherwise alert(0.00)
```

Figure 4. Machine diagram of the resultant DT selected by the SC technique.

4.5. Comparison of Performances

In this section we compare our technique of extracting a sure correct (SC) DT with MAP, and the maximum a posterior weight (MAPW). The comparison is made in terms of misclassification within 5 fold cross-validation. The misclassification rates of the above three techniques: SC, MAP, and MAPW are shown in Fig. 5. The left side plot shows the misclassification rates of the single DTs on the test data, and the right side plot shows its sizes.

In theory, the Bayesian averaging technique should provide lower misclassification rates than any other single DTs selected by the SC, MAP, and MAPW techniques. On the first run, we can observe that all the single DTs perform worse than the Bayesian ensemble of DTs which has misclassified 14.3% of the test data.

Comparing the misclassification rates of the SC, MAP, and MAPW shown in Fig. 5, we can see that the suggested SC technique more often out-performs the other two techniques, that is, the SC technique out-performs the MAP and MAPW techniques on the 4 runs.

Comparing the DT sizes on the right side plot, we can see that the SC technique has extracted shorter DTs than the MAP technique in 4 runs. At the same time, comparing the sizes of the SC and MAPW DTs, we can see that the SC technique has extracted shorter DTs in 2 runs only.

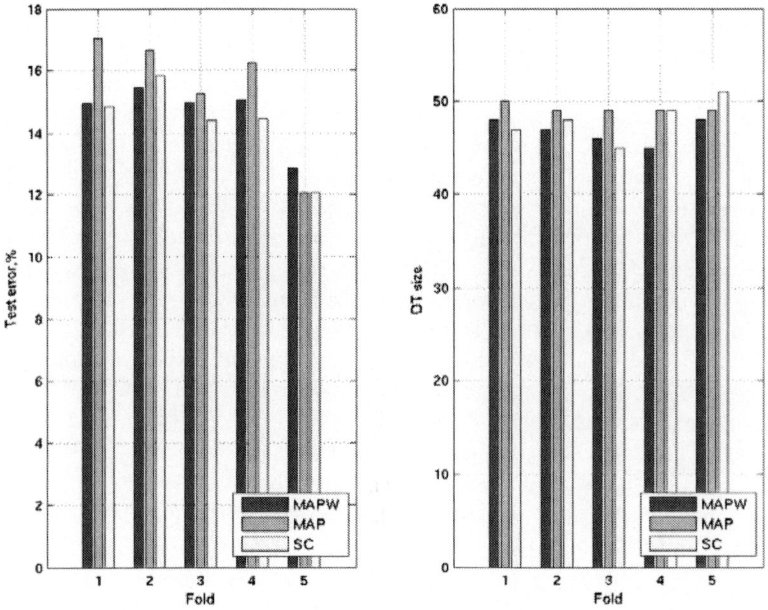

Figure 5. Comparison of test error and DT sizes within 5 fold cross-validation for the MAPW, MAP and proposed SC techniques.

5. An Application Scenario

In a general form an application scenario within our approach can be described as a sequence of the following steps.

1. Define a set of features considered by domain experts as the most important ones for the classification. For example, a domain expert can assume the velocities of aircraft in coordinates X and Y as the most important features for predicting the STCA.
2. Within a defined set of features, collect a representative set of data samples confidently classified by domain experts. For instance, a domain expert can arrange a set of radar data received from different pairs of aircraft in which some of the radar cycles were labeled as the STCA.
3. Analyze *a priori* knowledge and formulate priors within the Bayesian methodology. For example, domain experts can represent their knowledge in a form of decision tree asking specific questions. Such *a priori* information can be used within our approach in order to improve the performance.
4. Define parameters of DTs such as the minimal number of data samples, p_{min}, allowed to be in splitting nodes. By changing this parameter, a modeler can find in an *ad hoc* manner the number of splitting nodes providing the best performance of DTs.
5. Define parameters of MCMC such as the number of burn-in and post burn-in samples, proposal probabilities for the death, birth, change-split and change-rule, as well as the sampling rate. At this step a modeler can also specify suitable proposal distributions for the moves. However, if a modeler has no idea about the proposal distribution, a uniform distribution, known also as an "uninformative" prior, is used.
6. Specify a criterion for convergence of Markov Chain. Within the Bayesian MCMC technique, the convergence of a Markov Chain is usually achieved if the number of burn-in samples set enough large. Practically, the convergence is achieved if after approximately 1/3 of burn-in samples the likelihood values do not change significantly. This can be easily visualized by observing the log likelihood values plotted versus the number of post burn-in samples.
7. A modeler has to control the diversity of DTs collected during the post burn-in phase. The diversity is estimated implicitly by estimating an acceptance level which is the ratio of the number of the proposed and accepted DTs to the total number of proposed DTs. Practically, when the acceptance level is near 0.25, a set of collected DTs is optimally diversified.
8. Practically, the desired acceptance level is achieved by changing the following parameters: p_{min}, the variance of proposal distribution, and the expected number of splitting nodes.
9. To select a single DT providing the most confident classification, a modeler has to predefine a level of confident classification, γ_0. The value of γ_0 is dependent on the cost of misclassifications allowed in an application. Clearly, if the cost of classification is high, the value of γ_0 is defined close to 1.0, say 0.9990. This means that a decision is confident if no more than 10 classifiers from 10000 are contradictory. Otherwise, a decision is assigned uncertain.
10. Having obtained a confident DT, a modeler can observe a decision model and analyze the features used in this model. Additionally, a modeler can run the

Bayesian MCMC and DT selection techniques within *n*-fold cross validation in order to estimate the contribution of each feature to the classification. As a result, the contributions of all features can be ranked, and a modeler can find the most critical features in an application.

Related to the STCA problem, the application scenario used in our work is described as follows. The domain experts have defined 12 features listed in Table 1. From the collected 2500 data samples, 1250 were used for training and the remaining 1250 for testing. *A priori* information was not available in our case and, therefore, priors were given as "uninformative" except for a Gaussian for the proposal distribution with the variance set to 0.3. The number p_{min} = 15 was experimentally found to provide the best performance. The numbers of burn-in and post burn-in samples were set equal to 100 k and 10 k, respectively. The proposal probabilities for the death, birth, change-split and change-rules were set to 0.1, 0.1, 0.2, and 0.6, respectively. Every 7th DT was collected during the post burn-in phase, i.e., the sampling rate was 7. The convergence of the Markov Chain can be visually observed from the top right plot in Fig. 2 – from this plot we can see that after approximately 1/3 of burn-in samples the values of log likelihood do not change significantly. The acceptance level during the post burn-in phase was obtained equal to 0.22 that is close to 0.25 when the diversity of DTs is optimal. The level of confident classification, γ_0, was predefined to be 0.99. Fig 4 represents the machine diagram of the resultant DT selected under the given γ_0. Finally, the performance of the resultant DT is compared with the performances of the Bayesian DT technique as well as the MAP DT within the 5 fold cross-validation as shown in Fig. 5.

6. Conclusion

For estimating uncertainty of decisions in safety-critical engineering applications, we have suggested the Bayesian averaging over decision models using a new strategy of the RJ MCMC sampling for the cases when *a priori* information on the favored structure of models is unavailable. The use of DT models assists experts to interpret causal relations and find factors to account for the uncertainty. However, the Bayesian averaging over DTs allows experts to estimate the uncertainty accurately when *a priori* information on favored structure of DTs is available.

To interpret an ensemble of diverse DTs sampled by the RJ MCMC technique, experts select the single DT model that has maximum *a posteriori* probability. However in practice this selection technique tends to choose over-fitted DTs which are incapable of providing a high predictive accuracy.

In this Chapter we have proposed a new procedure of selecting a single DT. This procedure is based on the estimates of uncertainty in the ensemble of the Bayesian DTs. For estimating the uncertainty, the use of an Uncertainty Envelope technique has been advocated. As a result, in our experiments with the STCA data, the suggested technique outperforms the existing Bayesian techniques in terms of predictive accuracy.

Thus, we conclude that the technique proposed for interpreting the ensemble of DTs allows experts to select a single DT providing the most confident estimates of outcomes. These are very desirable properties for classifiers used in safety-critical systems, in which assessment of uncertainty of decisions is of crucial importance.

7. Acknowledgements

The work reported was largely supported by a grant from the EPSRC under the Critical Systems Program, grant GR/R24357/01.

References

[1] A. Haldar, S. Mahadevan, *Probability and Statistical Methods in Engineering Design*, Wiley, 1999.

[2] R. Everson, J.E. Fieldsend, Multi-Objective optimization of safety related Systems: An application to short term conflict alert, *IEEE Transactions on Evolutionary Computation* (forthcoming) 2006.

[3] L. Kuncheva, *Combining Pattern Classifiers: Methods and Algorithms*, Wiley, 2004.

[4] L. Brieman, J. Friedman, R. Olshen, C. Stone, *Classification and Regression Trees*, Belmont, CA, Wadsworth, 1984.

[5] D. Denison, C. Holmes, B. Malick, A. Smith, *Bayesian Methods for Nonlinear Classification and Regression*, Wiley, 2002.

[6] H. Chipman, E. George, R. McCullock, Bayesian CART model search, *J. American Statistics* **93** (1998), 935-960.

[7] P. Domingos, Bayesian averaging of classifiers and the overfitting problem, *International Conference on Machine Learning. Stanford, CA, Morgan Kaufmann* 2000, 223-230.

[8] V. Schetinin, J. E. Fieldsend, D. Partridge, W. J. Krzanowski, R. M. Everson, T. C. Bailey, and A. Hernandez, The Bayesian decision tree technique with a sweeping strategy, *Int. Conference on Advances in Intelligent Systems - Theory and Applications, (AISTA'2004) in cooperation with IEEE Computer Society,* Luxembourg, 2004.

[9] J.E. Fieldsend, T.C. Bailey, R.M. Everson, W.J. Krzanowski, D. Partridge, V. Schetinin, Bayesian inductively learned modules for safety critical systems, *Symposium on the Interface: Computing Science and Statistics*, Salt Lake City, 2003.

[10] P. Green, Reversible jump Markov Chain Monte Carlo computation and Bayesian model determination, *Biometrika* **82** (1995), 711-732.

[11] P. Domingos, Knowledge discovery via multiple models. *Intelligent Data Analysis* **2** (1998), 187-202.

[12] H. Chipman, E. George, R. McCulloch, Making sense of a forest of trees, *Symposium on the Interface*, S. Weisberg, Ed., Interface Foundation of North America, 1998.

Section II

Techniques, Frameworks, Tools and Standards

Integrated Intelligent Systems for Engineering Design
X.F. Zha and R.J. Howlett (Eds.)
IOS Press, 2006

Quantification of Customer Multi-Preference and Motivation through Data and Text Mining in New Product Design

Xiang LI [a,1], Junhong ZHOU [a,2] and Wen Feng LU [b,3]
[a] *Singapore Institute of Manufacturing Technology*
[b] *National University of Singapore*

Abstract. Effective collection and analysis of customer demand is a critical success factor for new product design and development. This chapter presents a set of customer requirement discovery methodologies to achieve broad and complex market studies for new products. The proposed approach uses data mining and text mining technologies to discover customer multi-preference and corresponding customer motivation. Using the proposed rule mining methodology, discovery rules can be flexibly defined, the complete customer multi-preference patterns are discovered and their statistic analysis of multi-preference can be conducted for new product design. With the proposed text mining methodology, the customer motivations are discovered and the percentage of surveyed customers with certain preference and the reason for this preference are presented. Combining the methodologies in text mining with rule mining, the customer motivations can be quantitatively described with statistic analysis results. A prototype system that allows on-line customer feedback collection, digitization of the language feedbacks, numerical descriptions of customer preferences and customer motivation of a product is developed to demonstrate the feasibility of the proposed methodologies. It is shown that the proposed work could significantly shorten the survey and analysis time for customer preference and is thus expected to help companies to reduce cycle time for new product design.

Keywords. Customer demand, association rule mining, text mining, new product design

Introduction

To successfully develop a new product, discovering customer demand preference and motivation are always important yet difficult [1]. In most cases, there are usually many ideas and options for selection during product conceptualization. One of the key factors for designers to consider during this selection is the preference and motivation

---

[1] Corresponding Author: Research Scientist, Singapore Institute of Manufacturing Technology, 71 Nanyang Drive Singapore 638075; E-mail: xli@simtech.a-star.edu.sg .
[2] Senior Research Engineer, Singapore Institute of Manufacturing Technology, 71 Nanyang Drive Singapore 638075; E-mail: jzhou@simtech.a-star.edu.sg .
[3] Associate Professor, Department of Mechanical Engineering/Design Technology Institute, Faculty of Engineering, National University of Singapore; Email: mpelwf@nus.edu.sg .

of customers. To investigate customer demands, one approach is to have a quantitative description and reasoning analyses of the preferences of different group of people. The most common practice for establishing such understanding is through survey with predefined questionnaires. New products or new features are listed and options are designed for customers to choose. Spaces are provided for customer to answer open-end questions or write reasoning description of their preferences. The feedbacks are tabulated and analyzed for product designers to use as references in their product design.

Although a good survey firstly depends on the design of questionnaire, the analysis methodology is also very critical as it will decide how much useful information can be extracted from a collection of feedbacks. It might be easy to get a preference percentage of the survey group for one particular feature. However, to get a statistical analysis on the preference of a combination of multiple features would be relatively difficult, even with the help of commonly available software tools such as MS Excel. It would more difficult, if not impossible, to manually get a statistical analysis of preferences for multiple feature combinations when the number of features in consideration is large and/or huge amount of feedbacks are received. Further more challenging to these is how to systematically discover customer motivations hiding in the feedback contexts. It is, therefore, desirable to have proper information handling methodologies for such survey collection, customer multi-preference and motivation reasoning analysis as well as linking customer motivation patterns with product design features. This chapter describes data and text mining technologies for these purposes.

Since 1995, there have been efforts to use data and text mining technologies to extract implicit, previously unknown and potentially useful knowledge from data and free text. The knowledge here refers relationships and patterns between data or text elements. Further, there have been quite a number of researches on data mining and knowledge discovery technologies to solve problems in marketing, planning, optimization and prediction [2]. In the area of engineering design, works have been done using data mining to detect aircraft component replacement [3], capturing the rationale for the detailed design process automatically [4], modeling knowledge to guide design space search [5], using incremental mining of the schema for semi-structured data [6] and so on. There are also new survey tools in the market, such as Host Survey by Hostedware Corporation [7], XPO Online Survey System by XPO Show Services Inc. [8], EZSurvey by David Hull & Associates Ltd [9], and Survey Pro Software by Survey Professional [10]. Most of them have the capability of being web-enabled, flexible documentation and powerful statistical analysis. SAS Data Mining Solution [11] uses data mining technology to reduce fraud, anticipate resource demand, increase acquisition and curb customer attrition. SPSS Text Analysis [12] could unlock the value contained in verbatim responses to open-ended survey questions. The tool creates categories or “codes” quickly and easily to support teams of people working on the same project. However, these tools have their limitations in addressing survey analysis issues as described in the previous paragraph. The work presented in this Chapter attempts to provide methodologies to fill the gap.

As for this chapter, Section 1 presents an approach for customer multi-preference pattern discovery with an associate rule mining methodology. It further explains the digitisation of survey feedbacks, and the algorithm for rule mining. In Section 2, a concept-based text mining method is proposed for reasoning analysis of customer preference to design features in a product as well as discovering the corresponding customer motivations. A software prototype with on-line customer feedback collection capability is developed based on these knowledge discovery methods and is described in Section 3. A scheme is also established to convert linguistic comments of customer feedbacks to digital values so that statistical analysis of customer preference can be performed. Section 4 describes a case study to illustrate the operational steps in using the proposed system for customer multi-preference pattern discovery. The system could significantly shorten the survey and analysis time and is thus expected to help companies not only to discover customer preferences, but also to reduce design cycle time for new product development. In addition, the proposed algorithms are generic and can be adapted for other scenarios that require customer preference and motivation analysis.

1. Customer Multi-Preference Patterns Discovery (CMPD) with Association Rule Mining

1.1. ARM Definition

Association rule mining (ARM) is one important method for knowledge discovery in databases. An ARM process searches for interesting relationships among items in a given data set [13] by finding a collection of item subsets (called item-sets) that frequently occur in database records or transactions, and then extracting the rules representing how one subset of items influences the presence of another subset [14]. For a given pair of confidence and support thresholds, the concept of mining association rules is to discover all rules that have confidence and support greater than the corresponding thresholds. For example, for a sale in a computer hardware shop, the association rule of "CD Writer $\Rightarrow$ Lens Cleaner" can mean that whenever customers buy CD writers, they also buy lens cleaner c% of the time and this trend occurs s% of the time. Therefore, an ARM operation performs two tasks [15] of discovery of frequent item-sets, and generation of association rules from frequent item-sets.

Many academic researchers have tackled the first sub-problem that is more computationally extensive and less straightforward [16]. These previous works designed many algorithms for efficient discovery of frequent item-sets. In the present work, we lever the ability of available ARM algorithms to rapidly discover frequent item-sets of customer preference patterns in new product design.

1.2. Approach for CMPD

There could be many ways to define questionnaire and handle customer feedbacks. In the present work, we adopted the most common practice that the survey questions are set with prescribed options. For example, if a company has a few new design options (A, B, C) for a mini-HIFI system, the survey questionnaire could appeare as follows:

- Which product is your preference among A, B, C?
- Where do you prefer to put the product in?
    a. personal room
    b. family room
    c. table top

d. other location

To discover customer preference patterns from this type of survey feedbacks is to find the frequent item-sets of association rules with customer preference patterns and its percentage of support (occurrence). In addition, designers might have one or few important particular products or product features in mind (feature of interest, or FOI) and would like to know customer preference patterns associated with the FOI.
Based on this perception on the CMPD tasks, we adopted an *ad hoc* network approach to handle the feedback collection and analysis. A three-layered network structure is designed as shown in Figure. 1, where

Q_n – *n*th question

p_k^n – *k*th option of *n*th question

$s_h\%$ – *h*th rule support percentage.

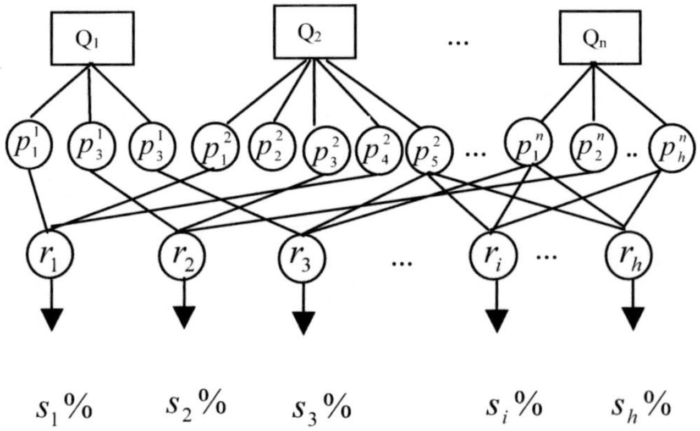

Figure 1 CMPD Network Structure

The nodes at the first layer are the input questions that are linguistic statement q_ks, such as 'my favorite product is', or 'I like to use the product in'. The task of layer one is to translate these linguistic statements to numeric numbers. The Q_is are their input numeric values, for example, Q_1=1; Q_2=2.

The nodes at the second layer are the question options, such as 'Product A', 'Product B', 'Product C', which are associated with certain questions. The function of this layer is to transfer the question numbers from layer one and translate the natural answers to binary format (explained in next section).

Layer 3 contains the rule nodes. The links here define the preconditions of the rule nodes, while the node outputs define their consequences. To determine the number of nodes in this layer, one should select one main node (main question) in the first layer relevant to the FOI. For each rule node in this layer there must have at least two links from the relevant nodes in the second layer and one of the links must go to an option node of the main node in the first layer. The operation of the network will find out the association rules of the options of other questions to the main question. The nodes at this layer perform the heuristic AND operations, and sometimes they also represent 'IF-AND-THEN' rules. The outputs of this layer are the percentage of supports s% to the rule. For example, an association rule can be described as follows:

Age (X, "20…29)∧Gender(X, Male)

⇒ Favorite(X, "Product B") ∧Location(X, "Own room")∧Function(X, "Connect with VCD/DVD") [support = 16.7%]

where X is a variable representing the choice of a customer.

1.3. Digitalization of Survey Feedback

Figure 2 illustrates the linguistic-to-numerical transformation method and process to digitalize the linguistic customer survey feedback to numerical data representation. The linguistic answer to a question is represented by a special formatted numbers, consisting of a decimal portion and a binary portion. The question number in the survey questionnaire is represented by using a decimal number in the decimal portion. The answer to the question is represented by binary digits in their sequence: "1" if the option is chosen and "0" if it is not chosen. For example, the output of Question #1 will be '1|010', which is corresponding to a customer answer of 'my favorite product is B'. Similarly, the output of second layer for question two will be '2|1010' if there are four options in question two and the customer's answer is option one and option three. To save storage space in database and further calculations, the binary digits can be further transferred into decimal numbers when store the customer data to the database.

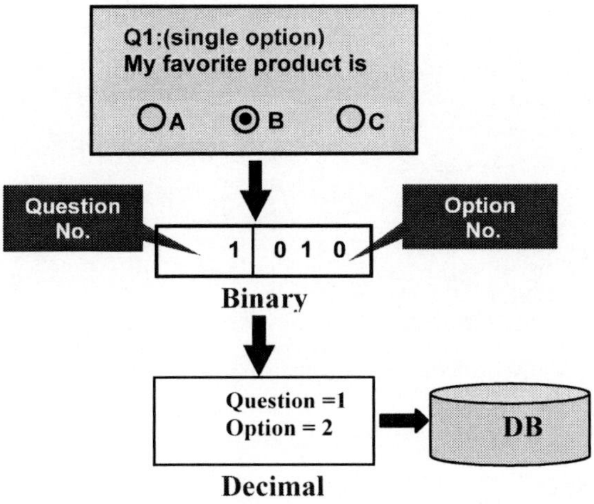

Figure 2 Illustration of the process for survey feedback digitalization

1.4. Rule Mining Algorithm

An algorithm is developed to find the percentage of support for all the customer preference patterns based on digitization mechanism of CMPD network structure described above and frequent item-sets.

Input: Database, D, of transactions; customer survey data include total number of questions n; survey question q_k (k = 1,2,…, n), sub-total number of option-item-set T(k); survey options p_j^k (j = 1,2,…,m) of kth question; minimum support threshold $s_{\min}\%$.

Output: Rule percentage of support s %, that is, the popularity of customer preference patterns percentage.

The algorithm is described as follows.

1. Convert customer questions q_k (k=1,2,…,n) and options p_j^k (j=1,2,…,m) to binary data format, and save them to the database D as decimal data format;
2. Transmit question values q_k to the layer one of CMPD architecture, forming binary data value of options p_j^k to the relevant nodes in layer two.
3. Select main question as q_1.
4. Work out the rule pattern collection and do statistic analysis for each rule with either of the following two ways:

(i) Count all of the question-option combination and determine the total rule numbers to identify total scenarios of customer preference patterns by assuming that all questions are defined with single choice. Then, for each rule pattern, the customer survey records will be fed through. The total number of rules can be calculated by Eq. (1)

$$R_{total} = \sum_{i=1}^{n-1}[\sum_{j_1=2}^{n-i+1}\sum_{j_2=j_1+1}^{n-i+2}\cdots\sum_{j_i=j_1+i-1}^{n} T(1)\cdot T(j_1)\cdot T(j_2)\cdot\ldots\cdot T(j_i)] \qquad (1)$$

(i = 1, 2, …, n-1 ; j_i = 1,2, …, i +1)

where

n — the total number of questions;

$T(1)$ — the numbers of option-itemset of main question;

$T(j_i)$— the numbers of option-itemset of j_i th question.

(ii) Generate customer rule patters by using the following recursive procedure: For each question option combination pattern, all customer survey records will be scanned and compared, and the pattern will be selected as a rule once a matching is found. Thus One customer selection pattern is a rule pattern. If the rule pattern already exists during the scanning, increase its rule counts. For the second methodology, the maxim rule number and loop times are equal:

$$R_{\max} = CU \sum_{i=1}^{n-1} C_{n-1}^{i}$$

where

n — total number of questions;

CU — total numbers of customer

Figure 3 describes the work flow to realize the question combination and pattern generation. As the option will join the recursive in the first method, usually, the second way is more effective. But when the number of customers is large and the questionnaire is simple, the first method may take less time.

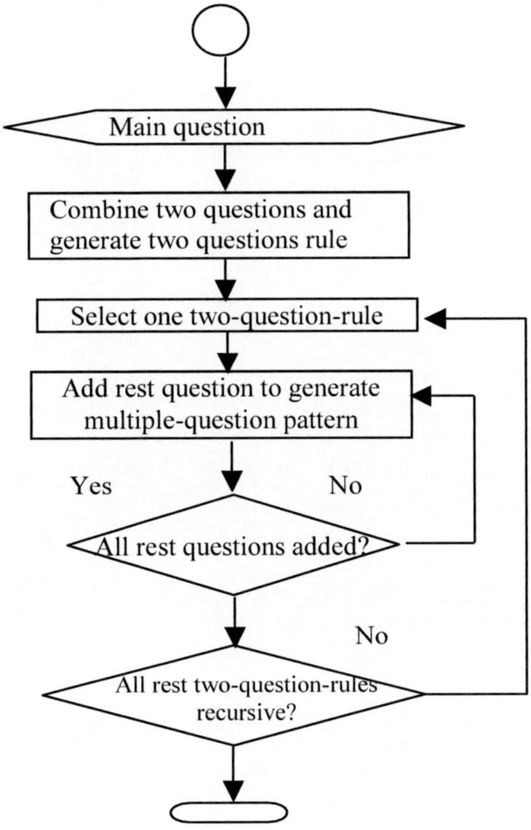

Figure 3 Question combination pattern generation flow chart

5. Filter out frequent rules R_u ($u = 1,2,\ldots,h$) at Layer 3. R_{total} is a superset of R_u, that is, its members may or may not be frequent rules, but all of the frequent rules R_u are included in R_{total}. R_{total} can be huge. To reduce the size of rule base and heavy computation, a scan of the database to determine the count of each candidate in R_{total} is carried out to determine of R_u. That is, all candidates having a rule count value s% not less than the minimum support count s_{min}% are frequent by definition, and therefore belong to R_u.
6. Result in the rule count value s% as the output of the node of the rule layer.
7. Save the frequent rules R_u and the rule support percentage s%, (the popularity of customer preference patterns percentage) to the rule base and database respectively.
8. Convert the discovered frequent rules R_u from digital format back to their original natural questions and options to make it easily understand to users.

2. Customer Motivation Discovery with Text Mining

With the CMPD module, it should have become easier to discover customer multiple preferences. However, more important and challenging to this would be the analysis of the reasons behind the preferences. It will be useful if designers could have understanding on why customers choose their preferred products and what are the main factors or product features that affect customers' selection. In this section, a concept-based text mining method is developed and employed to capture the useful information from huge data source of customer open-end answers.

2.1. Open-end Questionnaire Design

In customer survey questionnaire, some questions are open-end format which requests customer to answer by using free text. These questions are not well structured as the optional question types described in the Section 1. For example, if a company has a new design options (A, B, C) for a mini-HIFI system, the survey questionnaire could be as follows.

Question 1: Do you like Product A? Why? _________.
Question 2: Do you like Product B? Why? _________.
Question 3: Do you like Product C? Why? _________.

Customer 1 answers:
Answer 1: I like Product A because *the design is simple.*
Answer 2: I don't like Product B because *it's bulky.*
Answer 3: I don't like Product C because *it's ugly and heavy.*

Customer 2 answers:
Answer 1: I like Product B because *it's easy to carry.*
Answer 2: I don't like Product B because *it's expensive*.
Answer 3: I don't like Product C because *it's difficult to operate*.

The information carried by such free text answers are very useful to help designers to figure out customers' needs and create winning design ideas. Up to now, most of the companies manually read the answers gathered from customers, and manually figure out the useful information from these answers. It is can be a very time consuming exercise, especially when the amount of data is huge.

2.2. Concept-based Text Mining Method

Input: Database---Customer feedbacks with free texts.
Output: Percentage of surveyed customers with certain product preferences and the reasons.

The concept-based text mining method is described as follows and the work flow is shown in Figure. 4.

1. Identify the design key factors/features based on designer's knowledge or from previous keywords database.

2. Scan the text content of the answers from free text questions and identify keywords for each survey feedback.

3. Extract key words from these answers.

4. Add a key word into the keyword database if it does not exist in the database, otherwise increase its frequency count only.

5. If a key word's frequency is larger than the minimum support value, assign it to one of the categories describing important design key factors based on related domain knowledge (function, design, sound, size, etc) and further to either of the two sub-categories (positive impact and negative impact) for the factor category. The key words and design feature relationships are used to build up keyword database. This database will be continuously enriched with additional new key words.

6. Based on the information from the keywords database, the total positive and negative impact frequencies of the design features can be calculated.

7. The total positive and negative impact frequencies of the design features will be translated into the percentage of support. Save the support percentage s% to the database. This will provide the percentage of surveyed customers with certain product preferences and the reasons.

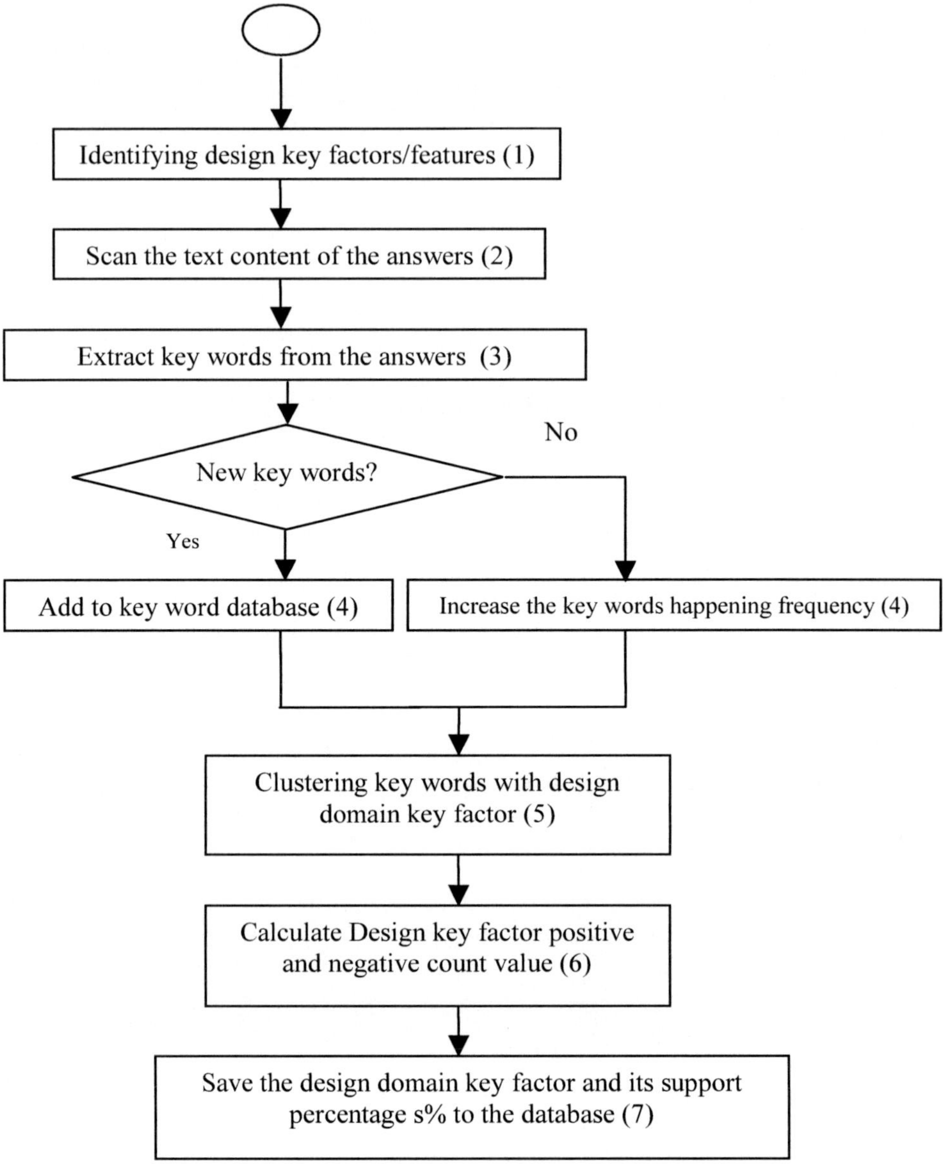

Figure 4 Concept-based Text Mining Procedure.

2.3. New Questionnaire Generation based on Reasoning Analysis

A good survey depends firstly on the design of questionnaire. At present, most of the companies manually create questionaires based on the designer's experience and the quality of the questions solely depends on the designer's knowledge. As such, the quality of the questions is not always guaranteed which will affect the outcome of the related surveys. Moreover, if all the questions are coming from designer's side, customers' perspectives might not be neglected. By combining the developed text mining method and customer multi-preference analysis method described above, generation of new questionnaires can be done from the results of reasoning analysis as well as knowledge pool?? . In this section, quantitative of customer motivation and new questionnaires generation from the quantitative analysis results will be discussed.

In Section 2.2, the methods for capturing customer motivations to create key word database were explained. Free text answers from customer are clustered into design features such as *appearance*, *size*, *function*, *brand*. These key words could further be clustered into more detailed categories of particular design features as shown in Figure 5. For example, design feature *appearance* can further cluster into *color*, *style*, etc. The detailed category can narrow the description of customer needs. The positive and negative impact frequencies of customer motivation can be calculated for the detailed category. Category clustering process can be repeated into further layers.

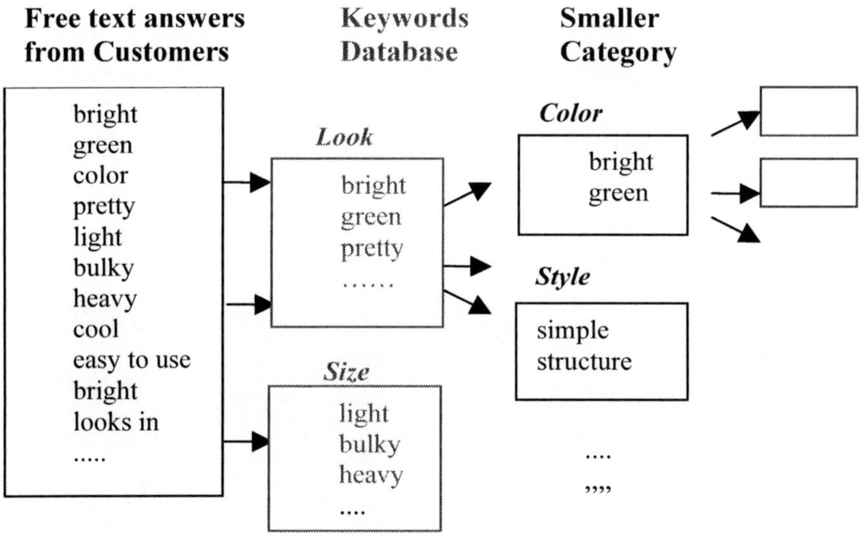

Figure 5 Illustration on clustering key words into design feature categories.

The relationship between a design feature and the corresponding detailed category can be treated as the question and its options. For example, *color*, *style* etc can be as the options for *appearance*. The new question created is as follow.

You like product A's look because

a. color
b. style
c. ......

This question is generated based on the feedback from answers of customers' free text. With such approach, designer could generate faster and more meaningful new questions that can represent customers' needs instead of designer's own thought. These new generated questions can be analyzed together with the original questions using the rule mining methods that were described in Section 1.

3. The Prototype System

The CMPD architecture and reasoning analysis methods described in the previous sections have been implemented in a prototype system with Java and VB languages. The system is designed to be web enabled for capturing customer's survey data and analyzing customer requirement with a standalone system. Figure 6 shows the system structure of the prototype. The core of the system is the knowledge discovery engine consisting of three components: statistical analyzer, multi-preference analyzer, and reasoning analyzer. The CMPD algorithm is embedded in the component of multi-preference analyzer while the concept-based text mining method built into the reasoning analyzer. The knowledge discovery engine is integrated with five supporting modules: User Account Manager, Data Manager, GUI controller, Database and Rule Base. In operation, the data manager receives customer data through Web-based survey form by GUI controller, and then converts the natural answers to binary formats and decimal formats to the database. The engine extracts the data and information from the database directly when performing a dictated task. Finally, the outcome, analysis results are presented to designers.

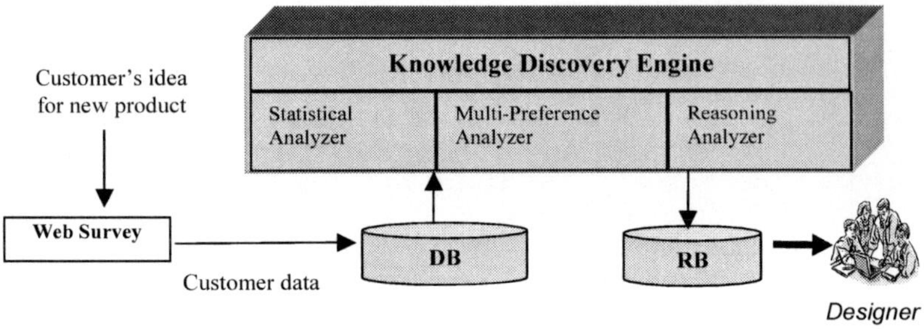

Figure 6 System Structure of the Prototype

4. An Illustrative Case Study

The prototype system has been tested in several practical applications. One such case is presented here to illustrate the quantification process and the working of the prototype system. The case is associated with the customer preference patterns discovery and reasoning analysis for a new HIFI system design in an electronic company. The designer in the company would like to have quantitative understanding on which product option was preferred, why customers like the products and how their customers would use their products. Three product options are used to conduct survey and more than 500 customers were involved in the survey exercise.

4.1. Customer Multi-Preference Analysis

Table 1 shows a few representative customer survey questions and their corresponding options. A total of 499 survey feedbacks are collected and their digitized results are represented in Table 2 (binary data format) and Table 3 (decimal data format). Based on the customer answer data, the CMPD architecture is built as shown in Figure. 7.

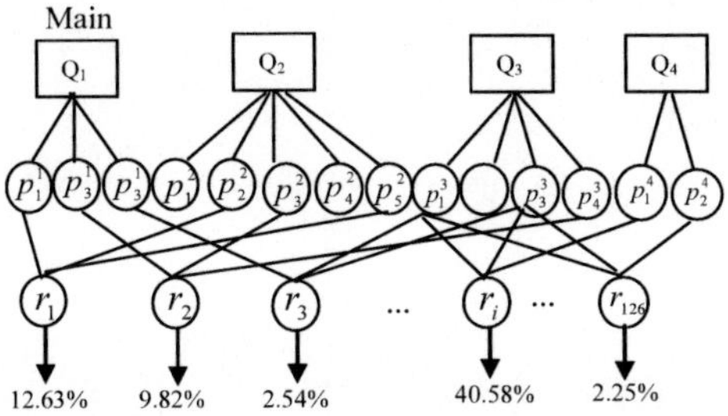

Figure 7 The network architecture of the case study

Table 1 Questions and options of customer survey for NPD

Question No.	Questions	Options (p_i^k)
1	My favorite product is	a. audio system A b. audio system B c. audio system C
2	I plan to put it in/on	a. my bed room b. family room c. bookshelf d. table top e. other location
3	I plan to use the audio system	a. stand alone b. connected to PC c. connected to game machine d. connected to VCD/DVD
4	I have at home a	a. VCD b. DVD

Both the values of question number, q_k (k = 1,2,3,4), and the binary data value of options p_j^k are assigned to the nodes in Layer 1, and to the relevant nodes in Layer 2, respectively. Taking the first question q_1 as the main question, the total number of rules is calculated as shown in Eq. (2):

$$
\begin{aligned}
R_{total} &= \sum_{i=1}^{n-1}[\sum_{j_1=2}^{n-i+1}\sum_{j_2=j_1+1}^{n-i+2}\cdots\sum_{j_i=j_1+i-1}^{n} T(1)\cdot T(j_1)\cdot T(j_2)\cdot\ldots\cdot T(j_i)] \\
&= T(1)\cdot[\sum_{j_1=2}^{4}T(j_1)+\sum_{j_1=2}^{3}\sum_{j_2=3}^{4}T(j_1)\cdot T(j_2)+\sum_{j_1=2}^{2}\sum_{j_2=3}^{3}\sum_{j_3=4}^{4}T(j_1)\cdot T(j_2)\cdot T(j_3)] \\
&= T(1)\cdot\{[T(2)+T(3)+T(4)]+[T(2)\cdot T(3)+T(2)\cdot T(4)+T(3)\cdot T(4)] \\
&\quad +[T(2)\cdot T(3)\cdot T(4)]\} \\
&= 3\times[5+4+2+4\times5+5\times2+4\times2+4\times5\times2] \\
&= 267
\end{aligned} \tag{2}
$$

Table 2 Customer feedback data in binary format

Customer ID / Question ID		Customer 1	Customer 2	Customer 3	Customer 4	…..	Customer 499
Q1	p_1^1	0	0	1	0	…..	0
	p_2^1	1	1	0	0	…..	1
	p_3^1	0	0	0	1	…..	0
Q2	p_1^2	1	1	0	0	…..	1
	p_2^2	0	0	1	1	…..	0
	p_3^2	1	0	1	0	…..	0
	p_4^2	0	1	0	1	…..	0
	p_5^2	0	0	0	0	…..	0
Q3	p_1^3	1	1	1	0	…..	1
	p_2^3	0	1	0	0	…..	1
	p_3^3	1	0	0	1	…..	0
	p_4^3	0	0	1	0	…..	0
Q4	p_1^4	1	1	0	0	…..	0
	p_2^4	0	0	1	1	…..	1

Table 3 Customer feedback data in decimal format

Customer ID / Question ID	Customer 1	Customer 2	Customer 3	Customer 4	…..	Customer 499
Q1	2	2	1	3	…..	2
Q2	20	18	12	10	…..	16
Q3	10	12	9	2	…..	12
Q4	2	2	1	1	…..	1

After gathering the answers from customers, corresponding binary data are generated as shown in Table 2. The first column in Table 2 represents question number, while the second column lists the options for each question. The remaining columns in Table 2 present the option selection in binary format for each question. Table 3 shows the result in decimal format converted from Table 2.

It can be seen that for such a relatively small number of questions with combinations of a few options, the R_{total} is as many as 267. The R_{total} can be huge if more questions and options are involved. To reduce the size of rule base as well as the computation loading, a rule filtering is carried out to pick out the frequent rules R_u (u = 1,2,…,h) at Layer 3 based on an assigned minimum rule support, such as 5% of the total feedbacks. Those rules with support of less than 1% are ignored in further calculation. In the illustrative case, the calculation results of the frequent rule R_u and their support s% are shown in Table 4.

It is not difficult to translate the rules back to natural language. For example, the rule number #39 can be explained as:

"The percentage of customers, who prefer product B, like to put it in their own bed room, like to use it as a stand alone unit and have a separate VCD at home, is 17%".

Figure 8 shows the user interface of the prototype system for customer multi-preference analysis.

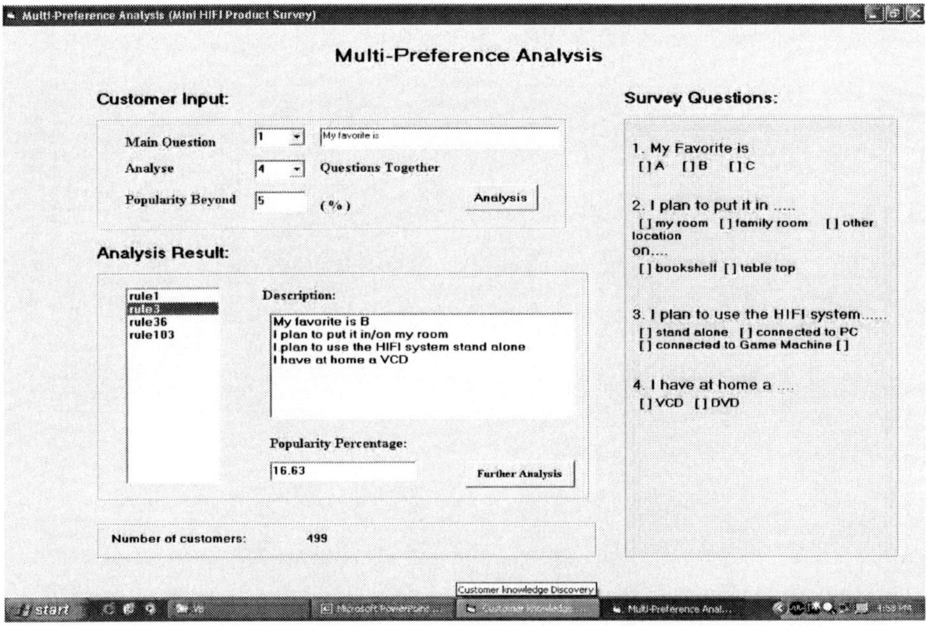

Figure 8 The Prototype User Interface of Customer Multi-preference Analysis

Table 4 Frequent rule results

Frequent Rule	**IF**	**THEN**	**Rule Support**
(R_u)	Option of main question	Option of other questions' combination	s (%)
#1	$p_1^1 \Rightarrow$	p_1^2	12.63
#7	$p_1^1 \Rightarrow$	p_1^3	9.82
#8	$p_1^1 \Rightarrow$	$p_1^2 \cap p_4^2$	2.43
#19	$p_2^1 \Rightarrow$	p_1^2	40.58
#20	$p_2^1 \Rightarrow$	p_1^3	35.68
#29	$p_2^1 \Rightarrow$	$p_1^2 \cap p_4^2$	3.81
#39	$p_2^1 \Rightarrow$	$p_1^2 \cap p_1^3 \cap p_2^4$	17.24
#42	$p_3^1 \Rightarrow$	$p_1^2 \cap p_4^2$	2.25
#50	$p_2^1 \Rightarrow$	$p_1^2 \cap p_1^3 \cap p_1^4$	8.93
#66	$p_2^1 \Rightarrow$	$p_1^2 \cap p_4^2 \cap p_2^4$	3.81
#67	$p_2^1 \Rightarrow$	$p_1^3 \cap p_1^4$	10.52
#73	$p_2^1 \Rightarrow$	$p_1^2 \cap p_2^4$	11.46
#104	$p_1^1 \Rightarrow$	$p_1^3 \cap p_1^4$	2.54
#111	$p_3^1 \Rightarrow$	$p_1^2 \cap p_1^3 \cap p_2^4$	4.73
#135	$p_3^1 \Rightarrow$	$p_1^2 \cap p_4^3 \cap p_1^4$	1.85
……	……	……	……
Total Frequent Rules	126	S_{min}	1.00

4.2. Customer Motivation Analysis

To capture the customer motivation, a reasoning analysis has been done using the concept-based text mining method. In this example, the total 210 customers' open-end answers were scanned and stored in the database for analysis. An example of the table structure is shown in Table 5. There are four columns in Table 5: Customer ID, Question ID, customer preference and the text answer from customers.

Table 5 Customer feedback information

Customer ID	Question_ID	Preference	Text Answer
1	1	Yes	design is simple
1	2	No	big
1	3	No	ugly and heavy
2	1	No	easy to carry
2	2	Yes	expensive
2	3	No	difficult to operate
--	--	--	--

The customer motivation analyzer starts from key words search. If it is a new key word that does not exist in the keyword DB, add this key word into the DB. If it is not a new key word, add one more to the frequency counter for this key word. The results of the example are stored in the database as shown in Table 6. If new key word frequency is greater than the minimum support value, the analyzer clusters them to the related key factors/features like function, design, sound, size, etc. with considering of the positive or negative impact of the words. The key word which is under the positive question, e.g. "Why do you like product A", is considered as the positive impact while the ones are the negative impact if the key words retreat from negative questions conversely. The relationship structure between key words and design factors/features is used to build key word dictionary. This dictionary will continuously enriched with additional new words.

Table 6 Key words table

Question_ID	Key words	Frequency
1	design simple	3
1	easy to carry	4
2	big	1
2	expensive	1
3	ugly	1
3	heavy	1
3	difficult to operate	1

The idea of how to cluster key words into design key word dictionary is shown in Figure 9. The key words, such as those in the example in Figure 9 (small, thin, bulky, heavy, fat, big, large, thick), are related to "Size" category in product design. Among these key words, small and thin is considered as the positive impact for size of mini HI-FI system. The other key words are categorized into the negative impact group. Based on the information from the key words database and key words dictionary, the total positive and negative impact frequencies of the design features can be calculated and the results fro the example are shown in Table 7.

The total positive and negative impact frequencies of the design feature will be translated to the percentage of support and save the percentage of support s% to the database.

Reasoning analysis is important for understanding the reasons why customers choose their preference products. Reasoning analysis can be realized by combining the proposed text mining method and customer multi-preference analysis method.

Key Words DB

Key words	Question_ID	Frequency
small	1	35
thin	1	19
bulky	2	31
heavy	2	28
fat	2	27
big	3	23
large	3	21
thick	3	20

Key Words Dictionary

Key words	Design Domain	Positive impact	Negative impact
small	size	√	
thin	size	√	
bulky	size		√
heavy	size		√
fat	size		√
big	size		√
large	size		√
thick	size		√

Figure 9 Clustering key words with design features

Table 7 Positive and negative impact frequency for product A

Question_ID	Design feature	Positive impact frequencies	Negative impact frequencies
1	Size	112	0
1	Design	46	3
1	Sound	4	35
1	Function	24	3
1	Others	37	1

With the total number of 210 customers, the percentages of customers that choose product A with positive impact are shown in Table 8:

Table 8 Positive impact reasoning analysis for product A

Question_ID	Design feature	Positive impact frequencies	% of Customers
1	Size	112	56
1	Design	46	23
1	Sound	4	2
1	Function	24	12
1	Others	37	18.5

Similarly, the reasoning and corresponding percentage of customers that do not choose product A is shown in Table 9.

Table 9 Positive Reasoning Analysis of Product A

Question_ID	Design feature	Negative impact frequencies	% of Customers
1	Size	0	0
1	Design	3	1.5
1	Sound	35	17.5
1	Function	3	1.5
1	Others	1	0.5

From Tables 8 and 9, the reasons of customer preference can be easily discovered with percentage of customers for each reason of the preference. 56% of surveyed customers like product A due to its small size while 17.5% of survey customers do not like product A because of its sound quality. Figure 10 shows an example of the user interface with analyzed results.

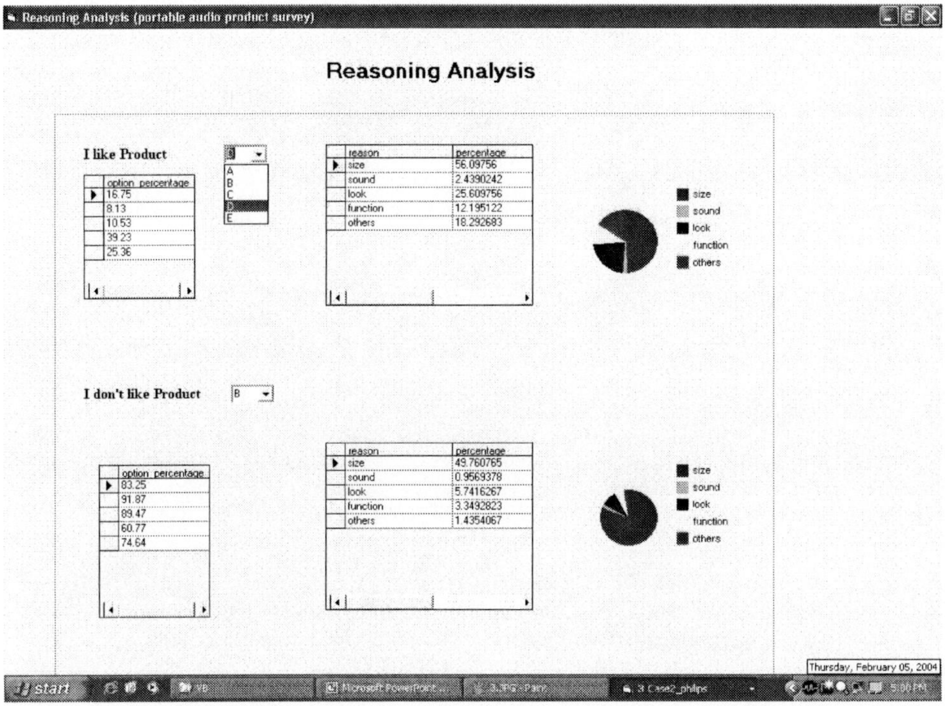

Figure 10 The Prototype User Interface of Customer Motivation Analysis

5. Concluding remarks

A new approach to discovery of customer multi-preferences and motivation analysis in new product design has been presented. An association rule mining algorithm and a concept-based? text mining method have been developed for customer demand discovery and analysis. Further, a software prototype system has been developed based on these methodologies . The working and effectiveness of the methodologies and prototype system are demonstrated with a case study. The prototype system allow online questionnaire design, online customer feedback collection, statistical analysis, and operations that cannot be easily realised with conventional tools. These include digitisation of language feedback, reasoning analysis and numerical description of customer preferences for one or more features of a product and linkage of customer preference motivation with product design features. The system could significantly shorten the survey and analysis time and is thus expected to reduce design cycle time for new product development.

Although the methodologies developed and illustrations are for product design applications, the methodologies are generic and can be adapted to other scenarios that require customer preference and motivation analysis.

References

[1] Crow, K. Voice of the Customer, *Product Development Forum,* DRM Associates, 2002.
[2] Adriaans, P. and Zantinge, D. *Data Mining*, Addison-Wesley Longman 1996.
[3] Sylvain Lourneau, Fazel Famili, Stan Matwin, Data Mining to Predict Aircraft Component Replacement, *IEEE Intelligent Systems*, **v.14 n.6**, p.59-66, November 1999.
[4] Karen L. Myers, Nina B. Zumel, Pablo Garcia, Automated capture of rationale for the detailed design process, *Proceedings of the sixteenth national conference on artificial intelligence and eleventh innovation applications of AI conference on Artificial intelligence and innovative applications of artificial intelligence*, July 18-22, Orlando, Florida, United States, (1999), p.876-883.
[5] Andrew Gelsey, Mark Schwabacher, Don Smith, Using modeling knowledge to guide design space search, *Artificial Intelligence*, **v.101 n.1-2**, May (1998), p.35-62.
[6] Zhou, A., Jin, W., Zhou, S., Qian, W., and Tian, Z., Incremental mining of the schema of semistructured data., *Journal of Computer Science and Technology.* **15(3)**, (2000), pp 241-248.
[7] Hostedware Corporation, *Hosted Survey*, USA, (2004). http://www.hostedsurvey.com
[8] XPO Show Services Inc., *XPO Online Survey*, USA, (2004). http://xposhow.com
[9] David Hull & Asociates Ltd., *EZSurvey*, Canada, (2002). http://www.survey-software.com.
[10] The Survey Professional Ltd, *Survey Pro Software*, USA, (2004). http://www.hostedsurvey.com
[11] SAS Institute Inc., SAS Data Mining Solution, USA, (2005). http://www.sas.com.
[12] SPSS Inc., SPSS Text Mining Analysis for Survey, USA, (2005). http://www.spss.com
[13] Han, J. and Kamber, M., *Data Mining: Concepts and Technique*, Morgan Kaufmann Publishers, San Francisco, 2001.
[14] Agrawal, R., Imielinski, T. and Swami, A.N., Mining association rules between sets of items in large databases, *Proceddings of the 1993 ACM SIGMOD International Conference on Management of Data*, Washington, D.C. 1993.
[15] Agrawal, R. and Srikant, R., Fast Algorithms for Mining Association Rules. In: *Proc. 20th Int. Conf. on Very Large Databases*. Santiago, Chile, 1994.
[16] Han, J. and Kamber, M., *Data Mining: Concepts and Techniques*, San Francisco: Morgan Kaufmann Publishers, 2001.

Integrated Intelligent Systems for Engineering Design
X.F. Zha and R.J. Howlett (Eds.)
IOS Press, 2006

An Approach to Software Design Reuse Using Case-Based Reasoning and WordNet

Paulo Gomes, Nuno Seco, Francisco C. Pereira, Paulo Paiva, Paulo Carreiro, José Ferreira and Carlos Bento
CISUC – Centro de Informática e Sistemas da Universidade de Coimbra
Departamento de Engenharia Informática – Polo II
Universidade de Coimbra – Portugal
pgomes@dei.uc.pt

Abstract: Reusing the knowledge gathered in the design phase of software development is an important issue for any software company. It enables software developers to work faster and make fewer mistakes, which decreases the development time due to the increased efficiency of the development team. In order to accomplish design knowledge reuse, we have developed an intelligent CASE tool that supports software design. Our system uses Case-Based Reasoning and WordNet, providing a framework for storage and reuse of design knowledge. This chapter presents our approach, which exploits a knowledge base and several reasoning mechanisms that reuse the stored knowledge.

Keywords: Case-Based Reasoning, Software Design and Reuse, UML, WordNet

Introduction

The reuse of software designs involves two complementary aspects: the management of the design knowledge repository, and the processes used to reuse this knowledge. The first aspect is responsible for the creation and maintenance of the design knowledge, while the second enables the knowledge's use in new situations by software designers. Having in mind these two aspects we developed a CASE tool named REBUILDER, which is the main focus of this chapter.

1. Motivations

The know-how acquired by the development teams is a valuable asset that software companies possess. Each engineer or programmer learns her/his own specific knowledge, which is then reused in other projects and tasks in which s/he participates. Experienced workers are very important to companies, in the sense that someone with experience has higher productivity. When an experienced engineer or other qualified employee leaves a company, the company loses the know-how and knowledge of these employees. In a high competitive market environment, where companies strive to be alive, this problem is crucial. Another relevant aspect is the importance of sharing know-how among the company employees increasing productivity and minimizing the losses when they leave the company. Inexperienced engineers would profit a lot from the stored knowledge, project development would need less resources and product quality would improve if a corporate memory and suitable retrieval mechanisms are available.

Software development [1] is a task involving a lot of expertise and know-how. The best software developers have years of experience and tackle new software projects based on the know-how gathered along these years. As in other areas of business, software development companies can take competitive advantage from sharing the knowledge acquired during project development. One possible solution to knowledge sharing is the creation of a central repository where this knowledge can be stored and indexed, ready to be reused in new projects. This has been performed in the past with software code, giving rise to code repositories and code reuse [2-6], where programmers can search for reusable functions and objects. But there are other types of software development knowledge that can be gathered and reused. In our research we are interested in the creation of a software design knowledge repository, and the mechanisms of reuse associated with such a knowledge base. There is a need for a design knowledge storage and usage methodology. Decisions made during the software design phase are going to condition the next phases; we consider design reuse a research field of strategic importance.

2. Our Approach

Case-Based Reasoning (CBR) [7, 8] is an Artificial Intelligence[9] field, based on reuse of experience. The reasoning framework used is based on the storage of experience episodes in the form of cases. Each case represents a specific situation or entity, and is stored in a case library ready to be reused when new situations arrive. Cases can be retrieved from the case library through a query defined by the designer. This query describes the current situation which needs to be solved or completed. The retrieval output is a list of cases ranked by similarity to the query. A CBR system can go a step further and adapt one or more retrieved cases to generate a new solution for the query. This new solution can then be stored in the library as a new case, thus closing a reasoning cycle [10] and enabling the CBR system to learn and evolve in time.

Our approach is based on the application of CBR to software design. One of the main aspects that lead us to use such an approach is the type of knowledge used by software designers. When a software designer is starting to develop a new project, s/he does not start from scratch. S/he makes use of old designs in which s/he was involved, in order to come up with a first approach, or to get design ideas. CBR provides an approach both to the design of knowledge repositories and to mechanisms for knowledge reuse. We have developed REBUILDER, which is an intelligent CASE tool that stores the corporation's design knowledge and enables its reuse. The software designer uses REBUILDER as a common UML [11] design tool, but s/he can also access the company's knowledge base. This knowledge base comprises cases, and other types of knowledge used in our system.

Another main aspect of our approach is the use of WordNet [12] as a general ontology, allowing REBUILDER to make semantic judgments about software objects. This ontology is an important aspect of our approach because it enables the classification of software objects based on semantic content. It also provides an indexing structure that improves case retrieval. WordNet can be seen as a broad-spectrum knowledge source, enabling other types of reasoning mechanisms, like analogy or verification algorithms. Thus our approach integrates within the CBR paradigm a general ontology capable of extending normal CBR mechanisms.

The next section describes related work, presenting systems with similar goals. Section 4 the architecture of REBUILDER. Section 5 describes the Knowledge Base format and content. Sections 6 to 10 present the various modules of the CBR engine used in REBUILDER. These modules are responsible for reusing the design knowledge. Finally section 11 concludes this chapter and presents future work on REBUILDER.

3. Related Work

The next subsections describe several software reuse systems, both with a CBR approach and other approaches.

3.1. CBR Software Reuse Systems

Fernández Chamizo [13] presented a CBR approach for software reuse based on the reuse and design of Object-Oriented code. Cases represent three types of entities: classes, methods and programming recipes, thus allowing the retrieval of these types of objects. Cases comprise a lexical description (problem), a solution (code) and a justification (code properties). It uses a lexical retrieval algorithm using a natural language query, and a conceptual retrieval using an entity and slot similarity measures.

Déjà Vu [14] is a CBR system for code generation and reuse using hierarchical CBR. Déjà Vu uses a hierarchical case representation, indexing cases using functional features. The main improvement of this system is the adaptation-guided retrieval, which retrieves cases based on the case adaptation effort instead of the similarity with the target problem.

CAESER [15] is another code reuse CBR tool. It works at the code level and uses data-flow analysis to acquire functional indexes. The user can retrieve cases from the case library using a prolog-like query goal, which is used by the system to retrieve similar functions.

Althoff and Tautz [16] have a different approach to software reuse and design. Instead of reusing code, they reuse system requirements and associated software development knowledge.

3.2. Software Reuse Systems

The RSL [3] is a software design system that combines several software design tools and library retrieval tools. It allows the reuse of code and design knowledge, and provides several software design tools to work the retrieved objects. RSL uses automatic indexing of components by scanning design documents and code files for specially labeled reuse component statements. It also provides a functional classification for components using a taxonomy of components. Component retrieval can be done using a natural-language query, or using attribute search. Component ranking is an interactive and iterative process between RSL and the user.

Prieto-Díaz [6, 17] approach to code reuse is based on a faceted classification of software components. Faceted classification has two different aspects according to Prieto-Díaz, it must classify components by the functions it performs and by the environment it works in. Conceptual graphs are used to organize facets, and a conceptual closeness measure is used to compute similarity between facets. He has identified six facets for software components: function, object, medium, system type, functional area and setting. Other work on facets is described by Liao [18], where he describes a hybrid similarity scheme using facets but where each facet can have multiple values.

Borgo [2] uses WordNet for retrieval of object oriented components. His system uses a graph structure to represent both the query and the components in memory. The retrieval mechanism uses a graph matching algorithm returning the identifiers of all components whose description is subsumed by the query. WordNet is also used for node matching. Helm [19] also presents a system for retrieval of Object-Oriented components based on the class source code, and the class documentation, using natural language queries.

4. REBUILDER's Architecture

REBUILDER has two main goals: create a corporative memory of software designs, and provide the software designer with a design environment capable of promoting software design reuse. This is achieved in our approach with CBR as the main reasoning process, and with cases as the main knowledge building blocks. REBUILDER comprises four different modules: Knowledge Base (KB), UML Editor, Knowledge Base Manager and CBR engine (see Figure 1).

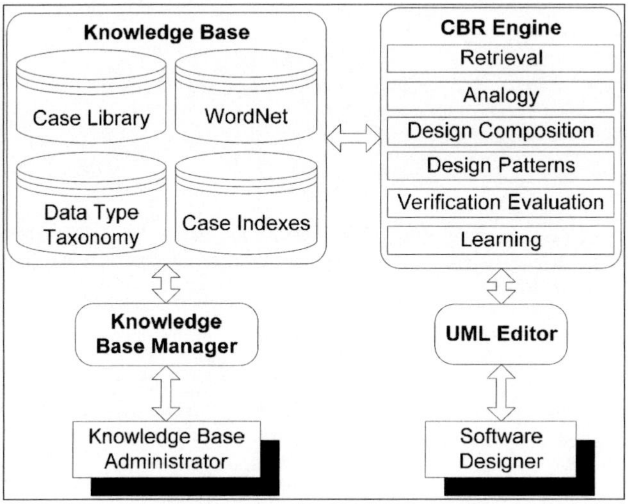

Figure 1. REBUILDER's architecture.

The UML editor is the front-end of REBUILDER and the environment where the software designer works. Apart from the usual editor commands to manipulate UML objects, the editor integrates new commands capable of reusing design knowledge. These commands are directly related with the CBR engine capabilities and are divided into two main categories:

- **Knowledge Base actions:** such as connect to KB and disconnect from KB.
- **Cognitive actions:** such as retrieve design, adapt design using analogy or design composition, verify design, evaluate design, and actions related with object classification using WordNet.

The Knowledge Base Manager module is used by the administrator to manage the KB, keeping it consistent and updated. This module comprises all the functionalities of the UML editor, and it adds case base management functions to REBUILDER. These are used by the knowledge base administrator to update and modify the KB. The list of available functions is:

- KB Operations: Creates, opens or closes a KB.
- Case Library Manager: Opens the Case Library Manager, which comprises functions to manipulate the cases in the case library, like adding new cases, removing cases, or changing the status of a case.
- Activate Learning: Gives the knowledge base administrator an analysis about the contents of the case library. REBUILDER uses several case base maintenance techniques to determine which cases should be added or removed from the case library (see section 10).
- Settings: Adds extra configuration settings which are not present in the normal UML Editor version used by software designers. It also enables the knowledge base administrator to fine tune the reasoning mechanisms.

The KB comprises four different parts: case library, which stores the cases of previous software projects; an index memory used for efficient case retrieval; data type taxonomy, which is an ontology of the data types used by the system; and WordNet, which is a general purpose ontology. This module is described in more detail in the next section.

The CBR Engine is the reasoning module of REBUILDER. As the name indicates, it uses the CBR paradigm to establish a reasoning framework. This module comprises five different parts: Retrieval, Design Composition, Analogy, Verification, and Learning. All these modules are detailed from section 6 through 10.

5. Knowledge Base

The KB stores all the corporate knowledge needed by the reasoning mechanisms and it consists of the WordNet ontology, a case library, the case indexes and the data type taxonomy. Each of these parts is described in the remaining of this section.

5.1. Case Library

In REBUILDER a case describes a software design, which is represented in UML through the use of Class Diagrams. Figure 2 shows an example of a class diagram representing part of an educational system. Nodes are classes, with a name, a set of attributes and methods. Links represent relations between classes. Conceptually a case in REBUILDER comprises: a name used to identify the case within the case library; the main package, which is an object that comprises all the objects that describe the main class diagram; and the file name where the case is stored. Cases are stored using XML/XMI (eXtended Mark-up Language), since it is a widely used format for data exchange. UML class diagram objects considered in REBUILDER are: packages, classes, interfaces and relations. A package is an UML object used to group other objects. A class describes an entity in UML and it corresponds to a concept described by attributes at a structural level, and by methods at a behavioral level. Interfaces have only method declarations, since they describe a protocol of communication for a specific class. A relation describes a relationship between two UML objects.

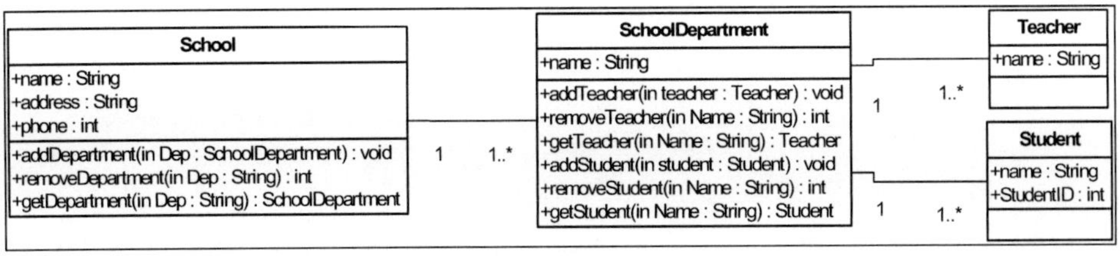

Figure 2. Example of an UML class diagram (*Case1*), the package classification is *School*.

5.2. WordNet

WordNet is used in REBUILDER as a common sense ontology. It uses a differential theory where concept meanings are represented by symbols (called synsets) that enable a theorist to distinguish among them. Symbols are words, and concept meanings are called synsets. A synset is a concept represented by one or more words. Words that can be used to represent a synset are called synonyms. A word with more than one meaning is called a polysemous word. For instance, the word mouse has two meanings, it can denote a rat, or it can express a computer mouse.

WordNet is built around the concept of synset. Basically a synset comprises a list of words and a list of semantic relations between other synsets. The first part is a list of words, each one with a list of synsets that the word represents. The second part, is a set of semantic relations between synsets, like *is-a* relations, *part-of* relations, and other relations. REBUILDER uses the word synset list and four semantic relations: *is-a*, *part-of*, *substance-of*, and *member-of*. Synsets are classified in four different lexical types: nouns, verbs, adjectives, and adverbs.

Synsets are used in REBUILDER for categorization of software objects. Each object has a context synset which represents the object meaning. In order to find the correct synset, REBUILDER uses the object name, and the names of the objects related with it, which define the object context. The object's context synset can then be used for computing object similarity (using the WordNet semantic relations), or it can be used as a case index, allowing rapid access to objects with the same classification. WordNet is used to compute the semantic distance between two context synsets. This distance is the length of the shortest path between the two synsets. Any of the four relation types can be used to establish the path between the synsets. This distance is used in REBUILDER to assess the type similarity between objects, and to select the correct synset when the object name has more than one synset. This process is called name disambiguation [20] and is a crucial task in REBUILDER. If a diagram object has a name with several synsets, then more information about this object has to be used to find which synset is the correct one. The extra information is the diagram objects that directly or indirectly are associated with it. In case of the object being a class, its attributes can also be used in the disambiguation process. This process is used when a case is inserted in the case library or when the designer calls the retrieval module.

5.3. Case Indexes

As cases can be large, they are stored in files, which make case access slower then if they were in main memory. To solve this problem we use case indexes. These provide a way to access the relevant case parts for retrieval without having to read all the case files from disk. Each object in a case is used as an index. REBUILDER uses the context synset of each object to index the case in WordNet. This way, REBUILDER can retrieve a complete case, using the root package of the case, or it can retrieve only a subset of case objects, using the objects' indexes. This allows REBUILDER to provide the designer with the possibility to retrieve not only packages, but also classes and interfaces. To illustrate this approach, suppose that the class diagram of Figure 2 represents *Case1*. Figure 3 presents part of the WordNet structure and some of the case indexes associated with *Case1*. As can be seen, WordNet relations are of the types *is-a*, *part-of* and *member-of*, while the index relation links a case object (squared boxes) with a WordNet synset (rounded boxes). For instance *Case1* has one package named *School* (the one presented in Figure 2), which is indexed by synset *School*. It has also a class with the same name and categorization, indexed by the same synset, making this class also available for retrieval.

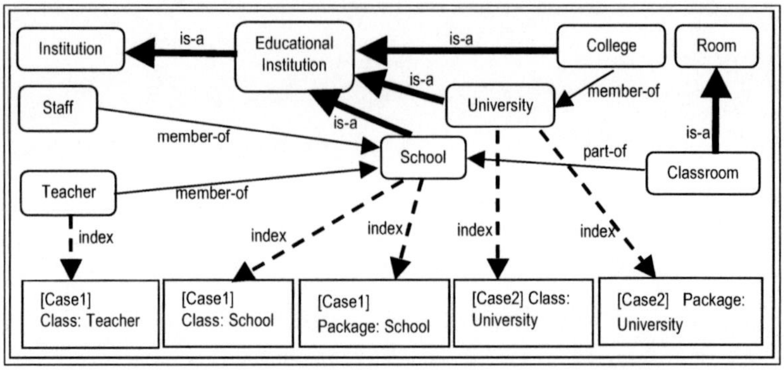

Figure 3. A small example of the WordNet structure and case indexes.

5.4. Data Taxonomy

The data type taxonomy is a hierarchy of data types used in REBUILDER. Data types are used in the definition of attributes and parameters. The data taxonomy is used to compute the conceptual similarity between two data types. Figure 4 presents part of the data taxonomy used in REBUILDER.

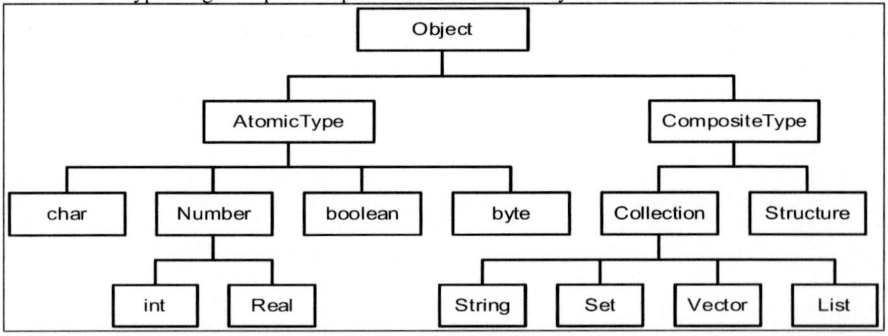

Figure 4. Part of the data taxonomy used in REBUILDER.

6. Retrieval Module

The case retrieval module can retrieve three types of objects: packages, classes or interfaces, depending on the object selected when the retrieval command is chosen by the designer. If the object is a package, the associated diagram will be considered as the query diagram (also designated as the target design or problem). The next subsections describe the retrieval algorithm and the similarity measures used in this module.

6.1. Retrieval Algorithm

The retrieval algorithm is the same for all three types of objects (packages, classes and interfaces), and is based on the object classification using WordNet. Suppose that the *N* best objects have to be retrieved,

QObj is the query object, and *ObjectList* is the universe of objects that can be retrieved (usually *ObjectList* comprises all the library cases). The algorithm is as follows:

```
ObjsFound  ← ∅
PSynset ← Get context synset of QObj
PSynsets ← {PSynset}
ObjsExplored  ← ∅
WHILE (#ObjsFound < N) AND (PSynsets ≠ ∅) DO
   Synset ← Remove first element of PSynsets
   ObjsExplored ← ObjsExplored + Synset
   SubSynsets ← Get Synset hyponyms / subordinates
   SuperSynsets ← Get Synset hypernyms / super ordinates
   SubSynsets ← SubSynsets - ObjsExplored - PSynsets
   SuperSynsets ← SuperSynsets - ObjsExplored - PSynsets
   PSynsets ← Add SubSynsets to the end of PSynsets
   PSynsets ← Add SuperSynsets to the end of PSynsets
   Objects ← Get all objects indexed by Synset
   Objects ← Objects ∩ ObjectList
   ObjsFound ← ObjsFound ∪ Objects
ENDWHILE
ObjsFound ← Rank ObjsFound by similarity
RETURN Select the first N elements from ObjsFound
```

Starting by *QObj* context synset, the algorithm searches for objects indexed with the same synset. If there are not enough objects, the algorithm uses the hypernyms[1] and hyponyms[2] of this synset to look for objects, going in a spreading activation kind of algorithm. When it has found enough objects, it stops and ranks them using the similarity metrics.

Object retrieval has two distinct phases. First the WordNet *is-a* relations are used as an index structure to find relevant objects. Then a similarity metric is used to select the best *N* objects. This process is a compromise between a first phase which is inexpensive from the computational point of view, and a second phase which is more demanding for computational resources but more accurate concerning the object selection and ranking. In the next subsection we present the similarity metrics.

6.2. Similarity Metrics

There are three similarity metrics concerning classes, interfaces, and packages. The class similarity metric is based on WordNet categorization and the object structure. The interface metric is equal to the class metric with the difference that it does not have attributes. The package similarity metric takes several aspects of the UML class diagram into account.

6.2.1. Class Similarity

The class similarity metric is based on three components: categorization similarity (also called type similarity), inter-class similarity, and intra-class similarity. The similarity between class C_1 and C_2, is:

$$S(C_1,C_2) = \omega_1 \cdot S(S_1,S_2) + \omega_2 \cdot S(Ie_1,Ie_2) + \omega_3 \cdot S(Ia_1,Ia_2) \qquad \textbf{(1)}$$

Where $S(S_1, S_2)$ is the categorization similarity computed as the distance, in *is-a* relations, between C_1 context synset (S_1) and C_2 context synset (S_2). $S(Ie_1, Ie_2)$ is the inter-class similarity based on the similarity between the diagram relations of C_1 and C_2. $S(Ia_1, Ia_2)$ is the intra-class similarity based on the similarity between attributes and methods of C_1 and C_2. Based on experimental work, we use 0.6, 0.1, and 0.3 as the default values of ω_1, ω_2 and ω_3 constants.

[1] The parents of a node in the *is-a* taxonomy.

[2] The children of a node in the *is-a* taxonomy.

6.2.2. Interface Similarity

The interface similarity metric is the same as the class metric with the difference that the intra-class similarity metric is only based on the method similarity, since interfaces do not have attributes.

6.2.3. Package Similarity

The similarity between packages Pk_1 and Pk_2 is:

$$S(Pk_1, Pk_2) = \begin{bmatrix} \omega_1 \cdot S(SPs_1, SPs_2) + \omega_2 \cdot S(OBs_1, OBs_2) \\ + \omega_3 \cdot S(T_1, T_2) + \omega_4 \cdot S(D_1, D_2) \end{bmatrix} \quad (2)$$

This metric is based on four items: sub-package list similarity - $S(SPs_1, SPs_2)$, UML class diagram similarity - $S(OBs_1, OBs_2)$, type similarity - $S(S_1, S_2)$, and dependency list similarity - $S(D_1, D_2)$. These four items are combined in a weighted sum. The weights used in our experiments are: 0.07, 0.4, 0.5, and 0.03, respectively. These values are supported on experimental work. From the weight values it can be seen that the most important factors for package similarity are the categorization similarity, and the class diagram similarity. The categorization similarity is the same as in the class similarity metric.

6.2.4. Object Categorization Similarity

The object type similarity is computed using the objects' context synsets. The similarity between synset S_1 and S_2 is:

$$S(S_1, S_2) = \frac{1}{\ln(Min\{\forall Path(S_1, S_2)\} + 1) + 1} \quad (3)$$

Where *Min* is the function returning the smaller element of a list. $Path(S_1, S_2)$ is the WordNet path between synset S_1 and S_2, which returns the number of relations between the synsets. *ln* is the natural logarithm function.

6.2.5. Inter-Class Similarity

The inter-class similarity between two objects (classes or interfaces) is based on matching of the relations in which both objects are involved. The similarity between objects O_1 and O_2 is:

$$S(O_1, O_2) = \frac{\sum_{i=1}^{n} S(R_{1i}, R_{2i})}{\# R_1 + \# R_2} - \frac{UM(R_1) + UM(R_2)}{2 \cdot (\# R_1 + \# R_2)} + \frac{1}{2} \quad (4)$$

Where R_i is the set of relations in object i, R_{ij} is the j element of R_i, n is the number of matched relations, $S(R_{1i}, R_{2i})$ is the relation similarity, and $UM(R)$ is the number of unmatched elements in R.

6.2.6. Intra-Class Similarity

The intra-class similarity between objects O_1 and O_2 is:

$$S(O_1, O_2) = \omega_1 \cdot S(As_1, As_2) + \omega_2 \cdot S(Ms_1, Ms_2) \quad (5)$$

Where $S(As_1, As_2)$ is the similarity between attributes, $S(Ms_1, Ms_2)$ is the similarity between methods, and ω_1 and ω_2 are constants, with $\omega_1 + \omega_2 = 1$ (default values are: 0.6; 0.4).

6.2.7. Sub-Package List Similarity

The similarity between sub-package lists SPs_1 and SPs_2 is:

$$S(SPs_1, SPs_2) = \frac{\sum_{i=1}^{n} S(SPs_{1i}, SPs_{2i})}{\# SPs_1 + \# SPs_2} - \frac{UM(SPs_1) + UM(SPs_2)}{2 \cdot (\# SPs_1 + \# SPs_2)} + \frac{1}{2} \quad \textbf{(6)}$$

Where n is the number of sub-packages matched, $S(SPs_{1i}, SPs_{2i})$ is the similarity between packages, $UM(SPsi)$ is the number of unmatched packages in SPs_i, SPs_{ij} is the j element of SPs_i, and $\#SPs_i$ is the number of packages in SPs_i.

6.2.8. UML Class Diagram Similarity

The similarity between lists of UML objects OBs_1 and OBs_2 is:

$$S(OBs_1, OBs_2) = \frac{\sum_{i=1}^{n} S(OBs_{1i}, OBs_{2i})}{\# OBs_1 + \# OBs_2} - \frac{UM(OBs_1) + UM(OBs_2)}{2 \cdot (\# OBs_1 + \# OBs_2)} + \frac{1}{2} \quad \textbf{(7)}$$

Where $\#OBs_i$ is the number of objects in OBs_i, $UM(OBs_i)$ is the number of objects unmapped in OBs_i, n is the number of objects matched, OB_{ij} is the j element of OB_i, and $S(OB_{1i}, OB_{2i})$ is the object categorization similarity.

6.2.9. Dependency List Similarity

Dependencies are an UML relation type expressing a dependence relation between two packages, for instance, package A depends on package B because a class in A uses a class from B. The similarity between dependency lists D_1 and D_2 is given by:

$$S(D_1, D_2) = \left[\omega_1 \cdot \frac{\left| \# ID_1 - \# ID_2 \right|}{Max\{\# ID_1, \# ID_2\}} + \omega_2 \cdot \frac{\left| \# OD_1 - \# OD_2 \right|}{Max\{\# OD_1, \# OD_2\}} \right] \quad \textbf{(8)}$$

Where ω_1 and ω_2 are constants, and $\omega_1 + \omega_2 = 1$ (default values are: 0.5; 0.5), ID are the input dependencies, and OD are the output dependencies.

6.2.10. Attribute Similarity

Attribute similarity is computed based on: the data type, the scope, and the default value. Data type similarity is assessed using the path distance between the data types being compared. This distance is found using the data type taxonomy. The scope similarity is based on a comparison table, which establishes the similarities between all possible scopes, as defined in UML and Java. The default value similarity is one if the default values are equal or if both do not exist, zero otherwise.

6.2.11. Method Similarity

Method similarity is based on: the scope, the input parameters, and the output parameters. The scope similarity is assessed as described before in the attribute similarity. The input and output parameter similarity is based on the similarity between parameters, which is the same as the attribute similarity, since attributes are treated as a special type of parameters.

7. Analogy Module

Analogical reasoning is used in REBUILDER to suggest class diagrams to the designer, based on a query diagram. The analogy process has three steps:

- Identify candidate diagrams for analogy.
- Map the candidate diagrams.
- Create new diagrams, by transferring knowledge between the candidate diagram and the query.

7.1. Candidate Selection

Cases are selected from the case library to be used as source diagrams. The selected candidates must be appropriate, otherwise the whole mapping phase can be at risk. Most of the analogies that are found in software design are functional analogies, that is, the analogy mapping is performed using the functional similarity between objects. Using the retrieval algorithm in the first phase of analogy enables this kind of analogies, since objects that are functional similar tend to be categorized in the same branch (or close to) of the WordNet *is-a* trees. Thus, the analogy module benefits from a retrieval filtering based on functional similarity (for more details see [21]).

7.2. Mapping Process

The second step of analogy is the mapping of each candidate to the query diagram, yielding an object list correspondence for each candidate. This phase relies on two alternative algorithms: one based on relation mapping, and the other on object mapping, but both return a list of mappings between objects.

7.2.1. Relation-Based Mapping

The relation-based algorithm uses the UML relations to establish the object mappings. It starts the mapping by selecting a query relation based on an UML heuristic (independence measure), which selects the relation that connects the two most important diagram objects. The independence measure is a heuristic used to assign each diagram object an independence value based on UML knowledge that reflects an object's independence in relation to all the other diagram objects. Then it tries to find a matching relation on the candidate diagram. After it finds a match, it starts the mapping using the neighbor relations, spreading the mapping through the diagram relations. This algorithm maps objects in pairs corresponding to the relation's objects. The relation-based mapping algorithm is:

```
Relations ← Get the best relation from the query diagram, based on the
independence measure of the relation
MappingList ← ∅
WHILE Relations ≠ ∅ DO
   PRelation ← Get best relation from Relations, based on the independence
                measure of the relation
   CRelation ← Get best matching relation from the base case relations that are
                matching candidates (structural constraints must be met)
   Mapping ← Get the mapping between objects, based on the mapping between
                PRelation and CRelation
   Remove PRelation from Relations
   Add Mapping to MappingList
   Add to Relations all the same type relations adjacent to PRelations, which
        are not already mapped (if PRelation connects A and B then the adjacent
        relations of PRelations are the relations in which A or B are part of,
```

```
        excluding PRelation)
ENDWHILE
RETURN MappingList
```

7.2.2. Object-Based Mapping

The object-based algorithm starts the mapping selecting the most independent query object, based on the UML independence heuristic. After finding the corresponding candidate object, it tries to map the neighbor objects of the query object, taking the object's relations as constraints into the mapping process. The object-based mapping algorithm is:

```
Objects ← Get object from the query diagram which has the higher independence
              measure value
MappingList ← ∅
WHILE Objects ≠ ∅ DO
   PObject ← Get best object from Objects, based on the object's independence
                 measure
   CObject ← Get best matching object from the base case objects that are
                 matching candidates (structural constraints must be met)
   Mapping ← Get the mapping between PObject and CObject
   Remove PObject from Objects
   Add Mapping to MappingList
   Add to Objects all the objects adjacent to PObject, which are not already
       mapped (an object (B) is adjacent to an object A, if there is a relation
       linking the two objects
ENDWHILE
RETURN MappingList
```

Both algorithms satisfy the structural constraints defined by the UML diagram relations. Most of the resulting mappings do not map all the problem objects, so the mappings are ranked by number of objects mapped (see [22]). An important issue in the mapping stage is: which objects to map? Most of the time, there are several candidate objects for mapping with the problem object. In order to solve this issue, we have developed a metric that is used to choose the mapping candidate. Because we have two mapping algorithms, one based on relations and another on objects, there are two metrics: one for objects, and another for relations. These metrics are based on the WordNet distance between the object's synsets, and the relative position of these synsets in relation to the most specific common abstraction concept (for details on these metrics see [22]).

7.3. Knowledge Transfer

The last step is the generation of new diagrams using the established mappings. For each mapping the analogy module creates a new diagram, which is a copy of the query diagram. Then, using the mappings between the query objects and the candidate objects, the algorithm transfers knowledge from the candidate diagram to the new diagram. This transfer has two steps: first there is an internal object transfer, and then an external object transfer. In the internal object transfer, the mapped query object gets all the attributes and methods from the candidate object that were not in the query object. This way, the query object is completed by the internal knowledge of the candidate object. The second step transfers neighbor objects and relations from the mapped candidate objects to the query objects, from the new diagram. This transfers new objects and relations to the new diagram, expanding it.

8. Design Composition Module

The Design Composition module generates new diagrams by decomposition/composition of case diagrams. The input data used in the composition module is an UML class diagram, in the form of a package. This is the designer's query, which usually is a small class diagram in its early stage of development (see Figure 1). The goal of the composition module is to generate new diagrams that have the query objects, thus providing an evolved version of the query diagram. Generation of a new UML design using case-based composition involves two main steps: retrieving cases from the case library to be used as knowledge sources, and using the retrieved cases (or parts of them) to build new UML diagrams. In the first phase, the selection of the cases to be used is performed using the retrieval algorithm described in section 6.1. The adaptation of the retrieved cases to the target problem is based on two different strategies: best case composition, and best complementary cases composition. The next subsections describe these two composition strategies.

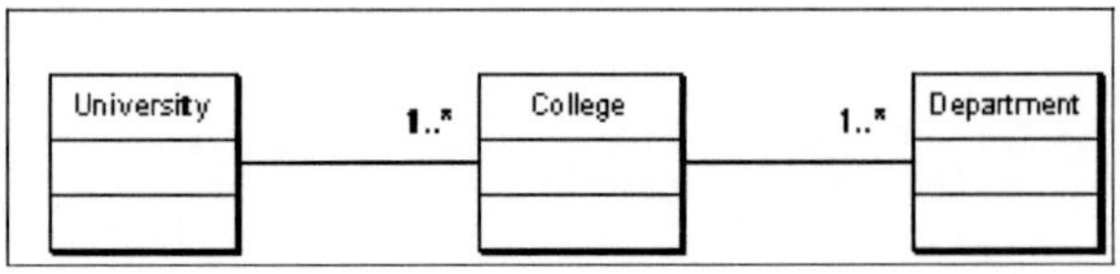

Figure 5. An example of a class diagram in the early stages of development.

8.1. Best Case Composition

In the best case composition, the adaptation module starts from the most similar case to the problem, mapping the case objects to the problem objects. The case mapped objects are copied to a new case. If this new case maps successfully all the problem objects, then the adaptation process ends. Otherwise it selects the retrieved case, which best complements the new case (in relation to the problem), and uses it to get the missing objects. This process continues while there are unmapped objects in the problem definition. Note that, if there are objects in the used case that are not in the problem, they can be transferred to the new case, generating new objects. The best case composition algorithm is:

```
RetrievedCases ← Retrieve cases from the Case Library using Problem
SelectedCases ← ∅
BestCase/Mapping ← Select the best case and map it to the problem
NewCase ← Use BestCase, Mapping and Problem to generate a new case
WHILE NewCase does not map all the Problem objects AND SelectedCases ≠ ∅ DO
   SelectedCases ← Search RetrievedCases for cases with unmapped problem objects
   SelectedCase ← Select the best case, the one wit more unmapped problem objects
   NewCase ← Complete NewCase with SelectedCase
   Remove SelectedCase from RetrievedCases
ENDWHILE
RETURN NewCase
```

8.2. Best Set of Cases Composition

The best complementary cases composition starts by matching each retrieved case to the problem, yielding a mapping between the case objects and the problem objects. This is used to determine the degree of problem coverage of each case, after which several sets of cases are built. These sets are based on the combined coverage of the problem, with the goal of finding sets of cases that globally map all the problem objects. The best matching set is then used to generate a new case. The best complementary set of cases composition algorithm is:

```
RetrievedCases ← Retrieve cases from the Case Library using Problem
FOR RetrievedCase in RetrievedCases DO
```

```
        Mapping ← Map RetrievedCase to Problem
END FOR
CaseSets ← Create the sets of complementary cases based on each case mapping
BestSet ← Select the best set from CaseSets, this is the set whose mapping has
   the best coverage of problem objects
NewCase ← Generate a new case using the BestSet
RETURN NewCase
```

9. Verification and Evaluation Module

The verification and evaluation module comprises two distinct phases: verifying a class diagram, and evaluating it. The next subsections describe each one of these functionalities.

9.1. Verification

The verification phase comprises checking the objects' names, relations, attributes, and methods. It's main purpose is to check the design coherence and correctness. The name checking is used whenever an object's name is changed. The other verification procedures are used when a verify action is requested by the designer. The verify action first checks the relations, attributes, methods, and sub-packages of the package being verified. The sub-package checking is a recursive call to the verify action.

Name checking is based on WordNet. Basically it uses WordNet to see if the object's name is one of the WordNet words. If the verification module finds no corresponding word, then the designer is presented with two alternatives: change the object's name, or browse WordNet to find a suitable name. This way it is assured that the system can find a suitable set of synsets. If the verification finds a word or a set of words (and a corresponding set of synsets) in WordNet for the object's name, there are two alternatives: the designer selects the appropriate synset, or the system can select it based on a word sense disambiguation algorithm.

Relation checking is based on: WordNet, design cases, or relation verification cases, which are cases describing successful and failure situations in checking relation validity. These cases have the following description: relation type {Association, Generalization, Realization, Dependency}, multiplicity {1-1, 1-N, N-N}, source object name, source object synset, destination object name, destination object synset, and outcome {Success, Failure}. The three knowledge sources can be used for checking the relation validity. For instance, if there is a relation in a design case or in WordNet connecting objects with the same synsets as the ones being checked, the system accepts this relation as being valid.

Attribute checking is based on: WordNet, design cases, and in Attribute Verification Cases, which are cases describing successful and failure situations in checking an attribute validity. These cases have the following description: object name, object synset, attribute name, and outcome {Success, Failure}.

Method checking is based on: a heuristic, the design cases, and in Method Verification Cases, which are cases describing successful and failure situations in checking method validity. These cases have the following description: object name, object synset, method name, outcome {Success, Failure}. The heuristic used is: if the method name has a word that is an attribute name or a class name in the same diagram, then the method is considered valid.

9.2. Evaluation

The evaluation phase provides an assessment of the design characteristics, which is based on software engineering metrics. Basically it provides the designer with some properties of the design diagram. They are divided in two categories: Software Metrics for Object Oriented Designs, and Design Statistics.

The used software metrics are:

- Class/Interface/Package complexity (class complexity is the number of methods and attributes.
- Interface complexity is the number of attributes.
- Package complexity is the sum of complexity of the objects it comprises, sub-packages, classes and interfaces).
- Number of children of a Class/Interface.

- Depth of inheritance tree of a Class.
- Coupling between Classes/Interfaces: number of non-inheritance related couples of an object with other objects.
- Package average of Class/Interface number of children.
- Package average of Class/Interface depth of inheritance tree.

Design statistics are (by package): number of sub packages; number of relations; number of classes; number of interfaces; average number of classes by package; average number of interfaces by package; average number of relations by package; average number of attributes by class; average number of methods by class; and average number of methods by interface.

10. Learning Module

Learning in REBUILDER is performed by storing new cases in the case library. Design cases can be a result of the software designer work, or they can be the result of the CBR process. In both situations, is up to the system's administrator to accept the case as a valid one and enter it into the case library. Several case-based maintenance strategies have been implemented in REBUILDER, which provide an advice to the software designer, whether a case should be added to the case library or not. The strategies are:

Frequency Deletion Criteria [23]: This criteria selects cases for deletion based on the usage frequency of cases.

Subsumption Criteria [24]: The subsumption criteria defines that a case is redundant if: it is equal to another case, or is equivalent to another case, or is subsumed by another case. In REBUILDER a case is considered subsumed when, there is another case with the same root package synset, and the package and objects structure is a substructure of the other case.

Footprint Deletion Criteria [25]: This criteria is based on two definitions: *case coverage*, or the set of problems that a case can solve; and *reachability set* of a case, which is the set of cases that can be used to provide solutions for a problem. Based on these two notions the case base is divided in three groups (initially in Smyth's work were four, but one of the subgroups can not be distinguished using our case representation and subsumption definition). Pivotal cases represent an unique way to answer a specific query. Auxiliary cases are those which are completely subsumed by other cases in the case base. Spanning cases are cases between pivotal and auxiliary cases, which link together areas covered by other cases. When the case library has too many cases the recommended order of deletion is: auxiliary, spanning, and pivotal.

Footprint-Utility Deletion Criteria [25]: This criteria is the same as the Footprint Deletion Criteria, with the exception that when there is a draw the selection is based on the case usage - less used cases are chosen for deletion.

Coverage Criteria [26]: The coverage criteria involves three factors: case base size, case base density, and case base distribution. A new case is added to the case library if its inclusion in the case base increases the case base coverage/case base number ratio.

Case-Addition Criteria [27]: The case-addition criteria involves the notion of case neighborhood, which in REBUILDER is defined as all the cases that have the same root package synset as the case being considered. This criterion uses the notion of benefit of a case in relation to a case set, which is based on the frequency function of cases. A new case is added to the case base if its benefit is positive.

Relative Coverage [28]: The goal of this criteria is to maximize coverage while minimizing case base size. The proposed technique for building case bases is to use Condensed Nearest Neighbor on cases that have first been arranged in descending order of their relative coverage contributions.

Relative Performance Metric [29]: This criteria uses the notion of relative performance to decide if a case should be added to the case base or not. The relative performance of a case is based on the sum of cost adaptation of the case being considered to all cases in its coverage set. If the relative performance of a new case being considered is higher than an established threshold then the new case is added to the case library.

Competence-Guided Criteria [30]: The competence-guided criteria extends previous works of Smyth, which use the notions of case competence based on case coverage and reachability. This criterion uses three ordering functions:

- Reach for Cover (RFC): uses the size of the reachability set of a case. The RFC evaluation function implements this idea: the usefulness of a case is an inverse function of its reachability set size.
- Maximal Cover (MCOV): uses the size of the coverage set of a case. Cases with large coverage sets can classify many target cases and as such must make a significant contribution to classification competence.
- Relative Coverage (RC): is defined in a previous criterion.

11. Conclusions and Future Work

This paper presents an approach to design knowledge reuse for corporate software development. This approach is based on CBR, which provides a reasoning framework for knowledge storage and the reasoning mechanisms that reuse this knowledge. The result was the development of REBUILDER, an intelligent CASE tool for software design and reuse. REBUILDER has three main advantages: (1) allows the design knowledge sharing throughout a company, (2) enables the reuse of this knowledge, and (3) provides cognitive mechanisms that support the software design task. These are some of the functionalities that an intelligent CASE tool should have, but there are others, like: natural language understanding, automatic verification and evaluation of design models, automating the code change when the design changes, and so on.

Future work in our system will consist on the creation of a new module in the CBR engine that allows the application of software design patterns as a design maintenance tool. Also being developed is a module that transforms Natural Language into UML, so that a first diagram can be created from the project specifications.

Acknowledgments

This work was partially supported by POSI - Programa Operacional Sociedade de Informação of Fundação Portuguesa para a Ciência e Tecnologia and European Union FEDER, under contract POSI/33399/SRI/2000, by program PRAXIS XXI. REBUILDER homepage is http://rebuilder.dei.uc.pt.

References

1. Boehm, B., *A Spiral Model of Software Development and Enhancement*. 1988: IEEE Press.
2. Borgo, S., et al. *Using a Large Linguistic Ontology for Internet-Based Retrieval of Object-Oriented Components*. in *9th International Conference on Software Engineering and Knowledge Engineering, SEKE'97*. 1997. Madrid, Spain: Knowledge Systems Institute, Illinois.
3. Burton, B.A., et al., *The Reusable Software Library*. IEEE Software, 1987. **4**(July 1987): p. 25-32.
4. Coulange, B., *Software Reuse*. 1997, London: Springer-Verlag.
5. Prieto-Diaz, R., *Status Report: Software Reusability*. IEEE Software, 1993(May).
6. Prieto-Diaz, R. and P. Freeman, *Classifying Software for Reusability*. IEEE Software, 1987(January).
7. Kolodner, J., *Case-Based Reasoning*. 1993: Morgan Kaufman.
8. Maher, M.L., M. Balachandran, and D. Zhang, *Case-Based Reasoning in Design*. 1995: Lawrence Erlbaum Associates.
9. Russel, S. and P. Norvig, *Artificial Intelligence: A Modern Approach*. 1995, New Jersey: Prentice Hall.
10. Aamodt, A. and E. Plaza, *Case-Based Reasoning: Foundational Issues, Methodological Variations, and System Approaches*. AI Communications, 1994. **7**(1): p. 39-59.
11. Rumbaugh, J., I. Jacobson, and G. Booch, *The Unified Modeling Language Reference Manual*. 1998, Reading, MA: Addison-Wesley.
12. Miller, G., et al., *Introduction to WordNet: an on-line lexical database*. International Journal of Lexicography, 1990. **3**(4): p. 235 - 244.
13. Fernández-Chamizo, C., et al. *Supporting Object Reuse through Case-Based Reasoning*. in *Third European Workshop on Case-Based Reasoning (EWCBR'96)*. 1996. Lausanne, Suisse: Springer-Verlag.
14. Smyth, B. and P. Cunningham. *Déjà Vu: A Hierarchical Case-Based Reasoning System for Software Design*. in *10th European Conference on Artificial Intelligence (ECAI'92)*. 1992. Vienna, Austria: John Wiley & Sons.
15. Fouqué, G. and S. Matwin. *Compositional Software reuse with Case-Based Reasoning*. in *9th Conference on Artificial Intelligence for Applications (CAIA'93)*. 1993. Orlando, FL, USA: IEEE Computer Society Press.
16. Althoff, K.-D., et al., *Case-Based Reasoning for Experimental Software Engineering*. 1997, Fraunhofer IESE: Berlin.
17. Prieto-Diaz, R., *Implementing Faceted Classification for Software Reuse*. Communications of the ACM, 1991(May).

18. Liao, S.Y., L.S. Cheung, and W.Y. Liu, *An Object-Oriented System for the Reuse of Software Design Items.* Journal of Object-Oriented Programming, 1999. **11**(8, January 1999): p. 22-28.
19. Helm, R. and Y.S. Maarek. *Integrating Information Retrieval and Domain Specific Approaches for Browsing and Retrieval in Object-Oriented Class Libraries.* in *Object-Oriented Programming Systems, Languages, and Applications.* 1991. Phoenix, AZ USA: ACM Press.
20. Ide, N. and J. Veronis, *Introduction to the Special Issue on Word Sense Disambiguation: The State of the Art.* Computational Linguistics, 1998. **24**(1): p. 1-40.
21. Gomes, P., et al. *Combining Case-Based Reasoning and Analogical Reasoning in Software Design.* in *Proceedings of the 13th Irish Conference on Artificial Intelligence and Cognitive Science (AICS'02).* 2002. Limerick, Ireland: Springer-Verlag.
22. Gomes, P., et al. *Experiments on Software Design Novelty Using Analogy.* in *European Conference on Artificial Intelligence ECAI'02 Workshop: 2nd Workshop on Creative Systems.* 2002. Lyon, France.
23. Minton, S., *Quantitative Results Concerning the Utility of Explanation-Based Learning.* AI, 1990. **42**: p. 363-391.
24. Racine, K. and Q. Yang. *Maintaining Unstructured Case Bases.* in *Proceedings of the 2nd International Conference on Case-Based Reasoning (ICCBR'97).* 1997: Springer.
25. Smyth, B. and M.T. Keane. *Remembering To Forget: {A} Competence-Preserving Case Deletion Policy for Case-Based Reasoning Systems.* in *Proceedings of the Fourteenth International Joint Conference on Artificial Intelligence (IJCAI 95).* 1995: Morgan Kaufmann.
26. Smyth, B. and E. McKenna. *Modelling the Competence of Case-Bases.* in *Proceedings of the 4th European Workshop on Advances in Case-Based Reasoning (EWCBR-98).* 1998: Springer.
27. Zhu, J. and Q. Yang. *Remembering to Add: Competence-preserving Case-Addition Policies for Case Base Maintenance.* in *Proceedings of the 16th International Joint Conference on Artificial Intelligence (IJCAI-99).* 1999: Morgan Kaufmann Publishers.
28. Smyth, B. and E. McKenna. *Building Compact Competent Case-Bases.* in *Proceedings of the 3rd International Conference on Case-Based Reasoning Research and Development (ICCBR-99).* 1999: Springer.
29. Leake, D. and D. Wilson. *Remembering Why to Remember: Performance-Guided Case Base Maintenance.* in *Proceedings of the European Workshop on Case-Based Reasoning (EWCBR-00).* 2000: Springer.
30. McKenna, E. and B. Smyth. *Competence-Guided Case Base Editing Techniques.* in *Proceedings of the European Workshop on Case-Based Reasoning (EWCBR-00).* 2000: Springer.

Integrated Intelligent Systems for Engineering Design
X.F. Zha and R.J. Howlett (Eds.)
IOS Press, 2006

Intelligent Process Planning Optimization for Product Cost Estimation

W.D. LI[1], S.K. ONG[2], A.Y.C. NEE[2], L. DING[1] and C.A. MCMAHON[1]
[1] *Department of Mechanical Engineering, University of Bath*
Bath BA2 7AY, U.K.
[2] *Department of Mechanical Engineering, National University of Singapore*
10 Kent Ridge Crescent, 119260, Singapore

Abstract. Manufacturing cost is crucial for the economic success of a product, and early and accurate estimation of manufacturing cost can support a designer to evaluate a designed model dynamically and efficiently for making cost-effective decisions. Manufacturing cost estimation is closely related to process planning problems, in which machining operations, machining resources, operation sequences, etc., are selected, determined and optimized. To solve the intractable decision-making issues in process planning with complex machining constraints, three intelligent optimization methods, i.e., Genetic Algorithm (GA), Simulated Annealing (SA) and Tabu Search (TS), have been developed to determine the optimal or near-optimal allocation of machining resources and sequence of machining operations for a process plan simultaneously, and a fuzzy logic-based Analytical Hierarchical Process technique has been applied to evaluate the satisfaction degree of the machining constraints for the process plan. Case studies, which are used to compare the three developed methods, are discussed to highlight their characteristics in the aspects of solution quality, computation efficiency and optimization result robustness.

Keywords. Process planning optimization, Genetic algorithm, Simulated annealing, Tabu search, Analytical hierarchical process, Manufacturing cost estimation

1. Introduction

Product development companies very often pursue "cheaper, faster and better" objectives when developing a product, and cost is one of the most crucial elements to decide the success of the product. Traditionally, product development companies calculate cost after the manufacturing of their products. The drawback of this practice is that a designer lacks a quick and accurate way to evaluate the cost of the products for making prompt and cost-effective decisions. Based on this situation, the research of cost estimation has been motivated. A cost estimation methodology can be regarded as a vital building block in a modern Concurrent Engineering (CE)-based product development environment to enable a designer to explore alternative plans and adjust design dynamically so as to avoid the high cost of design changes when process plans are already fixed and the production is in progress. For instance, an aircraft wing is designed according to the required performance of the aircraft. With a model to estimate the development cost of the wing based on its surface area [1], designers can attempt different design parameters and specifications effectively while the cost is well controlled and managed.

The cost of a product is driven or affected by many elements in design, manufacturing and services, such as geometry, material, tooling, process planning, labour, inventory, freight, logistics, etc. Among them, manufacturing is a crucial element and the estimation of the manufacturing cost involves a series of intractable reasoning and decision-making processes. Actually, the essence of the manufacturing cost estimation and computation for discrete-part manufacture is equivalent to that of a process planning problem, which is used to select and determine machining operations, machines, cutting tools, operation sequences, etc. An apt manufacturing cost model and a well-developed process planning optimization method can efficiently provide a designer with an optimal or near-optimal process plan and manufacturing cost for his or her designed product.

In this chapter, a developed process planning optimization method to support manufacturing cost estimation is presented. Users can identify some machining operations and resources for a designed model and a process plan can be generated intelligently and automatically. A Tabu Search (TS) method has been developed as an artificial intelligence-based optimization algorithm to determine the selection of machining resources and sequence of the machining operations in a process plan. Meanwhile, a fuzzy logic-based Analytical Hierarchical Process (AHP) technique has been applied to evaluate the satisfaction degree of the manufacturability for the process plan from the viewpoint of machining constraints. Based on the determined process plan, the manufacturing cost (as an optimal or near-optimal result) can be computed as the prediction result. Therefore, this method can allow users to estimate and optimize the manufacturing cost of a designed model dynamically and promptly during the product design stage so as to realize the CE-based product development philosophy.

2. Background

Cost estimation methods can be generally classified into two types: the parametric approach and the generative approach. The parametric approach is to predict cost through linking product parameters (e.g., weight, dimension, volume, size, process parameters, etc.) and cost information based on some physical relationships or statistical analyses of historical cost data. For instance, in the aerospace industry, computation models and systems have been developed to predict product cost based on some high-level costing parameters. A practical example is illustrated in [1]. The cost estimation of an aircraft wing depends upon two cost items: (1) a fixed $40,000 of wing cost not related to surface area (overhead cost), and (2) $1,000 per sq. ft. that is related to surface area to build one wing. Therefore, the surface area is an important parameter to decide the cost. For a wing with 200 square feet of surface area, the estimated cost is $40,000 + 200 sq ft x $1,000 per sq. ft. In the parametric Small Satellite Cost Model (SSCM) in the handbook, the input variables for estimating the development cost of a modern small satellite include mission type (communications, remote sensing, or space experiments), procurement environment (military, commercial or NASA), subsystem masses, performance measures and production schedule [1]. In the Whole Aircraft Parametric Cost Estimator (WAPCO) system developed by Airbus UK, volumes, key components and the general type of aircraft have been used as cost evaluation parameters [2]. Meanwhile, some heuristic rules have been developed to compare the similarity of the geometry or technical features of a new product to be predicted and a previous product, and from the similarity, the cost for the new product can be deduced

[3-5]. The major merits of the parametric approach include the rapid prediction of cost through several crucial variables, and good effects for some situations with well-defined cost formula and reference historical data. However, this approach has limitations in practice. For instance, accurate formulas become quite difficult to develop when the complexity of a product increases or a large amount of precise and concrete historical data is not in position. At the same time, it is not an easy job to define criteria to judge the similarity between a new product and a previous product. In the generative approach, the procedures for developing a product are broken down in detail and each procedure is associated with resources and the relevant cost [6-10]. Therefore, this approach can overcome some drawbacks of the parametric approach through depicting the cost drivers and resources in a more distinguishable way, and it is more adapted to modified resources, processes and products. Due to this characteristic, the generative approach has a higher requirement for intelligent reasoning, and AI techniques have been widely used to enhance this aspect.

In cost estimation, manufacturing cost and the corresponding manufacturing cost estimation are the important research topics to be investigated since they can drive bidding strategies, manufacturing resource management/optimization, production policies and the production-related crucial issues to ensure the economic success of products. In the manufacturing cost, process planning is one of the most significant factors affecting cost, and an optimized process plan can reduce the manufacturing cost dramatically. A generative process planning approach, which is the essential enabling technique for the aforementioned generative cost estimation approach, specifies how to organize the route and sequence of the machining operations of a product and how to allocate manufacturing resources in a generative way. During the past decade, a number of papers have appeared in this area and the most relevant to the present work are summarized as follows.

The approaches used in early published works are mainly based on knowledge-based reasoning [11-15] and graph manipulation [16-18]. The common characteristics of these approaches include: (1) Although the reasoning approaches can generate feasible solutions, it is very difficult to find global and optimal plans using the approaches; (2) In the reasoning processes, a challenging problem is how to manipulate some machining constraints between operations effectively; and (3) The reasoning efficiency is low in some complex machining environments.

Process planning objectives and machining constraints are often imprecise and can even be conflicting due to inherent differences in the feature geometry and technological requirements. Fuzzy logic is suitable for presenting such imprecise knowledge and several approaches using this technique have been developed to address process planning problems. Based on fuzzy membership, the objective of the work in [19, 20] is to minimize the dissimilarity among the process plans selected for a family of parts, and optimal process plans can be generated for each part family. A fuzzy logic-based approach reported in [21] has identified and prioritized important features based on the geometric and technological information of a part. Manufacturing cost is more closely correlated with the important features and their operations than with the less important features and operations. Hence, operations sequencing of important features is first carried out within a much smaller search space. The operations of the less important features can then be arranged easily due to reduced constraints.

Recently, evolutional and heuristic algorithms, which include Genetic Algorithms (GA), Neural Networks (NN) and Simulated Annealing (SA), have been applied to process planning research, and multiple objectives, such as achieving the minimum number of set-ups and tool changes, while applying good manufacturing practice, have been considered to achieve a global optimal or near-optimal target [22-29]. These algorithms are based on the heuristic and AI searching strategies to quickly find global optimization solutions in large, non-linear and multi-modal searching spaces. However, the following two issues are still outstanding and require careful consideration.

The first issue is the processing of precedence machining constraints in those evolutional and heuristic approaches. The efficiencies of the graph-based heuristic algorithm [22] and the tree traversal algorithm [23] are not high and the search is not global such that optimal plans might be lost during the reasoning processes. The test and generate method is to generate process plans randomly, and then test and select some feasible plans for further manipulation [24-26]. The fundamental problems of this approach include low efficiency and difficulty in generating reasonable initial plans for a complex part. Two-step manipulation [24, 27] is specific to the crossover operations of GA, and it is not applicable to SA and NN, or other operations in GA such as creating initial generations and mutations. In comparison, the penalty method [24, 28] and the constraint adjustment method [29] performed better in terms of computational efficiency and extensible search space. However, these methods cannot work well with conflicting constraints and evaluation criteria. New methods should be sought to handle the possibility of conflicting constraints, which can be classified according to their importance to the manufacturability of process plans, to achieve the better performance of an optimization algorithm in practical situations.

The second issue is that the multiple objectives in the unified optimization models create a very large search space. Much effort is needed to employ a suitable AI strategy and design a more effective optimization method to determine the optimal or near-optimal results with short reasoning iterations to meet the practical workshop situations with dynamically varying resources and workloads. It is also imperative to conduct comprehensive studies on the performance of the various optimization algorithms to highlight their characteristics.

3. Representation of Process Planning

A process plan for a part consists of machining operation types, applicable candidate machining resources, set-up plans, operation sequence, etc. A set-up can be generally defined as a group of operations that are machined on a single machine with the same fixture. Here, a set-up is specified as a group of features with the same Tool Approaching Direction (TAD) machined on a machine. For instance, in Fig. 1, a hole with two TADs is considered to be related with two set-ups [29].

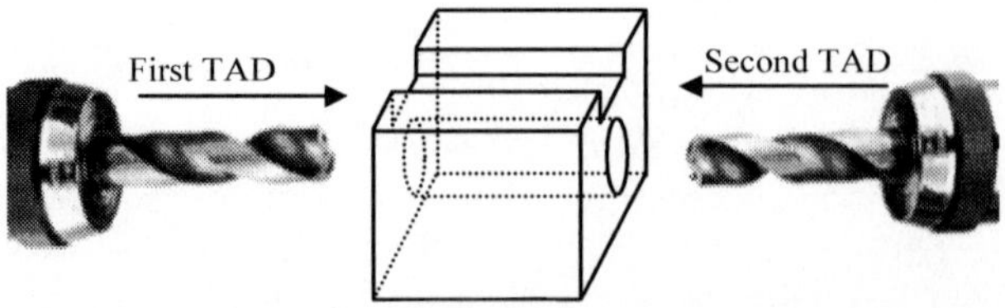

Fig. 1 A through hole with two TADs.

The features and their valid TADs can be recognized using a geometric reasoning approach [30]. In order to apply algorithms to the optimization of the process plan, the operations and their relevant machines, cutting tools and TADs are represented as a Process Plan (PP). For the purpose of applying an optimization approach, the PP can be represented by a bit sequence. That is, if there are n operations for machining a part, the PP will be n bits, and each bit represents an operation once and only once. In a PP, the sequence of the bits is the machining operation sequence of the plan. An operation has a set of candidate machines, tools, and TADs under which the operation can be executed. In an object-oriented description, a bit (an operation) in a PP can be represented as (illustrated in Fig. 2):

```
class PP_Bit
{
    int Oper_id;                         // The id of the operation
    int Mac_id;                          // The id of a machine to execute the operation
    int Mac_list[ ] = new int[20];       // The candidate machines for executing the operation
                                         // The maximum candidate number is assumed 20
    int Tool_id;                         // The id of a cutting tool to execute the operation
    int Tool_list[ ] = new int[20]       // The candidate tool list for executing the operation
                                         // The maximum candidate number is assumed 20
    int TAD_id;                          // The id of a TAD to apply the operation
    int TAD_list[ ] = new int[6];        // The candidate TAD list for applying the operation
                                         // The maximum candidate number is assumed 6
}
```

A PP can be described as an array - Oper[n], which is defined as:

```
PP_Bit Oper[ ] = new PP_Bit[ n ];        // Declare PP, and n is the number of operations
```

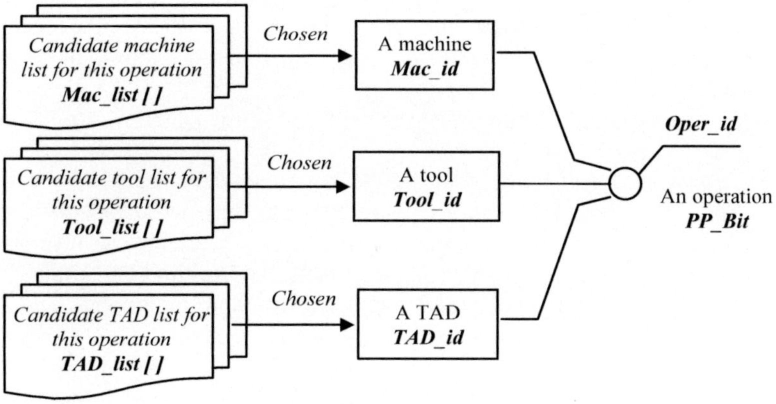

Fig. 2 An operation in a process plan.

The geometric and machining interactions between features as well as technological requirements in a part can be considered to generate some preliminary machining constraints between machining operations. These interactions and technological requirements can be summarized below [31]:

(1) Precedence constraints

- A parent feature should be processed before its child features.
- Rough machining operations should be done before finish machining operations.
- Primary surfaces should be machined prior to secondary surfaces. Primary surfaces are usually defined as surfaces with high accuracy or having a high impact on the design specifications, such as a datum plane. The rest of the surfaces are regarded as secondary surfaces, e.g., a threaded hole.
- Planes should be machined prior to holes and slots.
- Edge cuts should be machined last.

(2) Succession constraints

- Features or operations, which can be machined within the same set-up should be machined successively.
- Features to be machined with the same cutting tool should be machined successively.
- Operations of the same type, such as rough, semi-finish and finish machining, should be executed successively.
- Features with similar tolerance requirements should be machined successively on the same machine tool.

(3) Auxiliary constraints

- Annealing, normalizing and ageing operations of ferrous metal components should be arranged before rough machining or between rough and semi-finish machining.
- Quenching for ferrous metal workpieces should be arranged between semi-finish and finish machining or between rough and semi-finish machining if it is followed by high temperature tempering. Quenching for non-ferrous metals should be arranged between rough and semi-finish machining or before rough machining.
- Carburizing should be arranged between semi-finish and finish machining.

These constraints can be used to set the boundaries for the search in a smaller space. In a practical situation, not all constraints in a sequence of operations may be able to be met and there may be some conflicting cases arising. It is imperative to develop a strategy to consider the total effect of the constraints on the manufacturability of a designed model according to their importance and practical needs.

4. Determination of Cost Criterion

In the application of an optimization algorithm to a process planning problem, the performance criterion is considered as an essential ingredient since it indicates the degree of the objective satisfaction of a solution searched. Most of the previous research depends on an individual criterion, such as the minimum number of set-ups, the minimum machining cost or the shortest processing time, etc., and therefore the outcome solution is not optimal overall. In this research, an integrated optimization, which includes the overall cost consideration of the machine utilization, cutting tool utilization, number of machine changes, number of tool changes and number of set-ups, has been developed to find an optimal point satisfying the process constraints and minimum machining cost. The performance criterion is defined by the following expression:

$$F = f_c + w * f_m \quad (1)$$

where F is the performance criterion in terms of total cost,
f_c is the relative evaluating value for machining cost,
f_m is the degree of satisfaction of process constraints, and
w is the weight of f_m (it is used to convert f_m as a relative cost value, and the value is usually determined according to practical situations).

4.1 Determination of f_c

The contribution factors for f_c include all the machine utilization costs, tool utilization costs, machine change costs, set-up costs and tool change costs. These costs can be computed as given below [29].

- *Total Machine Cost (TMC)*

$$TMC = \sum_{i=1}^{n}(Oper[i].Machine\_id * MC[Oper[i].Machine\_id]) \quad (2)$$

where MC is the Machine Cost of a machine.

- *Total Tool Cost (TTC)*

$$TTC = \sum_{i=1}^{n}(Oper[i].Tool\_id * TC[Oper[i].Tool\_id]) \quad (3)$$

where TC is the Tool Cost of a tool.

- *Number of Set-up Changes (NSC), Number of Set-up (NS), and Total Set-up Cost (TSC)*

$$NSC = \sum_{i=1}^{n-1}\Omega_2(\Omega_1(Oper[i].Machine\_id, Oper[i+1].Machine\_id), \Omega_1(Oper[i].TAD\_id, Oper[i+1].TAD\_id)) \quad (4)$$

$$NS = 1 + NSC \quad (5)$$

$$TSC = \sum_{i=1}^{NS} SC \quad (6)$$

where $\Omega_1(X,Y) = \begin{cases} 1 & X \neq Y \\ 0 & X = Y \end{cases}$, $\Omega_2(X,Y) = \begin{cases} 0 & X = Y = 0 \\ 1 & otherwise \end{cases}$.

- *Number of Machine Changes (NMC) and Total Machine Change Cost (TMCC)*

$$NMC = \sum_{i=1}^{n-1} \Omega_1(Oper[i].Machine\_id, Oper[i+1].Machine\_id) \quad (7)$$

$$TMCC = \sum_{i=1}^{NMC} MCC \quad (8)$$

- *Number of Tool Changes (NTC) and Total Tool Change Cost (TTCC)*

$$NTC = \sum_{i=1}^{n-1} \Omega_2(\Omega_1(Oper[i].Machine\_id, Oper[i+1].Machine\_id), \Omega_1(Oper[i].Tool\_id, Oper[i+1].Tool\_id)) \quad (9)$$

$$TTCC = \sum_{i=1}^{NTC} TCC \quad (10)$$

- *The machining cost (f_c)*

$$f_c = TMC + TTC + TSC + TMCC + TTCC \quad (11)$$

4.2 Determination of f_m

In practical situations, it might be impossible to satisfy all constraints in a process plan. For instance, a high accuracy hole as a datum surface should be machined with a high priority according to the primary surfaces constraint, but it may be in conflict with the constraint of planes prior to holes and slots. The AHP technique [32], which is a systematic technique to decompose a complex problem in a hierarchical structure to make pair-wise comparisons, has been applied in this work to address the complex constraints and evaluate the degree of satisfaction of the manufacturability. In the developed AHP hierarchical structure, a set of fuzzy logic-based numerical weights has been incorporated to represent the relative importance of the constraints of a process plan with respect to a manufacturing environment. The relevant computation is depicted below.

Step 1: The constraints defined in Section 3 are organized in a hierarchy structure, which includes an overall objective (Level 1), three general constraint groups (Level 2) and rules under each constraint group (Level 3). This situation is illustrated in Fig. 3. For Level 2, a 3×3 pair-wise matrix (*R^0-matrix*) is created, where the number in the *i*th row and *j*th column, r_{ij}, specifies the relative importance of the *i*th group of constraints as compared with the *j*th group of constraints. For Level 3, three pair-wise matrices are created for each group of constraints (*R^1-matrix* (5×5) for Precedence constraints, *R^2-matrix* (4×4) for Succession constraints, and *R^3-matrix* (3×3) for Auxiliary constraints). Similarly, the number in the matrix (r_{ij})

specifies the relative importance of rules within each category of constraints. A *R-matrix* can be described as:

$$R = \begin{bmatrix} r_{11} & \cdot & r_{1i} & \cdot & r_{1m} \\ \cdot & \cdot & \cdot & \cdot & \cdot \\ r_{i1} & \cdot & r_{ii} & \cdot & r_{im} \\ \cdot & \cdot & \cdot & \cdot & \cdot \\ r_{m1} & \cdot & r_{mi} & \cdot & r_{mm} \end{bmatrix}$$

where $i = 1, 2, \ldots, m$ (m is the number of groups of constraints in Level 2 or the number of rules for each constraint group in Level 3),

$r_{ii} = 1$, and

$r_{ij} = 1/r_{ji}$.

Step 2: Evaluating criteria based on a 1-9 scale for the *R-matrices*, which are used to indicate the relative importance of two elements, are defined in Table 1. In order to get more neutral results, a group of experts is invited to fill in the four *R-matrices* according to their experience and knowledge.

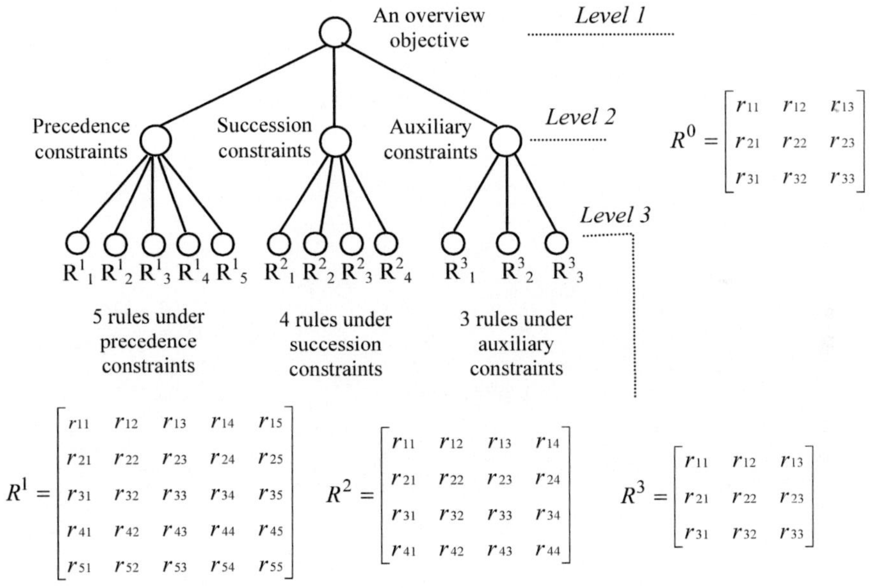

Fig. 3 A three-level hierarchy structure for the constraints.

For instance, considering two rules in the category of Precedence constraints - *Rule 2* and *Rule 4*:

Rule 2: Primary surfaces should be machined prior to secondary surfaces.
Rule 4: Planes should be machined prior to holes and slots.

From the perspective of an individual expert, if he thinks *Rule 2* is much more important than *Rule 4*, a weight of '7' is inserted in the juncture cell (r_{24}) of his filled R^l-*matrix*. On the contrary, the value in the juncture cell (r_{42}) is set to '1/7'.

Step 3: For Level 2 and Level 3, four weight vectors $W^0 - W^3$, which correspond to the four *R-matrices* respectively, are computed. The computation process consists of the following three steps.

Table 1 Evaluation criteria for *R-matrices*.

Definition	Intensity of importance (r_{ij})	Intensity of importance (r_{ji})
The *i*th rule and the *j*th rule have equal importance	1	1
The *i*th rule is slightly more important than the *j*th rule	3	1/3
The *i*th rule is more important than the *j*th rule	5	1/5
The *i*th rule is much more important than the *j*th rule	7	1/7
The *i*th rule is absolutely more important than the *j*th rule	9	1/9
Intermediate values between adjacent scale values	2, 4, 6, 8	1/2, 1/4, 1/6, 1/8

(1) Multiplication (M) of all elements in each row of a *R-matrix* is computed as:

$$M_i = \prod_{j=1}^{n} r_{ij} \qquad (12)$$

where j is the column index of elements, $j = 1, 2, \ldots, n$,
i is the index row of elements, $i = 1, 2, \ldots, n$, and
n is the number of the rows (columns) in a *R-matrix*.

(2) The *n*th root of M is calculated, that is:

$$\overline{w_i} = \sqrt[n]{M_i} \qquad (13)$$

where i is the row (column) number in a *R-matrix*, and $i = 1, 2, \ldots, n$.

Therefore, the relative importance weight vector can be built as follows:

$$\overline{W} = \left| \overline{w_1}, \ \overline{w_2}, \ \ldots, \ \overline{w_n} \right| \qquad (14)$$

Each element of the weight vector W ($|w_1, \ w_2, \ \ldots, \ w_n|$) is finally generated through a normalization operation.

$$w_i = \frac{\overline{w_i}}{\sum_{j=1}^{n} \overline{w_j}} \qquad (15)$$

For each W, it should be eventually denoted as $W^0 - W^3$ according to the individual computation process.

Step 4: There are totally 12 rules defined in this system (5 rules from Precedence constraints + 4 rules from Succession constraints + 3 rules from Auxiliary constraints). The element of a total weight vector for each rule - W^t ($|w^t_1, w^t_2, ..., w^t_{12}|$) can be generated as:

$$w^t_{1-5} = w^0_1 * w^1_{1-5}, \; w^t_{6-9} = w^0_2 * w^2_{1-4}, \; w^t_{10-12} = w^0_3 * w^3_{1-3} \qquad (16)$$

Step 5: A series of *V-matrices* are designed to record the situation of violating constraints for a process plan. For instance, for *Rule k*, its *V-matrix* is defined as:

$$V_k = \begin{bmatrix} v_{k11} & \cdot & v_{k1i} & \cdot & v_{k1n} \\ \cdot & \cdot & \cdot & \cdot & \cdot \\ v_{ki1} & \cdot & v_{kii} & \cdot & v_{kin} \\ \cdot & \cdot & \cdot & \cdot & \cdot \\ v_{kn1} & \cdot & v_{kni} & \cdot & v_{knn} \end{bmatrix}$$

where n is the number of operations in a process plan,

$$v_{ij} = \begin{cases} 1 & \text{if Operation i prior to Operation j is against Rule k} \\ 0 & \text{if Operation i prior to Operation j obey Rule k} \end{cases}, \text{ and}$$

$v_{ji} = 1 - v_{ij}$.

Step 6: The value to evaluate the manufacturability of a process plan is determined. f_m is finally calculated as:

$$f_m = \sum_{k=1}^{m} \sum_{i=1}^{n} \sum_{j=1}^{n} w_k{}^t v_{kij} \qquad (17)$$

where m is the total rule number of the constraints (here m =12).

5. Tabu Search Algorithm

5.1 Workflow of the algorithm

A typical TS algorithm can conduct a search process for a near-optimal or optimal solution through avoiding entrainment in cycles by forbidding or penalizing moves that take the solution, in the next iteration, to points in the solution space previously visited (hence "Tabu") [33]. It consists of three main strategies, i.e., the forbidding strategy, the freeing strategy and the aspiration strategy. During a search process, a Tabu list for recording the recently made moves is established and dynamically maintained. The global performance criterion of the algorithm is defined in Equation (1). Based on this algorithm, the workflow of the process planning optimization method has been designed (illustrated in Fig. 4), and the major elements of the algorithm are stated as follows.

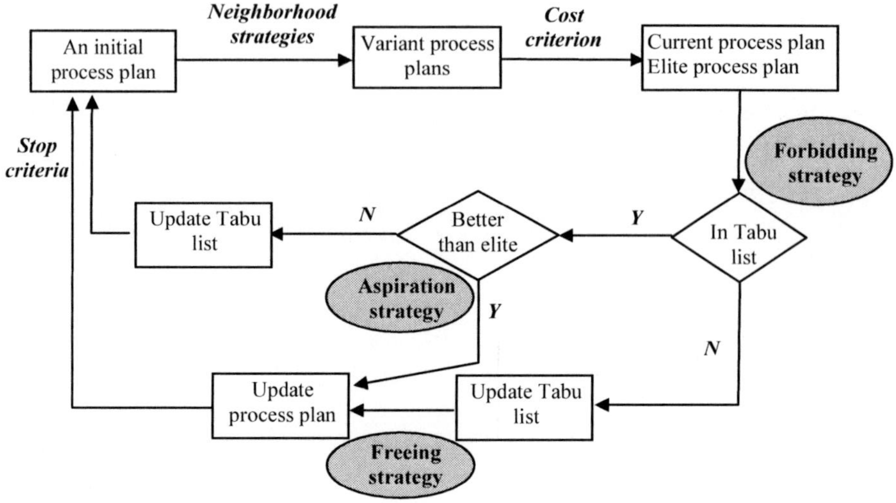

Fig. 4 Workflow of the TS-based optimization algorithm.

(1) *Initial plan, current plan and elite plan.* An initial plan is generated by randomly sequencing the machining operations and selecting the machine, tool and TAD for the execution of each operation in the plan from the corresponding candidate resource lists. A current plan is the best solution selected in each iteration and it is used to start a new iteration through generating some neighborhood trial solutions. An elite plan records the best solution found thus far.

(2) *Neighborhood strategies.* A set of process plans can be generated from a current plan for trials using neighborhood strategies. The neighborhood strategies include two basic manipulations, which are illustrated in Fig. 5. The first mutation manipulation randomly replaces the set of machine, tool and TAD from the candidate list for the operations of a plan. The second manipulation changes the sequence of two operations in a plan using shifting (selecting an operation from a plan to insert into a new position), swapping (selecting two operations for position exchange) or adjacent swapping (selecting two adjacent operations for position exchange) operations, which are the same as those described in [28].

(3) *Forbidding, aspiration and freeing strategies.* The forbidding strategy, which is used to manage the plans entering a Tabu list, can avoid cycling and local minimums by forbidding certain moves during the most recent computational iterations. The Tabu list is based on a "first-in-first-out" (FIFO) queuing rule to store recently searched plans. An aspiration strategy enables a plan that has been forbidden by the Tabu list to become acceptable if it satisfies a certain criterion, so as to provide some flexibility to the forbidding restrictions by leading the search in

a desirable direction. A common criterion is to override a tabooed plan if its machining cost is lower than that of the elite plan. The freeing strategy controls the plan that exits from the Tabu list and when it should exit. This strategy is applied in one of the following two cases: (a) When the Tabu list is full and a new plan needs to join, the earliest forbidden plan in the Tabu list should be freed so that it can be reconsidered in the future search; (b) When an evaluated current plan satisfies the above forbidding strategy as well as the aspiration criterion, this plan should be considered as admissible (in this case, the aspiration strategy is equivalent in function to the freeing strategy).

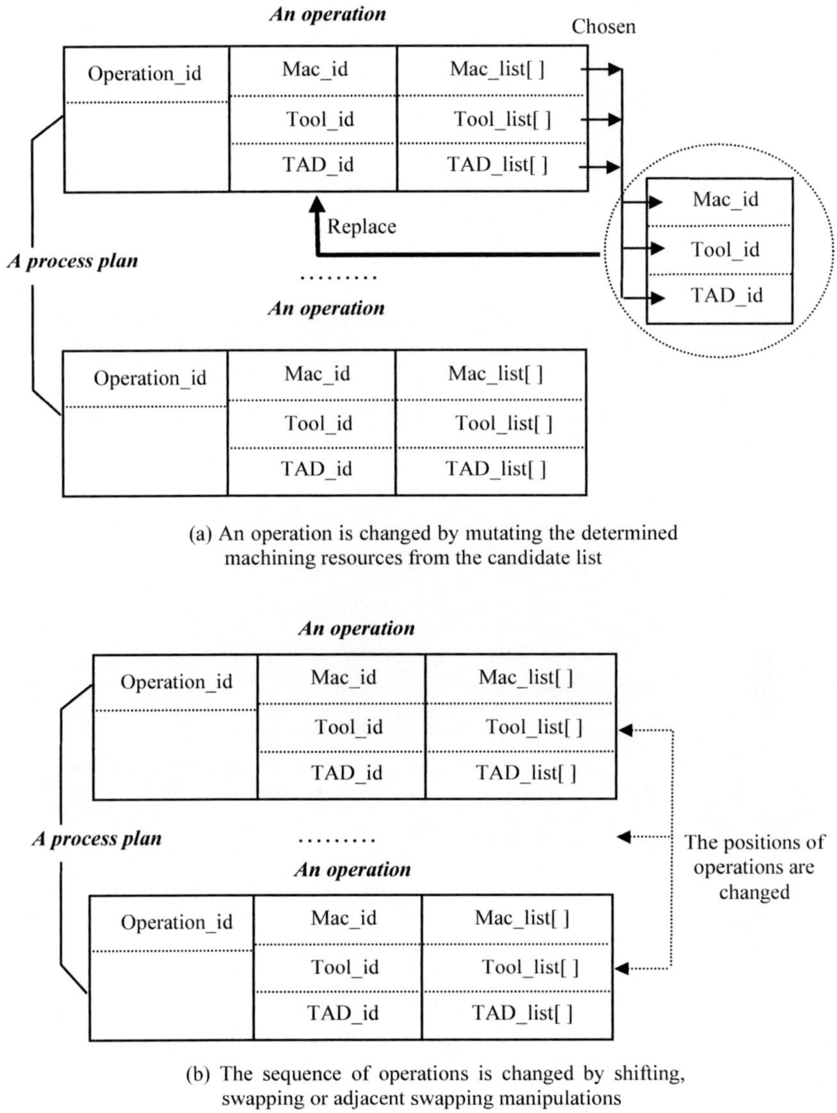

Fig. 5 Two basic manipulations to generate variant plans from a current plan.

(4) *Stopping criteria*. Termination conditions for the searching algorithm can be set using one of the following criteria: (a) the number of iterations reaches a pre-defined number; and (b) the elite plan remains unchanged for a pre-defined number of iterations.

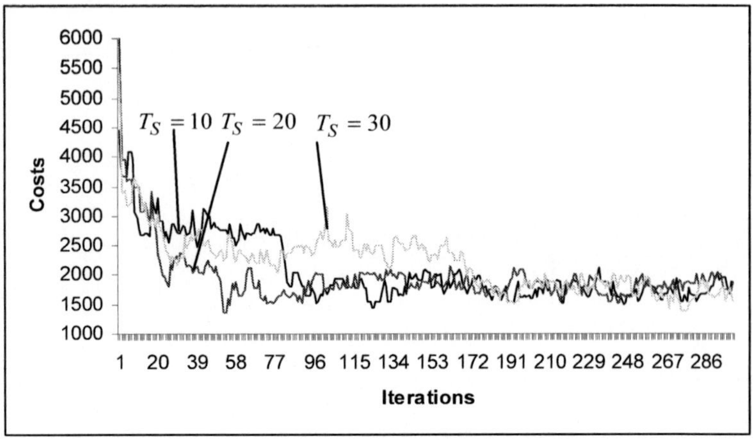

(a) Machining costs of current plans under different T_S

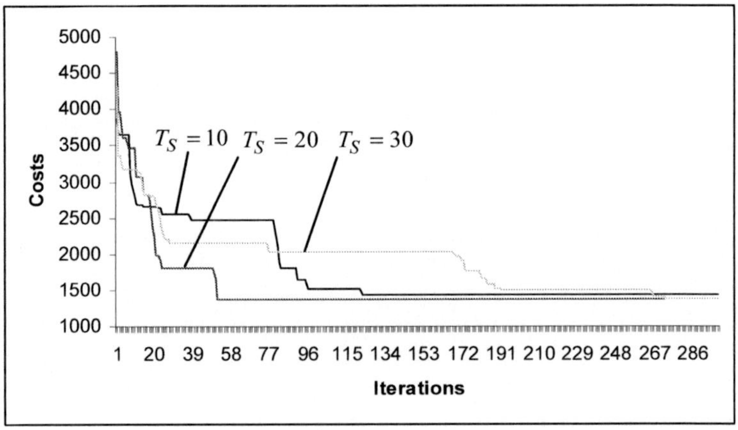

(b) Machining costs of elite plans under different T_S

Fig. 6 Determination of parameters of the TS algorithm.

5.2 Determination of parameters

The main parameters that determine the performance of the TS include the size of the Tabu list - T_S, the size of the variant plans from a current plan - S_{TS_N}, and the four probabilities of applying the shifting, swapping, adjacent swapping and mutation operations - P_{TS_sh}, P_{TS_sw}, P_{as} and P_{TS_mu}.

T_S plays an important role in the search for solutions. A small T_S might cause a high occurrence of cycling, and a large T_S might deteriorate the solution quality. Through trials, T_S was chosen as 20 and the comparison results are illustrated in Fig. 6 (a) and (b). A suitable S_{TS_N} can ensure good computational efficiency and algorithm stability. Similarly, through comparisons, $S_{TS\_N} = 30$ is a good choice for the algorithm to achieve good performance. After ten trials were conducted, the group of P_{TS_sh}, P_{TS_sw}, P_{as} and P_{TS_mu} were chosen as 0.85, 0.85, 0.5 and 0.85 respectively for the algorithm to yield good performance.

These comparisons are based on Part 1 (Fig. 7). Based on the trials for Part 2 (Fig. 8) and using the same chosen parameters, satisfactory results were obtained. Through more trials on other parts and the obtained results are also satisfactory. Thus, these parameters are generally acceptable.

5.3 *Comparison studies of TS, SA and GA*

Parts 1 and 2 are used to illustrate the performances of TS, SA and GA on the same optimization model to give a comprehensive understanding of their characteristics. Conditions (1) and (2) are chosen for the studies for both parts.

(1) All machines and tools are available, and all cost factors in Equation (1) are considered.
(2) All machines and tools are available, and only *TMC*, *TSC* and *TMCC* are considered.

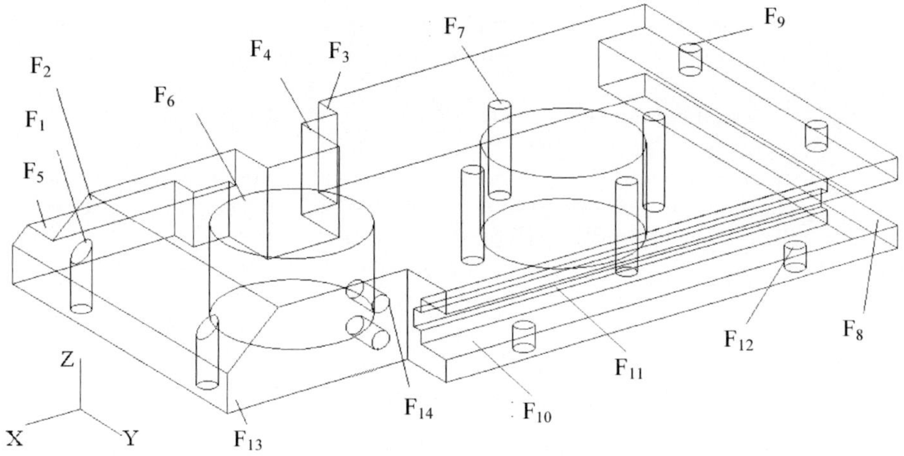

Fig. 7 A sample part with 14 features – Part 1.

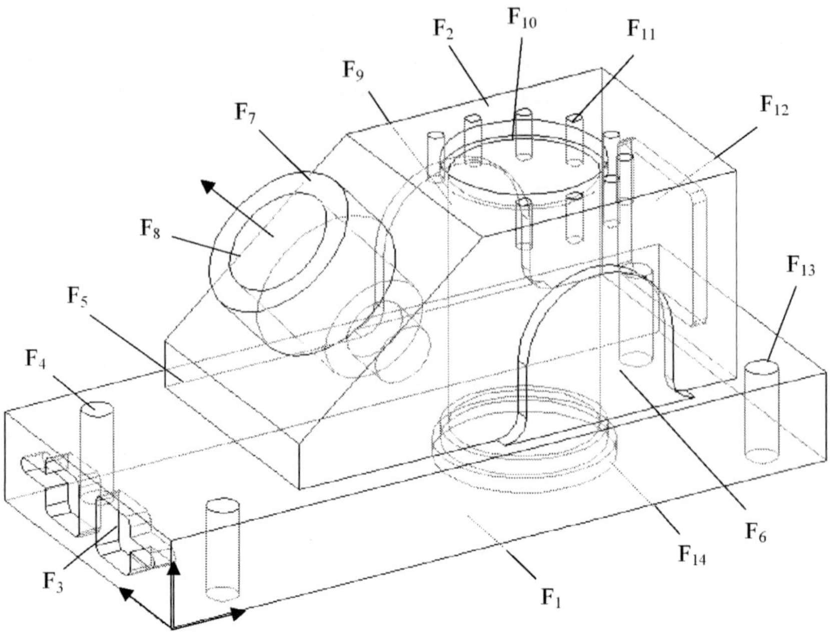

Fig. 8 A sample part with 14 features – Part 2.

The computations illustrated in Fig. 9 (for Part 1) and Fig. 10 (for Part 2) were performed for the two parts under Condition (1). The current plans of TS and SA at each iteration were used for generating the neighbourhood and next solutions, while in the GA, each refers to the best plan chosen from a population in a generation (an iteration). Each of the elite plans is the best plan at each iteration of the three algorithms.

It shows that the decreasing trends of the curves for TS and GA are smoother than that of SA. For SA, there are some "abrupt" decreasing points during its iteration process. Generally, TS and SA can achieve better (lower machining costs) solutions than GA, while TS can achieve a more stable performance than SA in terms of lower mean, maximum and minimum machining costs of the best process plans obtained. These observations are in accordance with the main characteristics of GAs, which are prone to "pre-maturity" (they converge too early and have difficulty finding optimal or near-optimal solutions), and SA, which is vigilant to the controlling parameters of the algorithm and specific situations [34]. In Tables 2 and 3, more thorough comparisons for the three algorithms on Part 1 under Condition (1), and on Part 2 under two conditions are made. The computations are based on ten trials for each algorithm under each condition respectively. Similar observations on the characteristics of the three algorithms can be obtained.

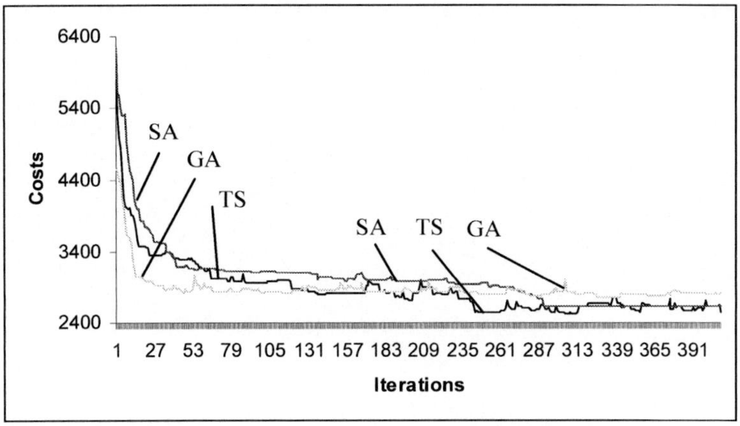

(a) Manufacturing costs of current plans for Part 1

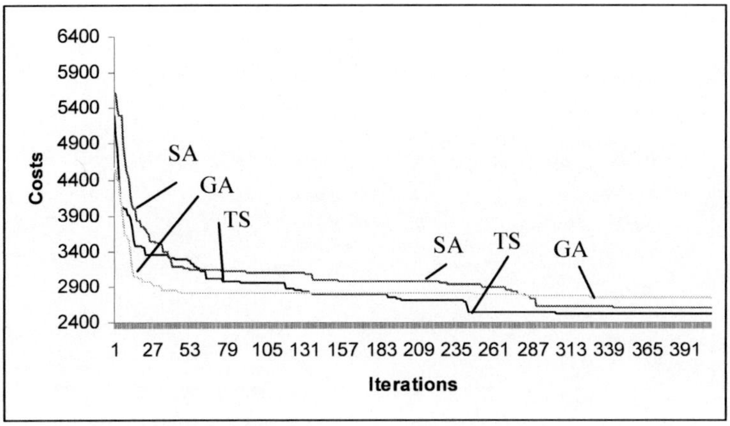

(b) Manufacturing costs of elite plans for Part 1

Fig. 9 Comparison studies of the three algorithms for Part 1.

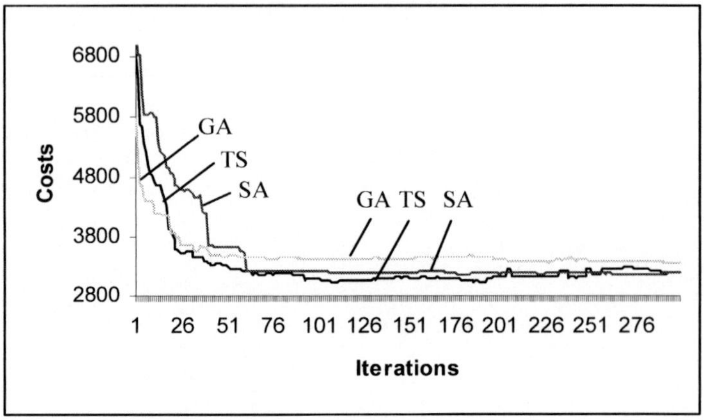

(a) Manufacturing costs of current plans for Part 2

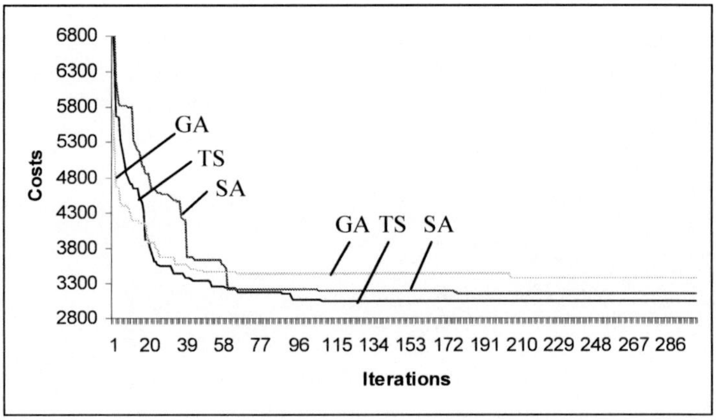

(b) Manufacturing costs of elite plans for Part 2

Fig. 10 Comparison studies of the three algorithms for Part 2.

Table 2 Comparison studies of the three algorithms for Part 1.

		TS	SA	GA
Condition (1)	Mean	1342.0	1373.5	1611.0
	Maximum	1378.0	1518.0	1778.0
	Minimum	1328.0	1328.0	1478.0

Table 3 Comparison studies of the three algorithms for Part 2.

		TS	SA	GA
Condition (1)	Mean	2609.6	2668.5	2796.0
	Maximum	2690.0	2829.0	2885.0
	Minimum	2527.0	2535.0	2667.0
Condition (2)	Mean	2208.0	2287.0	2370.0
	Maximum	2390.0	2380.0	2580.0
	Minimum	2120.0	2120.0	2220.0

6. Conclusions

Manufacturing cost estimation is one of the most important elements to build up in a CE-based product development environment, and this issue is closely related to the process planning optimization problem, in which machining operations, machines, cutting tools, operation sequences, etc. are selected and determined.

To solve this difficult problem with complex machining constraints, in this chapter, a TS method has been developed as an optimization algorithm to simultaneously determine the allocation of machining resources and optimization of machining operations for a process plan, and an AHP technique incorporating fuzzy logic has been applied to evaluate the satisfaction degree of the manufacturability of the process plan to address manufacturing constraints. Case studies to compare this method with GA-based and SA-based methods with the same evaluation criterion are presented, and the good performance of this method from the aspects of solution quality, computational efficiency and the robustness of the algorithm are highlighted. The major characteristics of the work include:

(1) The proposed method can generate optimal or near-optimal process plans for complex designed models with good computation efficiency based on a combined manufacturing cost criterion, which can conveniently simulate a practical dynamic workshop.
(2) Machining constraints are defined and classified. The developed fuzzy logic-based AHP technique can address a complicated constraint situation with conflicting constraints and achieve good computational performance.

7. Future Trend

Cost estimation is one of the important modules in product design. With this functional module, design concepts and models of products can be effectively adjusted in early design stages while cost can be optimized. In general, there are two approaches to estimate cost, i.e., parametric cost estimation and generative cost estimation. The parametric approach, which is used to estimate the cost of a product based on mathematical formulas of one or more relevant independent design variables, also known as cost driver (s), is more suitable in the very early stage of product design since this approach depends upon less design parameters. However, the drawback is that the estimation result is not very accurate. In contrast, the generative approach, which is used to forecast the cost of a product based on describing the relevant procedures of a product creation in a detailed work breakdown structure, is applicable when many design details of a product are determined and the estimated result is closer to the actual cost of the product. The current trend is to incorporate these two approaches as a hybrid approach, e.g., a parametric approach is first applied while the generate approach is then used for calibration and benchmarking, to enhance the accuracy and effectiveness of estimation.

On the other hand, lifecycle cost control and estimation is getting more important in product companies. Presently, CE and Design for X are being actively extended to Product Lifecycle Management (PLM) so as to further optimize the design of products from the perspective of the entire lifecycle. Most of the previously developed cost estimation works just consider product design and manufacturing, and so there lacks an effective method to evaluate the lifecycle cost. A new trend is to investigate the impacts of the lifecycle parameters of products on cost, and establish the relevant estimation computational models.

Acknowledgement

The authors would acknowledge the funding support from the Singapore Institute of Manufacturing Technology and the Engineering and Physical Sciences Research Council (EPSRC) of UK for this work.

References

[1] *NASA Parametric Cost Estimation Handbook*, Available at http://www.ispa-cost.org/PEIWeb/newbook.htm.

[2] J. Scanlan, T. Hill, R. Marsh, C. Bru, M. Dunkley and P. Cleevely, Cost modelling for aircraft design optimization, *Journal of Engineering Design*, 13(2002), 261-269.

[3] C. Ou-Yang and T.S. Lin, Developing an integrated framework for feature-based early cost estimation, *International Journal of Advanced Manufacturing Technology*, 13(1997), 618-629.

[4] E.M. Shehab and H.S. Abdalla, A design to cost system for innovative product development, *Proceedings of the Institute of Mechanical Engineers (Part B)*, 216(2002), 999-1019.

[5] A. Cardone and S.K. Gupta, Identifying similar parts for assisting cost estimation of prismatic machined parts, *ASME 2004 Design Engineering Technical Conferences*, DETC2004-57761, (2004).

[6] R. Stewart, R. Wyskidsa and J Johannes, *Cost Estimator's Reference Manual (2rd Edition),* Wiley, (1995).

[7] E.G. Welp and D. Giannoulis, Knowledge-based cost estimation of product concepts, *ASME 2004 Design Engineering Technical Conferences*, DETC2004-57766, (2004).

[8] H. Patwardhan and K. Ramani, Manufacturing feature based dynamic cost estimation for design, *ASME 2004 Design Engineering Technical Conferences*, DETC2004-57778, (2004).
[9] D. Ben-Arieh, Cost estimation system for machined parts, *International Journal of Production Research*, 38(2000), 4481-4494.
[10] J.Y. Jung, Manufacturing cost estimation for machined parts based on manufacturing features, *Journal of Intelligent Manufacturing*, 13(2002), 227-238.
[11] T.C. Chang, *Expert Process Planning for Manufacturing*, Addison-Wesley (1990).
[12] T.N. Wong and S.L. Siu, A knowledge-based approach to automated machining process selection and sequencing., *International Journal of Production Research*, 33 (1995), 3465-3484.
[13] C.C.P. Chu and R. Gadh, Feature-based approach for set-up minimization of process design from product design, *Computer-Aided Design*, 28(1996), 321-332.
[14] H.C. Wu and T.C. Chang., 1998, Automated set-up selection in feature-based process planning, *International Journal of Production Research*, 36(1998), 695-712.
[15] Y.J. Tseng and C.C. Liu, Concurrent analysis of machining sequences and fixturing set-ups for minimizing set-up changes for machining mill-turn parts, *International Journal of Production Research*, 39(2001), 4197-4214.
[16] C.L.P. Chen and S.R. LeClair, Integration of design and manufacturing: solving set-up generation and feature sequencing using an unsupervised-learning approach, *Computer-Aided Design*, 26(1994), 59-75.
[17] S.A. Irani, H.Y. Koo and S. Raman, Feature-based operation sequence generation in CAPP, *International Journal of Production Research*, 33(1995), 17-39.
[18] A.C. Lin, S.Y. Lin, D. Diganta and W.F. Lu, An integrated approach to determining the sequence of machining operations for prismatic parts with interacting features, *Journal of Materials Processing Technology*, 73(1998), 234-250.
[19] H.C. Zhang and S.H. Huang, A fuzzy approach to process plan selection, *International Journal of Production Research*, 32(1994), 1265-1279.
[20] M.K. Tiwari and N.K. Vidyarthi, An integrated approach to solving the process plan selection problem in an automated manufacturing system, *International Journal of Production Research*, 36(1998), 2167-2184.
[21] Z. Gu, Y.F. Zhang and A.Y.C. Nee, Identification of important features for machining operations sequence generation, *International Journal of Production Research*, 35(1997), 2285-2307.
[22] J. Vancza and A. Markus, Genetic algorithms in process planning, *Computers in Industry*, 17(1991), 181-194.
[23] D. Yip-Hoi and D. Dutta, A genetic algorithm application for sequencing operations in process planning for parallel machining, *IIE Transactions*, 28(1996), 55-68.
[24] S.V.B. Reddy, M.S. Shunmugam and T.T. Narendran, Operation sequencing in CAPP using genetic algorithms, *International Journal of Production Research*, 37(1999), 1063-1074.
[25] G.H. Ma, Y.F. Zhang and A.Y.C. Nee, A simulated annealing-based optimization algorithm for process planning, *International Journal of Production Research*, 38(2000), 2671-2687.
[26] D.H. Lee, D. Kiritsis and P. Xirouchakis, Search heuristics for operation sequencing in process planning, *International Journal of Production Research*, 39(2001), 3771-3788.
[27] L. Qiao, X.Y. Wang and S.C. Wang, A GA-based approach to machining operation sequencing for prismatic parts, *International Journal of Production Research*, 38(2000), 3283-3303.
[28] J. Chen, Y.F. Zhang and A.Y.C. Nee, Set-up planning using Hopfield net and simulated annealing, *International Journal of Production Research*, 36(1998), 981-1000.
[29] W.D. Li, S.K. Ong and A.Y.C. Nee, A hybrid genetic algorithm and simulated annealing approach for the optimization of process plan for prismatic parts, *International Journal of Production Research*, 40(2002), 1899-1922.
[30] W.D. Li, S.K. Ong and A.Y.C. Nee, A hybrid method for recognizing manufacturing features, *International Journal of Production Research*, 41(2003), 1887-1908.
[31] L. Ding, Y. Yue, K. Ahmet, M. Jackson and R. Parkin, Global optimization of a feature-based process sequence using GA and ANN techniques, *International Journal of Production Research*, 43(2005), 3247-3272.
[32] B.L. Golden, P.T. Harker and E.E. Wasil, *The Analytic Hierarchy Process: Applications and Studies*, Springer-Verlag, (1989).
[33] F. Glover, *Tabu Search*, Kluwer Academic Publishers, (1997).
[34] D.T. Pham and D. Karaboga, *Intelligent Optimization Techniques: Genetic Algorithms, Tabu Search, Simulated Annealing and Neural Networks*, Springer-Verlag, (2000).

Integrated Intelligent Systems for Engineering Design
X.F. Zha and R.J. Howlett (Eds.)
IOS Press, 2006

A Distributed Information System Architecture for Collaborative Design

Andrew FELLER, Teresa WU, and Dan SHUNK
Arizona State University

Abstract. Improving and streamlining the communication and decision processes used in Collaborative Product Development (CPD) requires a robust Distributed Information System that can enable intelligent notification, coordination and negotiation processes. In this chapter we review existing research and industry practice related to CPD information systems and propose an information framework called the Virtual Environment for Product Development (VE4PD) that is based on the integration of Web services and agent technologies to manage the CPD process. The VE4PD architecture is designed to support CPD functions such as design synchronization, timely notification, distributed control, role based security, support for distributed intelligent agents, and varying design rule standards. An implementation system including intelligent agents for design negotiation is also described that validates the application approach.

Keywords. Product Development, Collaboration, Information System, Web Services, Agents

1. Introduction

Modern design teams are often composed of globally dispersed staff based within different companies and facilities, using varying computer-based design tools, and supported by differing Information Technology (IT) applications and infrastructures. Still these teams must work together collaboratively to complete a design project under challenging time and cost constraints. Internet connectivity enables simple design data and file sharing schemes, however more advanced architectures are needed to address design process and application technology issues such as design synchronization, timely change notification, distributed control, role based security, support for distributed intelligent agents, and varying design rule standards. To meet these challenges, a number of Collaborative Product Development (CPD) practices and technologies have emerged including web-based distributed information systems. The need for more streamlined and efficient CPD technologies is intensifying due to highly competitive global markets, rapid technology advances and increasing customer demands. These forces compel companies to reduce product development cycle times in order to retain and grow market share. Success in this environment requires streamlined processes for design collaboration across geographically-dispersed internal teams and external partners participating in the product

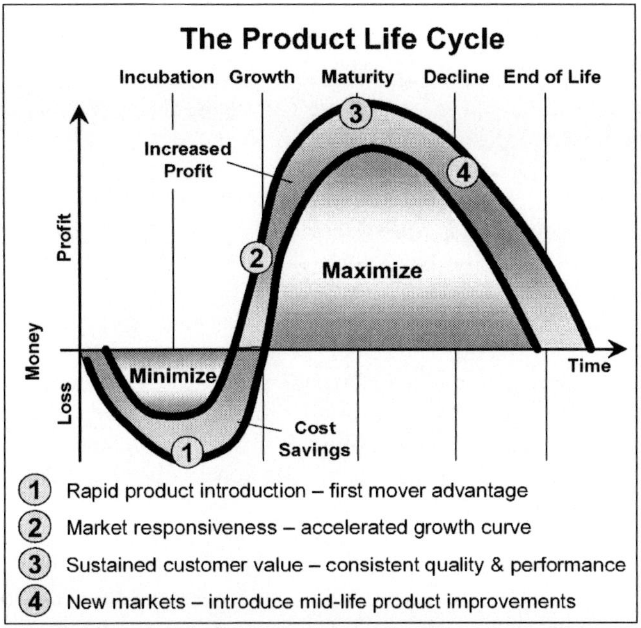

Figure 1. Benefits of Improved Product Development [1]

development value chain. The potential business impact from an improved product development process is significant across all stages of the product lifecycle [1]. It has been reported that over 70 percent of a product's manufacturing cost is influenced by decisions made during the product design stage [2]. Benefits of an effective product development process include a reduction in design changes [3], lower life cycle costs [4] and shortened product development time [5]. The relationship between product development time and market share shows that as time to market is decreased, potential market share increases [6]. The benefits from improved CPD are summarized in Figure 1. Given the importance of effective product development, there is considerable pressure to develop methods to improve communication and coordination of the design process. Such methods will facilitate decreased costs, decreased time to market and increased product innovation. One strategy is Collaborative Product Development (CPD), a process that involves not only coordination of dispersed product development functions, but also information management across multiple product development phases.

One of the fundamental requirements for enabling an effective CPD process is an information system architecture that is capable of seamlessly integrating the range of functions and technologies that must work together in a heterogeneous systems environment. Federated systems architectures are one approach for achieving this goal. In a federated architecture, information systems at individual sites and workstations operate under local control, yet are linked together by means of a common interface model. Another approach for enabling seamless integration across heterogeneous environments is to centralize the critical functionality and provide only thin-client access to centrally shared repositories. The prototype architecture that we present has leveraged both models, identifying some functions that are best supported by a federated model moderated using local agent technologies, and others that are best suited to run on a thin client interface.

Our overall purpose is to help lay the foundation for next generation intelligent collaborative design systems with an architecture for supporting distributed information systems. Newer approaches for distributed computing such as web services and agent technologies have emerged concurrently with increased demand for capabilities to support collaborative design. This creates an opportunity to build on these technologies to develop an application infrastructure that can provide critical design collaboration services. Intelligent design agents can then be built upon this architecture to implement specific CPD processes such as notification, coordination, and negotiation.

Reviewing existing research, we summarize a number of information system architectures that have been developed in research and industry environments, comparing them and identifying areas of opportunity for intelligent collaborative systems. We examine the research frameworks and enterprise Product Lifecycle Management (PLM) applications current in industry, focusing on the challenges and opportunities to be met by an enabling infrastructure such as design negotiation, management of central and local design files (check-in/check-out coordination), and design software integration.

In the proposed Virtual Environment for Product Development (VE4PD) framework [7] we provide an approach developed to meet the identified challenges, detailing a distributed information architecture that supports intelligent agents for collaborative design. The outline of VE4PD includes a shared information repository, a data management layer and a distribution layer. The repository consists of development file storage and a database for maintenance of design parameters. An open architecture, object-oriented data model is provided for the VE4PD data management layer. At the distribution layer we include the use of local agents for proactively synchronizing local engineering files for consistency and project agents for updating the shared information repository. An agent based approach is also detailed at the server level to monitor and synchronize changes, to identify required coordination and negotiation steps and to then notify the associated clients.

To demonstrate the approach, a set of example scenarios will be covered that takes an engineering design project through the initial stages of initializing a new project in the framework including file and data storage and distribution channels, establishing the design team including local repository setup and access, and maintaining a synchronized design base through an iterative development process.

For illustration, we've included a case study involving information and knowledge sharing between divisions of a company as well as between different companies. Typical CPD scenarios are implemented, providing application oriented examples for how the framework has been extended to enable collaborative design processes such as design synchronization and negotiation. The case provides details on the agents operating to maintain consistency between design files and design parameters, and notifying designers when conflicts may require negotiation.

Finally we provide an example of how intelligent agents are introduced into the framework. A distributed software agent mechanism is implemented for collaborative design negotiation. This agent based approach uses a principal-agent concept to impose penalties on design agents using a principal agent that manages the design negotiation to meet overall design constraints. In this approach, the design agents interact with the principal agent to implement a distributed multi-stage negotiation process called Penalty Induced Negotiation.

By proceeding from a general review of the supporting infrastructures being developed for CPD to a detailed example of intelligent collaborative agents executing a design negotiation in a newly proposed agent-based framework, we hope to stimulate continuing research into the development of Integrated Intelligent Systems for Engineering Design.

2. Background

Prior research has been conducted in areas related to CPD information frameworks to advance business and engineering disciplines in implementing critical CPD processes such as collection and analysis of product requirements and iterative product design and manufacturing [8]. Svensson and Barfod [9] emphasized the need to effectively manage product information flow in a supply network. Rupp and Ristic [10] stated that the lack of coordination and inaccurate information flows made production planning and control a difficult task. Over the last decade, a range of computer based information management technologies have been introduced to facilitate the application of CPD. These technologies are being used to control and support collaboration on product designs, manufacturing processes, and product knowledge management across product development phases. The development of standard communication protocols based on Internet technologies has eased the implementation of increasingly seamless distributed CPD information systems. As the Internet and distributed computing technologies progress, novel distributed platforms for CPD have evolved to enable globally distributed collaborative development environments [11, 12]. A recent field analysis of business processes at a firm with design sites in several countries investigated the collaboration technologies in use and found that different synchronous and asynchronous means are used including: phone and e-mail; intranet and groupware (e.g.: Lotus Notes); shared network directories; shared work spaces with whiteboards and shared editing (i.e.: NetMeeting); and (5) videoconferencing [13]. Shared visualization and virtual reality platforms have also come into use leveraging standards such as Virtual Reality Markup Language (VRML) [14] [15].

Most CPD platform development efforts have focused on enhanced collaboration by leveraging information technology and have emphasized the following issues:

- Definition and development of distributed information management and knowledge sharing platforms [8, 12, 16]
- Standards for information/knowledge exchange and presentation [8, 17]
- Standard processes across the CPD life cycle [18, 19, 20]
- Efficient information and knowledge base schemata for shared development and access among collaborators [21, 22]
- Design rules, strategies and applications to support a stable and secure information and knowledge sharing environment [23, 24, 25]
- Improving specific detailed functionality of a CPD system, such as project management [26, 27], portfolio management [28], visualization tools [29, 30, 31] and workflow management [32, 33]

Our initial focus on the first issue, developing an information/knowledge sharing platform, is motivated by the need to provide a suitable foundation on which to build distributed decision support capabilities for CPD [7]. Within the practice of CPD, the important challenge of implementing a suitable framework for collaboration and knowledge sharing is complicated by cultural and legal issues of multi-firm participation such as inter-enterprise trust and communication, negotiation and intellectual property ownership. There is significant potential for conflict to arise among different partners that have discrepant goals, strategies and roles. This situation creates the need for an information architecture that can support a design negotiation process. Negotiation processes during product development require access to design parameters, design files, and the design processes themselves maintained in a timely manner. The infrastructure we have developed provides an information sharing platform to support these design negotiation requirements. A variety of information frameworks have been developed for CPD ranging from web service based frameworks [34, 35, 36, 37] and remote service based frameworks [38, 39, 40, 41, 42], to remote repository based frameworks [43, 44]. These three primary architectures are summarized in Figure 2 as having a progressively heavier processing and information load on the client side, and a correspondingly decreasing load on the server side. They are each summarized in detail in the following sections. Note that the user interface represents the application's interface through which users access the distributed information system; data management represents the processing logic layer that executes business rules and makes system calls to retrieve files and data and generate responses for various system scenarios. Finally the information repository refers to the overall storage structure for product development information including design files and data.

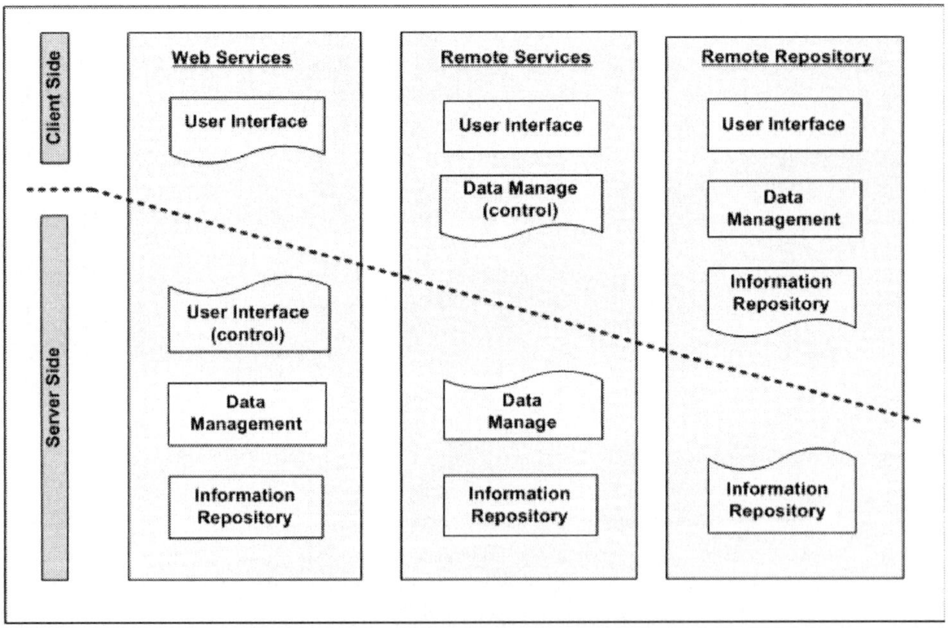

Figure 2. Three Distributed Information System Architectures for CPD [7]

Web service based frameworks only permit reactive information access and updates and have difficulty in the integration of both server and client side applications.

Remote service and repository based frameworks lead to severe challenges for system development and deployment efforts to keep the client and server systems correctly configured.

Due to these limitations in existing architectures, we found the need to explore and develop a hybrid web-client information framework for effective CPD that combines web service and agent based integration technologies to serve as a platform for design negotiation [7]. In this architecture, agent technology is applied to achieve information consistency and to overcome prior limitations in both server and client side integration.

2.1. Web Services Based Information Frameworks

Web service based frameworks take advantage of web technologies to support access and sharing of information from different locations. Web services based frameworks concentrate on server side programming. The user interface, data management system and information repository all function on the server side, allowing the client to interact with a simple application program via a Web browser. Web-PDM [25] uses the web services approach to structure a collaborative product data management system. KaViDo [35] concentrates on applying web services to record and monitor the product development process. The combination of Extensible Markup Language (XML) and Simple Object Access Protocol (SOAP) is used to construct a data communication, representation and standard storage layer. WebBlow [36] addresses the integration of product development control components on the server side using agent technology. Roy and Kodkani [37] propose a Web services based framework to share design data among multidisciplinary team members. E2open [46] uses Web services to simplify business process publication, consumption and discovery among trading partners to facilitate a many-to-many integration among multiple distributed companies. Agile PLM (Product Lifecycle Management) [47] employs Web services to achieve product and engineering collaboration, enable visibility and management of product information across global operations, and provide design collaboration and visualization support.

2.2. Remote Services Based Information Framework

Remote services are a traditional approach to designing distributed systems and are based on remote services technologies, such as Java Remote Method Innovation (RMI) and Common Object Request Broker Architecture (CORBA) to enable the communication among the clients and servers. Unlike Web services, remote services based information frameworks require development efforts spent on both the client and the server side. As shown in Fig. 1, the user interface and partial data management reside on the client side of the interface, while the server holds a part of the data management and the information repository. Urban et al. [42] provide an overview of remote services based technologies including CORBA, RMI to Enterprise JavaBeans, Jini Connection Technology, JavaSpaces, Java Messaging Service, and Java Transaction Service. A comparison of these different technologies is presented, discussing the similarities and differences, as well as the ways in which such technologies can be used for the product development. Gerhard [41] proposes a RMI based information framework to support collaborative design and manufacture event based processes. Another example is PRODNET [38, 39, 40], which aims to design and develop an open platform to support the communication among industrial virtual enterprises, especially small and medium enterprises. Among PLM vendors, MatrixOne has developed an architecture that can support both remote and web service oriented architectures [48].

2.3. Remote Repository Based Information Framework

Remote repository frameworks allow for central storage and versioning of information only. In this type of framework, most of the system's functionality resides on the client side of the interface, while the server only manages the storage of part of the common information. One example of a remote repository based information framework is Eclipse [43]. By applying plug-in technology on the client side, Eclipse builds an open, integrated architecture that supports a client-side development environment with a simple and clear information exchange structure. To our knowledge, remote repository based frameworks have been used to integrate software development, yet less effort has been spent on the CPD application. This might be due to the recent rapid development of Internet driven CPD, which focuses on server side development.

2.4. Summary of Existing Frameworks

In this section, we review and evaluate projects from the distributed system architecture point of view. Comparisons are summarized in Table 1.

Table 1. Comparison of Existing Architectures

	Web Service					Remote Service		Remote Repository
	KaViDo	**WebBlow**	**Web-PDM**	**E2open**	**Agile PLM**	**PRODNET**	**PRE-RMI**	**Eclipse**
Info Access Model	Remote	Remote	Remote	Remote	Remote	Loaded	Loaded	Local
Communication	HTTP/ FTP	HTTP/ FTP	HTTP/ FTP	HTTP/ FTP	HTTP/ FTP	RPC/HTTP	RPC	FTP/ RPC
Client Process	Thin	Thin	Thin	Thin	Medium	Medium	Medium	Thick
Server Process	Thick	Very Thick	Thick	Thick	Thick	Thick	Medium	Thin
Info Consistency	N/A	Server side	N/A	Server Side	Server Side	N/A	N/A	Client-server
Replication	N/A	N/A	Server Side	Server Side	Server Side	N/A	N/A	Client-server
System Extension	N/A	Server Side	N/A	N/A	Client/Server	N/A	N/A	Client side
Client Side Tech	HTML/ XSLT	Applet/ HTML	HTML	HTML/ RosettaNet/ EDI/XML	Java/Html	Java	Java/RMI	Java
Server Side Tech	J2EE/ XML	Servlet	CGI	J2EE/XML/ SOAP/UDDI /Perl	J2EE/ XML	Java/Corba	Java/RMI	Java

Table 1 data includes:

- Information Access Model (Remote, Loaded, Local): This refers to how clients access shared information. For Web services based architectures, the data is opened remotely using a Web browser and only transient files are stored on the client. For remote services, clients access the data locally after it is loaded from the server. For remote repository based systems, local information is saved on the client and can be accessed locally with only periodic updates from the repository.

- Communication (HTTP, FTP, RPC): This deals with protocols used for the client and server communication including HTTP, FTP and RPC (Remote Procedure Call). HTTP is the data communication protocol for most Web-based projects. FTP is sometimes added to HTTP to facilitate file transfers. RPC is a common protocol used between client and server in the absence of Web support.
- Client/Server Process (Thin, Medium, Thick): Thin and thick functionality refers to the level of complexity, business logic and control managed on one side of the client/server interface. A thick client is heavily loaded with processing and control software overhead, and operates more autonomously from the server. Since web service based frameworks locate all but the display processing on the server, they have a thin-client/thick-server architecture. At the other extreme, remote repository based frameworks have a thin-server/thick client architecture, while remote service based framework split the difference with business process and control functionality on both sides of the interface. This last architecture raises a set of interesting design decisions determining which functionality to place on each side of the interface.
- Information Consistency: This is a key design requirement for CPD information frameworks. Information consistency refers the system's ability to maintain consistent versions of design files and data across the client/server interface, and between multiple clients operating during a collaborative design process. Different approaches are used to address this requirement. For example, WebBlow provides a server side mechanism to check information consistency, while Eclipse provides a client side consistency checking mechanism.
- Replication: This item indicates whether the system provides a backup mechanism for the information repository. Of the systems reviewed, only Web-PDM and Eclipse provide server side replication and client side replication respectively.
- System Extension: The extension item indicates whether a system has the capability to integrate with other functional modules to make the framework extensible. WebBlow uses agent technologies to achieve the integration on the server side. Eclipse uses Java plug-ins to realize the integration on the client side.
- Client/Server Tech: This item lists the various technologies used to implement the interface architecture including Servlets, SOAP, XML, CGI, CORBA and Java

Analyzing these implementation approaches indicates that web service based frameworks are simpler to design, implement and deploy, however two drawbacks constrain their application to CPD. First, product development is a complex and comprehensive process consisting of many phases and involving a variety of engineering software tools running in different environments. Web service based frameworks have limited capability to support client-side integration of these design tools because the thin-client/thick-server approach uses internet/server resources through a virtual machine, and less capability is available to access the variety of computing and software resources residing on the client engineering workstation. WebBlow attempts to tackle this issue, however still leaves many open issues regarding client/server integration. One reason is that most advanced engineering software is too complicated and processor intensive to provide complete integration and client service using only server resources. Second, web service based frameworks follow a reactive process. That is, the system will not start an operation unless the user makes the request. Such a reactive process requires the user to update the information manually instead of initiating information consistency checking automatically or from within a client application.

Remote service based frameworks require medium size client and server side programs. Compared to Web service based frameworks, remote service based frameworks are difficult to deploy, maintain and even implement due to varying configurations of client hardware and software. There are also limitations related to remote service integration with technologies that support scalability and security [41]. For these reasons, most of the on-going research projects use remote service to support only part of the system functionality within an intranet/local-network scope. For example, the Web-PDM application provides an additional CORBA layer to map STEP translations to the standard data object to support clients without STEP APIs [32]. Ebbesmeyer et al. [49] provide a solution to build a special data transfer channel based on CORBA to satisfy the high security requirements of the Virtual Web Plant project.

Remote repository based frameworks are a promising approach to CPD due to their ability to achieve system integration with clear relationships between client/server repositories. One drawback is that the thick-client/thin-server approach can make system development and deployment a difficult task, especially in heterogeneous environments. Although Eclipse proposes an interesting methodology to integrate external software packages to enable information sharing, operation of the system is complicated for users and the architecture poses potential problems for the system implementation, installation and maintenance [43].

Many firms implementing commercial Product Lifecycle Management (PLM) systems take a significant step forward in establishing an information platform for CPD because these systems typically imbed an information architecture for managing distributed design files and data and electronic change control processes. A number of PLM systems have been widely implemented including Agile PLM from Agile Software, eMatrix from MatrixOne, Windchill from Parametric Technologies Corp., and TeamCenter from EDS [50]. It is interesting to note that most PLM systems are Web service based. While these commercial PLM systems are effective, there are a number of limitations:

1. Commercial PLM systems focus on product data management and design integration but neglect the steps and requirements of design negotiation. The PLM systems we have reviewed concentrate on how information is shared and controlled among different partners. We have not found a system that addresses the requirements of design negotiation. Apparently these systems assume that designed components are ready to be integrated when initially released, which is not the case in many CPD projects. Design negotiation is a critical aspect in collaborative product development.
2. Most commercial PLM architectures use a check-in/check-out protocol for design files built on an information architecture that does not support automated file integrity checking and integration. Though some attempts have been made for system integration, most models fall into the category of pessimistic checkout (copy and lock), requiring synchronization processes to be manually initiated by users. Local client manipulation of design files may go undetected, leading to delays in the CPD process. During these delays duplication and design conflict may occur. Therefore, consistency maintenance is an issue. The problem may

become exacerbated when considering the span of the lifecycle and the need to integrate up and down a design chain to have a seamless collaborative product development environment.

3. PLM solutions have come up with varied approaches to address client side integration with engineering tools and software on the client side. Due to heavy processing requirements, it is generally impossible to have all executable programs running off of the server. Therefore, many software applications run on the client, such as CAD, CAM, and CAE packages. There remain unresolved issues of how to integrate these software applications from the client side to truly integrate the system and move towards a proactive, optimistic checkout capability.

In summary, each of the approaches outlined in Table 1 has merits and drawbacks in building an effective information framework for CPD. Because CPD is a complex process spread across multiple companies in a heterogeneous computing environment, a key factor in selecting a CPD information architecture is the required development and deployment effort. In this regard, Web service based frameworks have advantages over remote service and remote repository based frameworks, because they are relatively simple to design, implement and deploy. Yet, the drawbacks of Web services need to be addressed. Our research builds a prototype solution using emerging software agent technology. Software agents are software programs that act in an intelligent and independent manner. In the following section we outline the development of a Virtual Environment for Product Development (VE4PD) based on web services with agent technology as a CPD information framework.

3. Proposed Approach: VE4PD

The VE4PD framework [7] is an architectural prototype for the information platform needed to support advanced collaborative product development processes such as design negotiation. Not all details of the framework have been fleshed out, however enough of the architecture has been built to validate the approach and its efficacy in supporting distributed processing and business logic in an agent based environment.

3.1. Overview

Information coordination and synchronization between distributed participants across the product development lifecycle is a primary function driving the requirements for an effective CPD information framework. The iterative nature of the product development process creates a dynamic environment in which information consistency needs to be maintained for the range of participants involved. In general, the design information that requires synchronized control can be represented by: (1) *design parameters*: the textual and structural (e.g.: Bill of Material) data, that can be controlled using database concepts; (2) *design files*: these files have a variety of formats such as design sketches, CAD files, and customer requirements documents and are typically controlled with versioning and revisioning strategies, and (3) the *linkages* between the design parameters and design files: the design parameters abstract specific elements from the design files and also provide the navigation scheme for file search/retrieval and change propagation and notification.. Based on the background review of CPD framework architectures and the need for standards for design information exchange and presentation, our approach addresses the requirements for maintaining information consistency using both web services and agent-based technologies for distributed computing. The architecture for VE4PD provides a flexible and reusable information framework to support communication and information sharing within the virtual enterprise context shown in Figure 3.

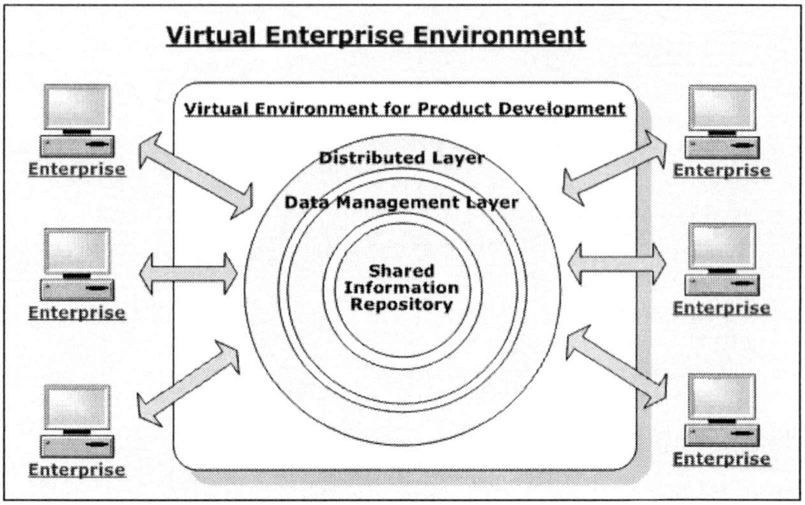

Figure 3. Outline of VE4PD

In VE4PD, Agent technology enables optimistic checkout procedures, which have been used extensively in software development environments but less so in applications for product development. The software agents in VE4PD will detect and notify the server of design changes made in design parameters and design files as well as business document objects, a feature that provides a foundation for design negotiation.

VE4PD consists of three main components: the shared information repository, the data management layer and the distributed structure layer. At the core of VE4PD, is the shared information repository which functions as the storage module to maintain product development information such as the product and process data and development knowledge. The next layer is the data management layer, which is the major operational control module for VE4PD. The third layer is the distributed structure layer. Software at this layer functions as an infrastructure manager handling facilities to share design information.

3.2. Architecture of VE4PD

To provide efficient development and deployment, a web services approach is applied as the foundation in developing VE4PD. On this foundation, the agent concepts are introduced to proactively assess and update information across the design chain. VE4PD is structured with two major segments as shown in Figure 4: (1) the functional agents include the project agent and local consistent monitor agent that are located on the client side of the interface, and the server consistent monitor agent located on the server; and (2) the operation layer that include standard and custom infrastructure elements including the data management system, web presentation layer, regular web browser and other product development tools available on the client side and the information repository (client & server side), which supports data and file storage.

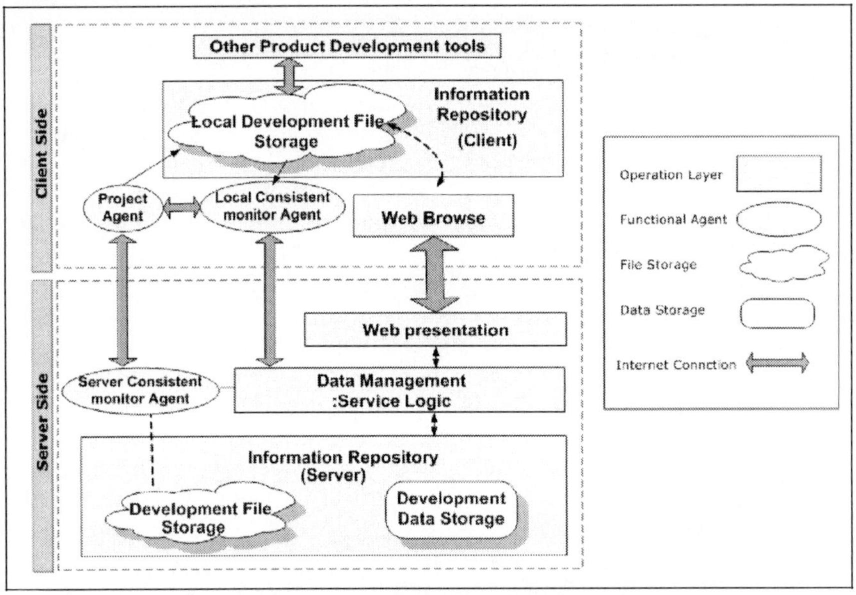

Figure 4. VE4PD System Architecture

A brief summary of critical functionality in the VE4PD architecture includes:

- The project agent on the client side structures and manages local storage. Including a client-side agent to manage a local repository enables development environment integration on the client side by controlling the development output files to ease the integration of various commercial product development engineering tools.
- The local consistent monitor agent on the client side enables the client workstation to access and update system information proactively. This agent detects changes in the local repository and notifies the data management component directly for the current development operations with the help of the project agent.
- Similar to the local consistent monitor agent, the server consistent monitor agent on the server side provides similar functions and also makes concurrent operational decisions based on the relationship between design parameters and design files. For instance this agent sends notifications to appropriate clients associated with related design changes from the server.
- Data management (service logic) is the primary operational module that maintains the business logic determining how the system respond to different design collaboration scenarios based on retrieved data and information. This layer enables integration with server-side development tools such as project management tools, portfolio management, visualization, workflow management and decision support.
- The web presentation layer builds a suitable interface for the end user.
- The information repository is the primary information storage component for the VE4PD architecture and is partially distributed. It consists of one development data storage database and two development file storage areas. The data storage database contains text and structure oriented design parameters, design file

structure management information and knowledge of the product development process. The two development file storage databases are located (1) in the server, which contains all the product development files and (2) in the client, which keeps partial product development files for the specified user.

In a traditional information system, the data structure of development data storage database is based primarily on relational database related architecture [15] and most of them only involve the design parameters and project management information. As discussed by Taner and Dennis [51] and Macgregor et al. [18], such a data structure has limitations. Without an open architecture data model, there could be a serious problem related to the reusability and flexibility of design information/knowledge. With the advances of computing technology and the research in product data standardization, many information systems have been developed for modeling different kinds of product development data [8, 35]. In order to provide a reusable, flexible and open architecture for the VE4PD framework, a simple CPD system data schema is introduced and named VE4PD Information/Knowledge Base Infrastructure. Three different kinds of data (design parameters, design file related data and development knowledge) and their relationships are modeled. This data schema is based on an object oriented data structure and can be implemented with either a relational database or an XML based object-oriented database. To save space, we will not explain the details of the VE4PD data schema. Interested readers may refer to [52]. An example of the information knowledge base is shown in Fig. 4.

The static working space in the upper left hand side of Figure 5 contains common product development knowledge such as general knowledge, domain specific knowledge, procedural or process knowledge as classified by Amrit et al. [21] and knowledge about collaborators such as virtual enterprise partners, supplier information and system user information. This space is independent of the particular product development project. When each project is initialized, the relationships are configured from this space for the particular product development information. According to Rezayat [53] and Lubell et al. [54], XML can play an important role in knowledge representation for a collaborative development environment and is becoming a standard for data transfer. In VE4PD, we have chosen to accommodate loading the necessary information (data + context) provided in XML format. This allows the information repository of VE4PD to potentially load data from external PLM systems and generate required relationships between the data, structure, and file management information.

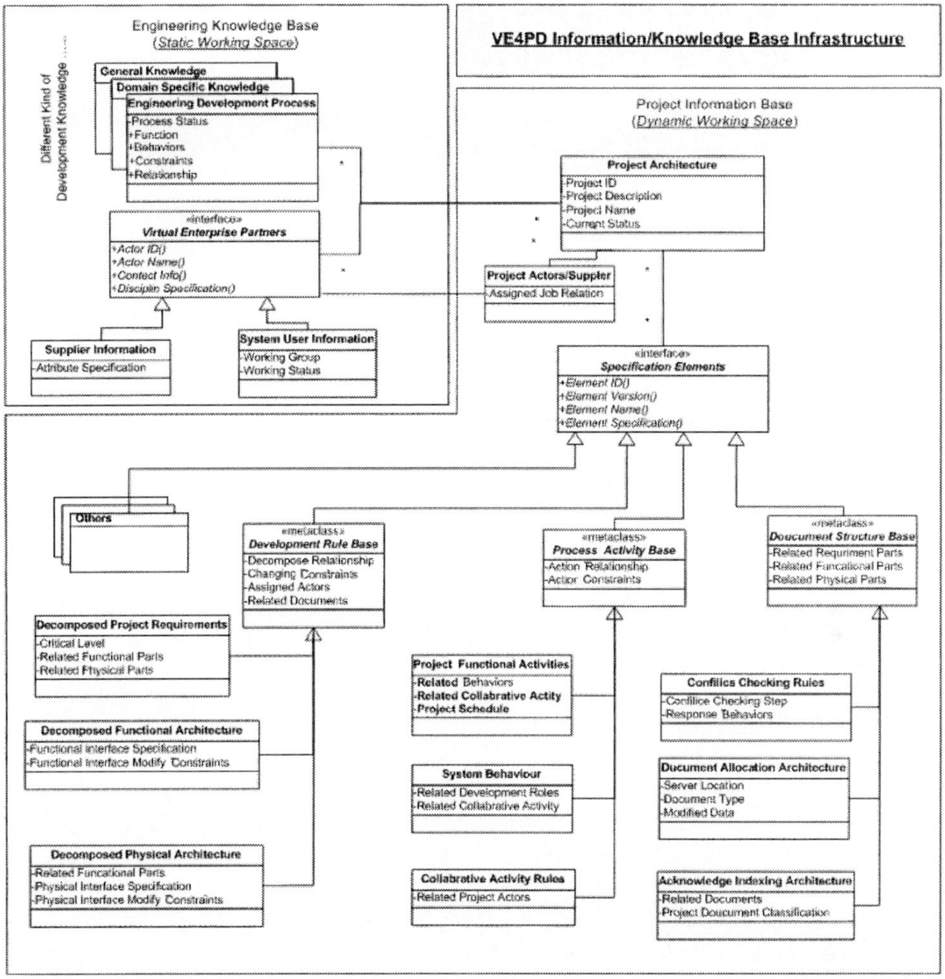

Figure 5. Abstract Representation of the VE4PD Information/Knowledge Base

The dynamic working space shown in Figure 5 is a project oriented data structure. For each product development project, an instance of the dynamic working space is created and the development information is stored using the provided data and knowledge base structures. The instantiated relationships between design parameters, design files and product knowledge are defined as the fundamental operational structures for the data management layer.

The complete data schema for a CPD system would be a very large and complicated structure. For example, the commercial Windchill Application Programming Interfaces (API) includes 80,000 methods and 1,700 classes with multiple data elements per class [55]. Figure 4 provides a conceptual snapshot to demonstrate how the data structure of VE4PD is configured to support critical relationships among the design parameters, design files and product/process knowledge. When a product development project is initialized in VE4PD, an instance of a “project architecture” class structure and it’s sub-nodes will be generated from the dynamic working space. The “development rule base” sub-node contains design parameters and is connected to domain specific knowledge from the static working space through attributes assigned to the specified data object.

For each of the decomposed functional modules of the product, an instance of the “decomposed functional architecture” sub-node is generated. This node maintains information related to the specific module as well as relationships between this itself and other physically or functionally related modules. When one module is modified in the iterative development process, all of the physical and functional connections among the modules will be detected based on their relationships and flagged for notification and modification accordingly. This goes beyond bill of material structure to provide a linkage model for propagating change data through a complex product architecture. The relationship between this model and related knowledge, such as specifications, design rules, constraints and rationale as pointed out by Szykman and Sriram [7], are established as well for future design change evaluation. In a similar fashion, other sub-nodes are instantiated to populate different kinds of development information and inter-relationships. For example, the “process activity base” sub-node is linked to the procedural or process knowledge component and maintains data on system activities, behaviors and their relationships. Instances of the “document structure base” node contain the information structure of the design file storage repository and the linkages between the design files and their related design parameters retrieved from the “development rule base” node.

3.3. VE4PD Process Scenario

In order to clarify the VE4PD system architecture and illustrate system activities, an abstract level System Sequence Diagram presented in Unified Modeling Language (UML) is shown in Fig. 5. A number of detailed scenarios are listed in Table 2 from this diagram that outline the steps involved in setting up a new project in VE4PD and beginning the design process. An application environment that implements these scenarios is described in a case study in the section 4.

Table 2: VE4PD Scenarios

Step 1 to 4	The Project Manager arrives and initializes the product development project. An instance of a VE4PD Project Information Base (Dynamic Working Space) will be created, the server side development file storage for the project and developer profiles and access channels will also be set up.
Step 5 to 10	The project developers arrive, sign up, download and set up their local project agents. The local repository structure and files are set up and downloaded from a server based on assigned access profiles. The project agents will start to monitor the local repositories.
Step 11 to 19	One particular project developer starts the iterative development process using their engineering tools. When a revision occurs, the project agent will detect the event and update the server side project information base. The project information base will trigger events based on any structure update and invoke a consistency checking process. The related files and developers will be detected based on the structure of the project information base and will be assigned to the particular consistency channel. The result will be broadcast to all developers based on the selected channel.
Step 20 to 27	The particular project developer will be notified when their project development files are updated on the server side and the consistency checking process will help the developer to apply these modifications and keep the developers working on a consistent design configuration.

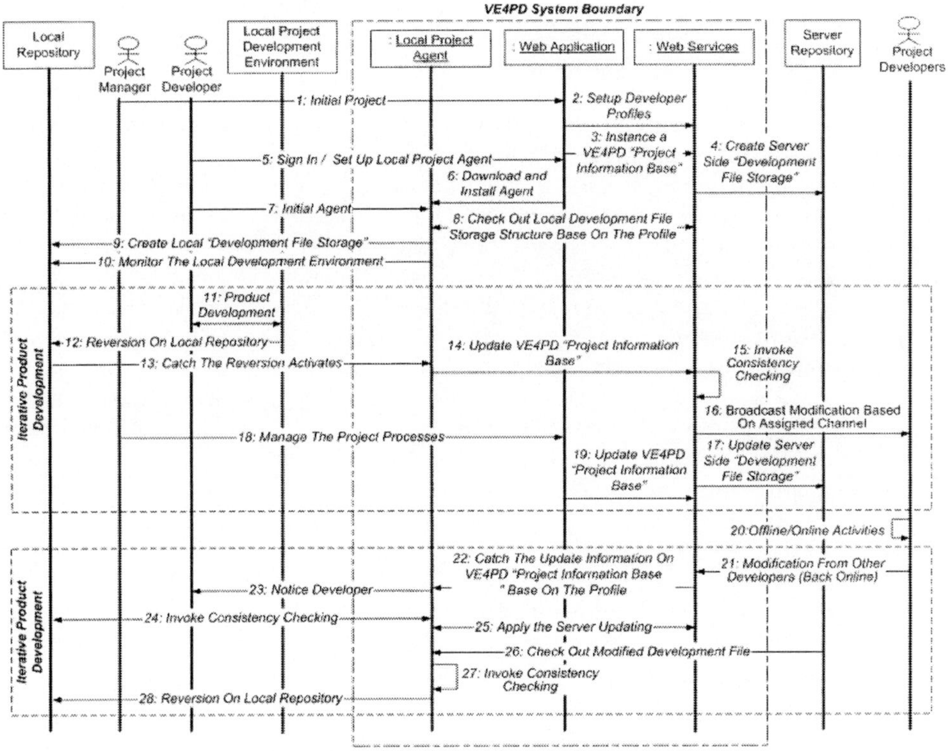

Figure 6. VE4PD System Sequence Diagram

Rather than relying on traditional check-in/check-out procedures, VE4PD takes advantage of remote local agents to achieve efficient remote synchronization of asynchronous design file updates. Since the remote agent can work as a windows service, even in the offline situation as shown in step 20 of Figure 6, the local agent will still capture and record the updated information within the local development repository. As long as the local system is connected back to the network as shown in the step 21 of Figure 6, the local agent will automatically link to VE4PD web services and invoke the consistency checking mechanism to achieve synchronization. Based on the data structure of VE4PD's Information/Knowledge base, the web service will set up a consistency channel by detecting related development files and design parameters. Then the web service will notify the appropriate remote local agents for selected developers through the consistency channel. This allows remote synchronization to be achieved and eliminates the problem of traditional check-in/check-out methods used in most PLM systems where information consistency must be controlled by the users.

Product development activities generate volumes of product data (e.g., engineering drawings, product assembly structures, etc.) and process data (e.g., partner selection, product fabrication, etc.). Streamlining this information flow through the requisite functions in the product realization process is a primary focus area for leading companies improving product development. Recent literature highlights the importance of information management in this area, however most work has focused on improving the operation of existing supply chains to reduce response times and costs for existing products rather than addressing collaboration considerations in the early design phase. The VE4PD architecture has been specifically targeted for inclusion of collaboration mechanisms to allow important CPD processes such as conflict resolution and design negotiation to be supported. We close this section considering a number of important collaboration scenarios and a brief set of derived requirements for supporting these collaboration processes.

Consider for example, when a supplier providing design services raises a serious issue with integration of their design into a higher level assembly? How will the other affected designers within the company or at other supplier sites with an interfacing component be identified and notified for conflict resolution. Or consider the process that a collaborative design team must go through when one supplier wants to change an approved design, perhaps for safety or manufacturability? How will the cost and lead time implications for other suppliers be identified and evaluated, and how will the change process be negotiated? To resolve these scenarios, a number of collaboration aspects are included in the VE4PD framework. VE4PD allows for negotiation processes with the following characteristics (a) the inclusion of detailed design information including parameters and files (b) conducted in a timely manner with synchronized data and (c) supporting cross-platform activities.

Considering these scenarios leads to a key design consideration for distributed information system architectures that support collaborative product development such as VE4PD. In general CPD information frameworks must support timely storage and retrieval of synchronized design parameters and files in a physically and functionally linked product structure. To do this, VE4PD uses agent technology and a product structure repository which enables proactive information access and notification to reduce the chance of time delay and human error in synchronizing local and shared repositories that can occur in traditional file-sharing systems. Finally, VE4PD uses web services to supports cross-platform activities and has an extensible architecture that can support storage and retrieval of design data providing the capability to handle complex relationships and business processes that occur in the CPD environment.

4. VE4PD Implementation and Case Study

In this section we present a prototype implementation and case study that illustrates and validates the proposed architecture.

4.1. VE4PD Implementation

In order to validate VE4PD, we have developed a prototype system implementation. This system includes the six components illustrated in Fig. 7(a): (1) a design parameter/file management/knowledge database is a relational database implemented with SQL Server 2000; (2) a server side repository is based on file management using the server's operating system and is supported by an FTP server; (3) a local client-side repository is based on the client's operating system file management; (4) a VE4PD web services component implements both the data management system and the server consistent monitor agent from the VE4PD architecture design. This component uses .NET Web-Services and performs most of the system's operations including building the application interface between the remote client and server database; (5) a VE4PD web server is based on ASP.NET for web page creation. (6) a VE4PD remote client is implemented using C#.NET windows forms.

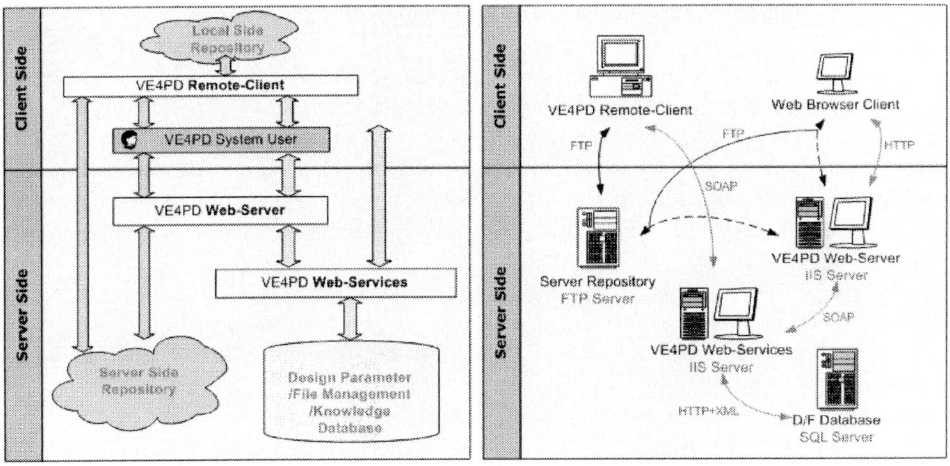

(a) *(b)*

Figure 7. VE4PD system Implementation and Deployment

Note that since the communication between the VE4PD remote client and the VE4PD web services is based on the standard SOAP protocol, other programming languages such as Java could also be used to develop a VE4PD remote client.

The implementation system can be deployed in six distributed nodes: the VE4PD Remote Client, the Web browser client, the server repository, the VE4PD Web services, the VE4PD Web server and database server, as illustrated in Fig. 7(b). If

warranted, this deployment can be simplified by allocating multiple logical nodes to one physical server.
Communication within the VE4PD prototype system is based on different protocols for various communication links as shown in Table 3.

Table 3. The Information Communication Protocols for the VE4PD

Communication	Protocol
Between System user and VE4PD Web server	HTTP
Between VE4PD Web server / VE4PD remote client and Web services	SOAP
Between VE4PD web services and the Database Server	HTTP+XML
All file transfers	FTP

4.2. VE4PD Case Study

The collaborative design of a water pump project was chosen as a case study to illustrate the design and implementation of the VE4PD system. The water pump is a small electro-mechanical system that can be decomposed into several parts for developers from different disciplines. The collaborative development process can be demonstrated by creating a set of test scenarios: one group of mechanical engineers designs the pump body, another group designs the transmission box, electrical engineers design the motor that can drive the pump through the transmission system, analysis engineers perform stress analysis of the shafts and the strength of the cover and send feedback to the designers, the manufacturing engineers evaluate the manufacturability of the components based on the domain knowledge and send feedback to all the designers, and procurement personnel are involved in purchasing components. This case can be used to explore the information and knowledge sharing among personnel from different divisions within a company as well as from different enterprises. Testing the VE4PD architecture focused on two different operating scenarios:

1. Scenario I: Using web services based functions to simulate the CPD process.
2. Scenario II: Taking advantage of remote agents to achieve an improved development environment with the capability to maintain information consistency and help with sever-client application integration.

Screen shots of the scenario implementations have been omitted due to space and print quality considerations. The authors may be contacted for screen prints or a demo. The first scenario begins with the basic structure of the product having been determined and uploaded to the VE4PD Web-Server. The development draft files and structure are initialized within an instance of the VE4PD Information/Knowledge Base. Based on this, the information and drawing files can be retrieved/download from the initialized VE4PD project. The information retrieval for this step is based on the login user's properties. In our example, a pump body is shown in a visualization window and the design parameters, design files and related knowledge of the decomposed product components are presented according to user properties. The relationships between and within the design parameters, design files and knowledge base are established using the VE4PD Information Knowledge Base infrastructure.

Finally, the Web browser client can update the design parameters and design files, and when the design files are changed, the related development concurrent operation services are invoked to check the related parameters, files and knowledge to maintain development data consistency. These operations are conducted by the VE4PD Web services and the results are sent back to the VE4PD Web server.

Scenario II introduces the agent concept and takes advantage the distributed framework of .Net to demonstrate the power of the proposed architecture. In our demonstration environment, the VE4PD remote client is built on an implementation of the remote agent and is downloadable from the server. A Windows based client has been implemented using the C# programming language. First, logging on to the VE4PD remote client assigns an information access channel. The first time launching a remote client, the user is prompted to initialize a local repository by copying the existing server side data structure. As the result of this initialization, a local repository is the created to match the server repository.

After the local repository is created, the VE4PD client starts to work as a monitor to check the remote information repository and capture any server notifications such as an information update message. The VE4PD client also detects any change on the local repository and sends a message to the server to update. As soon as a design file is renewed on the server repository, the VE4PD client will be notified and action will be taken to maintain consistency between the two repositories. When the update process is invoked, a concurrent development operation on the VE4PD Web services module will check the consistency of the file system and any related design parameters. If the update has no consistency problems, the new design file will be uploaded to the server, and the design version on the client and the server will be updated.

Finally, if information consistency problems arise, the client upload will fail based on a prototype design rule that assigns the design on the server a higher priority than designs from the clients to prevent inconsistent client updates. In this situation, the client invokes the VE4PD server for more information and two different designs of the pump body and related design parameters are illustrated. A decision support feature can then be invoked to determine how to update the server information. This approach bypasses the check-in/check-out method implemented in most PLM systems that rely on users to maintain information consistency, an approach that may not be acceptable for transacting real-time design negotiations which require timely notification of changes including duplications and conflicts to support the distributed decision support system. Using agent technology in VE4PD enables consistency to automatically be maintained by the agents. Even if the user is off-line, the client-side agent will keep track of the information and will align with the distributed decision support data. VE4PD includes the use of agents that monitor the local development repository to detect design file and parameter modifications. By combining a web service framework with agent technology the architecture provides a foundation for convenient and effective remote client platform solutions.

This case study demonstrates the VE4PD information framework for CPD. The web services framework makes the development and deployment straightforward, while the use of agent technology achieves information consistency. The proposed architecture allows for increased levels of collaboration across different product development entities and allows the design process to be effectively managed through product development phases.

5. Application Example: Intelligent Agents for Design Negotiation

In this final example we report on a distributed decision support prototype developed on the VE4PD platform that implements a Penalty Induced Negotiation (PIN) process based on multi-attribute utility theory. Details of the negotiation algorithm are described elsewhere [56]. This example is provided to validate an advanced collaboration mechanism using agent technology within the proposed framework. C# was used as the programming language and Lindo API acts as the optimization engine to solve the models. As shown in Figure 8, two types of agents are implemented:

1. **Design Agents** gather specific participant's design parameters and their related objective functions to establish/solve the local design optimization model, and
2. **Principal Agents** are implemented to monitor and control the global design constraint satisfaction and the iterative negotiation process.

The structure of the PIN Implementation consists of four parts: (1) **Principle Agent Functional Component** is a set of web services which realize the functions and optimization model of principle agent. (2) **Principle Agent User Interface** is the web portal for the principle agent. However, in some case, PA does not require any user interface and all the PA related activities are automatically controlled by PA Functional Component within the web logic layer. Two separate design agents are implemented based on an example of the design of a cylindrical pressure vessel adapted from (Martson 2000). (3) **Weight Agent** and (4) **Volume Agent** are implemented as windows applications that represent local private optimizations and interact with the Principle Agent Functional Component via its web services. The implementation approach of these two agents is identical but their optimization models and related design parameters are different. Consider a simple example, the design of a cylindrical pressure vessel, adapted from Martson 2000 to illustrate how the PIN mechanism and the implementation system work for distributed collaborative design. In order to clarify PIN implementation structure and illustrate the collaborative decision support process, an abstract-level System Sequence Diagram (SSD), represented in Unified Modeling Language (UML), is shown in Figure 9.

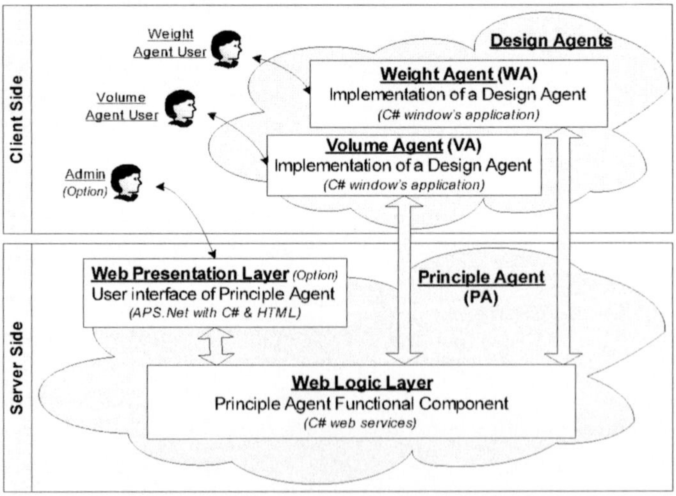

Figure 8. Structure of the PIN Implementation

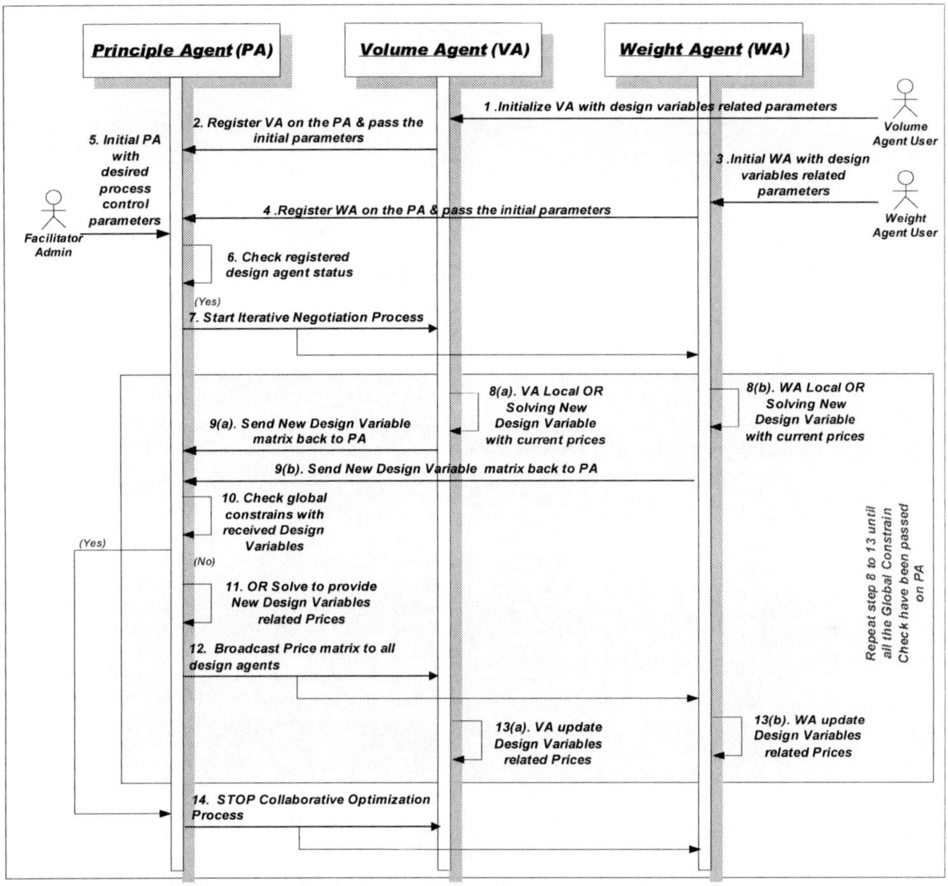

Figure 9. The PIN System Sequence Diagram

The system process is described in these steps:

Step 1: Initialization (1-7)

Step 1.1: Volume Agent initializes design variables and related parameters including initial price and minimum acceptable price for each design variable

Step 1.2: Volume Agent is registered on Principle Agent, initial parameters and utility function structure are sent to the Principle Agent

Step 1.3: Weight Agent initializes design variables and related parameters including initial price and minimum acceptable price for each design variable

Step 1.4: Weight Agent is registered on Principle Agent, initial parameters and utility function structure is sent to the Principle Agent

Step 1.5: Principle Agent collects control parameter such as the number of negotiation stages, checks the status of Volume and Weight Agent, and initiates the negotiation process.

Step 2: Negotiation (8-13)

Step 2.1: Volume Agent and Weigh Agent each determines the value of design variables based on local optimization model

Step 2.2: Volume Agent and Weigh Agent each sends the design variable information to the Principle Agent

Step 2.3: Principle Agent determines the new prices for each design variable based on the system constraint violations and announces the changed prices to Volume and Weight Agent

Step 2.4: Volume and Weight Agent each updates the design variables

Step 3: If all the global constraints are satisfied, negotiation stops (14); if the number of negotiation stages is reached, negotiation stops; otherwise, go back to Step 2.

This example demonstrates the use of intelligent agents to conduct a design negotiation process built on top of the proposed VE4PD architecture. It is easy to envision a number of agents operating in this environment that check designs for a variety of concerns such as conformance to material specifications, manufacturability constraints, environmental impacts, and so forth. Our goal in presenting this research is to stimulate further work in the direction of creating the foundation for next generation intelligent collaborative design systems built on a robust application architecture.

6. Future Trends

The development of internet based collaborative design systems is proceeding with many of the more novel as well as practical developments coming from industry. Virtual design reviews using 3-D visualization occurring with participants from across the globe are a commercial reality. Especially in the development of complex engineered systems, many disciplines must come together to evaluate and refine a system design through an iterative process. The creation of distributed application architectures on which to build intelligent, collaborative design systems for multi-discipline as well as multi-firm collaboration is a challenge that will continue being addressed in both research labs and in industry. Though promising, our approach needs further verification and validation with more case studies under various product environments. By continuing to enhance the VE4PD knowledge base we will contribute to the development of a general, reusable and standardized CPD platform.

7. Conclusion

We have proposed an information framework that addresses some of the drawbacks in current CPD systems that focuses on the needs of information consistency and client-server integration. Contributions of VE4PD includes the use of agent technology to improve consistency by proactively accessing information, and demonstration of the capability to support distributed decision support, particularly for design negotiation through an object oriented information/knowledge schema and the development of a consistency channel.

By sharing the right information to the right person at the right time, VE4PD has the potential to shorten the product development time, reduce the product development cost and increase the quality of the product. In addition, it allows for collaboration throughout the life cycle of the product.

References

[1] Boswell, B., 2005, PUTTING GLOBAL PRODUCT DEVELOPMENT TO WORK, Machine Design, 77(14), pp. 60-62.

[2] Andersen, M., Khler, S. and Lund, T. 1986, *Design for Assembly*, London, IFS Publications.

[3] Keys, K., Rao, R. and Balakrishnan, K., 1992, CONCURRENT ENGINEERING FOR CONSUMER, INDUSTRIAL PRODUCTS, AND GOVERNMENT SYSTEMS, IEEE Transactions On Components, Hybrids, and Manufacturing Technology, 15(3), 282 – 287.

[4] Shina, S., 1991, CONCURRENT ENGINEERING AND DESIGN FOR MANUFACTURE OF ELECTRONIC PRODUCTS, Van Nostrand Reinhold, New York.

[5] Zangwill, R., 1992, CONCURRENT ENGINEERING: CONCEPTS AND IMPLEMENTATION, IEEE Engineering Management Review, 20 (4), pp.40 – 52.

[6] Carter, T. and Baker, R., 1991. ACCURACY AND STABILITY OF A FINITE-ELEMENT PSEUDO-COMPRESSIBILITY CFD ALGORITHM FOR INCOMPRESSIBLE THERMAL-FLOWS, Numerical Heat Transfer Part B-Fundamentals, 20 (1), pp.1-23.

[7] Wu, T, Xie, N and Blackhurst, J, 2004, DESIGN AND IMPLEMENTATION OF A DISTRIBUTED INFORMATION SYSTEM FOR COLLABORATIVE PRODUCT DEVELOPMENT, Journal of Computing and Information Science in Engineering, 4(4), pp. 281-293.

[8] Szykman, S. and Sriram, R.D., 2001. THE ROLE OF KNOWLEDGE IN NEXT-GENERATION PRODUCT DEVELOPMENT SYSTEM, ASME Journal of Computation and Information Science in Engineering, 1(1), pp.3–11.

[9] Svensson, C. and Barfod, A., 2002, LIMITS AND OPPORTUNITIES IN MASS CUSTOMIZATION FOR "BUILD TO ORDER" SMES, Computers in Industry, 49 (1), pp. 77-89.

[10] Rupp, T.M. and Ristic, M., 2000, FINE PLANNING FOR SUPPLY CHAINS IN SEMICONDUCTOR MANUFACTURE, Journal of Materials Processing Technology, 10 (1), pp.390-397. X

[11] Grayson, P., 2001, DON'T BE LONELY AT THE TOP: BY FORMING A STRATEGIC COLLABORATION WITH SUPPLIERS OEMS CAN HAVE IT ALL, Product Design & Development, 56(6), pp.18.

[12] Pahng, F., Seninand, E. and Wallace, D., 1998, DISTRIBUTION MODELING AND EVALUATION OF PRODUCT DESIGN PROBLEMS, Computer-Aided Design, 30(6), pp.411-423.

[13] Dustdar, S., Gall, H., 2003, ARCHITECTURAL CONCERNS IN DISTRIBUTED AND MOBILE COLLABORATIVE SYSTEMS, Journal of Systems Architecture 49, pp. 457-473.

[14] Kan, H. Y., Guffy, V. G. And Chuan-Jun, S., 2001, AN INTERNET VIRTUAL REALITY COLLABORATIVE ENVIRONMENT FOR EFFECTIVE PRODUCT DESIGN, Computers in Industry 45(2), pp. 197-213.

[15] Li, W. D., 2005, A WEB-BASED SERVICE FOR DISTRIBUTED PROCESS PLANNING OPTIMIZATION, Computers in Industry, 56(3), pp. 272-288.

[16] Yang, H., Xue, D., 2003, RECENT RESEARCH ON DEVELOPING WEB-BASED MANUFACTURING SYSTEM: A REVIEW, International Journal of Production Research, 41(15), pp.3601-3629.

[17] Ray, S. R., 2002, INTEROPERABILITY STANDARDS IN THE SEMANTIC WEB, Journal of Computing and Information Science in Engineering, 2(1), pp.65-69.

[18] Biggs, S. and Smith, S., 2003, A PARADOX OF LEARNING IN PROJECT CYCLE MANAGEMENT AND THE ROLE OF ORGANIZATIONAL CULTURE, World Development, 31(10), pp.1743-1757.

[19] Macgregor, S. P., Thomson, A. I. and Juster, N. P., 2001, INFORMATION SHARING WITHIN A DISTRIBUTED, COLLABORATIVE DESIGN PROCESS: A CASE STUDY, ASME 2001 Design Engineering Technical Conference and Computers and Information in Engineering Conference. Pittsburgh, Pennsylvania, September 9-12.

[20] Rezayat, M., 2000, THE ENTERPRISE-WEB PORTAL FOR LIFE-CYCLE SUPPORT, Computer-Aided Design, 32, pp.85–96.

[21] Erik, H. and Anders, T. 2001, INFORMATION MODELING FOR SYSTEM SPECIFICATION REPRESENTATION AND DATA EXCHANGE, Proceedings of the 8th IEEE International Conference and Workshop on the Engineering of Computer-based System, pp.136-143.

[22] Amrit, T. and Balasubramanian, R., 2001, A DESIGN KNOWLEDGE MANAGEMENT SYSTEM TO SUPPORT COLLABORATIVE INFORMATION PRODUCT EVOLUTION, Decision Support Systems, 31, pp.241-262.

[23] Olivero, N. and Lunt, P. 2004, PRIVACY VERSUS WILLINGNESS TO DISCLOSE IN E-COMMERCE EXCHANGES: THE EFFECT OF RISK AWARENESS ON THE RELATIVE ROLE OF TRUST AND CONTROL, Journal of Economic Psychology, 25(2), pp.243-262.

[24] Schwartz, R. A. and Kraus, S., 2004, STABLE REPEATED STRATEGIES FOR INFORMATION EXCHANGE BETWEEN TWO AUTONOMOUS AGENTS, Artificial Intelligence, 154 (1-2), pp. 43-93.
[25] Yong, E.T., 2004, THE ROLE OF E-MARKETPLACES IN SUPPLY CHAIN MANAGEMENT, Industrial Marketing Management, 33(2), pp.97-105.
[26] Anderson, D. and Merna, T., 2003, PROJECT MANAGEMENT STRATEGY—PROJECT MANAGEMENT REPRESENTED AS A PROCESS BASED SET OF MANAGEMENT DOMAINS AND THE CONSEQUENCES FOR PROJECT MANAGEMENT STRATEGY, International Journal of Project Management, 21(6), pp.387-393
[27] Mahaney R. and Lederer, A., 2003, INFORMATION SYSTEMS PROJECT MANAGEMENT: AN AGENCY THEORY INTERPRETATION, Journal of Systems and Software, 68(1), pp.1-9.
[28] Tang, T., Winoto, P., Niu, X., 2003, I-TRUST: INVESTIGATING TRUST BETWEEN USERS AND AGENTS IN A MULTI-AGENT PORTFOLIO MANAGEMENT SYSTEM, Electronic Commerce Research and Applications, 2(4), pp.302-314.
[29] Kim, N. K., Kim, Y. and Kang, S. H., 1997, SUBDIVISION METHODS OF CONVERTING STEP INTO VRML ON WEB, Computers & Industrial Engineering, 33(3-4), pp.497-500.
[30] Hao, J. P., Yu, Y. L. and Xue, Q., 2002, A MAINTAINABILITY ANALYSIS VISUALIZATION SYSTEM AND ITS DEVELOPMENT UNDER THE AUTOCAD ENVIRONMENT, Journal of Materials Processing Technology, 129(1-3), pp.277-282.
[31] Jezernik, A. and Hren, G., 2003, A SOLUTION TO INTEGRATE COMPUTER-AIDED DESIGN (CAD) AND VIRTUAL REALITY (VR) DATABASES IN DESIGN AND MANUFACTURING PROCESSES, International Journal of Advanced Manufacturing Technology, 22 (11-12), pp.768-774.
[32] Huang.,G.Q., Huang., J. and Mak, K.L., 2000, AGENT-BASED WORKFLOW MANAGEMENT IN COLLABORATIVE PRODUCT DEVELOPMENT ON THE INTERNET, Computer-Aided Design, 32, pp.133-144.
[33] Krishnan, R., Munaga, L. and Karlapalem, K., 2002. XDOC-WFMS: A FRAMEWORK FOR DOCUMENT CENTRIC WORKFLOW MANAGEMENT SYSTEM, Lecture Notes on C.S., 2465, pp.348-362.
[34] Xu, X. W. and Liu, T., 2003, A WEB-ENABLED PDM SYSTEM IN A COLLABORATIVE DESIGN ENVIRONMENT, Robotics and Computer Integrated Manufacturing, 19, pp.315–328
[35] Tamine, O. and Dillmann, R., 2003, "KaViDo—a Web-based system for collaborative research and development processes". Computers in Industry, 52 , pp.29–45
[36] Wanga, Y. D., Shena, W. and Ghenniwab, H. 2003, "WebBlow: a Web/agent-based multidisciplinary design optimization environment". Computers in Industry, 52, pp.17–28.
[37] Roy, U. and Kodkani, S.S., 1999, PRODUCT MODELING WITHIN THE FRAMEWORK OF THE WORLD WIDE WEB, IIE Transactions, 31(7), pp. 667–677.
[38] PRODNET, http://www.uninova.pt/~prodnet/
[39] Camarinha-Matos, L.M. and Lima, C.P., 1998, A FRAMEWORK FOR COOPERATION IN VIRTUAL ENTERPRISE, Proceedings of DIISM'98 - Design of Information Infrastructures Systems for Manufacturing, Fort Worth, Texas.
[40] Garita, C., Afsarmanesh, H. and Hertzberger, L.O., 1999, THE PRODNET COOPERATIVE INFORMATION MANAGEMENT FOR INDUSTRIAL VIRTUAL ENTERPRISES, Journal of Intelligent Manufacturing, 12(2), pp.151-170.
[41] Gerhard, J. F., Rosen, D., Allen, J. K. and Mistree, F., 2001, A DISTRIBUTED PRODUCT REALIZATION ENVIRONMENT FOR DESIGN AND MANUFACTURING, Journal of Computing and Information Science in Engineering, 1(3), pp.235-244.
[42] Urban, S. D., Dietrich, S. W., Saxena, A. and Sundermier. A., 2001, INTERCONNECTION OF DISTRIBUTED COMPONENTS: AN OVERVIEW OF CURRENT MIDDLEWARE SOLUTIONS, Journal of Computing and Information Science in Engineering, 1(1), pp.23-31.
[43] Eclipse Research Community, http://www.eclipse.org/technology/research.html.
[44] Lüer, C., 2002, EVALUATING THE ECLIPSE PLATFORM AS A COMPOSITION ENVIRONMENT, 3rd International Workshop on Adoption-Centric Software Engineering, Portland, Oregon.
[45] Tanenbaum, A. S. and Steen, M., 2002, DISTRIBUTED SYSTEMS - PRINCIPLES AND PARADIGMS. Upper Saddle River, NJ, Prentice Hall, pp.50-52.
[46] E2open Software, http://www.e2open.com
[47] Agile PLM Platform, http://www.agile.com
[48] MatixOne ,2005, THE MATRIX PLM PLATFORM, Datasheet, http://www.matrixone.com/pdf/ds_prod_plmplatfom.pdf

[49] Ebbesmeyer, P., Gausemeier, J., Krumm, H., Molt, T. and Gruß, T., 2001, VIRTUAL WEB PLANT: AN INTERNET-BASEDPLANT ENGINEERING INFORMATION SYSTEM, Journal of Computing and Information Science in Engineering, 1, pp.257-261.
[50] Burkett, J., Kemmeter, J, and O'Marah, K., 2002, PRODUCT LIFECYCLE MANAGEMENT: WHAT'S REAL NOW, AMR Research Report, http://www.amrresearch.com/Content/View.asp?pmillid=14830&docid=650
[51] Taner, B. and Dennis, R., 1997, PRODUCT DATA MANAGEMENT SYSTEM: STATE OF THE ART AND THE FUTURE, ASME Design Engineering Conference, September 14-17, 1997, Sacramento, California.
[52] Nan, X., 2003, IMPLEMENTATION OF DISTRIBUTED INFORMATION SYSTEM FOR CPD, Technical Report, VCIE lab, Arizona State University.
[53] Rezayat, M., 2000, KNOWLEDGE-BASED PRODUCT DEVELOPMENT USING XML AND KCS, Computer-Aided Design, 32, pp.299–309.
[54] Lubell, J., Peak, R., Srinivasan, V. and Waterbury, S., 2004, STEP, XML, AND UML: COMPLEMENTARY TECHNOLOGIES, Proceedings of ASME 2004 Design Engineering Technical Conference, Utah.
[55] AMR Research Vendor Profile - PTC, 2002, http://www.ptc.com/WCMS/files/30579en_file1.pdf
[56] Ganguly, S., Wu, T. Xie, N. Blackhurst, J, and Zha, X. F., 2005, A DISTRIBUTED INFORMATION SYSTEM FRAMEWORK FOR COLLABORATIVE PRODUCT DESIGN NEGOTIATION, ASME Transactions: Journal of Computing & Information Science in Engineering, Special Issue: Computer-Supported Collaborative Product Development (to appear).

Integrated Intelligent Systems for Engineering Design
X.F. Zha and R.J. Howlett (Eds.)
IOS Press, 2006

Towards an Evolvable Engineering Design Framework for Interactive Computer Design Support of Mechatronic Systems

Zhun FAN[a], Mogens ANDREASEN[a], Jiachuan WANG[b], Erik GOODMAN[c], and Lars HEIN[a]

[a] *Department of Mechanical Engineering, Technical University of Denmark, Denmark*
[b] *United Technologies Research Center, Systems Department, USA*
[c] *Michigan State University, Department of Electrical and Computer Engineering, USA*

Abstract. The paper proposes an integrated evolutionary engineering design framework that integrates the chromosome model in the domain theory, the evolutionary design, and human interaction. The evolvable chromosome model helps the designer to improve creativity in the design process, suggesting them with unconventional design concepts, and preventing them from looking for solutions only in a reduced solution space. The systematic analytical process to obtain a chromosome model before running evolutionary design algorithms also helps the designer to have a complete view of design requirements and intentions. Human interaction is integrated to the framework due to the complex and dynamic nature of engineering design. It also helps the designer to accumulate design knowledge and form a design knowledge base. An example of vibration absorber design for a typewriter demonstrates its feasibility.

Keywords. chromosome model, evolutionary synthesis, interactive evolutionary computation, domain theory, morphology

1. Introduction

Application of evolutionary computation (EC) as a tool for design search, exploration and optimization has achieved rapid progress in both academy and industry during the past ten years. EC as optimal information gatherers in designing and decision-making has the following major advantages: (1) it is a good global optimization tool, capable of processing both discrete and continuous variables, and is not sensitive to the shape of the landscape of the objective function. (2) As a population-based approach, it has strong capability to search and explore the design space, and be able to find innovative design candidates which may not be obvious to human designers' intuitions. (3) In a well-designed framework, human designers can interact with the EC in the designing process to gather information and thus foster design insights.

Research of developing a framework of an interactive evolutionary design system (IEDS) is very promising because EC can provide the level of search and exploration required across the very ill-defined, uncertain problem spaces that most likely involve multiple objectives, constraint and high modality. While much current research focus on human evaluation of the fitness of solutions generated from evolutionary search where quantitative evaluation is difficult or impossible to achieve, this paper aims to encompass a larger spectrum of the information generated during a design process ranging from design specifications to detailed information like geometry models in the IEDS. For instance, in traditional function based evolutionary design synthesis, the encoding of the design is decided by the designer, and no formal process of encoding is provided in most research in this line. As a result of limitations in the encodings typically used, evolutionary design synthesis often produces only small-scale designs – that is, relatively few components are used. And because they fail to cover the full spectrum of user needs, and frequently do not consider manufacturing processes, costs and constraints, the design results are often not very useful to industry and only end up as 'new ideas' generated to satisfy academic curiosity. This paper tries to extend the algorithmic design process of traditional evolutionary synthesis approaches, integrating the product development process and broader customer needs. In particular, we made the first attempt to integrate the IEDS with a product modeling framework, a chromosome model based on the domain theory described in [1]. From the viewpoint of the chromosome model, this approach is an extension of the model towards an evolvable chromosome model within the domain theory.

An evolvable chromosome model is reasonable because designs are intrinsically evolutionary and no design should be static. The evolvable chromosome model can also facilitate computer-aided conceptual design in an interactive evolutionary design system, thus pushing the research in functional-based evolutionary design synthesis a further step towards industrial-oriented applications.

The remainder of the paper is organized as follows. Section 2 introduces the research of evolutionary engineering design and discusses some important considerations in applying EC in the engineering design process. Section 3 provides some basics of the chromosome model as a product modeling framework. Section 4 discusses the integration of the chromosome model with evolutionary engineering design. A case study of vibration damper design for a typewriter system is discussed in section 5 to explain the utilization of the IEDS proposed in this paper. Conclusions and discussions of some future research directions are provided in section 6.

2. Evolutionary engineering design

Highly automated function-based synthesis methods have emerged in recent years [2]. Among them approaches based on evolutionary computation (EC) appear to be one of the most promising groups. During the past ten years, the effort of integrating evolutionary computation with engineering design has rapidly increased, taking advantage of EC's search ability to explore the design space. Many important research advances and results of evolutionary engineering design have been reported, including [2]-[8].

2.1. Generation of morphology using evolutionary approaches

In the design community, one of the most powerful systematic methods for creating conceptual designs is Morphology. The core idea of the Morphology Method is that there exist sets of important characteristics that are believed to be common in all desired solutions. Each characteristic can be varied and a certain number of alternative solutions for satisfying the characteristic can be established. Then if we can identify that set of characteristics, any combination of each sub-solution will be a potential solution or design candidate.

EC refers to a class of general-purpose search algorithms based on (admittedly very incomplete) abstraction of principles of biological evolution and natural selection. These algorithms implement biologically inspired computations that manipulate a population of candidate solutions (the "parents") to generate new variations (the "offspring"). At each step (or "generation") in the computation, some of the less promising candidates in the population are discarded and replaced by new candidates ("survival of the fittest").

In summary, EC is very relevant to the core principles of design methodology, namely, to create several concepts, and to select the best one based upon criteria that mirror what is believed to represent high quality in a solution. Design models based on evolutionary computation start, like the Morphology Method, with generating a population of design concepts candidates, and then according to evaluation criteria set forth by the designer, each design candidate in the population will be evaluated and assigned a value that represents its 'goodness' / fitness for the design. With this, EC uses certain mechanisms, such as crossover and mutation that are analogous to mechanisms Nature uses to evolve its creatures, to gradually evolve / reconfigure the population of design candidates so that in each offspring generation, the population of design concepts as a whole is superior to its parental generation, according to evaluation criteria set forth by the designer. In this way, the population of design concepts is guided towards better designs in each generation, and after a number of generations of improvement / evolution it will converge to one or a set of 'good' candidates that the designer can select from or use to make further trade offs. Figure 1 shows an overflow of the design process based on evolutionary computation.

2.2. Topology exploring and parameter optimization in engineering design

Two most widely used types of evolutionary computation techniques include genetic algorithm (GA) and genetic programming (GP).

GA is a very simple, straightforward, yet a powerful approach for global search of the parameter design space [6]. GA usually represents/encodes an individual design candidate with a string that concatenates parameters (both real and binary) considered important in the design. The design topology, in most cases, is fixed, so that the length of the string for each individual is the same, facilitating the crossover and mutation operations towards strings.

Genetic programming is an extension of the genetic algorithm, and it uses evolution to optimize actual computer programs or algorithms to solve some task, typically involving a tree-type (or other variable-length) representation, thus lending itself very well to explore topology design space [7]. Because GP (genetic programming) can manipulate variable-sized strings, it is especially useful for representing developmental processes and processing topological information. Most design methods based on GP require a preliminary design, or a design embryo, which need not contain all of the necessary components, or the necessary number of components, but only enough information to allow specifying the behaviors desired of the system (defining objectives and variables constrained, for example).

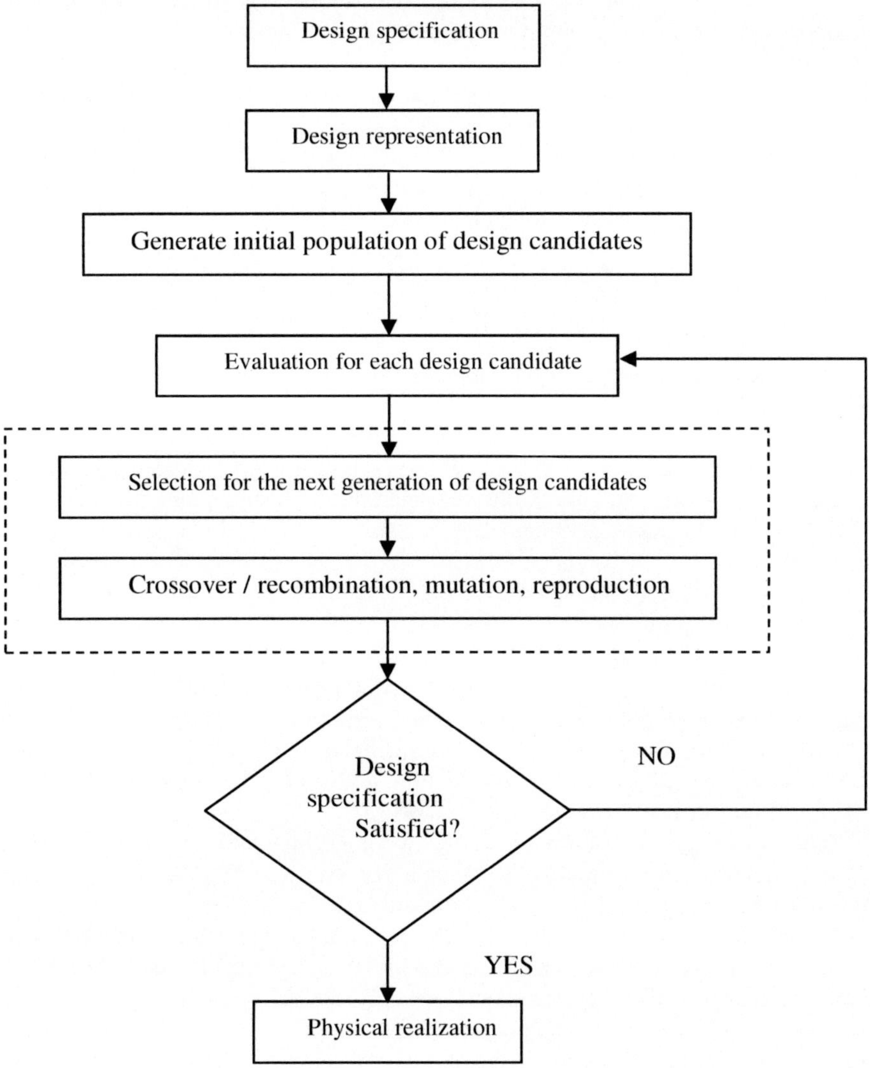

Figure 1. An overflow of the evolutionary engineering design process

Genetic algorithm is widely used to optimize parameters, but lacks the ability in exploring topology search space. Compared with genetic algorithm, genetic programming is an intrinsically strong tool in open-ended topological exploration, because the tree structure of the GP chromosome is flexible in generation and reconfiguration, with constraints of maximum depth and maximum nodes only imposed by practical implementation considerations. In addition, functions used in GP, rather than rules used in specially designed GA, allow the designer to explore design regions (in the whole design space) with which he or she is not familiar. In practice, GP and GA may be used together in an evolutionary design system to explore both the topology and paramter design space.

It is important to point out that in the conventional design environment, designers' decision-making is biased by both the capabilities of simulation tools and the designer's experience and intuition [4]. It is hard for the designer to make an "imaginative jump or creative leap" from one design candidate to another. But design tools based on evolutionary approaches can free designers from this kind of "design fixation" and the limitations of conventional wisdom, allowing them to explore a huge number of possible candidates for a design problem, and increasingly, the probability to discover novel designs uncharted before by human exploration.

3. Chromosome model

Unfortunately, despite the significance of research break-throughs in academia, the reported results of evolutionary engineering design are still not ready to be used widely in industry because they cannot link user aspects and design intent to a structural product model, so they only cover part of our current function vocabulary and/or design process, and generally lack the ability to generate realizable geometries and structural topologies. As a consequence, the identification of important characteristics is also an ad-hoc process. It is highly recommended that a richer representation language adding process, function, organ and geometric issues of product should be used to 'spell' the product so that storage and reuse of design knowledge becomes possible. A complete product definition is necessary in supporting engineering designers in their design activities. According to [9], four attributes are relevant in defining a product: characteristics, inherent properties, relational properties, and qualities.

Characteristics are a class of design attributes that the designers can determine directly during design. They may include structure (both behavioral and physical), form, dimension, surface quality, material and so on.

Inherent properties describe the behaviors of a design and can be determined by the design characteristics and the environment. They can also be determined at high-level behavorial models by the designers in a top down design process.

Relational properties are design attributes that describe the behavior of the so-called meetings between the design and the life phase system. Relational properties are causal determined by the characteristics of the design, the life phase system characteristics, and the meeting characteristics. Examples on relational properties are costs, throughput time, flexibility, etc.

Quality, meaning pride of ownership, can also be considered as the stakeholder's reactions on inherent and relational properties. Determining quality requires a person observing and reacting, and there is no causality between properties and quality.

But how can these attributes be identified? A very promising approach is the so-called chromosome model, which has a structure in accordance with the domain theory [1]. The basic idea is to model the product from four hierarchical viewpoints, based on strict identification of structural and behavioral aspects of a product:

- A process view, with a structure of activities related to the product, for example the use process, the product life cycle etc. In this viewpoint, to understand how the transformation of materials, energy, and information of the product are related to their use or functions is central.

- A functional view, with a structure of the desired functions or effects. These functions must be able to facilitate the necessary transformations.
- An organ view, with a structure of functional carriers or solutions which create the desired functions or effects of the product. The result of design considerations is an organ structure.
- A part view, with a structure of parts and their assembly relations. By determine materials, form, tolerances and surface quality of each part and relations between the parts, the necessary conditions for the organs and their functionality are created.

The graphical representation of the chromosome model is show in Figure 2.

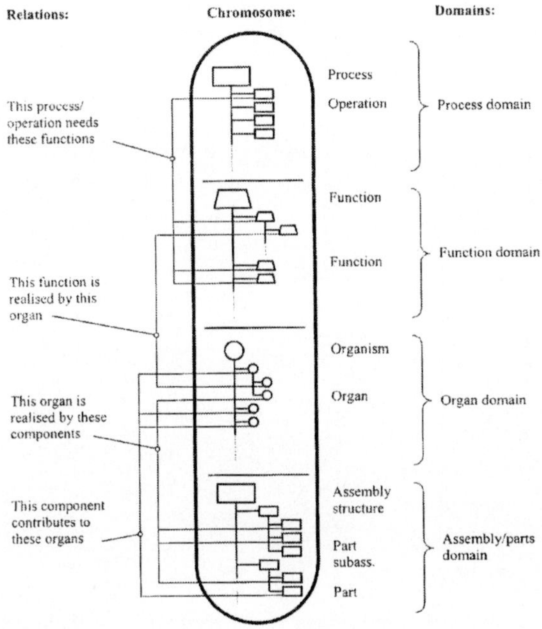

Figure 2. The chromosome product model, adapted from [1]

According to the theory of technical systems the design can be modelled from two constitutive viewpoints: organ and part viewpoints. The two constitutive viewpoints are necessary for explaining the behavior of a design and the physical realization.

The other two viewpoints, process view and functional view, provide a systematic way of analysing the design requirements and intentions and relate them to the organ and part viewpoints. The chromosome model links the structures with causal relationships: the processes determine the functions, the functions are created by the organs, and the organs are materialized by the components. The causal relationship constituting the genetic information of the system makes the chromosome model well suited for developing a design history system based on a tool called Function-Means-tree. An example of airbag application using Function-Means-tree can be found in [10]. In summary, the chromosome model provides an extended and hierarchical view of product configuration that relates a substantial part of all the data, information and knowledge about the product to the product model. It offers a general framework of modeling the various aspects of technical systems and is also strongly linked to design methodology. This view can be used to extend previous research in evolutionary engineering design that only encoded the part domain or perhaps organ domain of the chromosome model.

4. Integrated evolutionary engineering design

4.1. Integrating chromosome model with evolutionary design

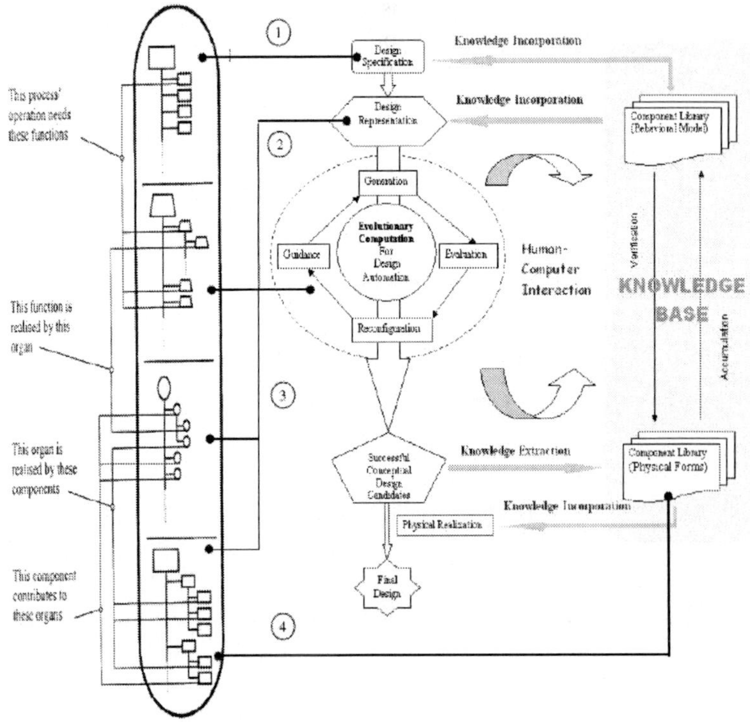

(1) The design specification motivates analysis in the process domain of the chromosome model. This will then in turn decide the functions needed to realize the process in the function domain, and the organs needed to realize the functions in the organ domain, and components that contribute to the organs in the part domain.
(2) The design representation is collectively determined by the decisions made in the function domain, the organ domain, and the part domain.
(3) The function domain, the organ domain, and the part domain of the chromosome model can all influence how evolutionary design may be carried out, e.g. in the stage of encoding design characteristics, and defining proper EC operators like GP functions.
(4) Parts domain relates to the physical forms of the component library used in the design.

Figure 3. The integrated evolutionary engineering design framework with an evolvable chromosome model

To record and reuse a large spectrum of the information generated during a evolutionary design process, the framework of Integrated Evolutionary Design System (IEDS) is developed and shown in Figure 3. The overall procedure of IEDS starts with the design specification, including design objectives, design constraints, and design preferences, etc. The design specification motivates analysis in the process domain of the chromosome model. This will then decide the functions needed to realize the process in the function domain, and the organs needed to realize the functions in the organ domain, and components that contribute to the organs in the part domain. With a complete chromosome model, we can make a decision on the how to represent design candidates. It is important to point out that in many engineering design cases, a design need to be represented in multiple levels of abstractions. The different levels of abstractions may correspond to the part domain, or the organ domain, or even the function domain in the chromosome model.

After we have design representations, we can move on to the next step of the Integrated Evolutionary Design framework – to run evolutionary computation for design automation. A preparatory step for this includes several more issues to be determined, e.g. the encoding of design characteristics, and defining of proper EC operators like GP functions if needed. Again, they will be collectively determined by previous decisions we made on part domain, organ domain, and even function domain.

The automated design loop of evolutionary computation includes four steps of generation, evaluation, reconfiguration, and guidance. Design candidates of engineering systems are often represented in several levels of abstractions [10] in the conceptual design level. The automated design loop of evolutionary computation may also take place in different cycles, leading to design results in corresponding representations. After the step of conceptual design, we move to the following step of detailed design. In this step, physical realization transforms the conceptual design to its final physical structure according to the physical forms of the component library used in the design, i.e. decided in the part domain of the chromosome model. It is noted that physical realization may be a comprehensive procedure itself.

4.2. Integrating human interaction with evolutionary design

Due to the characteristics of uncertainty, multi-objectives, severe constraints and high-modality related to real world designs, it is almost impossible for EC to evolve strictly realizable designs in an efficient manner if we use EC merely as a set-and-run tool. EC can perform much better to play a supporting role to enhance design insight and assist decision-making, rather than to act merely as a terminal optimizer that gives customers a final result [5] [8].

It is hoped that a design knowledge base can be created in this interactive evolutionary design process. For example, the multiple diversity solutions of engineering designs obtained through evolutionary computation can provide valuable information to the user to foster a better insight of the problem domain and help to identify best direction for future investigation. In addition, the knowledge acquired in the process may assist the designer to refine design objectives and modify design representations. In the process, Knowledge incorporation and knowledge extraction are two major forms of knowledge interaction.

In summary, human-computer interactions may happen in various forms that include but are not limited to the following aspects.

Specification of design objective and constraints

Specifications of design objectives and constraints are the input to the interactive evolutionary design framework. They are provided by the user at the beginning stage of the design, and specified at the process domain of the chromosome model. It is obvious that specifying design objectives is a process that incorporates domain knowledge and human preferences. What is more, because many assumptions about objectives and constraints may not be correct at the beginning, and are subject to changes such as the market condition, human's interactions with the computer are desired to reflect the corrections and modifications.

Identification and encoding of significant design features

Encoding is a critical step in the designing of an evolutionary algorithm. Generally speaking, designers should be able to identify significant design features and encode them in the chromosome of design candidates, so that they can be evolved in the running process of the evolutionary algorithm. In different stages of the design, the significant design features that the designers concern may change. For instance, in the early conceptual design phase, the significant features may be represented as functional building blocks or organs. But in a later detailed design stage, the significant features may well be the dimensionalities in the parts domain. How to identify and encode the significant design features in the design process obviously needs human designer's involvement and interaction with the evolutionary design engine.

Design of EC operators

Take the definition of GP functions for example. Executing the GP tree can accomplish the collective tasks that the user embedded in the functions. All functions in GP tree belong to a function set. Designing the function set is therefore one of the most significant steps in setting up GP run. Because most functions in the function set deal with the configuration of building blocks/organs of the design, in practice, it is important to first decide the selection of building blocks, or the component library, of the proposed design. Although the component library should be decided in the chromosome model, at organ domain and part domain respectively, it happens often that the designer finds it necessary to change the contents of the organ and component library in the design cycle. In this case, the designer's expertise knowledge should be incorporated into the chromosome model and accordingly modify the evolutionary design process in an interactive manner.

Definition and modification of design evaluation

Design evaluation involves defining an objective or fitness function against which each design candidate is tested for suitability for matching the design specifications under various design constraints. Because the purpose of engineering design is to make products for a changing world, engineering design is an interactive process of integrating new information, new technologies and new biases from the marketplace. As a result, the fitness function should be able to take feedback from designers and customers constantly, enabling it to reflect changing market environments or user preferences.

It is speculated that with the above procedures to implement an interactive evolutionary design system, a unique, rich knowledge system that can not only gather and relate data, but also process and evolve data will be created. In addition, the chromosome model in this knowledge system is now not a static one, but more a dynamic one that can evolve during the design process, continuously adapting to changing environments including market demands, customer preferences, or technology advances, etc. In this way the designers and the computer can work as a symbiotic, interacting team to tackle design problems.

5. An example

We are going to use an example of typewriter redesign to illustrate the Integrated Evolutionary Design method.

The problem was presented by C. Denny and W. Oates of IBM, Lexington, KY, in 1972. The original typewrite system includes electric voltage source, motor and mechanical parts (see Figure 4). The problem with the design is the position output of the load has intense vibrations (see Figure 5). The design specification is to reduce the vibration of the load to an acceptable level, given certain command conditions for rotational position. In particular, we want the settling time to be less than 70ms when the input voltage is stepped from zero to one.

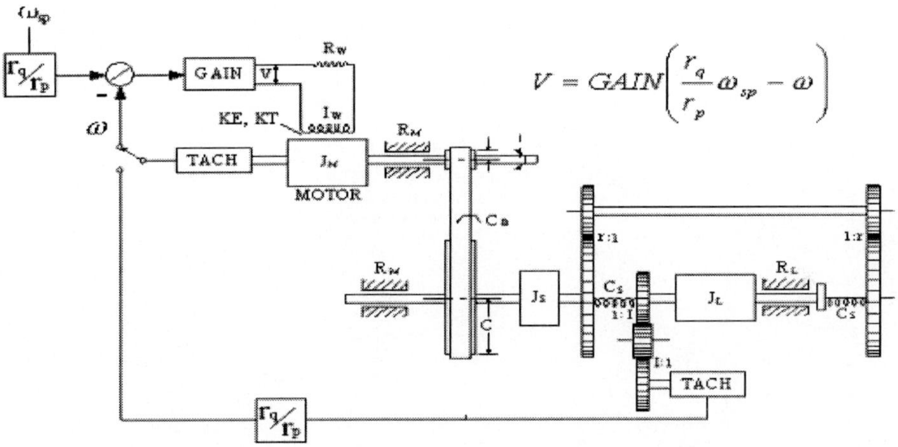

Figure 4. Schematic of the typewriter drive system

Given the design specification, we use a Function-Means-tree to analyze the system to decide the organs we are going to use in the IEDS. The Function-Means-tree is shown in Figure 6, and explained as the following.

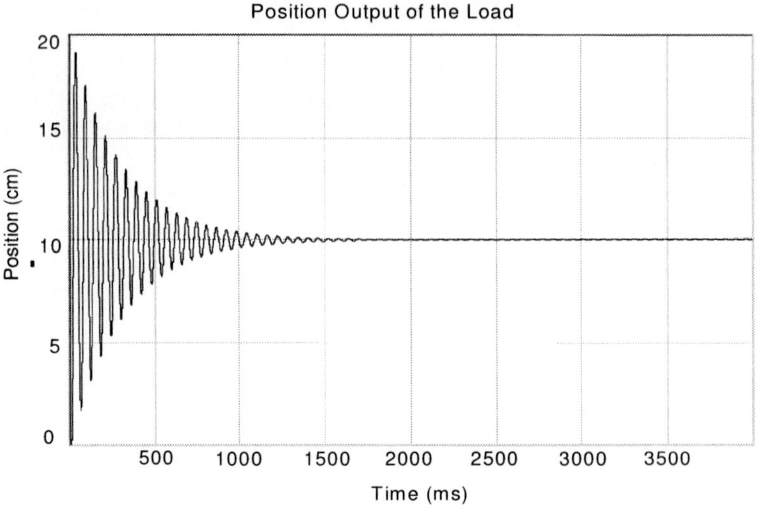

Figure 5. Positional vibration of the load

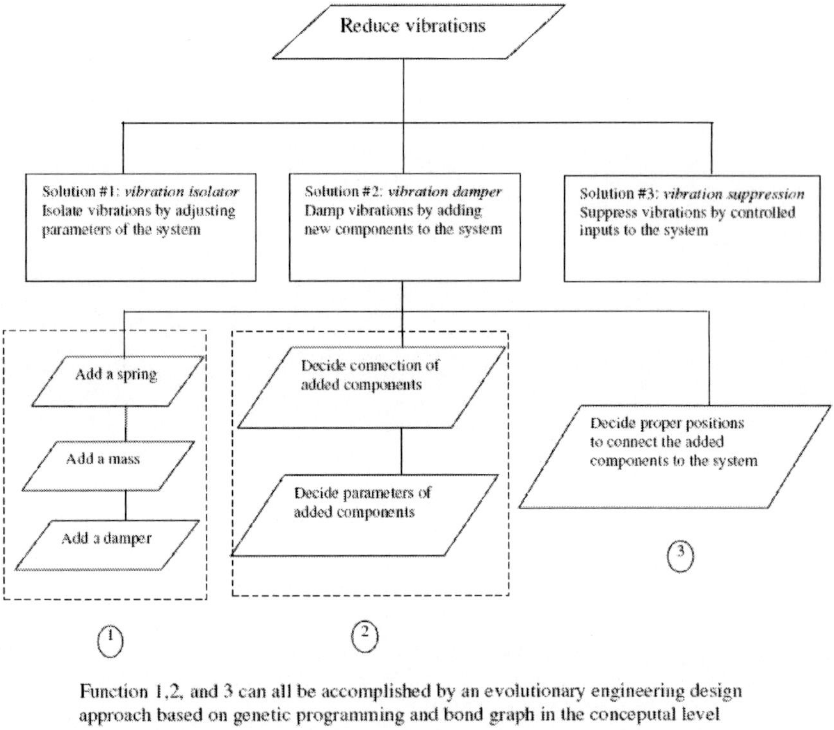

Figure 6. A Function-Means tree for vibration damper of the typewriter system

To reduce vibrations, there are basically three ways: vibration isolation, vibration absorber / damper, or active vibration suppression.

Vibration isolation reduces vibrations through adjusting parameters (e.g. the stiffness and damping) of the existing system to cause its vibration response to behave in a desired fashion. After analysis, we decided not to take this method because the original design has fixed the materials so that it is difficult to change the mass and stiffness of the system more than a few percent.

Active vibration suppression uses an external adjustable (or active) device, called an actuator (e.g. a hydraulic piston, a piezoelectric device, and an electric motor) to provide a force to the device, structure, or machine whose vibration properties are to be changed. Again, it is not attractive to us because the added components of actuators and sensors may be too expensive for a typewrite design.

Therefore, we are left with the choice of vibration absorber / damper, which basically inserts new components to take vibrations from the primary system that is to be protected from the vibrations.

The types of organs that we can use to add to the system include a spring, a damper, and a mass. They can be considered as the functional building blocks of the evolutionary algorithm used in IEDS. Because we not only need to decide their parameters, but also need to determine the topological configurations of them, genetic programming is a suitable selection in the toolbox of evolutionary algorithms.

We therefore need to identify a proper embryo for genetic programming. By analyzing the model, we conclude that the critical part for the design is a subsystem that involves the drive shaft and the load (see Figure 7). The input is the driving torque, T_d, generated through the belt coupling back to the motor. This subsystem was deemed a logical place to begin the design problem.

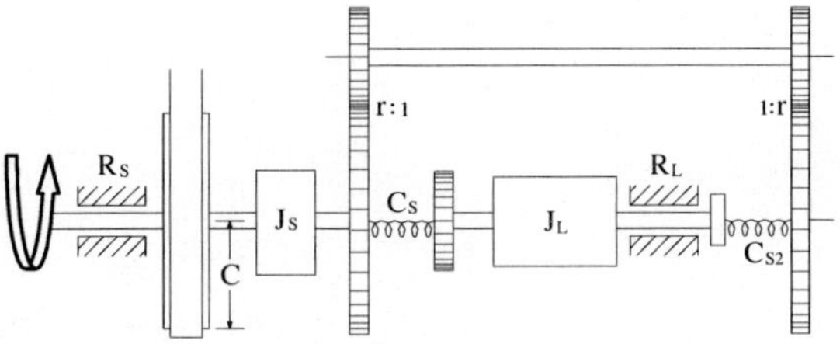

Figure 7. The focused subsystem

Then several questions need to be answered, e.g. what types of components need to be inserted, how should we connect the components before inserting them to the primary system, at which positions should the components be inserted, and how to decide parameters for the components, etc. Because our IEDS framework specially designed for mechatronic systems uses a search engine that combines both genetic programming and bond graphs to explore the design space of mechatronic systems, we represent the focused subsystem using a bond graph model as shown in Figure 8.

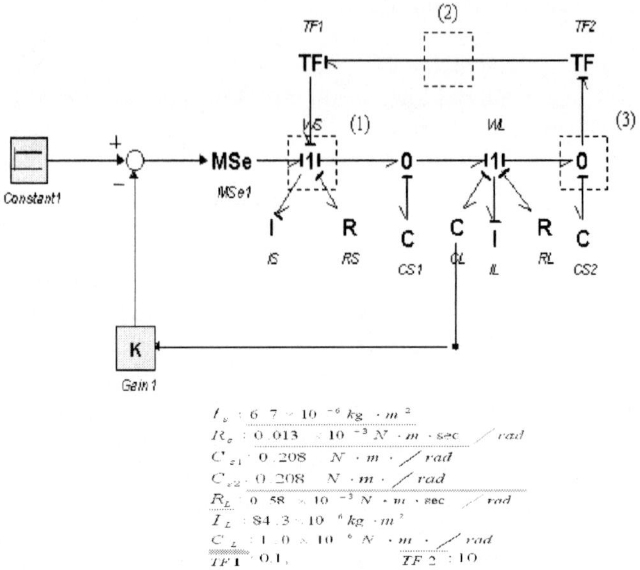

Figure 8. the bond graph model of the focused subsystem with modifiable sites specified

Given several modifiable sites that indicate potential locations that the added components may be inserted, the evolutionary design approach described in this research can answer the above questions in a highly automated manner and suggest several designs (with different topologies) that damp the vibrations successfully. Two competing design candidates with different topologies, as well as their performances, are provided in Figure 9 to Figure 14 (evolved components are circled).

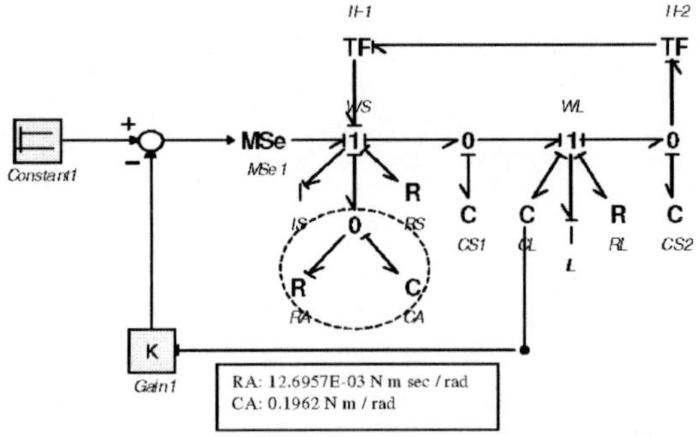

Figure 9. The evolved bond graph model I

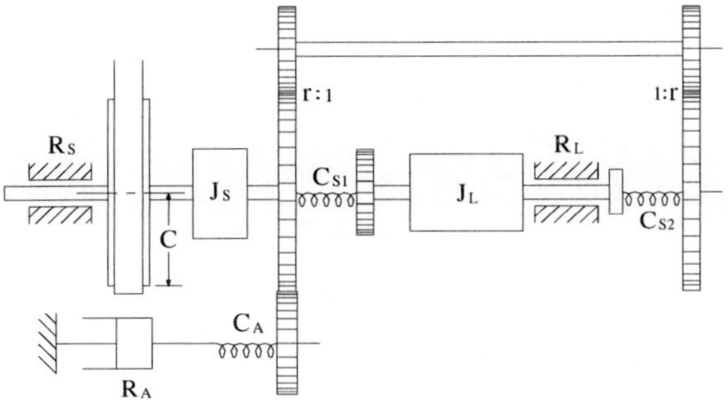

Figure 10. The physical realization of evolved bond graph model I

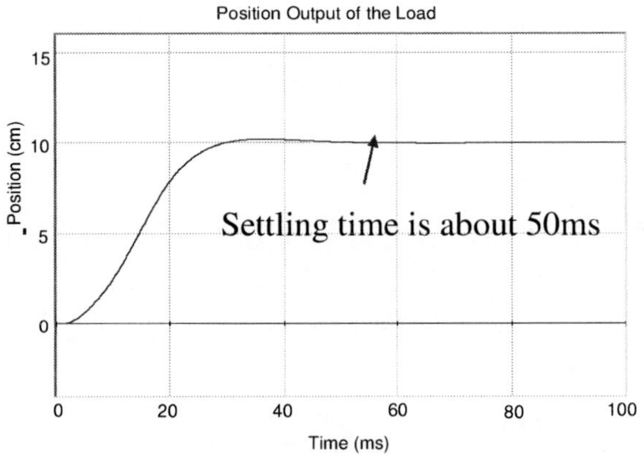

Figure 11. Simulation result of evolved bond graph model I

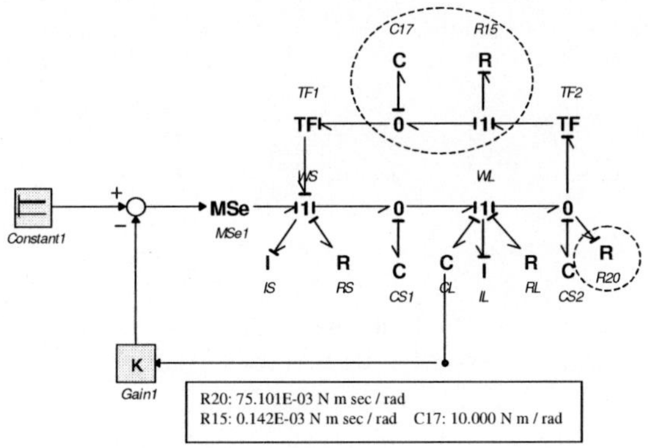

Figure 12. The evolved bond graph model II

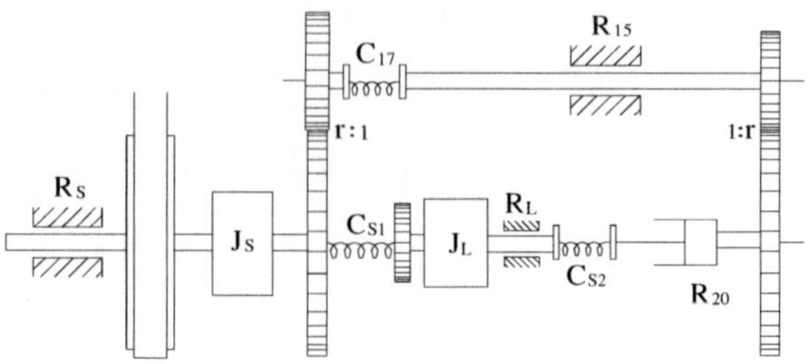

Figure 13. The physical realization of evolved bond graph model II

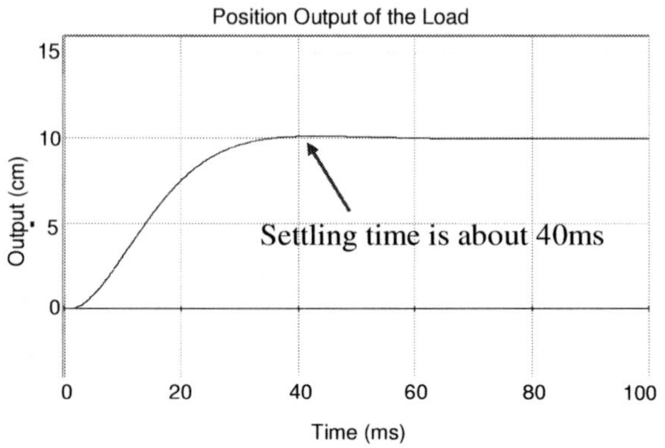

Figure 14. Simulation result of evolved bond graph model II

One of the designs has three elements, one each of 0-junction, C, and R, added to modifiable site 1 of the focused subsystem model (Figure 9). Dashed circles highlight the newly evolved components in the bond graph figures. The performance of this model is shown in Figure 11. The position response for step function input quickly converges in about 50msec, which was an acceptable timeframe. One possible physical realization of the bond graphs model is shown in Figure 10. A spring and a damper are added and coupled to the original printer subsystem as shown in Figure 10.

Another design is shown in Figure 12. Four elements, 0-junction with C, 1-junction with R are added to modifiable site 2 and one R is added to modifiable site 3. One possible physical realization of the design is shown in Figure 13. Figure 14 displays the performance of this model.

We can see from the output rotational position responses that they all satisfy the design specification of settling time less than 70ms. Note that the time scale of the plots is 100 ms. Now the designer has the possibility to make a trade-off decision – if fewer inserted components is preferred, a decision can be made to select design candidate I; if shorter settling time is a higher priority, design candidate II may be the choice.

As an example, we selected candidate I and made a final physical realization for it (in the parts domain). Figure 15 shows the inserted components and their physical realization for design candidate I.

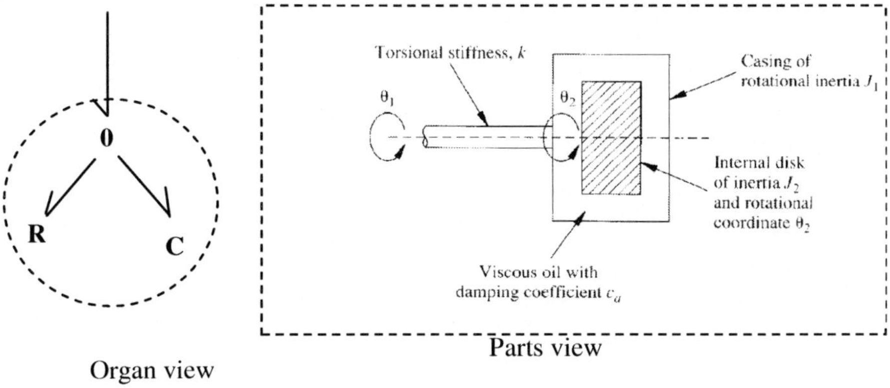

Figure 15. the inserted components and their physical realization

6. Conclusions

The paper proposes an integrated evolutionary engineering design framework that integrates the chromosome model in the domain theory, the evolutionary design, and human interaction. In this framework, the chromosome model is not a static one, but dynamic and evolvable with the help of evolutionary design process. This evolvable chromosome model helps the designer to improve creativity in the design process, suggesting them with unconventional design concepts, and preventing them from looking for solutions only in a reduced solution space. The systematic analytical process to obtain a chromosome model before running evolutionary design algorithms also helps the designer to have a complete view of design requirements and intentions. By taking advantage of genetic programming as a search method for competent designs and the bond graph as a representation for mechatronic systems, we have created a design environment in which open-ended topological search can be accomplished in a semi-automated and efficient manner and the design process thereby facilitated. By incorporating specific design considerations the method can be used to explore design space of special types of mechatronic systems. Human interaction can be integrated to the framework due to the complex and dynamic nature of engineering design. It also helps the designer to accumulate design knowledge and form a design knowledge base.

Although a simple example of vibration damper design for a typewriter system demonstrates its feasibility, many issues still need to be addressed before the framework can be utilized widely in industry. For instance, an extended implementation of the chromosome model based on Function-Means-tree reasoning can be developed so that it can integrate the evolutionary design framework in a more complete software environment. With a better record of the design knowledge and design history, reuse of design knowledge can be enhanced by techniques such as information extraction and data visualization.

References

[1] Andreasen M. M., "The Role of Artefact Theories in Design", in Universal Design Theory, H. Grabowski, S. Rude, G. Grein (ed.), 1998, pp. 47 – 56

[2] Antosson E. K. and Cagan J. (ed.), "Formal Engineering Design Synthesis", Cambridge University Press, 2001

[3] Gero J. S., "Computers and Creative Design,", in M. Tan and R. Teh (eds), The Global Design Studio, National University of Singarpo, 1996, pp. 11-19

[4] Clement S., Jordan A., Vajna S., "The Autogenetic Design Theory – an Evolutionary View of the Design Process", International Conference on Engineering Design, ICED 03, Stockholm, August 19 – 21, 2003

[5] Parmee I. C. (ed.), "Engineering Optimization, Special Issue on Evolutionary Search, Exploration and Optimization for Engineering Design", Taylor & Francis Group, Vol 36, No.2, 2004

[6] Deb K. , "Optimization For Engineering Design: Algorithms and Examples", Prentice Hall of India, 2004

[7] Koza J, "Genetic Programming: On the Programming of Computers by Means of Natural Selection", The MIT Press, 1992

[8] Kamalian, R.H., Agogino, A.M., and Takagi, H., "The Role Of Constraints and Human Interaction in Evolving MEMS Designs: Microresonator Case Study", Proceedings of DETC'04, ASME 2004 Design Engineering Technical Conference, Design Automation track, Paper # DETC2004-57462, CD ROM, ISBN # I710CD.

[9] Mortensen, N. H., "Design Modelling in a Designer's Workbench – Contribution to a Design Language", dissertation, Department of Control and Engineering Design, Technical University of Denmark, 1999

[10] Summers J., Hernandez N., Zhao Z., Shah J., Lacroix Z., "Comparative Study of Representation Structures for Modeling Function and Behavior of Mechanical Devices", ASME Computers in Engineering Conf., Pittsburgh, DETC01/CIE-21243, Sept. 2001.

[11] Malmqvist, J. "Improved Function-Means Trees by Inclusion of Design History Information", Proceedings of ICED-95, 10th International Conference on Engineering Design, Praha, Czech Republic, 1995b, pp 1415-1423

Integrated Intelligent Systems for Engineering Design
X.F. Zha and R.J. Howlett (Eds.)
IOS Press, 2006

Integrated Intelligent Design for STEP-based Electro-Mechanical Assemblies

Xuan F. Zha
University of Maryland & NIST, USA

Abstract. Assemblability analysis and evaluation plays a key role in assembly design, operation analysis and planning. In this paper, we propose an integrated intelligent approach and framework for evaluation of assemblability and assembly sequence for electro-mechanical assemblies (EMAs). The approach integrates the STEP (STandard for the Exchange of Product model data, officially ISO 10303) based assembly model and XML schema with the fuzzy analytic hierarchy process for assembly evaluation. The evaluation structure covers not only the geometric and physical characteristics of the assembly parts but also the assembly operation data necessary to assemble the parts. The realization of the integration system is implemented through a multi-agent framework. Through integration with the STEP-based product modeling agent system, CAD agent system and assembly planning agent system, the developed assembly evaluation agent system can effectively incorporate, exchange, and share concurrent engineering knowledge into the preliminary design process so as to provide users with suggestions for improving a design and also helping obtain better design ideas. The proposed approach has the flexibility to be used in various assembly methods and different environments. The applications show that the proposed approach and system are feasible.

Keywords: assemblability, evaluation, STEP, intelligent CAD, design for assembly, multi-agent-based integration, fuzzy analytic hierarchy process (AHP) approach

1. Introduction

Design for assembly (DFA) or assembly oriented design (AOD) concerns assemblability analysis and evaluation and integrates the specific domain knowledge of product design, manufacturing and assembling, and decision-making automation in assembly process (Molloy et al. 1991). The essence of DFA is to evaluate and rationalize the parts and assembly processes (Boothroyd and Dewhurst 1989, Boothroyd & Alting 1992). The analysis of assembly properties of a product is needed during the initial design stage in order to identify potential assembly problems that may affect product performances in the later stages of its life cycle. To produce a low-cost product, designers need to know whether the designed product can be assembled and how difficulty the assemblability of its components is. Assembly evaluation is a means to recognize design quality in terms of assemblability or feasibility. The information feedback from such an analysis and evaluation process in the early design stage is a key to improving design quality for better assemblability. On the other hand, an effective and efficient evaluation method for assemblability should indicate the cause of design weakness by identifying the tolerances, form features, and geometries of assembly parts, rather than simply provide an evaluation score for the assembly parts or assembly operations. During the assembly design process a set of solution alternatives is evaluated, and a solution is suggested based on the degree of satisfaction resulting from selection of alternative functional requirements. In view of assembly design as "generate and test", the most important step in design decision making is the comprehensive evaluation and justification that comes out a final solution from a set of alternatives determined by multiple factors or suggested by many evaluators.

From the literature review, there are some *ad hoc* methodologies developed for assemblability or feasibility analysis and evaluation using detailed design information. The existing methodologies can be classified into three categories: design heuristics, design ratings, and assembly geometric analysis. Based on these methods, there have some academic/research systems developed and a few efforts on the integration of assemblability knowledge with current CAD systems, but the research into assemblability analysis and evaluation during the initial design stage is limited, and the implementation of integration of assembly design, analysis and evaluation remains difficult due to interoperability limitations of current CAD and DFA systems. There is still a gap between the standardized representation of product data and information and the general-purpose assembly analysis and evaluation method. To bridge the gap, this work aims to propose an integrated fuzzy analytic hierarchy process (AHP) approach to quantitatively analyzing and evaluating assemblability for design of electro-mechanical assemblies (EMAs) based on the STEP. An agent-based integrated intelligent framework will also be provided.

This Chapter is organized as follows. In the following sections, related work into assemblability or feasibility analysis is first reviewed in Section 2; Section 3 presents an assembly evaluation integration scheme and data flow. Section 4 provides a STEP-based EXPRESS/XML schema assembly model; Section 5 discusses the use of the STEP-Based EXPRESS/XML schema model for assembly evaluation. Section 6 presents a fuzzy AHP approach for evaluating the assemblability and the assembly sequence; Section 7 proposes a multi-agent framework for integrated assembly evaluation; Section 8 provides a case study to verify and illustrate the proposed approach; and Section 9 gives a summary and some concluding remarks.

2. Review of Related Work

In this section, we give an overview of the related work on both assemblability analysis and evaluation methods and systems. However, this is not a complete survey (Abdullah et al. 2003, Lim et al. 1995, Zha et al. 1998). Most of the early work in the analysis of assemblability was rule based. The design attributes of the components, the assembly operations and relationships between components were used to estimate the ease or difficulty of the assembly. Plan-based evaluation systems were later developed to address the effects of sequencing assembling components on assemblability. The pioneering work of Boothroyd and Dewhurst (1989) in developing the design-for-assembly guidelines has resulted in several automated assembly evaluation and advisory systems (Jakiela 1989,

Jakiela and Papalambros 1989, Zha et al 1999, Zha 2001). Swift (1981) also presented a knowledge-based DFA technique with procedures similar to that of the Boothroyd and Dewhurst approach. Sturges and Kilani (1992) developed a semi-automated assembly evaluation methodology that attempts to overcome some limitations for the scheme proposed by Boothroyd and Dewhurst (1989). Although this system lacks geometric reasoning capabilities, it serves as an interactive environment to study the effect of various design configurations on assembly difficulty.

Li and Hwang (1992) developed another semi-automated system, which closely follows the Boothroyd-Dewhurst methodology. Their assembly difficulty analysis and cost-estimation modules were direct computer implementations of the DFA rules. Their method considers the multiple assembly sequences, calculates the time for all feasible sequences, performs limited feature recognition for assembly, and interactively obtains the non-geometric information that will affect the assembly. The final result is a table similar to a manual assembly worksheet. It is argued that assembly information developed quickly and in a proper format gives the designer enough input to perform further analysis for design modification.

The Hitachi Assemblability System (Miyakawa et al 1990) has served as a basis for the development of an automated assemblability system. It is based on the principle of one motion per part, with symbols for each type of assembly operation, and penalties for each operation based on its difficulty. The method computes an assembly evaluation score and assembly-cost ratio. An assembly-cost ratio gives an indication of the current assembly cost to the previous cost. The methodology is common for manual, automatic and robotic systems.

Miles and Swift (1992) also developed an assembly evaluation method in which parts are grouped according to functional importance: "category A" parts are those required to fulfill the design specification, and "category B" parts are the accessories. The goal is to eliminate as many type-B parts as possible through redesign. Analyses of feeding and fitting were carried out on the parts, with both results combined into a total score. The total is then divided by a number of type-A parts to obtain a final score. A proposed assembly sequence is used to perform fitting analysis. Warnecke and Bassler (1988) studied both functional and assembly characteristics. Parts with low functional value but high assembly difficulty receive low scores, while parts with high functionality and low assembly cost receive high scores. The scoring is used to guide the redesign process.

Assembly information models usually provide the input for assemblability evaluation methods. An assembly information model contains information regarding parts and their assembly relationships. One of the earlier works on assembly modeling was reported in (Lee and Gossard 1985). There are some academic systems that offer some facilities to represent assembly information. One such system developed by Whitney and Mantripragada (1998) represents the high-level assembly information as the "key characteristics". The chains of dimensional relationships and constraints in a product are handled by the so-called Datum Flow Chain concept (Whitney 2004). The system reported in van der Net (1998) focuses on designing assemblies taking into account requirements from the assembly process planning phase, in order to prevent design errors, reduce lead times, and be able to automate process planning. These requirements are captured in the assembly by specifying geometric, assembly and tolerance –specific relations on and between the assembled parts. Noort et al. (2002) worked on the integration of the views supporting parts design and assembly design of the whole product. Callahan and Heisserman (1997) proposed a strategy for evaluating, comparing, and merging design alternatives. Assembly features have also been subject to many studies (Shah 1991, Shah and Regers 1993, van Holland and Bronsvoort 2000, Coma et al. 2003, Chan and Tan 2003).

The need for the integration of assemblability knowledge with current CAD systems has been motivated by the fact that DFA methods have the greatest impact on a product design when they are incorporated into the preliminary design stage (De Fazio et al. 1997). In these integrated CAD-DFA systems, a proposed design is evaluated and recommendations for improvements are presented based on the results of the evaluation. Liu and Fisher (1994) use a STEP-based mechanical system data model as the assembly evaluation information source. This organizes the assembly-related information in a feature-based fashion. The proposed general-purpose assembly evaluation method is built by adopting the basic concepts of multi-attribute utility theory. The feature of this method is the linkage of the STEP product definition to the assembly evaluation method. Jared et al. (1994) presented a DFA system that performs geometric reasoning based on mathematical models for assembly operations. This reduces the user input requirement. Their system calculates a manufacturability index for individual components and a fitting index between the components. Zha (2002) proposed an integrated assembly evaluation method for STEP-based electro-mechanical systems.

Artificial intelligence (AI) techniques, such as knowledge-based expert systems fuzzy set theory, and case-based reasoning, may be used for the integration of DFA and CAD (De Fazio et al. 1997, Abdullah et al. 2003). Several methodologies and systems, such as IDAERS (integrated design for assembly evaluation and reasoning system), intelligent CAD-DFA, DFAES (design for assembly expert system) (Zha et al. 1999), the neuo-fuzzy system (Zha 2001), have been developed for integrated intelligent design for assembly (IIDFA). IDAERS can provide feedback on the estimated time required for assembling a product. Automatic identification of assembly attributes from a CAD description of a component has been investigated (Li and Hwang 1992). Jakiela and Papalambros (1989) developed an intelligent CAD system by encoding the Boothroyd DFA knowledge with feature-based representation. The system is able to provide users with suggestions in order to improve a design and also to help obtain better design ideas.

3. Assembly Evaluation Integration Scheme and Data Flow

Generally, assembly evaluation can benefit the improvement of a design in two ways, if the evaluation method can: (1) feedback high resolution information of design's weak points and (2) accept abstract or high level product information to evaluate the product assemblability in the earliest design stage. On the one hand, an effective, efficient evaluation method can indicate the cause of design weakness by identifying the tolerance(s), form feature(s), and geometry(ies) of assembly parts that cause the problem, rather than simply provide an evaluation score for the assembly parts or assembly operations. On the other hand, accepting abstract or conceptual product information as evaluation inputs is valuable, because suggested design changes can be evaluated and implemented at an early design stage.

In nature, an assembly can be considered as an application/process to be implemented through some sub-applications/processes, including design, planning and evaluation (Zha and Du 2002). Each application/process may include some other sub-applications/processes, for example, the evaluation includes assemblability evaluation and assembly sequence evaluation. The assembly operation planning may include sequence generation, cost estimation and evaluation, time scheduling, and resource management, etc. Therefore, there is a need to exchange data from assembly modeling to assembly planning, or more generally from the design phase to the assembly planning phase. In addition, even for various sub applications/processes, there are data exchange needs between them. Thus, assembly planning must be integrated into assembly modeling in a design environment, which enables the product model data to be directly exchanged from the modeling process to the evaluation and planning processes.

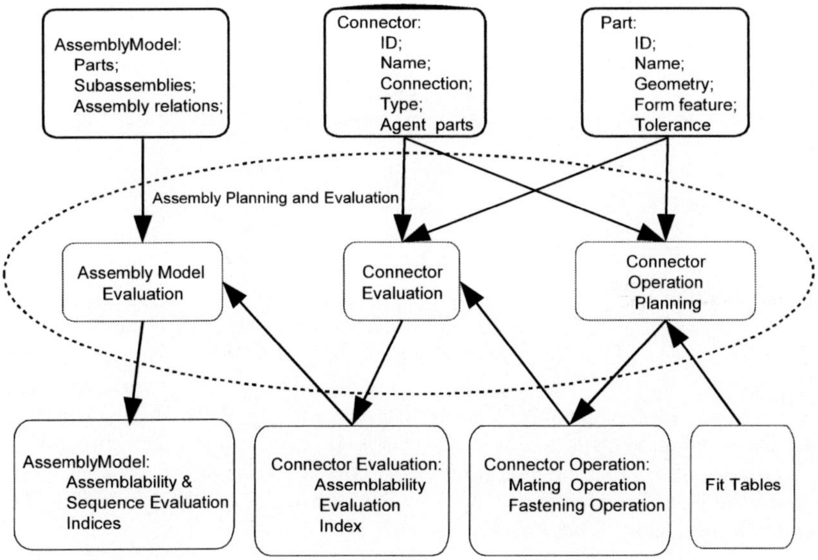

Figure 1: Agent-based assembly application integration scheme and data flow

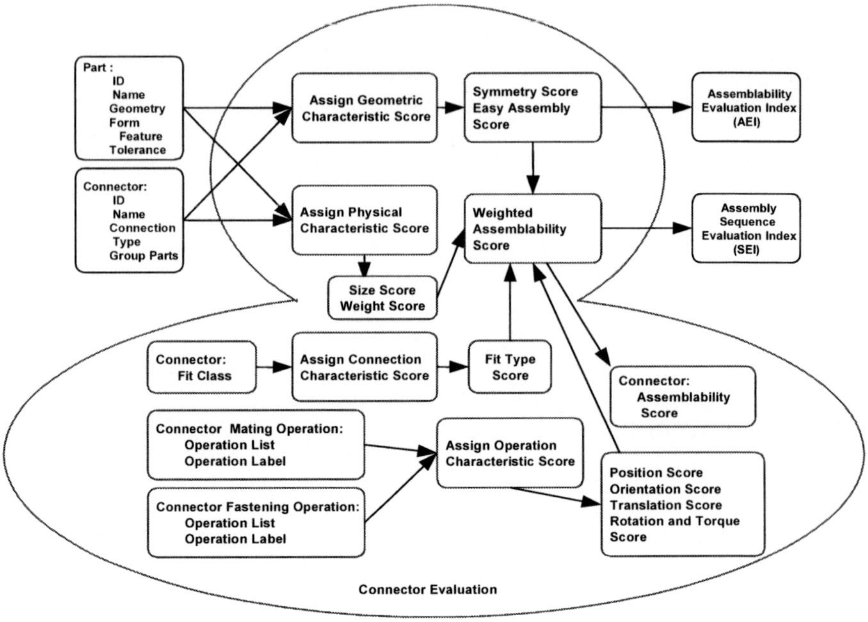

Figure 2: Data flow at connector evaluation

The overall integration scheme and data flow for the assembly evaluation application are shown in Figure 1, in which the needed input data and the output data generated by the assembly application are given (Zha 2004). To describe more detailed data flow of the assembly application, the data flow can be broken down into more specific descriptions. Sub-applications/processes of the assembly are introduced to implement the corresponding object methods for assembly design, planning and evaluation. To further understand the data flow within each assembly sub-application/process, such as connector process planning and evaluation process, at least one more level of the data flow diagram is required. Figure 2 illustrates the data flow of the connector evaluation process. In practice, if necessary, the data flow diagram with more detailed levels can be specified by breaking down a process into many sub-processes.

4. STEP-Based Assembly Model and EXPRESS/XML Schema

As discussed above the assembly model play a crucial role in the assembly evaluation integration scheme and data flow. The hierarchical relation model was used to describe assembly structure model and can be used to simplify the assembly evaluation and the generation of assembly sequence plans (Zha and Du, 2002, Zha 2004). A generic product/assembly data model can be defined using EXPRESS/EXPRESS-G data languages.

The main descriptive elements of EXPRESS are type (ID, name), entity, algorithms (function, procedure), and rule (Liu 1992). In this work, the STEP-based hierarchical assembly model consisting of product entity, subassembly entity, part entity and connector entity is described as an EXPRESS-G shown in Figure 3.

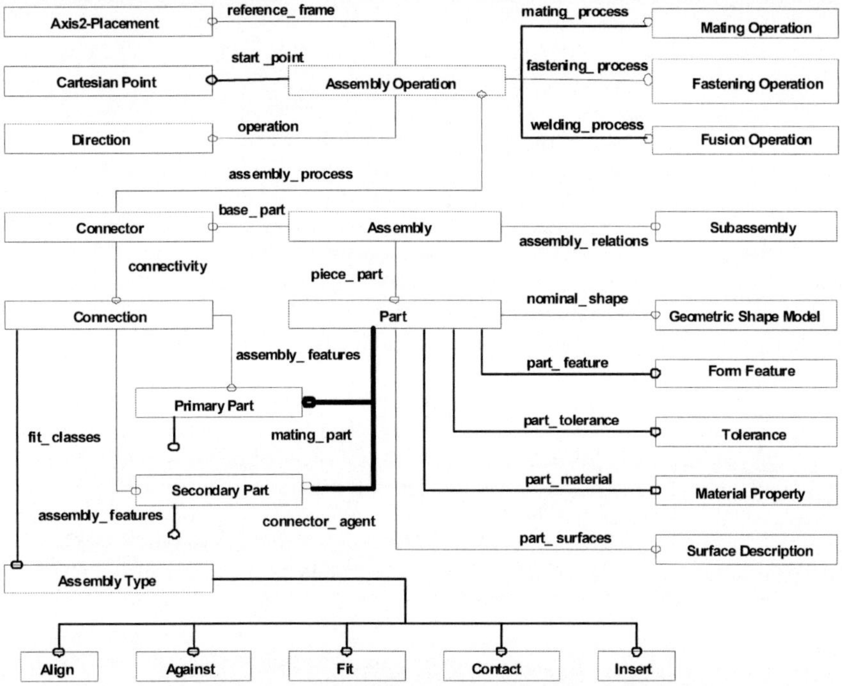

Figure 3: EXPRESS-G of assembly model

The entity *assembly* represents the product; it is the abstraction of common characteristics of products, which consists of several attributes including id, name, description, size, weight and subassemblies, parts, assembly relations. A brief explanation of the product is stored in the attribute description. The attributed subassemblies, parts and assembly-relations have the type of *subassembly*, *part* and *connector*. Since the history of product design should be recorded for the integrated system, the instances of the entity *product-version* keeps the track of the product forms which consists of the attributed id, description, make-or-buy, and of-product (Part 41) of STEP.

The entity *subassembly* is defined as the subtype of *assembly*, so that it can inherit the attributes of product without redefined. The inheritance mechanism provided by STEP simplifies the coding process and enhances the systems maintainability. The only difference between subassembly and product is that the subassembly is not a final product. The entity connector or joint should further express its upper structure and assembly relations with other parts or subassemblies.

The entity *part* provides detailed information about a part. A part in a mechanical system is a solid entity that has specific geometry and material properties. Its attributes are id, name, code, nominal-shape, part-features, part-tolerances, material properties, etc. The nominal-shape, part-features and part-tolerances correspond to three components of STEP: geometric model (Part 42), form features (Part 48) and tolerance (Part 47). A form feature adds detailed geometric characteristics to the geometric model to precisely define the shape of a part. Precision features such as tolerances and surface texture describe additional geometric characteristics of the final product design information for manufacturing and assembling such as assembly process and assembly method.

The entity *connector* is defined based on the mating conditions and kinematic constraints between parts in the global product definition. From an assembling viewpoint, a connector is an ordered sequence of assembly operations and specifies assembly operations and mating conditions between parts. According to the way that parts are assembled, a connector can be an operational connector, a fastener connector, or a fusion connector, in which a fastener connector contains additional information, i.e., its connector agent(s) with a designed part (such as a pin) or a standard mechanical part (such as a bolt and nut, screw, or rivet) used as a medium to assemble parts.

The information used by design and manufacturing can be classified into different features based on the information type. Form and precision features are defined in part object above. Assembly features are particular form features that affect assembly operations, which are defined by connectors. The attributes of connector include id, name, priority, connectivity and assembly process. Connectivity consists of primary part, assembly type and secondary part. Primary part and secondary part are subtypes of part; therefore, their necessary assembly features can be found in part object. Meanwhile, assembly process is classified to mating operation, fastening operation and fusion operation. By the mating operation the mated parts have certain assembly relationship in position. The fastening operation refers to operations to fix the agents and mated parts. The fusion operation joins the contact parts.

Therefore, the STEP integrated product information model comprises not only geometry but also form feature and product structure information. The EXPRESS schema for the STEP-based assembly model was provided in (Zha and Du 2002). From the representation/transformation of XML from EXPRESS schema and data (Part 28), the generated XML schema for representing the above generic assembly model is described as follows:

EXPRESS Schema

```
SCHEMA assembly;
TYPE Connector_type: ENUMERATION OF [operational, fastener];
END_TYPE;

ENTITY assembly_model;
sub_assemblies: OPTIONAL LIST [0:#] OF assembly_model;
piece_parts: OPTIONAL LIST [0,#] OF part;
assembly_relations: LIST [0,#] OF (LIST [1:#] OF Connector);
END_ENTITY;
```

```
ENTITY part;
name: STRING (80);
id: STRING (80);
nominal_shape: geometric_shape_model;
part_features: OPTIONAL LIST [0,#] OF form_feature;
part_tolerances: OPTIONAL LIST [0:#] OF tolerance;
part_material: OPTIONAL LIST [0,#] OF material;
part_surface: OPTIONAL LIST [0,#] OF surface;
END_ENTITY;

ENTITY Connector SUPERTYPE OF (operational_Connector AND fastener_Connector);
name: STRING (80);
id: STRING (80);
type: connector_type;
connectity: connection;
END_ENTITY;

ENTITY connection;
base_part: primary_part;
mating_part: LIST [1:#] OF secondary_part
END_ENTITY;

....

END_SCHEMA
```

XML Schema

```
<?xml version="1.0" encoding="UTF-8"?>
<xsd:schema xmlns:xsd="http://www.w3.org/2001/XMLSchema">
 <!--XML Schema created by E2XS-->
 <xsd:element name="express_data">
  <xsd:complexType>
   <xsd:sequence minOccurs="1" maxOccurs="unbounded">
    <!--SCHEMA assembly-->
    <xsd:element name="Assembly">
     <xsd:complexType>
      <xsd:sequence minOccurs="0" maxOccurs="unbounded">
       <xsd:element ref="Assembly_Model" />
       <xsd:element ref="Part" />
       <xsd:element ref="Connector" />
       <xsd:element ref="Connection" />
       <xsd:element ref="Primary_Part" />
       <xsd:element ref="Secondary_Part" />
        ......
       <xsd:element ref="Operational_Connector" />
       <xsd:element ref="Fastener_Connector" />
       <xsd:element ref="Assembly_Operations" />
```

```
            <xsd:element ref="Assembly_Label" />
            ......
           </xsd:sequence>
          </xsd:complexType>
         </xsd:element>
        </xsd:sequence>
       </xsd:complexType>
      </xsd:element>
      <!--ENTITY Assembly_Model-->
      <xs:element name="Assembly_Model">
            <xs:complexType mixed="true">
                <xs:sequence minOccurs="0" maxOccurs="unbounded">
                    <xs:element ref="subassemblies"/>
                    <xs:element ref="piece_parts"/>
                    <xs:element ref="assembly_relations"/>
                </xs:sequence>
            </xs:complexType>
        </xs:element>
     <xs:element name="assembly_relations">
            <xs:complexType>
                <xs:sequence>
                    <xs:element ref="Connector"/>
                </xs:sequence>
            </xs:complexType>
        </xs:element>
     <xs:element name="Part">
        <xs:complexType mixed="true">
            <xs:sequence minOccurs="0" maxOccurs="unbounded">
                <xs:element ref="name"/>
                <xs:element ref="nomial_shape"/>
                <xs:element ref="part_features"/>
                <xs:element ref="part_material"/>
                <xs:element ref="part_surface"/>
                <xs:element ref="part_tolerances"/>
            </xs:sequence>
            <xs:attribute name="id" type="xs:string"/>
        </xs:complexType>
     </xs:element>
     ......
     </xsd:schema>
```

---

5. STEP-Based EXPRESS/XML Schema Model for Assembly Evaluation

The standardization efforts support information exchange between different design, analysis, planning and evaluation systems. An integrated information model is the kernel for various kinds of applications in which features are used as the key integration elements.

In this section, a STEP-based integrated EXPRESS/XML schema model as an information source for assembly evaluation is discussed.

5.1. Role of the STEP-Based EXPRESS/XML Schema Model in Assembly Evaluation

The STEP-based assembly EXPRESS/XML Schema model has been developed as a basic building block for integrating assembly applications. For example, the STEP based data model is used to define a procedure for feature-based assembly operation planning in (Zha 2004). The proposed data model defines a set of assembly information entities, called joints or connectors, to work with each component part's geometry, form feature, and tolerance definition. The connector is used to (1) define the assembly relations in an assembly model and (2) carry the corresponding assembly process data.

The STEP form features play a key role in product information model in which the form features are used as design features in CAD and assembly features in assembly operation planning and evaluation. A form feature can be one of, some of, or all of the above application features, depending on the characteristics and functionality of the form feature. For example, a design feature, say hole_A, can also be a machining feature that requires a drilling operation. Furthermore, hole_A can be an assembly feature when used to receive a cylindrical part that forms a mated or joined connection of an assembly.

The design information of component parts is necessary but not enough to perform an accurate assemblability evaluation. At least two additional information sources, assembly system configuration and assembly relationships of component parts, are needed. In the proposed model, the STEP-based part definition, connector entities, and assembly model, can be used as the information source for an assembly evaluation method. A connector carries both the assembled parts' connectivity information and the assembly operation data. The connectivity information, decided by the designer, contains mated or joined part identifications and their assembly features. The assembly process data as the result of assembly process planning carry the fit-class, assembly operation list, and assembly operation label. An example of the connector data structure is shown in Figure 4, which shows the roles of connectors in linking the assembly parts' form features to specify the assembly relations of component parts.

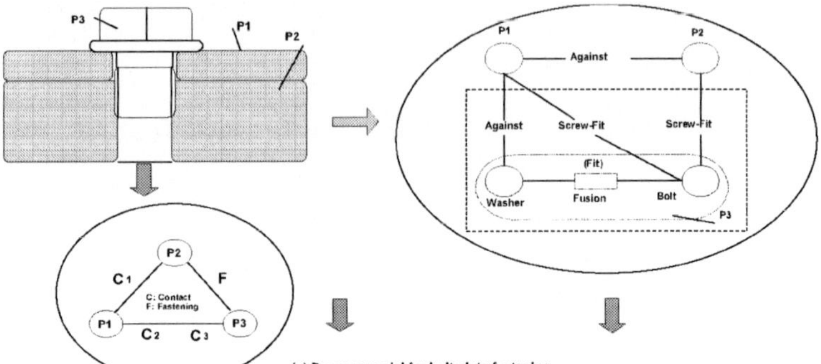

(a) Process model for bolt plate fastening

Connector	ID			1101	1102	1103	1104
	Name			C1	F	C2	C3
	Connectivity	Base Part	ID	P1	P2	P1	P1
			Priority	0	1	1	1
			Assembly Features	Surface	Thread Hole	Surface	Hole
		Mating Part	ID	P2	P3	P3	P3
			Priority	1	0	0	0
			Assembly Features	Surface	Thread Pin	Surface	Pin
	Assembly Process	Fit Classes		Against	Screw Fit	Against	Fit
		Operation Type		Mating	Fastening	Mating	Mating
		Requirement		-	-	-	-

(b) The instances of connector for bolt plate fastening

Figure 4: Fastener connector model instance

The advantages of using the STEP-based model for assembly evaluation are described as follows (Liu and Fisher 1992, Zha 2004):

(1) At an early design stage, evaluation can be performed when only the abstract connectivity information is provided, e.g., consider only the major shape of part geometry, assembly related form features, and tolerances that are involved in the assembled parts. Feature-based design environments allows designers to prototype a design by combining primitives, e.g., block, cylinder, or sphere, and/or using profile sweeping and ruling to create the pre-existing solid model of a part, then attach necessary form features to the solid model to fulfill the part's functional requirements. The parameters that describe a primitive or a form feature are grouped under the type of primitives or the type of form features. This concept not only allows a designer to change the parameters of a primitive or form feature easily, but also makes a clean separation of abstract information (e.g., block, cylinder, hole, slot, or pocket) and the detailed parameter set (x,y,z) dimension of a block, location, orientation, diameter, and depth of a hole, position, depth, width and length of slot).

(2) When the connectivity information provides enough details to generate the assembly process data, the evaluation method can consider the information when making an operation based estimation, in which time and cost for each assembly operation

procedure are considered as evaluation factors.

(3) The cause of low assembly performance can be linked to the assembled parts' definitions, which give high resolution to the indicated problem, e.g., how a designed part's shape, form features, and tolerances parameters affect the assembly operations or the assemblability.

5.2. Data Derivation for Assembly Evaluation

The data and information about connectors and parts of an electro-mechanical system, represented by the hierarchical assembly model, can be used for assembly evaluation. The connectors are defined to be common information set that carries both the connectivity information of mated parts and the assembly process data of mating and joining operations. To fully represent connectors' behaviors, a set of algorithms need to be developed to describe how the assembly process data can be derived from the connectivity of connectors and the standard product definition of the involved parts. These algorithms are based on the relationships between parts' form features, dimensions and tolerances, and assembly. The connector object class carries all the assembly features of a product and the part object class carries form features and precision features. The shape, location, and orientation of form features and precision features involved in mating parts determine the process and degree of difficulty of an assembly operation. In this connection, the data for assembly evaluation can be derived. Details are discussed below.

To derive the assembly operation information from design information, certain rules established on the types of connectors are used. For example, three general rules regarding the structure of connector attributes are listed below:

Rule 1: IF the connector type is operational
THEN the connections of primary and secondary parts do not have joining features
AND the assembly process only involves mating operations

Rule 2: IF the connector type is fusion
THEN the connections of primary and secondary parts do not have joining features
AND the assembly process involves both mating and fusion (welding) operations

Rule 3: IF the connector type is fastening
THEN the connections of primary and secondary parts have mating features and joining features
AND the connector agent is involved
AND the assembly process involves both mating operations and fastening operations

The relationships between connector types, the assembly operation list, fit class, and attributes of operation label can be obtained. For example, forces for the assembly operation can be determined from mating or joining form features and fit classes derived from the tolerances of the form features. Given assembly features and precision features of mating parts, the operation list, fit class, and contents of operation label can be determined. Thus, the attributes of assembly operation for the assembly planning and evaluation can be determined.

Because form features in STEP are defined to incorporate with the shape variational tolerance information model, extracting the assembly features' dimensions and tolerance ranges are relatively simple. Thus, the classification of fit can be done. Subsequently, the force and operation types that are needed for the connector can be decided. The following two algorithms can be used to assign assembly process data, such as: mating_operation and fastening_operation, and to determine the fit_class and assign the operation_label (Liu 1992).

Algorithm 1: Determine fit_class

```
{
        get assembly feature sizes and tolerances;
        calculate the min clearance and max clearances;
        if (min clearance >=0) then
            compare the min and max clearances with the limits of clearance in Clearance fit table to decide the fit type;
            else if (max clearance <=0) then
        compare the min and max clearances with the limits of Interference in Interference fit table to decide the fit type;
        else
            compare the min and max clearances with the fit ranges in Transition fit table to decide the fit;
            end_if
          end_if
        end.
}
```

Algorithm 2: Assign operation label;

```
{
        assign assembly_operation reference frame;
        assign starting point based on the origin of assembly feature reference frame of the secondary part
              or agent which is referred to the assembly operation reference frame;
        assign operation direction;
        assign operation movement distance;
        if (assembly features involves thread form features)
              then assign operation rotation degree;
              assign torque;
         end_if
    end.
}
```

Figure 4 illustrates a simple process model and the instances of the connector for bolt plate fixing. It shows the data derivation for an assembly process from the defined generic assembly model.

6. Fuzzy Analytic Hierarchy Process for Assembly Evaluation

Due to the uncertainty and fuzziness of design specifications and technical requirements, and the above parameters with different degrees of importance on the overall difficulty of assembly, it is difficult to assess the assemblability of the design using the traditional approaches reviewed in Section 2. In this work, a method for assembly evaluation presented is constructed using fuzzy analytic hierarchy process (FAHP) approach to multi-order fuzzy justification and evaluation (MFJE) problem (Zimmermann 1987). Figure 5 shows the fuzzy evaluation process for assemblability and assembly sequence. In this section, we discuss in detail the fuzzy evaluation for assembly difficulty.

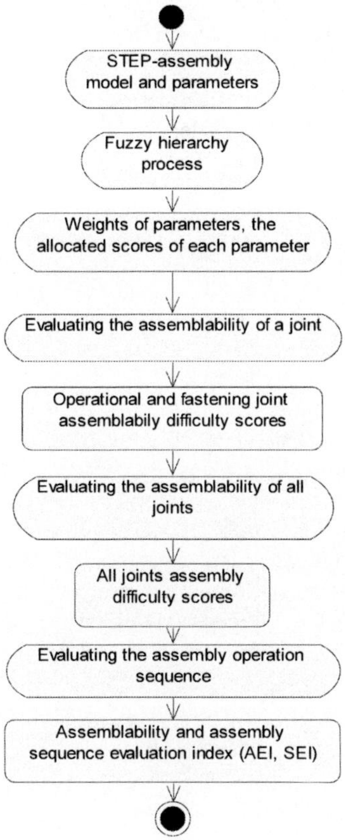

Figure 5: Fuzzy Evaluation of Assemblability and Assembly Operation Sequence

6.1. Fuzzy Analytic Hierarchy Process

The AHP mechanism proposed by Satty (1991) is known as an effective tool to support the multi-attribute decision-making. Its versatility in dealing with qualitative factors, multiple objectives, and decision makers has resulted in an impressive array of applications such as energy planning, conflict resolution, banking, architecture, etc.. It is a compositional approach in which a multi-attribute problem is first structured into a hierarchy of interrelated elements, and then a pairwise comparison of elements in terms of their dominance is elicited. The weights are given by the eigenvector associated with the highest eigenvalue of the reciprocal ration matrix of pairwise comparisons. Currently, most of research efforts compose AHP comparison matrix A according to user's individual and flexible preference. In this work, fuzzy membership functions are combined with the AHP to pursue the preference of user/agent dynamically (for collaborative design and evaluation), and as a result, the fuzzy comparison matrix A can be obtained. The details of the AHP algorithm are described as the following four steps (Kim 2003):

Step 1: Pairwise comparison matrix *A*

In an AHP, it is possible to decide the weights by comparing the importance of two criteria subjectively. We first define the pairwise comparison matrix A as

$$A = \begin{bmatrix} a_{11} & a_{12} & a_{13} & \cdots & a_{1m} \\ a_{21} & a_{22} & a_{23} & \cdots & a_{2m} \\ \vdots & \vdots & \vdots & \cdots & \vdots \\ a_{m1} & a_{m2} & a_{m3} & \cdots & a_{mm} \end{bmatrix} \tag{1}$$

where, a_i, i =1, 2, ...m, represents the criterion; a_{ij} indicates how much more important the ith criterion is than the jth criterion to construct the column vector of importance weighting of criteria. For all i and j, it is necessary that a_{ii} = 1 and $a_{ij} = 1/a_{ji}$. The possible assessment value of a_{ij} with the corresponding interpretation can be represented either with linguistic representation or graphical representation (Table 1). The graphical representation can be transformed from the linguistic representation with numeric values. The pairwise comparison ratio which is comparison of the importance of criterion i and criterion j, that is w_i and w_j, is defined as:

$$a_{ij} = w_i / w_j \tag{2}$$

Table 1: Linguistic and graphic representation and numeric value of relative importance

Value of a_{ij}		
Linguistic representation (term)	*Numeric value*	*Interpretation*
Equivalent	*1*	Criterion i and j are of equal importance
A little strong	*3*	Criterion i is weakly more important than criterion j
Strong	*5*	
Very strong	*7*	Criterion i is strongly more important than

Absolutely strong	*9*	criterion *j* Criterion *i* is very strongly more important than criterion *j* Criterion t *i* is absolutely more important than criterion *j*
-	*2, 4, 6, 8*	*Intermediate values*
Graphic representation 1) i=j, i *equivalent to* j 2) i>j, i *absolutely strong* j 3) i↑, j↑, i,j *=1→9, prefer* i *to* j		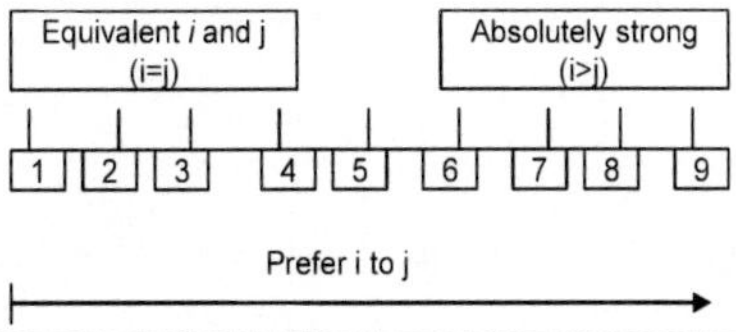

Step 2: Generalization of pairwise comparison matrix A

In this step, each entry in column i of $\boldsymbol{A}$ is divided by the sum of all the entries in column i. This yields a new matrix A_w, in which the sum of the entries in each column is 1.

$$A_w = [a_{ij} / \sum_{i=1}^{m} a_{ij}]$$

$$(i=1, 2, \ldots, m; j=1,2,\ldots,m) \tag{3}$$

Step 3: Average vector C

Compute c_i as the average of the entries in row i of A_w to yield column vector C.

$$C = [c_i] = [\sum_{j=1}^{m} (a_{ij} / \sum_{i=1}^{m} a_{ij}) / m]$$

$$(i=1, 2, \ldots, m) \tag{3}$$

where, c_i represents the relative degree of importance for the ith criterion in the column vector of importance weighting of criteria. In addition, c_i represents the evaluation score that the ith candidate alternative is assessed for a particular criterion to make the optimal decision.

Considering a pairwise comparison matrix $A = [a_{ij}]$ and an importance index (weight) vector $W = [w_i]$, their relationship can be described as:

$$AW = nW \tag{4}$$

When A is given, the vector W and n are calculated as an eigenvector and an eigenvalue of A.

Step 4: Consistency check for A and C

The pairwise comparison matrix should be examined whether consistency is reliable. To check for consistency in a pairwise comparison matrix, we need to do the following required sub steps:

(i) Define and calculate c_i , x_i, and δ :

$$A \bullet C = \begin{bmatrix} a_{11} & a_{12} & a_{13} & \cdots & a_{1m} \\ a_{21} & a_{22} & a_{23} & \cdots & a_{2m} \\ \vdots & \vdots & \vdots & \cdots & \vdots \\ a_{m1} & a_{m2} & a_{m3} & \cdots & a_{mm} \end{bmatrix} \begin{bmatrix} c_1 \\ c_2 \\ \vdots \\ c_m \end{bmatrix} = \begin{bmatrix} x_1 \\ x_2 \\ \vdots \\ x_m \end{bmatrix} \tag{5}$$

$$\delta = \frac{1}{m} \sum_{i=1}^{m} \frac{x_i}{c_i} \tag{6}$$

(ii) Determine and check the consistency index (CI)

$$CI = \frac{\delta - m}{m-1}, \; CR = \frac{CI}{RI} \tag{7}$$

where, CI is the consistency index; CR is the consistency ratio; RI is the random index; the average random consistency index = m-1

(iii) Determine if the degree of consistency is satisfactory through comparing CI to the random index (RI) for the appropriate value of m. If CI is sufficiently small, the decision maker's comparisons are probably consistent enough to give useful estimates of the weights for the objective function. For example, if $CI/RI < 0.10$, the degree of consistency is satisfactory, but if $CI/RI > 0.10$, serious inconsistencies may exist, and the AHP may not yield meaningful results. In this case, the matrix needs to be reset by comparing the importance again. The reference values of the RI for different numbers of m are shown in Table 2 below (Chuang 2001):

Table 2: The reference values of the *RI* for different numbers of *m*

The reference values of the *RI* for different numbers of *m*									
m	2	3	4	5	6	7	8	9	10
RI	0	0.58	0.90	1.12	1.24	1.32	1.41	1.45	1.51

6.2. STEP-based Assembly Model, Assemblability Parameters and Weights

To adopt the fuzzy AHP approach for assembly evaluation, the assemblability factors/parameters and their weights must be first identified from the STEP-based assembly model. This will be discussed as follows.

Form the STEP-based assembly model as discussed above, we classify the factors/parameters that affect the assemblability into two categories: geometry-based parameters and non-geometric parameters. Four types of characteristics of the parts and operations involved are of significance: geometry characteristics (related to parts' geometry), physical characteristics, connection characteristics (related to the type of contact between the components), and operation characteristics. They are described as an evaluation factor tree. In practice, these parameters are defined in terms of fuzzy linguistic descriptors, which can correspond to a range of actual parameter values (e.g., length), or to a qualitative description of a value of the parameter (e.g., interference). The above various parameters have different degrees of importance, which means that they have various degrees of influence on the overall difficulty. The widely used methods to find the relative importance of each parameter are: pairwise comparison; block distance model; and rank reciprocal rule (Ben-Arieh 1994). In this research, the fuzzy hierarchical process discussed above is used.

The acquisition of fuzzy quantities or knowledge is mainly referred to as acquiring the weights of assemblability factors. The fuzzy contribution of each parameter can be acquired based on expert advice, time study analysis, or even on experimentation with the various values of each parameter and analysis of the added difficulty. To describe these parameters, fuzzy values are used, and the values of the parameters are represented by linguistic variables with corresponding membership functions. For example, the amount of interference expected in the assembly can be described as "low," "low-medium," "medium," "medium-high," and "high." Each such descriptor implies a certain degree of difficulty that is described as a triangular fuzzy number. For example, a fit of type "pressure fit" with "high" amount of force required implies a basic difficulty of (26, 35, 55). A "push fit" with "low" force required contributes a difficulty of (16,20,26) to the assembly operations. The range of difficulty levels is from 0 to 100 with 100 representing an impossibly difficult operation.

As there are many factors involved, multi-order (2-order) model is required in ranking them for comprehensive fuzzy evaluation and justification for assembly. The first-order factors set can be described as: (u_{11}, u_{12}, u_{13}, u_{21}, u_{22}, u_{31}, u_{41}, u_{42}, u_{43}, u_{44}, u_{45}) =(α-symmetry, β-symmetry, number of ease of assembly form features, size, weight, fit type, position, orientation, translation, rotation, force/torque). The second-order factors set is described as: (u_1, u_2, u_3, u_4)=(geometric factor, physical factor, connection factor, operation factor). The degree of importance for the first-order factors on assemblability can be described using a linguistic variables set - (very important, important, medium important, almost not important, no relation), while the degree of importance for the second-order factors on assemblability is described as a linguistic variables set - (almost not important, medium important, very important). Thus, expert advice can be collected to elaborate the contribution of each factor to assemblability. Using the consistence function, the relative values of linguistic variables can be determined and normalized. Hence, the weight of each first-order factor can be obtained as (0.0875, 0.0875, 0.175, 0.080, 0.040, 0.18, 0.065, 0.133, 0.042, 0.065, 0.042), and the weight of each second-order factor as (0.35, 0.12, 0.18, 0.35).

6.3. Models for Fuzzy Evaluation of Assemblability

The assemblability can be evaluated through fuzzy value measurement using analytic hierarchical decision analysis. One model for fuzzy evaluation of assemblability, i.e., fuzzy additive aggregation model is discussed in this section below.

6.3.1. Fuzzy Hierarchical Evaluation Model

To evaluate a design, each factor makes a different contribution. This can be represented by the membership function defined in the universe of discourse of linguistic evaluation variable set E=(v_1, v_2, …, v_l), e.g., E= =(Low, Medium, High). Thus the voting matrix or evaluation matrix can be derived as a form/table. The second-order and first-order voting matrices, *R* and *r*, can be shown as follows, respectively. After carrying out the analysis and statistics, we can obtain the percentage r_{ij} $(i = 1,...,(m_1 + m_2 + ..., m_n)$, j=1,…, l) and z_{ij} (i=1,…,n; j=1,…,l) of the evaluated values of each factor and its item with respect to the evaluation linguistic variable matrix E. Let an evaluation vector be *Z*, weight vector be *W*, and evaluation matrix be *R*. As there may be many hierarchical-level factors to be considered in a complex design problem, it is reasonable to adopt a multi-order model to comprehensively evaluate the performance of an object. This is dependent on the hierarchical classification of the evaluation factors as described above. From the evaluation factors set above, we can define two-order evaluation models as such that are composed of U=(u_i,i=1,…,n) and $u_i = (u_{ij}, j = 1,..., m_i)$. For the first-order model, the value matrix and its voting matrix are $w_i = (w_{ij}, j = 1,..., m_i)$ and r_{ij}, respectively. Based on the definition of evaluation matrix, we have

$$z_i = w_i \circ r_i \tag{8}$$

and

$$z_i = \vee_{j=1}^{m_i} (w_{ij} \wedge r_{ij})(i = 1,..., n) \tag{9}$$

where, $\vee$ and $\wedge$ are union and intersection operators. For example, typical operators in the fuzzy set theory are maximum and minimum, "+" (addition) and "-" (minus). For the second-order model, considering the first-order evaluation results, its voting matrix and value matrix can be represented as $R = (z_i, i = 1,..., n)$ and $W = (w_i, i = 1,..., n)$ respectively. Thus, the final evaluation results are determined as follows

$$Z = W \circ R \tag{10}$$

Figure 6 is the fuzzy hierarchical evaluation model.

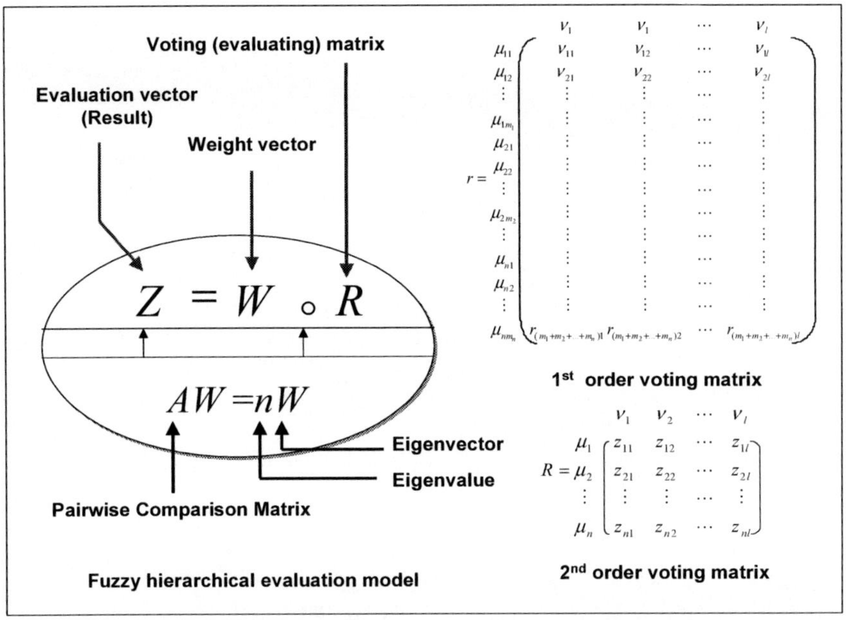

Figure 6: Fuzzy hierarchical evaluation model

6.3.2. Additive Aggregative Model

For a specific design or in a specific assembly environment, each assemblability factor has a reasonable range. Specifying a reasonable range for each factor constrains the evaluation method to provide a better response. For example, a 20 kg part may be considered heavy for manual assembly, but light for robot assembly. In fuzzy theory, a variable v can belong to more than one set, according to a given membership function $\mu_X(v)$. The possible values of a linguistic variable are not numbers but so called linguistic terms, such as *S* (small), *M* (medium) and *L* (large). Standard membership function types as Z, λ, π and *S* type can be mathematically represented as piecewise linear functions. For example, the variable v=2.5 belongs to *S* with a membership grade $\mu_S(v)$=0.75 and to M with $\mu_M(v)$=0.25. As such, the acceptable range of each assemblability factor can be categorized by the grade. The grade of each assembly factor is subject to change for refining the resolution of an evaluation.

Based on the weight of parameters, the allocated score of each parameter can be obtained, and used to evaluate the assemblability in design. This can be accomplished by evaluating the assemblability of a joint/connector. With respect to the STEP-based model, the following two types of joints/connectors are considered: the fastener joint (with joint agent, e.g., screw, pin, bolt and nut) and the operational joint (without joint agent):

(1) For an operational joint, the secondary part is mated into the primary part, and the

assembly difficulty score as an Assembly Evaluation Index (AEI) is calculated using the following equation:

$$AEI\ (J) = \frac{1}{100} \sum_{i=1}^{n} ds_i(x_i) \tag{11}$$

where, $ds_i(x_i)$is the relative difficulty score of the joint for the *i* assembly factor. AEI(J) is the assembly difficulty score of Joint J, which is regarded as an assemblability evaluation index of Joint J.

(2) For a fastener joint, the primary part and secondary parts are mated together first, and then the joint agent part(s) is used to join the mated parts. Assuming all the assembly characteristics among the mated parts and the joint agent parts are equally important, the assembly score for a fastener joint is calculated as follows:

$$AEI\ (J) = \frac{1}{100p} \sum_{j=1}^{p} \left(\sum_{i=1}^{n} [ds_i(x_i)]_j \right) \tag{12}$$

where, p is the total number of secondary parts and agents involved in the fastener joint.

6.4. Assembly Operation Sequence Evaluation

The choice of the assembly sequence in which parts or subassemblies are put together can drastically affect the efficiency of the assembly process. For example, one feasible and reasonable sequence may require less fixturing, less changing of tools, and include simpler and more reliable operations than others. Assembly sequence generation plays an important role in designing and planning the product assembly process and assembly system. In order to select the optimal assembly sequence that most nearly meet the needs for a particular purpose within the available resources, it is essential to develop some procedures to reduce the sequence count. The generation of assembly sequence and its econo-technical justification and evaluation was discussed in detail (Zha et al 1998). Using the proposed integrated knowledge-based approach, all the feasible assembly sequences can be generated, and through Petri net modeling and simulation the optimal assembly sequence can be obtained.

In the previous section, we discussed the fuzzy assemblability evaluation based on the degree of difficulty of assembly operations. As a matter of fact, once the mating operation is evaluated, the entire sequence of operations can be evaluated. The evaluation of the entire sequence needs to support comparison and selection of a preferred one; therefore, the aggregate measure of difficulty for the entire sequence is represented as a fuzzy number between 0 and 1. Suppose that the following notation is used: S_i= sequence *i*, *i*=1, ..., *n*; n_i = number of operations in sequence S_i, S_{ij}= operation *j* in sequence *i*,*j*=1,...,n_i; ds_{ij}= assembly difficulty score that represents the degree of difficulty of operation j in sequence *i*. For the entire sequence, the assembly difficulty scores for the sequence i are calculated using the following equation:

$$SEI\ (S_i) = \frac{1}{100n_i} \left\{ \sum_{i=1}^{n_{i_1}} ds_i(x_i) + \frac{1}{p} \sum_{k=1}^{p} \left(\sum_{j=1}^{n_i - n_{i_1}} [ds_j(x_j)]_k \right) \right\} \tag{13}$$

where, *SEI* (S_i) is the sequence evaluation index of sequence *i*; $n_i = n_{i1} + n_{i2}$, n_{i1} is the number of operational joints in sequence *i*, and n_{i2} is the number of fastener joints in sequence *i*; $ds_i(x_i)$ is the relative difficulty score of the joint for the *ith* assembly factor; *p* is the total number of secondary parts and agents involved in the fastener joint. Based on Eq.(13), the preferred sequence is chosen as the one with the lowest sequence evaluation index.

7. An Overview of Multi-Agent-based Integrated Assembly Design Framework

Multi-agent technology has been emerging as a promising approach for dealing with integration in distributed information system applications. In this section, a multi-agent framework as shown in Figure 7 is proposed for integrated assembly evaluation. The framework is a multi-agent environment, including a feature-based CAD agent, data and information modeling agent, product modeling agent, assembly planning agent, and assembly evaluation agent. These agent systems correspond to the integration models described in Section 3.

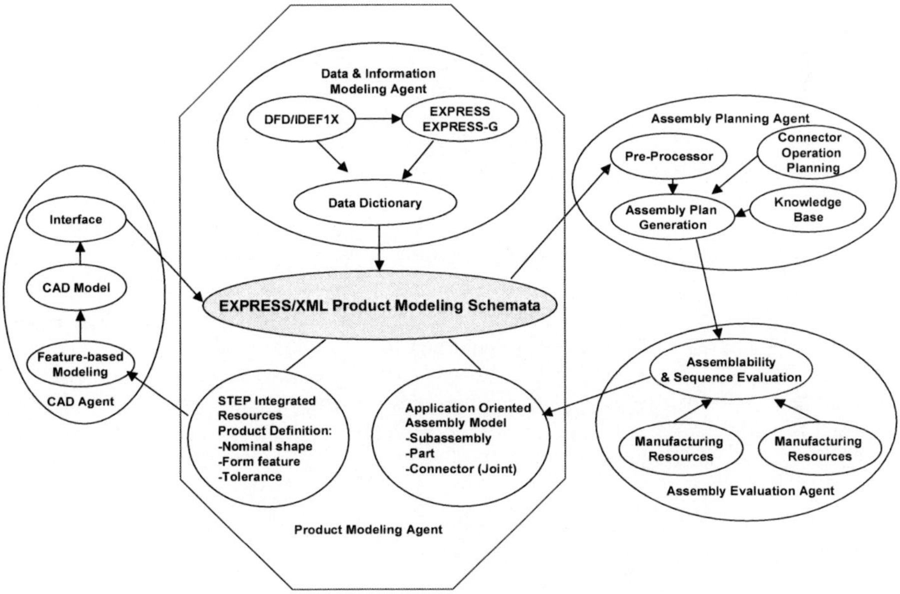

Figure 7: An overview of multi-agent-based integrated assembly design framework

1) The purpose of product modeling agent system is to provide mechanisms for representing, managing and exchanging product data using STEP. It is the central piece

of the framework. The assembly-oriented product model is defined as numerous STEP entities from integrated resources (IR) written in EXPRESS and XML to meet the need of assembly design and planning. Once a product or parts of the product are designed, the product data, for example, hierarchical structure of assembly and assembly relations, are generated by a feature-based CAD agent system. They are stored in the product model as instances of STEP entities.

2) The feature-based CAD agent system can also accept the imported CAD files of individual components and assemblies from DXF (e.g. AutoCAD) and STEP-based modeling system, and organize them into an assembly representation. Using feature recognition techniques, the assembly editor can differentiate connectors between parts and assembly features on individual parts.
3) The assembly operation planning agent system obtains the necessary information from the product model through preprocessor and generates the feasible assembly sequences (Zha 2004). The assembly plan generation agent is implemented by incorporating several subagents, such as the geometric checking and reasoning agent for interference and collision detection, precedence generation, user input constraints, a sequence generator, a Petri net tool for representing, analyzing and searching, simulating and visualizing as well as optimizing assembly sequences (Zha et al. 1998). The interference detection for disassembly operation is required for assembly sequence generation. The geometry checking employs a collision detection algorithm for every component in every candidate assembly direction. The objective is to determine the set of components that would obstruct the assembly operation if they were already in their final position, similarly, consider the disassembling of the final product. The precedence generation subagent determines the precedence relationships among parts with a precedence graph. There are many constraints considered when planning assembly sequence. Allowing the user to input constraints or criteria on which assembly sequences are chosen helps to prune the number of feasible sequences. The sequence generator via a disassembly approach performs generation and construction of assembly plans.
4) The assembly evaluation agent system is used to evaluate the design and planning results, i.e., the assemblability. As discussed above, the assemblability is evaluated in terms of the minimization of total assembly time or the assembly evaluation index (AEI) and assembly sequence evaluation index (SEI) with the proposed FAHP. The results of the evaluation feed back to then redesign stage through the product-modeling agent.

8. Case Study

To verify and illustrate the proposed approach, the optic lens assembly with eight parts, as shown in Figure 8, is simulated. The product consists of eight parts labeled: O_1-doublet 1, O_2-spacer, O_3-doublet 2, O_4-lock ring, and O_5-subassembly 1 (which is composed of sp_1, sp_2, sp_3 and sp_4, and pre-assembled.) Doublet 1 (O_1), spacer (O_2), doublet 2 (O_3), and lock ring (O_4) are connected with contact fits. Lock ring (O_4) also connects sp_1 with a screw fit.

The assembly is accomplished by a robotic system. The assembled product is a module/subassembly that will be mated to a large assembly.

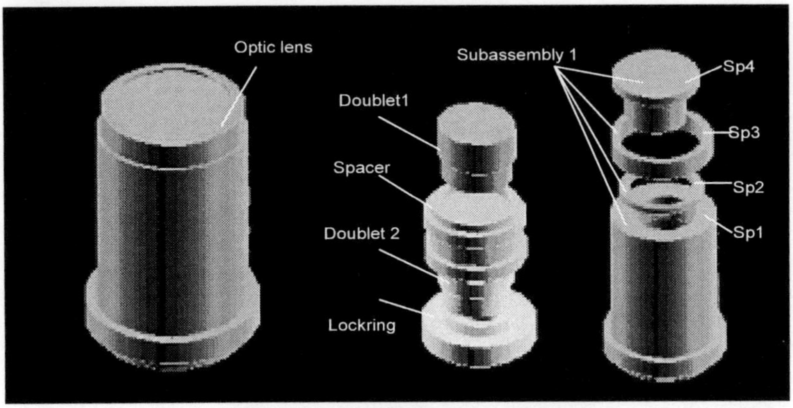

Figure 8: Optic lens assembly

As discussed above, the unified description of the feature-based models of both an assembly and single piece components can be obtained through the data abstraction of components and connectors on various levels. Therefore, the feature model for the optic lens in an assembly can be thought of as consisting of the shaft and a set of connectors: against, chamfer, face, cylinder, screw fit, and several cases of fix_fit, all of which are features as usual. In terms of the hierarchical model, part descriptions are form feature oriented, and product assembly structure descriptions are hierarchical multi-level graphs with feature-links. All these data and information comprise the main parts of the STEP model of the optic lens. For instance, the entity definition and its XML for the lock ring can be described as follows:

```
part 4(O4) {
                Name                Lock_ring
                ID                  1001
                nominal_shape       step_cylinder
                part_features       [chamfer, fixfit, cylinder, screw_fit]
                part_tolerances     [0.01]
                part_material       [aluminum]
                part_surface        [cylinder]
           }
```

XML:


```
<part id="part 4(O4)">
    <name> Lock ring</name>
    <nomial_shape>
      <geometric_shape_model>step cylinders </geometric_shape_model>
```

```
        </nomial_shape>
        <part_features>
            <form_feature id="00013">
                <name>CLD 1</name>
                <type> cylinder</type>
                <diameter>24</diameter>
                <height>10</height>
            </form_feature>
             <form_feature> chamfer </form_feature>
             <form_feature> screw_fit</form_feature>
        </part_features>
        <part_tolerances>
             <tolerance>0.01</tolerance>
        </part_tolerances>
        <part_material>
             <material>aluminium</material>
        </part_material>
        <part_surface>
             <surface>cylindric surface</surface>
        </part_surface>
    </part>
    ......
```

After all connectors are evaluated, the total assembly difficulty scores can be obtained by summing all of the evaluated scores of these joints. As different assembly sequence requires different assembly operations, the total assembly difficulty scores are therefore different. For the sequence: doublet 2 → spacer → doublet 1 → lock ring → subassembly 1, the total assembly difficulty score and assemblability evaluation index (AEI) after training and learning are 11.93 and 0.1193 (0.12), respectively. The larger the value of AEI the more difficult the assemblability is. Table 3 shows an assembly difficulty evaluation for a mating operation in the assembly process. With the fuzzy hierarchical evaluation method, the evaluation result is (0.42, 0.71, 0.91), which means that the assemblability is High, i.e., this assembly operation is easy. Table 4 lists partial results of assembly evaluation.

Table 3: One mating operation assembly difficulty score and evaluation

Mating Operation	Assembly Factors	u_{11}	u_{12}	u_{13}	u_{21}	u_{22}	u_{31}	u_{41}	u_{42}	u_{43}	u_{44}	u_{45}
	Data	360	0	2	5	2	LN	clear	vertical	4	0	0
	Score	4.66	0.59	1.75	0.64	0.40	4.5	0.0	1.67	0.60	0.00	0.00
	Total assembly difficulty score=11.93, assembly evaluation index $\approx$ 0.12											

Table 4: Weights and partial assemblability ratings in fuzzy hierarchical evaluation

Criterion No.	Criterion Item	Criterion Weight	Partial Assemblability Rating		
1st order		Weight Value	Linguistic Term	Fuzzy Number	Rating Value
1	u_{11} (α-symmetry)	0.085	Medium	(0.4,0.5,0.5,0.6)	0.500
2	u_{12}(β-symmetry	0.085	High	(0.7,0.8,0.8,0.9)	0.800
3	u_{13} (number of ease assembly features)	0.175	High	(0.7,0.8,0.8,0.9)	0.800
…	…	…	…	…	…
11	u_{45} (force/torque)	0.042	Very Low	(0.0, 0.0,0.1,0.2)	0.075
2st order		Weight Value	Linguistic Term	Fuzzy Number	Rating Value
1	u_1 (geometry)	0.35	Medium	(0.4,0.5,0.5,0.6)	0.500
2	u_2 (physical)	0.12	High	(0.7,0.8,0.8,0.9)	0.800
3	u_3 (fit type)	0.18	Very Low	(0.0, 0.0,0.1,0.2)	0.075
4	u_4 (operation)	0.35	Fairly High	(0.5, 0.6, 0.7, 0.8)	0.650

Evaluation Results:

Assembly Sequence	Evaluation Index (AEI)	Rankings
1-2-3-4-5	0.5212	4
3-2-1-4-5	0.1204	1
4-3-2-1-5	0.2822	2
5-3-2-1-4	0.3944	3

All the feasible sequences for the optic lens assembly can be generated using the approach proposed in (Zha et al 1998). There are 12 feasible assembly sequences remaining after considering hard and soft constraints for linear assembly. Here, hard constraints are the geometric and physical constraints related to the generation of the assembly sequence, and soft constraints are imposed by assembly planners while evaluating and selecting an assembly sequence. Through the assembly sequence evaluation discussed above, the fuzzy evaluation of difficulty score and assembly sequence evaluation index (SEI) can be obtained. The assembly difficulty scores of four different assembly sequences (1 → 2→ 3→ 4→ 5, 3→ 2→ 1→ 4→ 5, 4 →3 →2 →1 →5, 5→ 3→ 2→ 1→ 4) are (0.52, 0.12, 0.28, 0.39), as shown in Figure 9. Therefore, among these four assembly sequences, the optimal one is 3→ 2→ 1→ 4→ 5.

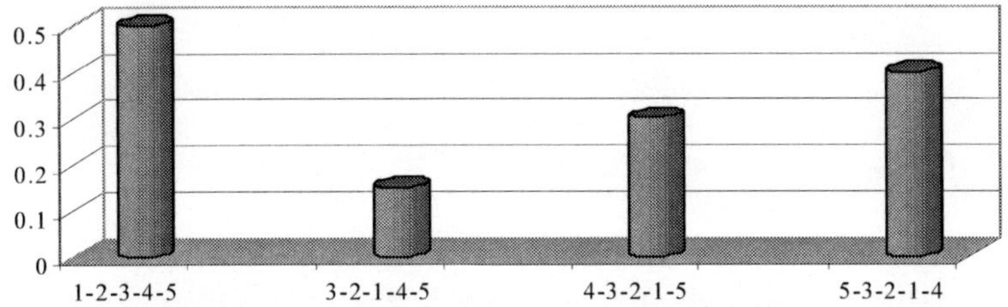

Figure 9: Assembly difficulty score under different sequences

9. Conclusions

This paper presented a multi-agent integrated fuzzy AHP approach to evaluating assemblability and assembly sequence for STEP-based electro-mechanical assemblies (EMAs). The fuzzy AHP approach could be used as an alternative to the Boothroyd/Dewhurst approach of a straight numeric (linear) metric. The approach uses the STEP-based data model represented by EXPRESS and XML schema as the assembly evaluation information source. The evaluation structure covers not only the assembly parts' geometric and physical characteristics, but also takes into account the assembly operation data necessary to assemble the parts. The weight of each assemblability factors is subject to change to match the real assembly environments based on expert advice through the fuzzy AHP approach. The approach was designed for general-purpose assembly evaluation, which can find wide application in developing a knowledge-based expert system for design and planning of assemblies. This approach has the flexibility to be used in various assembly methods and different environments. The developed multi-agent integrated intelligent framework can provide users with suggestions in order to improve a design and also help obtain better design ideas. The concurrent engineering knowledge can be effectively incorporated into the preliminary design process through the use of the developed framework. The contributions of this work can be summarized as three-fold: 1) the STEP-based EXPRESS/XML schema assembly model; 2) the STEP-based EXPRESS/XML schema assembly model tailored for assembly evaluation; 3) the fuzzy AHP approach for evaluating the assemblability and the assembly sequence; 4) the multi-agent integrated assembly evaluation framework.

Disclaimer

The bulk of the work reported here was conducted at Nanyang Technological University and Institute of Manufacturing Technology, Singapore. No approval or endorsement by the National Institute of Standards and Technology is intended or implied.

10. References

Abduliah, T.A., Popplewell, K., and Page, C.J., (2003), A review of the support tools for the process of assembly method selection and assembly planning, *International Journal of Production Research*, 41(11), pp.2391-2410

Boothroyd, G. and Dewhurst, P. (1989), Product Design for Assembly. Boothroyd Dewhurst Inc.

Boothroyd, G. and Alting, L. (1992), Design for assembly and disassembly, *Annals of the CIRP*, 41 (2), pp. 625-636

Ben-Arieh, D. (1994), A methodology for analysis of assembly operations' difficulty, *International Journal of Production Research*, 32 (8), pp.1879-1895

Ben-Arieh, D. (1994), A methodology for analysis of assembly operations' difficulty, *International Journal of Production Research*, 32 (8): 1879-1895

Chan, C.K. and Tan, S.T., Generating assembly features onto split solid models, Computer-Aided Design, 2003, 35(14):1315-1336

Coma, O., Mascle, C. and Veron, P., Geometric and form feature recognition tools applied to a design for assembly methodology, Computer-Aided Design, 2003, 35(13): 1193-1210

Chang, E. and Li, X. and Schmidt, L.C., The Need for Form, Function, and Behavior-based Representations in Design, DATLab, University of Maryland, 2000

Callahan, S., Heisserman, J., Pratt, M.J. , Sriram, R. D. and Wozny, M.J., Product Modeling for Computer Integrated Design and Manufacture, A Product Representation to Support Process Automation, 285-296, Chapman and Hall, 1997

Chuang, P.T. (2001), Combining the Analytic Hierarchy Process and Quality Function Deployment for a Location Decision from a Requirement Perspective, *International Journal of Advanced Manufacturing Technology*, 18, 842-849.

De Fazio, T. L., Rhee, S. J. and Whitney, D. E., (1997), A design-speci.c approach Design For Assembly (DFA) for complex mechanical assemblies, *Proceedings of the IEEE International Symposium on Assembly and Task Planning*, CA, USA, pp. 152–158

Hsu, W.,Lee, C.S.G. and Su, S.F. (1993), Feedback approach to design for assembly by evaluation of assembly plan, *Computer-Aided Design*, 25(7): 395-410

ISO TC184/SC4/WG3, STEP Part 42, Geometric and topological representation

ISO TC184/SC4/WG3, STEP Part 43, Representation structure

ISO TC184/SC4/WG4, STEP Part 21, Exchange of product model data

ISO TC184/SC4, STEP Part 11, Description methods: The EXPRESS language reference manual.

ISO TC184/SC4, STEP Part 28: XML representation of EXPRESS schemas and data

ISO TC184/SC4, STEP Part 203, Configuration controlled design

Jakiela, M.J. (1989), Intelligent suggestive CAD system, *Proceedings of MIT-JSME Workshop*, MIT,

Cambridge, pp.411-441, USA

Jakiela, M.J. and Papalambros, P. (1989), Design and implementation of a prototype intelligent CAD system, *ASME Journal of Mechanisms, Transmission, and Automation in Design*, 111(2), pp. 252-258

Jared, G.E.M., Limage, M. G., Sherrin, I. J., and Swift, K.G. (1994), Geometric reasoning and design for manufacture, *Computer-aided Design*, 26(9): 528-536

Kim, J.S. (2003), Negotiation support in electronic commerce using fuzzy membership functions and AHP, *Proceedings of the 6th Pacific Rim International Workshop on Multi-Agents (PRIMA) 2003*, Seoul (Korea), pp.93-104.

Lee, K. and Gossard, D. C., A hierarchical data structure for representing assemblies: Part-1, CAD, 1985, 17(1): 15--19

Li, R.K. and Hwang, C.L., (1992), A framework for automatic DFA system development, *Computers in Industrial Engineering*, 22(4): 403-413

Lim, S.S., Lee, I.B.H., Lim, L.E.N., & Ngoi, B.K.A., (1995), Computer aided concurrent design of product and assembly processes: A literature review, *Journal of Design and Manufacturing, 5*, 67–88

Liu, T.H. and Fischer, G.W. (1994), Assembly evaluation method for PDES/STEP-based mechanical systems, *Journal of Design and Manufacturing*, 4, pp.1-19

Liu, T.H., (1992), An object-oriented assembly applications methodology for PDES/STEP based mechanical systems, *PhD Thesis*, The University of Iowa, USA

Miles, B.L. and Swift, K.G. (1992), Working together, Manufacturing Breakthrough

Miyakawa, S., Ohashi, T. and Iwata, M. (1990), The Hitachi new assemblability evaluation method, *Transactions of the North American Manufacturing* Research Institute, SME

Molloy,E., Yang, H. and Brown, J. (1991), Design for assembly with concurrent engineering, *Annals of CIRP*, 40(1), pp.107-110

Noort, A., Hoek, G. F. M. and Bronsvoort, W. F., Integrating part and assembly modeling, Computer-Aided Design, 2002, 34(12): 899-91

Saaty, T.L. (1991), The analytic hierarchy process, McGraw-Hill, New York

Shah, J. J., Assessment of features technology, Computer-Aided Design, 1991, 23(5): 331-343

Shah, J. J. and Rogers, M. T., Assembly Modeling as an Extension of Feature-based Design, Research in Engineering Design,1993, 5: 218-237

Sturges,R.H. and Kilani, M.I. (1992), Towards an integrated design for an assembly evaluation and reasoning system, *Computer-Aided Design*, 24(2):67-79

Swift, K.G. (1981), Design for Assembly Handbook, Salford University Industrial Center, UK

Van der Net, A, Designing and manufacturing assemblies, Eindhoven University of Technology, 1998

van Holland W. and Bronsvoort W. F., Assembly features in modeling and planning, Robotics and Computer Integrated Manufacturing, 2000, 16(4): 277-294

Warnecke, H.J. and Bassler, R. (1988), Design for assembly - part of the design process, Annals of the CIRP, 37(1): 1-4

Whitney, D. E. and Mantripragada, R, The Datum Flow Chain: A systematic approach to assembly design and modeling, ASME Design Engineering Technical Conferences and Computers in Engineering Conference, 1998, ASME

Whitney, Daniel E. Mechanical Assemblies: Their Design, Manufacture, and Role in Product Development, Oxford University Press, 2004

Zha, X.F., Lim, S.Y.E. (1998), Integrated knowledge-based assembly sequence planning, *International Journal of Advanced Manufacturing Technology*, 14(1), pp. 50-64

Zha, X.F., Lim, S.Y.E., Fok, S.C. (1999), Integrated knowledge-based approach and system for product design for assembly, *International Journal of Computer Integrated Manufacturing*, 12(3), pp. 211-237

Zha, X.F., (2001), Neuro-fuzzy comprehensive assemblability and assembly sequence evaluation for assembly, *Artificial Intelligence for Engineering Design, Analysis, and Manufa*cturing, 15(4), pp. 367-384

Zha, X.F., and Du, H., (2002), A PDES/STEP-based model and system for concurrent integrated design and assembly planning, *Computer-Aided Design*, 34(12), pp. 1087-1110

Zha, X.F. (2002), Integrating the STEP-based assembly model and XML schema with the fuzzy analytic hierarchy process (AHP) for assembly evaluation, NTUIR, Singapore

Zha, X.F., (2004), Planning for STEP-based electro-mechanical assemblies: an integrated approach, *International Journal of Computer Integrated Manufacturing*, 17(4): 305-326

Zimmermann, H.-J. (1987), Fuzzy Sets, Decision-Making, and Expert Systems, *Kluwer Academic Publishers*, Boston

Section III

Applications

Integrated Intelligent Systems for Engineering Design
X.F. Zha and R.J. Howlett (Eds.)
IOS Press, 2006

Adaptive Tabu Search and Applications in Engineering Design

Sarawut SUJITJORN [a,1], Thanatchai KULWORAWANICHPONG [a], Deacha PUANGDOWNREONG [b] and Kongpan AREERAK [a]
[a] *School of Electrical Engineering, Institute of Engineering, Suranaree University of Technology, Nakhon Ratchasima, Thailand, 30000*
[b] *Department of Electrical Engineering, Faculty of Engineering, South-East Asia University, Bangkok, Thailand, 10160*

Abstract. This chapter presents detailed step-by-step description of an intelligent search algorithm namely Adaptive Tabu Search (ATS). The proof of its convergence, and its performance evaluation are illustrated. The chapter demonstrates the effectiveness and usefulness of the ATS through various engineering applications and designs in the following fields: power system, identification, and control.

Keywords. Adaptive tabu search, convergence, performance evaluation, identification, neuro-tabu-fuzzy control.

1. Introduction

Engineering design can range from the design of an individual component to an entire system. Conventionally, precise mathematical expressions are required in every part of the works. Problem formulation is the first step that one may regard it as the most important procedure when an engineering work is committed. A difficulty of this procedure depends on the degree of model complexity used to characterize the system. During the design stage, engineers often use simplified models sometimes called design models, and attempt their designs via trial-and-error, whilst there is a large number of engineers and researchers working with optimization. In today competitive world, engineering design becomes increasingly more complicated and subjected to many variables and constraints. To attack such a design problem needs a very efficient algorithm for number crunching on a highly complex model sometimes called truth model, which accurately describes the system behaviors. Many old fashion techniques of mathematical programming are unable to find a good enough solution due to variable complexity, for example. Fortunately, for about two decades, tabu search [1] has been proposed for combinatorial optimization in which a searching space is discontinuous. With a long history of tabu search development, adaptive tabu search (ATS), one of its modified versions, has been released. The ATS has been proved to be top of the most efficient search algorithms. Its features are suitable for practical engineering design works. The capability of handling discrete variables distinguishes

[1] Corresponding Author.

the ATS from the others. Imagine that when engineers design some things, e.g. a conveyor belt, the final decision is to answer very simple questions like these: how many motors are required to drive the system and what are their ratings? Undoubtedly, a total number of motors to be installed and the motor ratings provided by manufacturers are discrete and strongly associated with investment cost. What would be the engineers' decision when a solution from their calculation is 13.6! Can we round it up? If the answer is yes, how can we trust this round-up answer because there might be some integers with a certain degree of probability in which their costs might be less than the round-up one! Interestingly, the ATS gives flexibility to handle such a problem with satisfactory execution time consumed. With a specified search space, an integer that gives the minimum cost can be found efficiently. By adjusting variable step sizes, a fine search space can be generated, and a decimal solution can be also obtained.

In 1986, Glover [1] proposed the tabu search method for solving combinatorial optimization problems. Its principles are the neighborhood search and the tabu list (TL). To obtain the best solution, the method repeatedly moves from a current solution to a better one nearby. The method also utilizes various techniques to escape from local solution locks, and to reach the global solution. These techniques are aspiration criteria, deterministic recency, and frequency approaches [2,3,4]. Successful applications of the tabu search method are found in many fields such as power system [5], transportation [6], food processing [7], flow shop [8], etc. Despite the success, the simplistic tabu search, of the type sometimes applied in the literature, cannot completely escape from a local minimum lock. This can be readily verified by applying the method to search for the global minimum of the Bohachevsky's function [9]. To avoid the difficulty of local solution locks, some researchers proposed modified versions of the tabu search such as reactive tabu search [10], and probabilistic tabu search [11]. We have proposed a novel version of tabu search so called adaptive tabu search (ATS) that is more efficient than the conventional method.

The chapter's presentation begins with the review of the conventional tabu search as a background for the readers. Then, it presents detailed step-by-step explanation of the ATS. The readers will be able to follow the algorithms easily. The convergence of the ATS is guaranteed by our theorem proofs. The chapter also presents detailed evaluation of the ATS performance. The usefulness of the algorithms is described as applications in power system, model identification, and control. A novel control structure so called neuro-tabu-fuzzy control is also inclusive. The chapter is closed by giving a discussion on future trends, and conclusion, respectively.

2. Naïve Tabu Search

Naïve tabu search (NTS) is an extension of local search methods. Glover proposed this attractive search tool to overcome local minima almost twenty years ago [1,2,3], especially in combinatorial optimization problems. The earliest and simplest release of the tabu search method as we call "Naïve tabu search" has four basic elements, which are i) search space, ii) neighborhoods, iii) search memories, and iv) aspiration criteria [12]. Later versions including ours, namely adaptive tabu search, are technically based upon these four features. Their combination gives one, among the very efficient intelligent search algorithms, able to provide many good solutions close to the global optimum. Besides, it is capable of handling some problem difficulties to achieve closer solutions to the global one much better than other intelligent search methods are. Basic structure and algorithm to perform the search according to the conceptual framework of minimization are reviewed as follows.

The backbone of the NTS lies on the theme of local search methods in combination with search memories. The algorithm starts with an initial guess solution in the search space. A neighborhood of the initial solution is thus generated as a local search. The best among the neighborhoods is selected to be the initial solution for neighborhood generation in the next iteration. A repetitive process will guide the most recently updated solution to a local minimum. This strategy, however, does not always succeed. Frequently, it misleads the iteration to a local trap cycling. Any generated solutions will occasionally be locked in this loop. When the search process goes on and on, it seems to us that there is no further significant improvement to the best solution found thus far, and it jumps to the conclusion that the global solution would have been successfully obtained. To prevent this failure, a short-term memory block called tabu list (TL) is employed. It is used to store the sequence of a few previously visited solutions. Previous solutions in the TL are regarded as forbidden moves. In other words, they will not be included in a set of generated neighborhoods for later search rounds.

Use of the TL is effective to prevent a local trap to some extent. Nevertheless, in some applications, the forbidden moves can cause another serious consequence. With too large memory size of the TL, the search process may spend a substantial amount of its searching time to reach the global minimum or in the worst case it may experience a failure. Therefore, an additional feature referred to as aspiration criteria (AC) for cancellation of such forbidden moves is committed. However, it is just a conceptual mechanism. There is no precise rule for constructing this feature. It is only postulated that the AC is a tool to revoke tabus when the improvement of the objective function of the recently updated solution is stagnated or many attractive moves are prohibited, although there is no sign of any solution cycling. By this reason, the use of the TL enhances the local search algorithm significantly in such a way that the move continues in the right direction, e.g. descent direction in the minimization problem, and its solution converges to the global minimum at the end of the process.

To terminate the search process, there must be at least one termination criterion to justify whether or not the search has reached the global solution. The most commonly used termination criteria found in literature [6,7,12,13,14,15] are i) a fixed number of iterations or a fixed amount of CPU time (it must be large enough), ii) no further improvement in the objective function within a certain number of iterations, and iii) a pre-defined threshold value of the objective function.

The above explanation reflects the earliest and simplest version of the tabu search. When a problem complexity arises, the NTS often fails to achieve the global minimum. Many researchers over the last decade have foreseen this weakness. Thus, various works [10,16,17,18,19,20, 21,22,23,24] have been released to enhance the tabu search algorithm and to bring this technique back onto the top of the league. Although several techniques have been used as supplements, there can be categorized into two essential strategies: i) Intensification, and ii) Diversification [12,18]. Briefly, the Intensification is a strategy to guide the search process to focus on only a small portion of the search space, which would probably contain the best solution. While the Diversification is a mechanism forcing the search process to jump to other unexplored areas of the search space such that a deadlock could be broken out. These two optional features can be embedded into every tabu search process. If either one or both are applied, the search might be named by its originators, e.g. reactive tabu search [10,16,17], probabilistic tabu search [18,19] or even our adaptive tabu search [20,21,22,23,24], of which details can be found in the following section.

3. Adaptive Tabu Search: Structure and Algorithm

Although to enhance the NTS in practice has a long history and can be achieved by various approaches based on the two optional features, we particularly devote our several research works over half a decade focusing on two distinctive features, namely adaptive search radius and back-tracking mechanisms. Besides both features, we also notice that, during each iterative search round, a set of generated neighborhoods is not necessary to be entirely explored. Without providing an evidence at this moment, we presume that only a portion of a whole neighborhood is sufficient to overcome a local minimum. The ideas underlying these supplements are to completely eliminate local traps, to avoid tedious computation in creating recursive neighborhoods, and to speed up the search process to reach the global minimum as quickly as possible. The structure of the ATS can be explained as follows.

We start the search from an initial guess solution as common. In the ATS, only a portion of all the neighborhoods is randomly chosen as shown in Figure 1. The best solution among the obtained samples is selected as the initial for the next iteration. Also, the best solution from the samples at each iteration is put into a first-in-last-out memory block, so-called a local tabu list (TL). It must be noted that we define two TLs in our proposed algorithm, i) short-term or local TL, and ii) long-term or global TL. Although it consists of two TLs, each will be used for different purposes. The local TL is to store the sequence of the local solution candidates found consecutively during any descent movements toward one local minimum. It implies that each iteration is sampled as shown in Figure 2. To reduce the size of the local TL, it is reset every time when a local minimum has been successfully found. As the NTS, the local TL is a device to prevent solution cycling. Whereas, the global TL has a lower sampling rate that must be greater than one iteration. The global TL will be used in conjunction with our back-tracking mechanism and will be explained later. It is clear that there is no confusion between both TLs. So, we will call them "the TL" throughout this chapter.

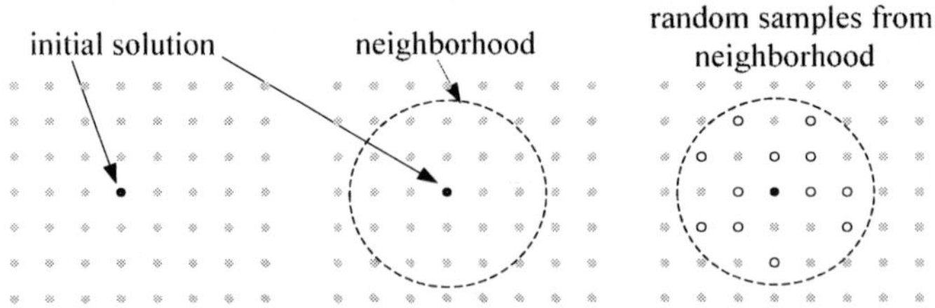

Figure 1. Neighborhood and its random samples

Random sampling from neighborhood at each iteration may cause some problems. Some potential candidates might not be chosen, and a loss of good offsprings could occur. With uniform random and a sufficient numbers of samples, the best solution movement can keep its descent direction toward a local minimum. However, fast search time to reach the minimum cannot be guaranteed. To overcome this problem, a refinement of the next iteration search is necessary when the movement is close to the minimum. In the ATS, the adaptive search radius is introduced. Particularly, we use this feature as the Intensification strategy. At the beginning or after reaching a local minimum, the search process begins with a provided initial solution. Neighborhoods are defined as solutions around the initial within a circle of a regular search radius. The search process is then performed iteratively with the regular search radius until the recent move is sufficiently close to the local minimum. The adaptive search radius mechanism is therefore activated, and subsequently adjusted in a heuristic manner. The search radius is usually decreased. With no limitation, we may invoke the search radius adaptation as many times as we prefer. After performing this adaptation, the search process will eventually reach the local minimum. As mentioned previously, the search radius will be reset to the regular value for the next local search. The concept of adaptive search radius can be represented by the diagram in Figure 3.

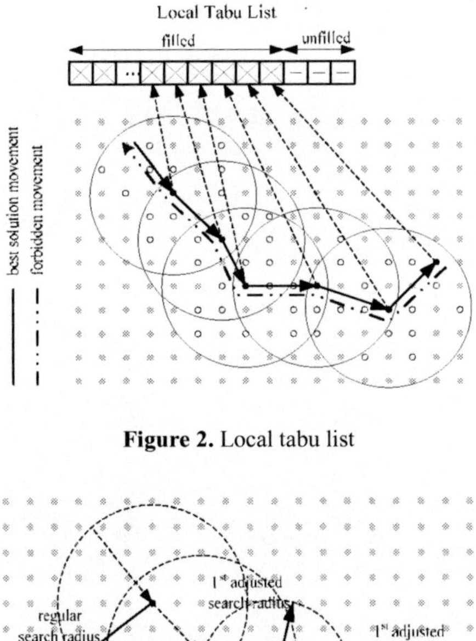

Figure 2. Local tabu list

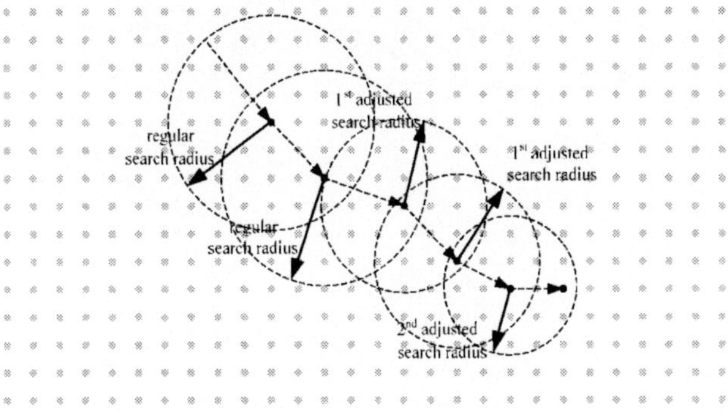

Figure 3. Adaptive search radius

We also use the Diversification strategy, called back-tracking, to reinforce our algorithm such that it is able to jump across local solutions effectively to an unexplored area in the search space. This leads to an expectation of a new local minimum found at the end of the next local search. To achieve this goal, another memory stack, the global TL, is required to store some already visited solutions. An explored solution in a set of samples can be put into this stack whether it is the best solution in the current iteration or not. This idea gives a variety of initial solutions to be used as the starting point for the next local search round. The string of sub-search-spaces must be sampled less frequent than this is done with the local TL, e.g. 3 iterations, 5 iterations, or as large as its allocated memory space, as shown in Figure 4. However, the best solution as well as other explored solutions can also be placed into the global TL. When the back-tracking with the global TL are active, the search is allowed to generate a new path to trace another local minimum as graphical representation shown in Figure 5.

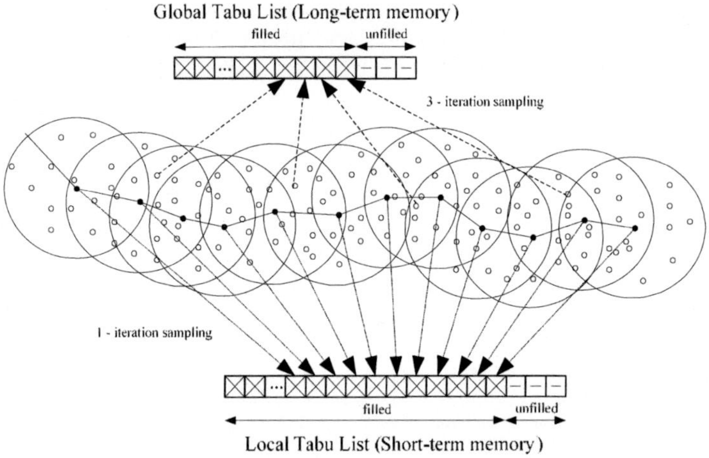

Figure 4. Global tabu list

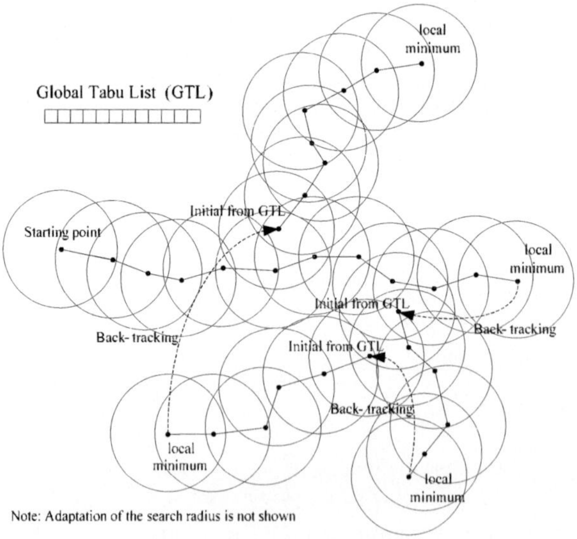

Figure 5. Multiple local minima obtained by the back-tracking scheme

As explained, the algorithm of the ATS can be summarized, step-by-step, as follows.

1. Initialize the Tabu Lists (TL = ϕ) and reset all program counters.
2. Randomly select an initial solution x_0 from the search space Ω and assign it as the global minimum, x^*.
3. Create a sub-space $\Psi \subset \Omega$ from the neighborhood of the initial. Evaluate the objective function of $\forall x \in \Psi$. A solution that gives the minimum objective function among them is given as x'.
4. If $x' < x_0$, keep x_0 in the local TL and set $x_0 = x'$. Otherwise put x' in the local TL instead to prevent cycling. Update the global TL (if any).
5. Update the global minimum. $x^* = x_0$ if $x_0 < x^*$.
6. Check the termination criteria (TC) and the aspiration criteria (AC), respectively
   - Go to step 7 if TC is satisfied, otherwise repeat step 3.
   - Activate the adaptive search radius (AR) mechanism if necessary to speed up the searching process.
   - Activate the back-tracking (BT) mechanism if a local minimum trap occurs and repeat step 3.
7. Terminate the search process. The last updated x^* is the global minimum found.

As can be seen, only a few numbers of solutions in Ω would be visited randomly and it is sufficient to locate the global minimum.

4. Convergence of the ATS

The ATS is more efficient than the NTS in the senses that the ATS can reach the global minimum faster than the NTS can. The following proofs of theorems confirm the claim.

Theorem 1. If a sub-space Ψ has a large member of its total members, m, to give good representatives of a neighborhood, a local minimum in the vicinity can be found by generating a few successive sub-spaces.

Proof : Let $x_{t=0}$ be an initial solution in a sub-space $\Psi_t \subset \Omega$ to generate a sequence of solutions x_{t+1}. The successive property has to be considered to guarantee that the search converges to a local minimum solution. For a current search, two cases occur: (i) The search can find a better solution than the previous one, i.e. $f(x_{t+1}) < f(x_t)$; or (ii) The search cannot find a better solution, i.e. $f(x_{t+1}) \geq f(x_t)$.

During the search process, a set of nearby solutions, $N_\rho(x_t)$, to the current solution, x_t, will form and possess N members. Concurrently, a sub-search-space, $\Psi_{t+1} \subset N_\rho(x_t)$, will form randomly and possess m total members, where m is constant and $m < N$. Only some u members belonging to $N_\rho(x_t)$ result in $f(x) < f(x_t)$, $x \in \Psi_{t+1}$.

Let the probability of finding a better solution be $P = P(f(x) < f(x_t))$. Hence,

Case I : $(m > N-u)$

$P = 1$ and at least one of the m members provides a satisfied condition of $f(x) < f(x_t)$.

Case II : $(m \leq N-u)$

Probability of not being able to find any better solution is

$$\bar{P} = \frac{\binom{N-u}{m}}{\binom{N}{m}} = \frac{(N-u)!(N-m)!}{(N-u-m)!N!}$$

Each search round, the ATS updates the current solution with a better solution if it exists. As time passes, the search moves toward a local solution, i.e. $\lim_{t\to\infty} u(t) = 0$. It can be said that the search cannot find any better solutions than the current one. That is to say

$$\lim_{t\to\infty} P(t) = \lim_{t\to\infty} \bar{P}(t) = \lim_{t\to\infty} \frac{\left(N-u(t)\right)!(N-m)!}{\left(N-u(t)-m\right)!N!} = 1$$

and a local minimum is found.

This completes the proof.

Theorem 2. The AR mechanism accelerates the search to reach the minimum. The search rapidly converges to the local minimum by suitably adjusting the search radius ρ.

Proof : The search radius of a sub-space Ψ is $\rho = \mu \cdot \gamma$ in which $\mu = 1.0$ at the beginning of the search. When a current solution moves closer to $\hat{x}$, i.e. $\gamma < \rho$, in other words $\hat{x} \in N_\rho(x_i')$, one of the total N members becomes the local minimum $\hat{x}$. Hence, probability of finding the local minimum, $\hat{x}$, is

$$P = \frac{\binom{m}{1}}{\binom{N}{m}} = \frac{m \cdot m!}{N(N-1)\cdots(N-m+1)}$$

where N is the total number of members of $N_\rho(x_t')$. From theorem 1, $m < N$ means that a longer search radius (N is large) results in a lower probability P. Thus, an appropriate way to adjust the search radius is to reduce the radius of the sub-space gradually, i.e. $0 < \mu \leq 1$ selectively. Therefore, for $\gamma < \rho' < \rho \wedge (N > N' \in \mathrm{I}^+)$

$$\left[P_{\rho'}(\hat{x}) = \frac{m \cdot m!}{N'(N'-1)\cdots(N'-m+1)}\right] > \left[P_\rho(\hat{x}) = \frac{m \cdot m!}{N(N-1)\cdots(N-m+1)}\right]$$

where ρ' is a reduced radius, N and N' are the total number of members of $N_\rho(x_t')$ and $N_{\rho'}(x_t')$, respectively. Since a shortened search radius results in a greater probability of finding the local minimum, the search can reach the solution more rapidly.

This completes the proof.

Theorem 3. With the BT mechanism, the search process obtains multiple local minima, one of which is the global minimum.

Proof: The search sometimes fails to improve the current solution x_0 leading to an unchanged initial solution for the next search, and eventually an entrapment of solutions. Based on a random process, the next search may direct a new search to the vicinity of the boundary of $\Lambda(x_0)$ and the convex region nearby. $N_\rho(x_0)$ may overlay with some convex regions, i.e. $N_\rho(x_0) - \Lambda(x_0) \not\subset \Lambda(x_0)$. This property enables the search to escape from solution locks to some extent. In fact, the escaping process may fail because of the ineffectiveness of the conventional algorithms. To increase the possibility of successful solution-lock run away, some candidate solutions historically stored in the TL can be selectively used as an initial solution for the next search. Using this tactic, a jump from one convex region to another occurs to escape a deadlock made by a local minimum.

Given *n_re* be a counter for a solution cycling. In this context, "solution cycling" means that the search is trapped by the just visited local minimum, so the next search will be attracted to that just visited solution. The *n_re* counter is increased by one at any time the final solution of the search being equal to the one previously visited and already stored in the TL. Given *n_re_max* be the maximum allowance of the solution cycling. Therefore, the ATS invokes the BT mechanism according to the following conditions. If $n\_re \geq n\_re\_max$, then perform the BT, else continue the search whether or not it can escape the solution lock. The condition "$n\_re \geq n\_re\_max$" is a kind of AC. Once it is met, one historic candidate in the TL is retrieved to be a new initial for creating the next sub-space Ψ. That is to say the BT selects a solution $x_h \in$ TL such that $x_h = \underset{x_i \in \mathrm{TL}}{M} \left\| x_i - x_0 \right\|$, and the condition $f(x_0) < f(x_h)$ holds. Then, a new initial for the next search is $x_0 = x_h$ Therefore;

(i) Providing the local minimum $\hat{x}$ is obtained, and the length of TL is finite, there exists at least one solution very close to the boundary of $\Lambda(x_0)$. Hence,

$$\mathrm{length(TL)} >> 1 \rightarrow \exists x \in \mathrm{TL} \wedge \left\| x - x_B \right\| < \delta$$

where x_B is a boundary point, and δ is the maximum allowance.

(ii) Providing the statement (i) is true, and the radius ρ is long enough to reach some solutions outside $\Lambda(x_0)$, the best solution of a current Ψ can be situated outside $\Lambda(x_0)$ with high probability. That is

$$\left(\Psi(x_0) - \Lambda(x_0)\right) \subset \left(N_\rho(x_0) - \Lambda(x_0)\right) \not\subset \Lambda(x_0) \rightarrow \exists x \notin \Lambda(x_0)$$

(iii) Providing the statement (ii) is true, a new search could reach a new local minimum of a new search region nearby. Repeating the process could search k different local minima within finite search time $p \cdot m \cdot Time(x) < \sum_{i=1}^{n} w_i \cdot Time(x)$, one of the k minima is the global minimum.

This completes the proof.

5. Performance of the ATS

To evaluate its performance, the ATS was coded in MATLAB™ for running on a Pentium IV, 1.6 GHz, 256 Mbytes SD-RAM. The task was to find the global minimum against two functions: the Bohachevsky's function (BF), Eq. (1), [9] and the circle function (CF), Eq. (2), [25]. Their surfaces are shown in Figure 6 and Figure 7, respectively. The reason for choosing these functions is that finding their global minima is not possible using some conventional methods. Some unconventional methods may hit the global minimum only by chance, but commonly fail to find it, and they are usually trapped by local solutions. Both functions have the global minimum at $x = y = 0$ with $f(0,0) = 0$. $\varepsilon = 1\times10^{-5}$ is used to approximate zero and set as the TC.

$$f(x,y) = x^2 + 2y^2 - 0.3\cos(3\pi x) - 0.4\cos(4\pi y) + 0.7 \tag{1}$$

$$f(x,y) = \left(x^2 + y^2\right)^{1/4}\left(\sin^2\left(50\left(x^2 + y^2\right)^{1/10}\right) + 0.1\right) \tag{2}$$

There are five parameters affecting the search performance of the ATS as follows (i) the initial search radius (R), (ii) the number of neighborhood members (n), (iii) the number of repetitions of a solution before invoking the back-tracking mechanism (n_re_max), (iv) the kth backward solution selected by the back-tracking mechanism (k[th] backward selection), and (v) the amount (in %) of search radius reduction compared to the previous radius.

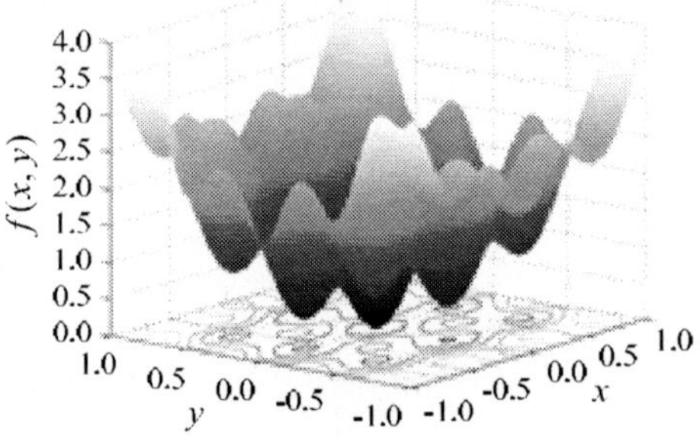

Figure 6. Surface of BF

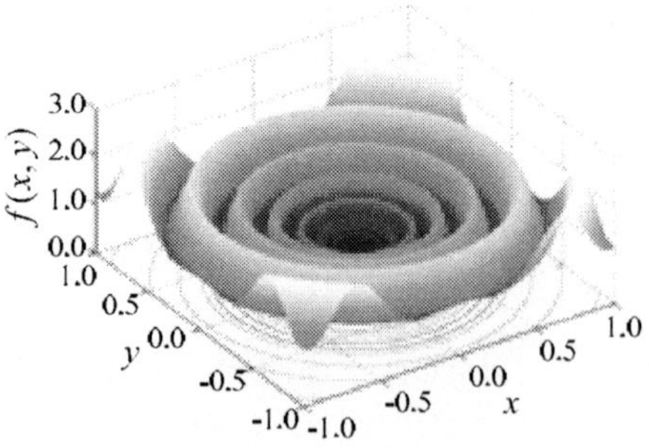

Figure 7. Surface of CF

The performance tests were conducted 1,000 trials against each parameter's values. Each trial starts with a random solution. The ATS stops when either of the following stop criteria is met: (i) the maximum search round of 10,000, or (ii) the cost value $\leq \varepsilon$. According to the test procedures, the ATS method is classified as ATS^1 and ATS^2 that denote the ATS having only the back-tracking mechanism, and that having both the back-tracking and the adaptive search radius mechanisms, respectively. The performance tests of the ATS^1 were conducted first. The following values were assigned to search parameters as follows: R = 2.5, 5.0, 7.5, 10.0, 12.5, 15.0, 20.0 % of the search space radius, n = 10, 20, 30, 40, 50, 60, n_re_max = 5, 10, 15, 20, 25, and k^{th} backward selection = -1, -2, -3, -4, -5. The satisfactory search parameters obtained were transported to the ATS^2 for further tests. The tests of the ATS^2 resulted in suitable percent reduction of the search radius providing three steps were used. The adaptive radius scheme can be expressed as: if [cost value < 10^{-1}] then [R_a=R/DF, R= R_a]; if [cost value < 10^{-2}] then [R_a=R/DF, R= R_a]; if [cost value < 10^{-3}] then [R_a=R/DF, R= R_a], where R_a means adapted radius. DF stands for decreasing factor that provides the current search radius of 10, 15, 20, 25, and 30% of the previous radius, respectively. The small constants appear in the conditions of the if-then clauses were derived from local solution locks.

Table 1. Effects of the search parameters.

Search parameters		BF			CF		
		Average search rounds	Average search time (sec)	No. of trials solution found	Average search rounds	Average search time (sec)	No. of trials solution found
R (%)	2.5	7923.3	48.51	203	9341.4	52.43	214
	5.0	7156.6	41.18	296	6610.2	38.87	405
	7.5	3876.6	20.95	757	3742.3	20.63	885
	10.0	1353.1	6.06	1000	4878.8	25.40	816
	12.5	2263.7	10.60	988	5955.1	33.03	700
	15.0	3071.4	14.91	954	6796.2	37.60	548
	20.0	4832.4	24.59	808	8038.0	47.15	363
n	10	4135.1	10.33	893	7134.4	21.82	526
	20	2203.3	7.72	987	4802.4	20.23	804
	30	1353.1	6.06	1000	3742.3	20.63	885
	40	1089.1	6.37	1000	3029.3	19.37	910
	50	904.1	6.49	1000	2928.5	25.06	873
	60	802.5	6.70	1000	2304.2	22.31	908
n_ re_ max	5	1310.9	5.91	1000	3029.3	19.37	910
	10	1322.1	5.95	1000	3279.4	22.15	871
	15	1353.1	6.06	1000	3438.3	23.91	858
	20	1518.8	6.61	998	3466.4	24.12	851
	25	1438.8	6.51	1000	3360.0	23.37	862
k^{th}	-1	1498.4	6.33	998	3455.1	24.09	866
	-2	1488.4	6.68	998	3364.4	23.65	871
	-3	1478.4	6.90	998	3253.7	22.21	867
	-4	1586.3	7.31	998	3115.4	21.45	884
	-5	1462.8	6.21	999	3029.3	19.37	910
Re-duced R (%)	10	24.4	0.09	1000	1195.4	8.70	892
	15	26.2	0.10	1000	1200.9	10.52	887
	20	30.2	0.13	1000	600.9	4.20	942
	25	38.4	0.16	1000	601.1	4.42	940
	30	64.2	0.31	1000	802.9	7.15	914

The results obtained by the performance tests are shown in Table 1. From the ATS[1] tests, it can be noticed that the R = 7.5-12.5%, n = 30-40, n_re_max = 5-15, and k^{th} backward selection = -5 give a satisfactory search performance in terms of solution convergence and search time. The search parameters obtained by the ATS[1] tests rendering good performance were transported to the ATS[2] test as follows: BF–{R = 10.0%, n = 30, n_re_max = 5, and k^{th} = -5} and CF–{R = 7.5%, n = 40, n_re_max = 5, and k^{th} = -5}. The results of the performance test of the ATS[2] are also summarized in the Table 1. Although very high rate of solution convergence can be assured for both functions, it can be concluded that 20-25% reduction of the radius R gives very good performance in terms of speed and solution convergence.

The recommendation for the use of the ATS method is provided for parameter selection to obtain an efficient search as follows:

(i) the initial search radius, R, should be 7.5-15.0% of the search space radius,
(ii) the number of neighborhood members, n, should be 30-40,
(iii) the n_re_max should be 5-15,
(iv) the kth backward selection should be equal or close to the n_re_max,
(v) the adaptive search radius should employ 20-25% of radius reduction, and
(vi) a well educated guess of the search space that is wide enough to cover the global solution is necessary.

However, the proposed ATS method is still problem-dependent like other AI search techniques, although it performs well for minimization of both tested functions. So, understanding the problem domain well, and selecting an appropriate form of the objective function are essential for a successful application of the method.

6. Applications

6.1 Power Systems

A 5-bus power system as shown in Figure 8 is chosen to demonstrate a relay setting example using the ATS. Although there are many possible fault locations and types, in this demonstration only two fault conditions are situated at buses 2 and 5. Currents and voltages during faults can be obtained from the fault calculation software. These two fault cases are used to set up the objective function for relay coordination.

To achieve the optimum time relay grading, the system objective is given as follows.

$$\text{minimize} \qquad F_{obj} = \sum_{k=1}^{m} \left(\sum_{i=1}^{n} \frac{\alpha_i \times TDS_i}{\left(\frac{I_{i,k}}{I_{S,i}} \right)^{\gamma_i} - 1} \right)$$

$$\text{Subject to} \qquad t_{u,j} - t_{d,j} > T_{gm} \qquad ; j = 1,2,\ldots,J$$

where

F_{obj} is the system objective function,

α_i are γ_i are arbitrary constants of relay i,
TDS_i is the time-dial setting of relay i,
$I_{S,i}$ is the pick-up current of relay i,
$I_{i,k}$ is the fault current seen by relay i for case k,
T_{gm} is the time-grading margin allowance,
$T_{u,j}$,$T_{d,j}$ are the operating time of upstream and downstream relays of pair j,
n, m and J are the total number of designed relays, fault cases, and relay pairs, respectively.

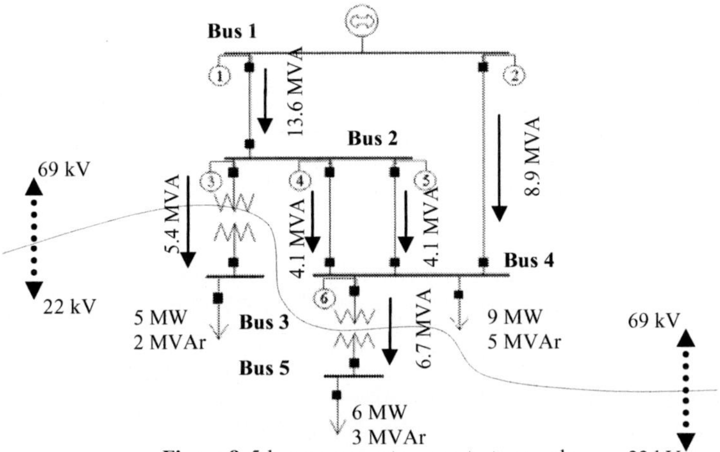

Figure 8. 5-bus power system as a test example

In this case study, all the digital relays are characterized by $\alpha = 0.14$ and $\gamma = 0.02$ (standard inverse) from the IEEE Std C37.112-1996 and 100% pick-up current setting. Also, all selected CT ratios are shown in Table 2. To minimize the objective function, the search space of the problem is defined by [0.05,1.00] for the time-dial setting. Optimizing the objective function by using the ATS method, the obtained optimal solution (TDS) is shown in Table 2 together with the operating time of the two fault cases. In addition, the grading graph interpreted from the obtained optimal solution is shown in Figure 9(a). Note that relay 3 is not involved with the fault cases. Moreover, the convergence of the search process is presented in Figure 9(b).

Table 2. Optimal solution resulting from the ATS method for digital relay coordination.

Relay number	TDS	CT ratio	Operating time (s)	
			Fault at bus 2	Fault at bus 5
1	0.15	150/5	0.71	0.93
2	0.17	100/5	1.27	0.95
4	0.12	50/5	0.86	0.52
5	0.12	50/5	0.86	0.52
6	0.05	50/5	not operate	0.12

Note: relay 3 is not involved with these two fault cases

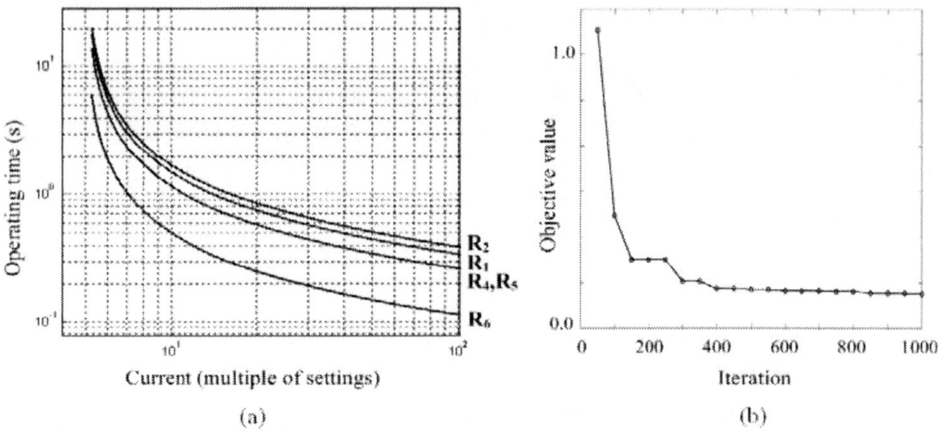

Figure 9. Coordination result and convergence
(a) grading graph of digital relaying and (b) convergence of the search process

6.2 Identification

6.2.1 Identification of Static Nonlinear Models

A. Saturation Characteristic

A two-inertia system with mechanical resonance is represented by the block diagram in Figure 10. The plant, input, and feedback compensators are of third-order transfer functions. The plant's parameters are obtained from the conventional ARMAX identification. The mechanical resonance has been compensated for. Under linearity assumption, both compensators having three poles and three zeros are conventionally designed [26]. In practice, a nonlinear characteristic of saturation type appears in the system. It represents the protective mechanisms of drive amplifier and compensators. We could acquire the speed responses according to over-, normal-, and under-drive situations. With the ATS, the saturation characteristic could be extracted from these recorded responses; however, this is not possible to do with any conventional identification methods. Figure 11 shows the plot of the model representing the system in Figure 10 against the measured data. With the stop criteria, $f(x,y) \leq 1.0$ and maximum search round = 10,000, we have obtained m = 1.145, x_u = 2.759, and x_l = 1.916 with sum squared error = 0.5.

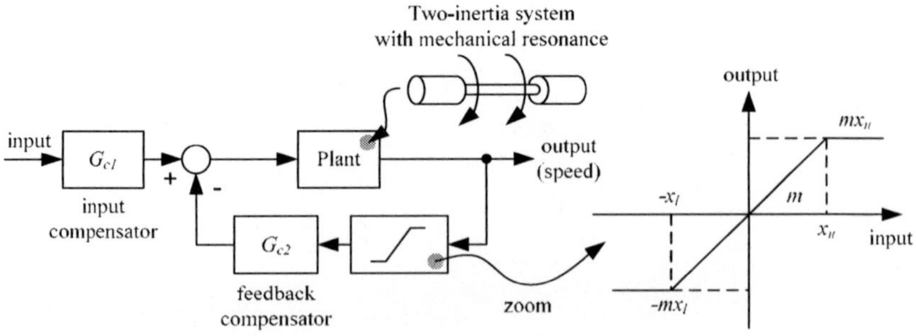

Figure 10. Block diagram representing a compensated two-inertia system

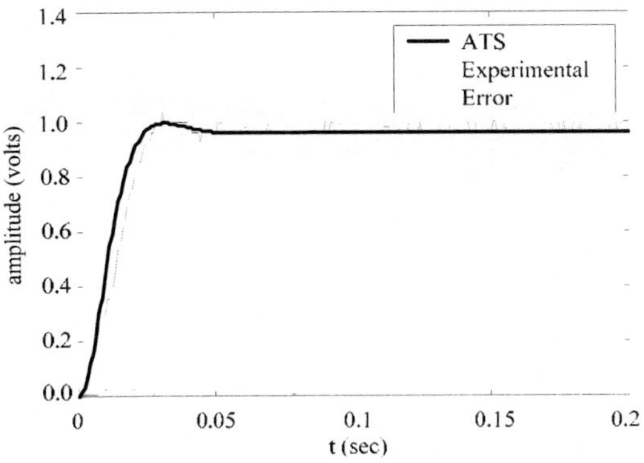

Figure 11. Model plot against measured data

B. Nonlinear Friction Model

It has been known for many years that stick-slip phenomenon occurs when a solid mass translates on a solid surface at very low velocity. This phenomenon is influenced by nonlinear friction described macroscopically as Stribeck's effect [27]. Figure 12 depicts the recorded waveforms of position (②), and speed (④) of a linear slide bed exhibiting stick-slip. Figure 13 shows the nonlinear friction curve in which F_{s+} and F_{s-} = static friction (N), F_{c+} and F_{c-} = Coulomb friction (N), F_{v+} and F_{v-} = viscous friction coefficients (N·s/mm), $+vss$ and $-vss$ = critical speeds (mm/s) (remarks: plus and minus signs indicate positive and negative directions of motion, respectively.) The friction model is expressed by Eq. (3) where v = velocity of mass (mm/s), and F_{in} = force inserting on mass (N).

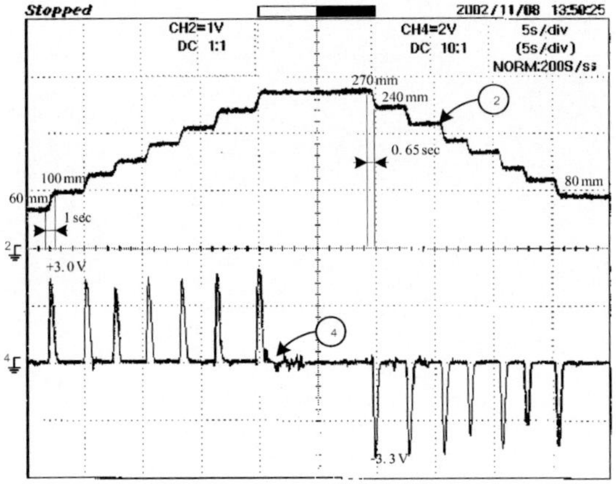

Figure 12. Recorded position (curve ②), and speed (curve ④)

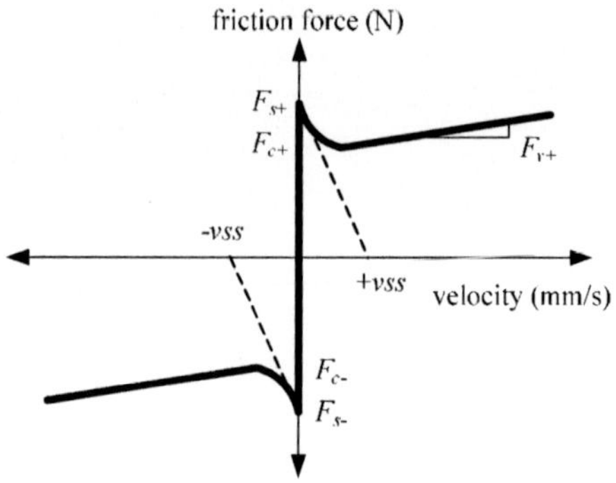

Figure 13. Nonlinear friction with Stribeck's effect

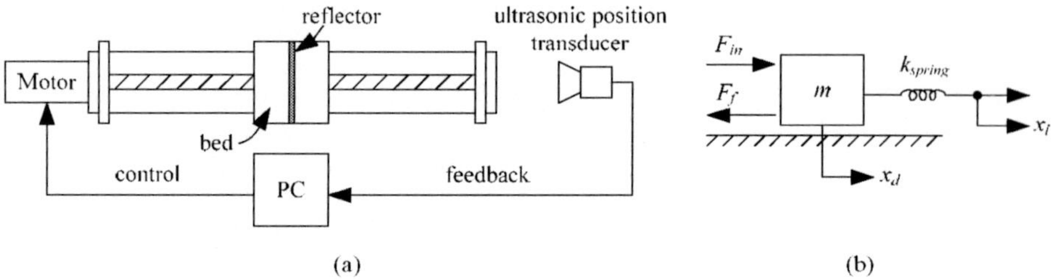

Figure 14. The slide bed
(a) a linear slide bed controlled by a PC and (b) mass-spring system representing the slide bed

$$F_f(v,F_{in}) = \begin{cases} F_{c+} + (F_{s+} - F_{c+})e^{-\left|\frac{v}{vss}\right|} + F_{v+}|v| \ , & v > 0 \\ F_{s+} \ , & v = 0 \text{ and } F_{in} > 0 \\ F_{s-} \ , & v = 0 \text{ and } F_{in} < 0 \\ F_{c-} + (F_{s-} - F_{c-})e^{-\left|\frac{v}{-vss}\right|} + F_{v-}|v| \ , & v < 0 \end{cases} \tag{3}$$

$$m\ddot{x}_d = k_{spring}(x_i - x_d) - F_f(v, F_{in}) + F_{in} \tag{4}$$

Many observations were made on a feedback control system of a linear slide bed represented by the diagram in Figure 14(a). Using position feedback control, the bed followed the reference speed of ±5 mm/s. However, it did not move smoothly as shown by the waveforms in Figure 12. For the purpose of parameter identification, the following quantities were recorded: bed position and speed, control signal, motor voltage and current. The sliding range of the bed was 400 mm. Some careful signal scaling was necessary during the measurement. Our linear slide bed can be represented by the mass-spring system as shown in Figure 14(b). Eq. (4) describes the motion where m = mass of the bed (kg), x_d = bed displacement (mm), F_f and F_{in} already declared. During the ATS search process, it was necessary to compare the experimental results with the simulation results based on the motion and friction models such that sum squared errors (SSE) could be calculated. The objective of the ATS was to minimize the SSE. The stop criterion was $J_{SSE} \leq 2{,}700$ mm^2. With the following constants $m = 10.9$ kg, $k_{spring} = 0.37$ N/mm, and $vss = 3.198$ mm/s, the following model parameters (average values) were obtained: $F_{s+} = 130.771$ N, $F_{c+} = 47.750$ N, $F_{v+} = 0.8$ N·s/mm for positive direction of motion, and $F_{s-} = -123.902$ N, $F_{c-} = -48.230$ N, and $F_{v-} = -0.8$ N·s/mm for negative direction of motion.

6.2.2 Pendulum-on-Cart System

A pendulum-on-cart system practically represents a crane lifter that is intrinsically nonlinear and unstable. The diagram in Figure 15 represents the pendulum-on-cart system having a non-uniform force, *f*, exciting the cart. This force is generated by a motor and a flexible belt. The belt motion consists of at least three modes: longitudinal (left-right), up-down swing, and sway depending upon the motor input, *u*. So, the pendulum swings unpredictably.

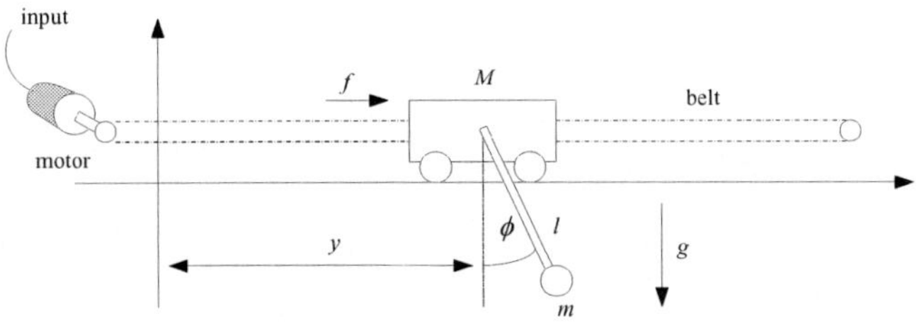

Figure 15. Pendulum-on-cart system

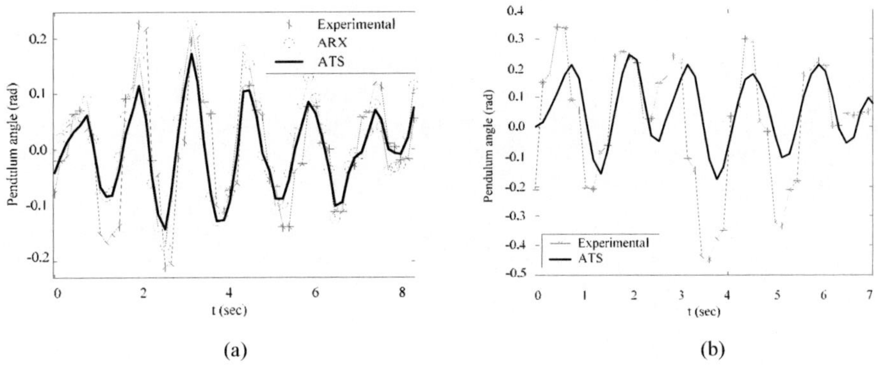

Figure 16. Plot of the models against observed data (a) linear model and (b) nonlinear model

With small angles of oscillation, ϕ, the system can be considered linear and described by the 5th-order black-box model. We have applied both the classical ARX method and the ATS for comparison of their identification results. The observed angles of oscillation, ϕ, and the plots of the 5th-order models of both methods are depicted in Figure 16(a). The sum squared errors of the ARX model and the ATS method are 10.1412 and 7.7107, respectively.

With larger angles of oscillation, ϕ, the pendulum system becomes nonlinear and can be described by the Eqs. (5) and (6) [28]. In both equations, y = cart position, M = cart mass (not known), f = force exerted by belt (not known), m = pendulum mass = 0.251 kg., l = length of pendulum rod = 0.4 m, and g = gravity = 9.81 m/s^2. The pendulum rod is assumed to be weightless. The non-uniform force, f, is transmitted from the motor axle through a flexible belt. The seventh order polynomial representation of the force, f, in terms of the motor input, u, is assumed. The ATS could provide the coefficients of the force expression, and the mass, M. The stop criteria for the search are the cost $J_{SSE} \leq 1.32$ or the maximum search rounds of 10,000. The tests were conducted 10,000 trials with random initial solutions. The ATS provided the solutions with an average $J_{SSE} = 1.3188$, average search rounds of 841.55, and consumed 83.72 seconds of average search time. The search results in $M = 1.0738$ kg, and $f = 4.595u^7 + 6.714u^6 - 1.476u^5 - 5.714u^4 - 5.621u^3 - 1.867u^2 + 2.330u + 0.0083$. Figure 16(b) illustrates the model plotted against the observed angles of oscillation of the pendulum. The model provides satisfactory information on modes of oscillation although the amplitude errors are still great. The amplitude errors can be decreased by attempting to model the force more accurately.

$$\ddot{\phi} = \frac{f\cos(\phi) + 0.5ml\sin(2\phi)(\dot{\phi})^2 + (M+m)g\sin(\phi)}{l\left[m\cos^2(\phi) - (M+m)\right]} \tag{5}$$

$$\ddot{y} = \frac{f + 0.5mg\sin(2\phi) + ml\sin(\phi)(\dot{\phi})^2}{\left[(M+m) - m\cos^2(\phi)\right]} \tag{6}$$

6.2.3 Vibrating Tube System

The diagram of the vibrating tube system is shown in Figure 17(a). The tube positions are depended on the force, f, applied by the motor's shaft. With motor's random input, the tube motion consists of at least three modes: clockwise-counterclockwise rotation, up-down swing, and sway. The dynamic behavior of this system in 3D space can be represented by the PDE models as shown in the Eqs. (7) and (8). The vibrating string and membrane models form the basis of these two equations [29].

$$\frac{\partial^2 z(x,y,t)}{\partial t^2} = \frac{f}{\mu}\left(\frac{\partial^2 z(x,y,t)}{\partial x^2} + \frac{\partial^2 z(x,y,t)}{\partial y^2}\right) - g_y \tag{7}$$

$$f = K_D K_M u \tag{8}$$

Referring to the Eqs. (7) and (8), $z(x,y,t)$ = tube position in z axis, μ = mass of tube per unit length (not known), u = motor input, K_M = motor gain (not known), K_D = driver gain (not known), and gravity $g_y = 9.81$ m/s^2. The tube position, $z(x,y,t)$, varies with the positions in x axis (x), y axis (y), and time (t).

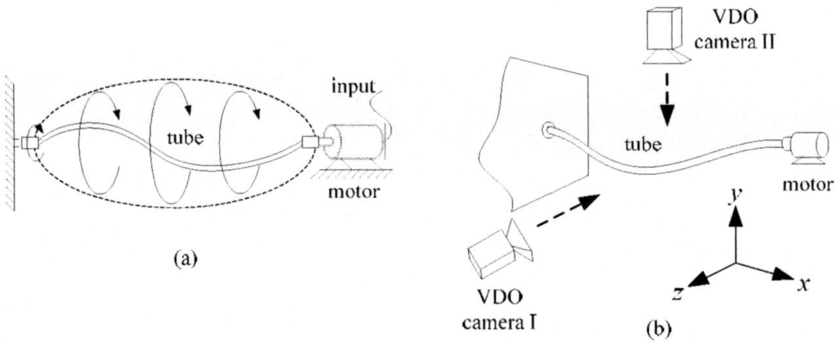

Figure 17. Vibrating tube system
(a) system diagram and (b) experimental set up

It is not possible to install sensors on our experimental tube since the tube is small and soft. To acquire the tube dynamic, we develop a novel identification technique via image processing approach. The diagram in Figure 17(b) illustrates our experimental set up. The VDO cameras require some careful calibrations [30], and synchronization. The data representing the tube dynamic in response to the motor's random excitation can be extracted from the recorded images. Then, the ATS is applied to search for the model parameters. The stop criteria are the maximum search rounds of 1,000 or the cost $J_{MSE} \leq 23.26$ representing accumulated mean squared errors (MSE) between the experimental and the model quantities. The tests were conducted 1,000 trials with random initial solutions. The ATS provided the solutions with an average $J_{MSE} = 23.25$, average search rounds of 10.19, and consumed 8.83 seconds of average search time. The following model parameters are obtained: $\mu = 0.0064$ kg/m, $K_D = 0.022$, and $K_M = 0.0243$. Figure 18 illustrates the model plotted against the observed positions of the tube. The results obtained are highly satisfactory.

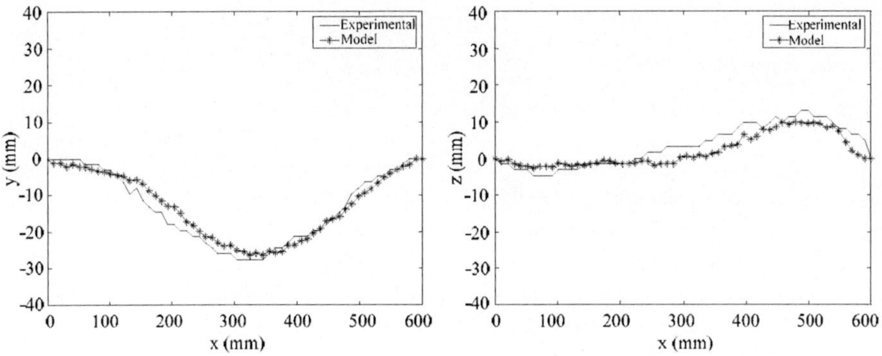

Figure 18. Plot of the models against observed data

6.3 Control

6.3.1 Input-Weight Optimization for PID and PIDA

The use of a PID controller in a feedback control system for industrial applications was first introduced in 1939 [31], while a PIDA controller was first introduced in 1996 [32]. Due to harsh industry environment, a real plant is prone to performance degradation, and parametric perturbations. In this context, Eitelberg [33] introduced a method to maintain satisfactory system response by leveling of input signals, called input weighting factors. The original control was proposed for manual operation subjected to an operator's experience. This type of control is suitable for an AI-based controller implemented online to optimize system performance. Moreover, an offline optimization is an alternative. Regarding this, the ATS method is to provide an offline optimization of the input weighting factors for the PID and PIDA controllers as shown in Figures 19(a) and 19(b), respectively. The cost function *J*, errors between the reference *R*(*s*) and the actual response *C*(*s*), is fed back to the ATS block, and minimized to obtain a set of input weighting factors, *Fs*, driving the system to produce the optimum response.

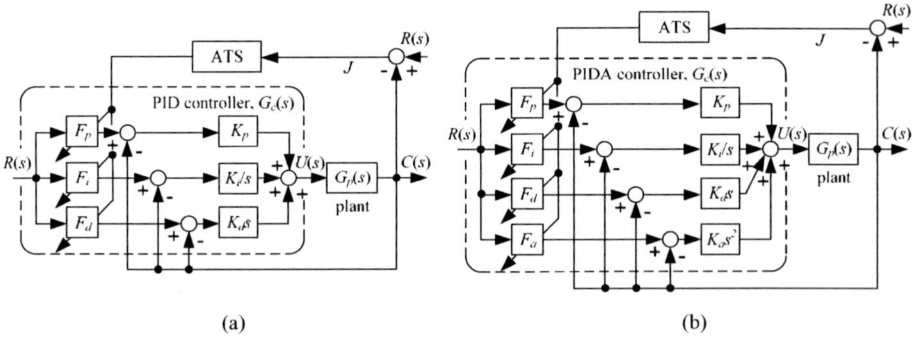

Figure 19. Input weighting optimization by ATS
(a) PID controller and (b) PIDA controller

For the PID case, a DC servo motor is used as a plant. Its transfer function is expressed by Eq. (9). K_p = 25.6, K_i = 282.35, and K_d = 0.1 are the PID parameters obtained by a conventional design method. Parameters of the plant and the controller are assumed to be variant to some extent due to aging and working environment. In this case, the following desired specifications are $T_r \leq$ 0.01 s, $P.O. \leq 20\%$, $T_s \leq 0.03$ s, and $E_{ss} \leq 0.001$. They are set as inequality constraints of the problem. After termination of the search, the input weighting factors are successfully obtained by the ATS. Figure 20(a) displays step responses of three cases for comparison with the *Fs* reported in the figure.

$$G_p(s) = \frac{1}{(1+8.5\times10^{-2}s)(1+1.77\times10^{-3}s)(1+5.55\times10^{-3}s)} \quad (9)$$

$$G_p(s) = \frac{8.60\times10^{11}}{s^5+1590s^4+9.44\times10^5 s^3+1.24\times10^8 s^2+5.38\times10^{10}s+8.42\times10^{11}} \quad (10)$$

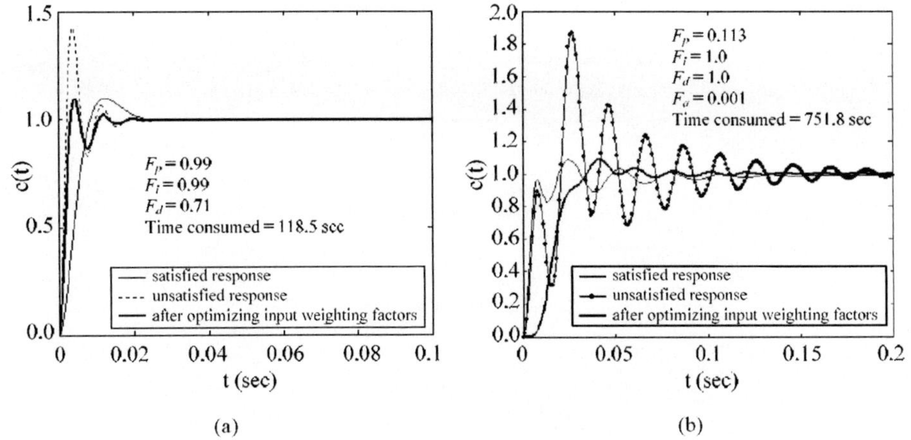

Figure 20. Step responses
(a) DC servo motor control system and (b) two-mass rotary control system

For the PIDA case, a two-mass rotary system that exhibits torsional resonance akin to the one described in the topic 6.2.1(A) is used as a plant. The transfer function of the system is stated in Eq.(10). K_p = 15, K_i = 120, K_d = 0.0056 and K_a = 0.0003 are obtained by the ATS such that a preliminarily satisfactory response is achieved. The ATS has been applied because the conventional PIDA design approach results in a response with too light damping as can be seen in Figure 20(b). System parameters are assumed to be variant. The desired specifications are given by $T_r \leq 30$ ms, $P.O. \leq 10\%$, $T_s \leq 0.1$ s, and $E_{ss} \leq 0.001$, and set as inequality constraints for the ATS to seek for the optimum *Fs*. Figure 20(b) displays step responses in a similar manner to the Figure 20(a) with the *Fs* embedded in the figure. It can be noticed that the systems with the optimum *Fs* obtained from the ATS render the most satisfactory responses.

6.3.2 Neuro-Tabu-Fuzzy Control

The single input rule module (SIRM) fuzzy controller has been proposed [34,35] to control single and double inverted pendulum systems. This controller requires the dynamic importance degrees (DIDs) in its structure. These values indicate the relative importance among rules' inputs in terms of weights. Basically, they are obtained from trial-and-error. Each rule can be read as " *if* $x = A$ *then* $u = C$ " in which A and C are the membership functions of input and output, respectively. We propose the neuro-tabu-fuzzy (NTF) control structure represented by the block diagram in Figure 21. The OSF in the diagram represents the output scaling factor block. With the NTF control, tuning of the DIDs can be done via backpropagation learning of the feedforward multilayer neural network (NN) whose optimum initial weights and biases are obtained from the ATS. While NN helps on learning and reliefs the need for trial-and-error task, the optimum parameters provided by the ATS effectively reduce stabilization time.

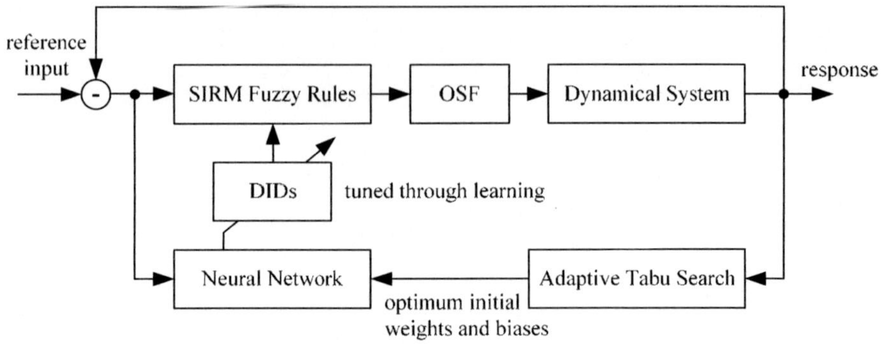

Figure 21. Neuro-tabu-fuzzy (NTF) control structure

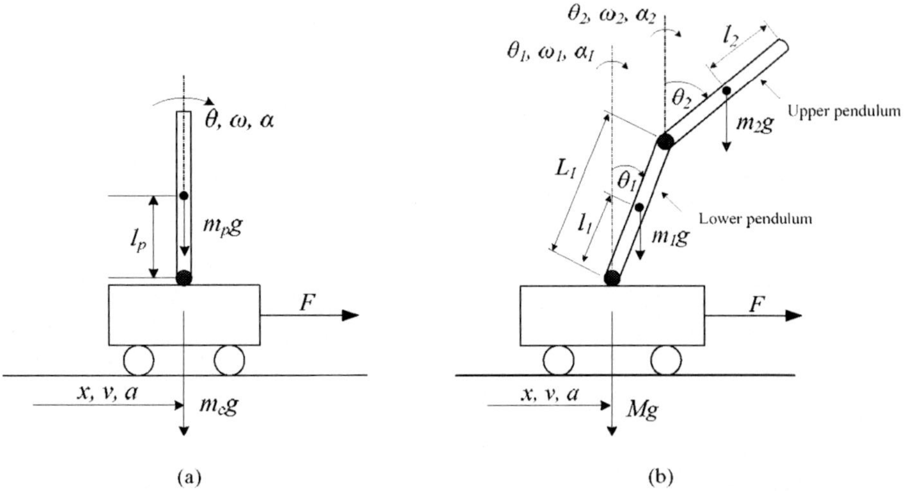

Figure 22. (a) Single inverted pendulum system and (b) Double inverted pendulum system

We compare the performances between the NTF control and the SIRM fuzzy control by having them stabilize the single and the double inverted pendulum systems (Figures 22(a) and (b)) to upright positions. Both controllers have the same set of fuzzy rules. Some simulation results are shown in Figure 23. We conducted thousands of simulation cases with various initial conditions of pendulum angles, pendulum lengths, and cart positions. To summarize, the NTF controller could stabilize the single inverted pendulum system with shorter stabilization time of about 30-55% than the SIRM fuzzy controller could, and 15-38% shorter for the case of double inverted pendulum.

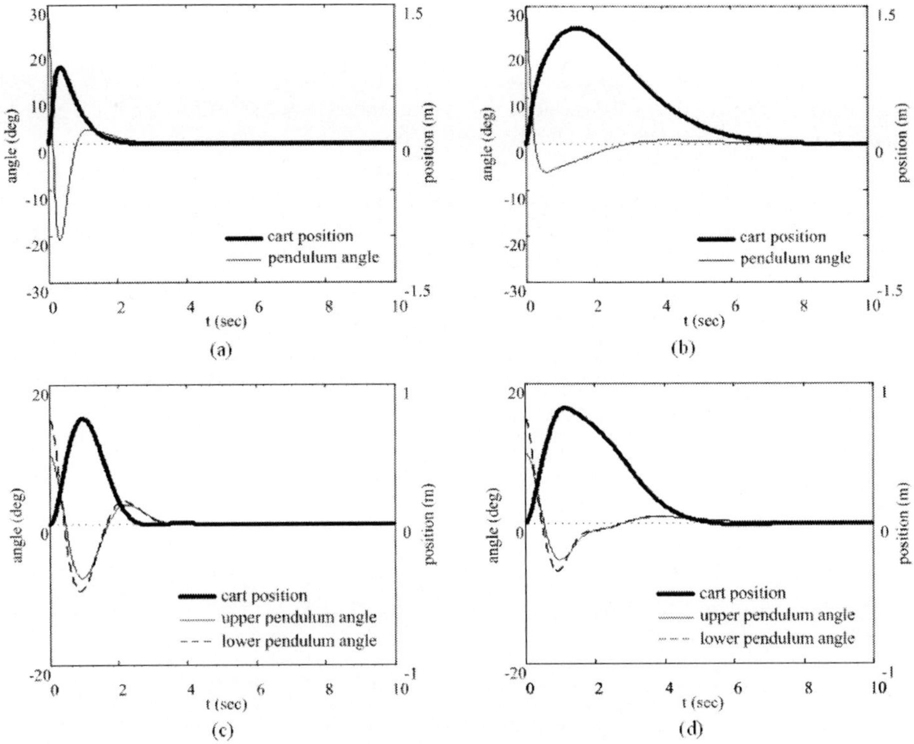

Figure 23. Simulation results of stabilization of
(a) single pendulum by NTF controller (b) single pendulum by SIRM controller
(c) double pendulum by NTF controller and (d) double pendulum by SIRM controller

7. Future Trends

Since the emergence of the ATS, successful applications in various areas have been demonstrated. Those include our innovations on identification via image processing, and neuro-tabu-fuzzy control. Despite advantages of the ATS over the NTS, faster solution finding is still a challenging goal. Our future contributions will be parallel ATS (PATS) and multi-path ATS (MATS). Both share the same principle of dividing a search space into many sub-spaces. With PATS, those sub-spaces will be searched simultaneously on parallel platforms. This means that PATS needs more than one computer. With MATS, searches will begin with different initial solutions for those sub-spaces. Many sets of the ATS algorithms will perform the tasks on a single platform. It may seem slow at the start of the search. During the search process, the ATS algorithms on different sub-spaces will compete each other. Based on some criteria of convergent rate and successful avoidance of local solution lock, the best among them will be singled out. In this, "the best" means a particular sub-space on which the ATS currently performs with the highest probability of finding the global solution. Convergence analysis for PATS and MATS will be based on either probabilistic or Markov chain approach.

8. Conclusion

This chapter presents the detailed explanation of the adaptive tabu search (ATS) with recommendations for users. Since the ATS is an enhanced version of the conventional or naïve tabu search (NTS), the chapter gives a review of the NTS also. To ensure the potential users about the performance of the ATS, the performance evaluation and the convergence proofs are elaborated rigorously. The readers can find the recommendations for selecting search parameters in the section describing performance evaluation. To confirm the usefulness and the effectiveness of the ATS, the chapter presents case studies in the following areas: power system, model identification, and control. Among those cases, the readers can find our innovations on system identification via image processing, and neuro-tabu-fuzzy control. One optimistic future of the ATS will be the emergence of multi-path ATS (MATS) and parallel ATS (PATS) with a variety of technical applications.

Acknowledgment

The authors' thanks are due to Dr.Arthit Srikaew for his valuable discussions, Kittiwong Suthamno, and Sudarat Khwan-on for performing some computing parts of the works.

References

[1] F. Glover, Future paths for integer programming and links to artificial intelligence, *Computers and Operations Research* **13**(1986), 533 549.
[2] F. Glover, Tabu search – Part I, *ORSA Journal on Computing* **1**(1989), 190 206.
[3] F. Glover, Tabu search – Part II, *ORSA Journal on Computing* **2**(1990), 4 32.
[4] F. Glover and M. Laguna, *Tabu Search*, Kluwer Academic Publishers, Norwell, 1997.
[5] A.H. Mantawy, Y.L. Abdel-Magid and S.Z. Selim, Unit commitment by tabu search, *IEE Proc.-Gener. Transm. Distrib* **45(1)** (1998), 56 64.
[6] J-F. Cordeau and G. Laporte, A tabu search heuristic for the static multi-vehicle dial-a-ride problem, *Transportation Research Part B: Methodological* **37(6)** 2003, 579 594.
[7] G. Zhang, W. Habenitch and W.E.L. SpieB, Improving the structure of deep frozen and chilled food chain with tabu search procedure, *Journal of Food Engineering* **60(1)** (2003), 67 79.
[8] E. Nowicki and C. Smutnicki, A fast tabu search algorithm for the flow shop problem, *European Journal of Operational Research* **91**(1996), 160 175.
[9] I.O. Bohacheevsky, M.E. Johnson and M.L. Stein, Generalized simulated annealing for function optimization, *Technometrics* **28(3)** (1986), 209 218.
[10] R. Battiti and G. Tecchiolli, The reactive tabu search, *ORSA Journal on Computing* **6(2)** (1994), 126 140.
[11] Y.A. Kochetov and E.N. Goncharov, Probabilistic tabu search algorithm for multi-stage uncapacitated facility location problem, *Operations Research Proceeding,* Springer (2000), 65 70.
[12] M. Gendreau, An introduction to tabu search, *98' INFORMS Conference: Bridging Continents & Cultures* (1998), 26 29.
[13] E.L.D. Silva, J.M.O. Areiza, G.C.D. Oliveira and B. Binato, Transmission network expansion planning under a tabu search approach, *IEEE Trans Power Systems* **16(1)** (2001), 62 68.
[14] T. Kulworawanichpong and S. Sujitjorn, Optimal power flow using tabu search, *IEEE Power Engineering Review* **22(6)** (2002), .40 37
[15] B. Lin and D.C. Miller, Application of tabu search to model identification, *AIChE Annual Meeting* (2000)
[16] P. William Nanry and J. Wesley Barnes, Solving the pickup and delivery problem with time windows using reactive tabu search, *Transportation Research Part B: Methodological* **34(2)** (2000), 107 121.

[17] Y. Fukuyama, Reactive tabu search for distribution load transfer operation, *IEEE Power Engineering Society Winter Meeting* **2** (2000), 1301 1306.
[18] Y. Rochat and E.D. Taillard, Probabilistic diversification and intensification in local search for vehicle routing, *Journal of Heuristics* (1995), 147 167.
[19] U. Faigle and W. Kern, Some convergence results for probabilistic tabu search, *ORSA Journal on Computing* **4(1)** (1992), 32 38.
[20] D. Puangdownreong, K-N Areerak, A. Srikaew, S. Sujitjorn and P. Totarong, System identification via adaptive tabu search, *Proc. IEEE Int. Conf. on Industrial Technology (ICIT'02)* **2** (2002), 915 920.
[21] T. Kulworawanichpong, K-L. Areerak, K-N. Areerak and S. Sujitjorn, Harmonic identification for active power filters via adaptive tabu search method, *Lecture Notes in Artificial Intelligences LNAI 3215 Part III* (2004), 687 694.
[22] T. Kulworawanichpong, K-N. Areerak and S. Sujitjorn, Moving towards a new era of intelligent protection through digital relaying in power systems, *Lecture Notes in Artificial Intelligences LNAI 3215 Part I* (2004), 1255 1261.
[23] T. Kulworawanichpong, K-L. Areerak, K-N. Areerak, P. Pao-la-or, D. Puangdownreong and S. Sujitjorn, Dynamic parameter identification of induction motors using intelligent search techniques, *IASTED International Conference on Modelling, Identification and Control (MIC2005)* (2005), 328 332.
[24] D. Puangdownreong, K-N. Areerak, K-L. Areerak, T. Kulworawanichpong and S. Sujitjorn, Application of adaptive tabu search to system identification, *IASTED International Conference on Modelling, Identification and Control (MIC2005)* (2005), 178 183.
[25] S.D. Miller, J. Marchetto, S. Airaghi and P. Koumoutsakos, Optimization based on bacterial chemotaxis, *IEEE Trans. Evol. Comput.* **6** (2002), 16 29.
[26] S. Sujitjorn, C. U-Thaiwasin and Y. Prempraneerat, Torsional resonance suppression via pole-zero assignment, *IASTED International Conference on Modelling, Identification and Control (MIC2000)* (2000), 288 292.
[27] B. Armstrong-Helouvry, Stick slip and control in low-speed motion, *IEEE Trans. AC* **38(10)** (1993), 1483 1495.
[28] A.M. Bloch, N.E. Leonard, and J.E. Marsden, Controlled Lagrangians and the stabilization of mechanical systems I: the first matching theorem, *IEEE Trans. AC* **45(12)** (2000), 2253 2270.
[29] A.P. French, *Vibrations and waves*, MIT Introductory Physics Series, Chapman & hall, 1992.
[30] J. Heikkila and O. Silven, A four-step camera calibration procedure with implicit image correction, *IEEE Int. Conf. on Computer Vision and Pattern Recognition* (1997), 1106 1112.
[31] S. Bennett, Development of the PID controller, *IEEE Control System Magazine* (1994), 58 65.
[32] S. Jung and R.C. Dorf, Analytic PIDA controller design technique for a third order system, *Proc. of the 35th Conf. on. Decision and Control* (1996), 2513 2518.
[33] E. Eitelberg, A regulating and tracking PI(D) controller, *Int. J. Control* **45(1)** (1987), 91 95.
[34] J. Yi and N. Yubazaki, Stabilization fuzzy control of inverted pendulum systems, *AI in Engineering* **14(2)** (2000), 153 163.
[35] J. Yi, N. Yubazaki, and K. Hirota, Stabilization control of series-type double inverted pendulum systems using the SIRMs dynamically connected fuzzy inference model, *AI in Engineering* **15(3)** (2001), 297 308.

Integrated Intelligent Systems for Engineering Design
X.F. Zha and R.J. Howlett (Eds.)
IOS Press, 2006

Intelligent Experimental Design Using an Artificial Neural Network Meta Model and Information Theory

Shi-Shang Jang[1,a], David Shan-Hill Wong[a] and Junghui Chen [b]
[a] *Chemical Engineering Department, National Tsing-Hua University*
Hsin Chu, Taiwan 300
[b] *Department of Chemical Engineering, Chung-Yuan Christian University*
Chung-Li, Taiwan 320

Abstract. Ability to rapidly design products and their manufacturing process is a key to being competitive in a dynamic market environment. Traditional methods of design of experiment development are unsatisfactory when applied to design problems with large number of input variables and nonlinear input-output relation. A meta-model driven experimental design scheme is developed. The approach uses artificial neural network as the meta-model, and a combination of random-search, fuzzy classification, and information theory as the design tool. An information free energy index is developed which balances the needs for resolving the uncertainty of the model and the relevance to finding the optimal design. The procedure involves iterative steps of meta-model construction, designing new experiments using meta-model and actual execution of designed experiments. The effectiveness of this approach is benchmarked using a simple optimization problem. Three industrial examples are presented to illustrate its applicability to a variety of design problem.
Keywords: Information theory, artificial neural network, experimental design

Introduction

In the competitive market, speediness in product or process development is the key to success due to short product life cycles. Due to the shortness of the development stage and process shelf life, it is unlikely that a process be understood thoroughly so that sophisticated first-principle model be developed and used for optimization and design. Finding recipes and designing new processes are basically empirical. Getting experimental data, if not difficult, is time-consuming and costly. Traditionally, a systematic methodology that includes statistical data analysis and decision making is known as experimental design [1], [2], [3] is used for to minimize the number of experiments and direct process development. However, such methods become unsatisfactory when the number of design variables becomes very large and the input-output relation is highly nonlinear with multiple local minima

Product and process development are regarded as learning experiences that have been the focus of many artificial intelligence researchers. For example, Fukunaga [4] described the process of classifier design or statistical pattern recognition in a series of iterative steps: data gathering, registry, analysis, classifier design, and testing. The logic does not differ from that of experimental design, except that the tools employed are more suitable for problems with high dimensionality and nonlinearity. Saraiva and Stephanopoulos [5] demonstrated that with existing plant data, top-down induction of decision trees can be used to explore process-improvement opportunities. Pattern recognition or other machine learning methods mine information from data, and validation techniques usually concern with how the model, i.e. synopsis of information can be validated, without regards of how the model will be used. The effect of model accuracy on large combinatory optimization problems have been investigated by Ho and co-workers (e.g., [6], [7]). They showed that if one is concerned with finding an acceptable design but not the optimal design, then even a very rough model with limited accuracy, i.e. a meta-model, would very helpful.

In this chapter, we shall present a novel meta-model driven experimental design scheme that uses the artificial neural network (ANN) as the meta-model. Information theory is used to characterize the uncertainty of the meta-model. The basic philosophy of this work is that predictions of optima and uncertainty are both important features of the model that should be validated. During early stages of development, the emphasis should be placed on resolving uncertainty, while predicted optima should only be pursued only when the model is relatively accurate. This strategy is implemented by

[1] Corresponding Author

performing "importance sampling" on the meta-model with regional random search and fuzzy classification.

The rest of this chapter is organized as the following: in the next section, the theoretical background is introduced. In section 3, the complete algorithm together with the flow chart is presented. An illustrated numerical example and two industrial applications are demonstrated in section 4. Some new development of this approach is discussed in section 5. Conclusive remarks are given in the last section.

1. Theoretical Development

This work aims at exploring the optimal recipe. An optimal design architecture that integrates the neural network and information induction to model the process and search the potential optimal recipe is presented using the following steps:

(1) Using the historical data, an ANN based response surface model RS is constructed. The major advantage of ANN model is not limited such that the polynomial model is only a special case.

(2) Using the above ANN model, two information induction modules: regional optimal search and fuzzy information analysis, are proposed to create some optimal regions and possible optimal operating conditions.

(3) Based on the optimal conditions, the final step, feedback control, suggests the input parameters should be changed to see if the output performance is improved. If not, it is necessary to regenerate the response surface and classification information induction. The model is updated based on the concept of "importance sampling" developed by information theory and fuzzy clustering to explore new design conditions. The iterative procedure keeps running until a satisfied optimal recipe is found.

1-1 ANN Meta-Model Construction

Consider a process system S that generates the experimental data with an input vector $\boldsymbol{x} \in R^N$, unknown disturbance vector $\boldsymbol{z} \in R^P$, and a system output (quality indices) vector $\boldsymbol{y} \in R^M$, such that

$$\boldsymbol{y} = S(\boldsymbol{x}, \boldsymbol{z}) \tag{1}$$

It is our objective to find an optimal recipe that a defined plant profit can be maximized:

$$\max_{\boldsymbol{x}} P(\boldsymbol{x}, \boldsymbol{y}) \tag{2}$$

Since the real input-output relation is unknown, assume that a set of experimental data $\Omega = \{(\tilde{\boldsymbol{x}}_1, \tilde{\boldsymbol{y}}_1), \cdots (\tilde{\boldsymbol{x}}_N, \tilde{\boldsymbol{y}}_N)\}$ are available, where $(\tilde{\boldsymbol{x}}_i, \tilde{\boldsymbol{y}}_i)$, $i=1...N$ denote vectors of the measured control variables and responses. A response surface model for system S:

$$\hat{\boldsymbol{y}} = RS(\boldsymbol{\Xi}, \boldsymbol{x}) \tag{3}$$

can be constructed by regression of the model parameters $\boldsymbol{\Xi} \in R^L$

$$\min_{\boldsymbol{\Xi}} \sum_{i=1}^{N} (\tilde{\boldsymbol{y}}_i - \hat{\boldsymbol{y}}_i)^2 \tag{4}$$

Artificial neural networks are known to be a powerful tool to approximate complex multivariable functions [8], [9]. In this work, standard one-layer feed-forward neural network (Figure 1) is used as the response surface:

$$\begin{aligned} \boldsymbol{h} &= \boldsymbol{W}^h \bullet \boldsymbol{x} + \boldsymbol{b}^h \\ z_i &= a(h_i) \\ \hat{\boldsymbol{y}} &= \boldsymbol{W}^o \bullet \boldsymbol{z} + \boldsymbol{b}^o \end{aligned} \tag{5}$$

A hyperbolic tangent function is used as the activation function a [10]. To obtain the parameters $\boldsymbol{W}^h$, $\boldsymbol{b}^h$, $\boldsymbol{W}^o$ and $\boldsymbol{b}^o$, the pseudo-Gauss-Newton method [11], [12] is used for training. Due to the small number of training data, a statistical technique called the leave-one-out (LOO) cross-validation scheme is used [13].

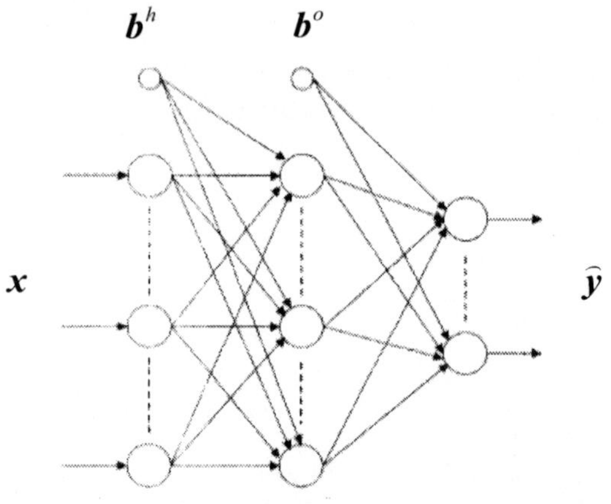

Figure 1: Architecture of a feedforward neural network

1-2 Regional Optimal Search

In product and process development, the feature of interest is the condition that satisfies the desired optimal operation. However, multiple local optima are frequently encountered. It is often necessary to rate alternative local optima based on secondary objectives, such as robustness, safety, etc. Therefore, non-gradient based search procedure is implemented here to extend the area of search on the ANN model (meta model). The four steps are as follows:

Step 1:Use existing experimental point $(x_i,y_i)\in\Omega$ as starting points. Define a local search region as hypersphere around these experimental points.

Step 2:Generate a set of N_r random points at each star point and evaluate the objective function at these points.

Step 3:Extract the best N_s points. Define a local search as a hypersphere with a volume equal to the total search space divided by N_s. Reset N_s as new starting points.

Step 4:Repeat step 2 and 3 until the average performance of the N_s points has no significant change.

1-3 Fuzzy Classification

In reality, it is may be costly to perform the new experiments at all N_s points found in the regional optimal search. Fuzzy classification is used to select the most representative candidate points from these N_s points. The classification algorithm we use is an unsupervised fuzzy classification algorithm (FCM) [14]. The data-clustering problem is to find out C clusters in a set of N_s points: $\{\boldsymbol{x}_1,\boldsymbol{x}_2,\cdots\cdots,\boldsymbol{x}_{N_s}\}$. The cluster structure can be defined by a set of cluster centers $\{\boldsymbol{c}_1,\boldsymbol{c}_2,\cdots\cdots,\boldsymbol{c}_C\}$ and the membership matrix $\boldsymbol{U}=\{u_{ij},i=1\cdots c,j=1\cdots N_s\}$. They can be determined using the following steps:

Step 1. Randomly initialize the membership matrix between 0 and 1 with the constraints

$$\sum_{i=1}^{C} u_{ij} = 1 \qquad \forall j = 1,\ldots,N_s \tag{6}$$

Step 2. Calculate centroids ($\mathbf{c}_i$) by

$$\boldsymbol{c}_i = \frac{\sum_{j=1}^{N_s} u_{ij}{}^m \boldsymbol{x}_j}{\sum_{j=1}^{N_s} u_{ij}{}^m} \qquad i = 1,2,\cdots,C \tag{7}$$

where $m \in [0,\infty)$ is a weighting exponent.

Step 3. Compute dissimilarity between centroids and data points

$$J(U, c_1, c_2, ..., c_C) = \sum_{i=1}^{C}\sum_{j=1}^{N_s} u_{ij}^{m} \left\| x_j - c_i \right\|^2 \quad (8)$$

with $\left\| x_j - c_i \right\|$ being the Euclidean norm between i^{th} centroid c_i and j^{th} data point x_j. Stop if its improvement over the previous iteration is below a threshold.

Step 4. Compute a new membership matrix

$$u_{ij} = \frac{1}{\sum_{k=1}^{C}\left(\frac{\left\| x_j - c_i \right\|}{\left\| x_j - c_k \right\|}\right)^{2/(m-1)}} \quad (9)$$

and go to Step 2.

1-4 Information Index

The purposes of the clustering process are to distill a certain number of homogeneous clusters or classes from a large data set and to classify a concise representation of the individual local optimal behavior. Experiments will then be performed only at the clustering centers. However, the total number of cluster centers is a variable and can be determined by the information index derived in this section. The information theory discussed in this section is to extract K experimental points to be performed in the next stage.

In the above algorithm, the total number of cluster centers is a variable. Information theory is used to determine an appropriate number of clusters. According to Shannon's definition [15] for a variable x, which can randomly take values from a set X, the information entropy of the set X is:

$$S(x) = \sum_{x \in X} p(x) \ln[p(x)] \quad (10)$$

This concept can be extended to measure how clearly the ith cluster is classified:

$$S_i = \sum_{j=1}^{N_s} p(x_j \mid c_i) ln[p(x_j \mid c_i)] = \frac{\sum_{j=1}^{N_s} u_{ji} \, ln(u_{ji})}{N_i} - N_i \quad (11)$$

with N_i being a fuzzy number of data in the ith cluster

$$N_i = \sum_{j=1}^{N_s} u_{ji} \quad (12)$$

And the entropy of the entire classification can be measured:

$$S \equiv \sum_{i=1}^{C} \frac{N_i}{N_s} S_i = \frac{1}{N_s}\left(\sum_{i=1}^{C}\sum_{j=1}^{N_s} \mu_{ij} \ln \mu_{ij} - \sum_{i=1}^{C} N_i \ln N_i\right) \quad (13)$$

An information energy that is the expected value of the performance index is defined as:

$$U = \sum_{i=1}^{C} \frac{N_i}{N} f[\hat{y}(c_i)] - f_{min} \quad (14)$$

f_{min} is the value of the minimum performance index recorded in the optimal search and $f[\hat{y}(c_i)]$ is the value of the performance index at the cluster centers. The information energy is a measure of the relevance of the messages generated by the clustering analysis to the optimization procedure. Provided we have full confidence in our model, it is most desirable that only one cluster center with objective function close to global minimum is generated.

The indices of entropy and energy are measures of how well a set of cluster means classifies the data points and how well a set of cluster means performs if is chosen as the next set of experiments, respectively. To balance them a composite information index the information free energy is defined:

$$F = U - TS \tag{15}$$

The temperature defined is a normalization factor

$$T = \frac{f_{max} - f_{min}}{ln\,N} \tag{16}$$

where f_{max} is the maximum f of all surviving points in the regional optimal search; and N is the total number of existing experiments. During the procedure of determining the number of clusters, temperature remains constant. This is analogous to the thermodynamic equilibrium criterion under the isothermal condition that the free energy is minimized. During the initial phase of the search, when N is small, the data are relatively scattered and f_{max}-f_{min} is relatively large. We would put more emphasis on obtaining the shape of the performance rather than finding the optimum. As the data accumulates with more new experiments, the result of region optimal search will concentrate toward global optima, and f_{max}-f_{min} would decrease. Emphasis should be put less on categorizing information and more on optimization.

1-5 The Algorithm

A flow chart of the entire experimental design procedure is illustrated in Figure 2. There are two ways to implement this criterion. If the subject of investigation is a recipe of a new product, many tests can be conducted simultaneously in a laboratory environment. The cost of a single experiment is of little concern. We can first decide the maximum number of experiments that can be performed in a single batch, and then calculate the information free energy of each classification. The number of experiment in the next batch should be the one that minimizes information free energy. In the early stages of the search, a relatively large number of experiments will be collected before we update the ANN model, but the total number of batches can be reduced. If the subject of investigation in each single experiment is an expensive step, we can start the classification procedure with just one cluster. Calculate the change of information free energy if another cluster is added. If there is an increase in information free energy, the addition of the cluster is rejected. Experiments are then performed at the existing cluster centers. If there is a decrease in information free energy, the addition of the cluster is accepted, and possibility of adding another cluster is investigated again. When data are scarce, the results of regional optimization will be scattered. The information free energy is likely to decrease when we try to add another cluster. The number of experiment in each batch will be kept small. The ANN model is updated more frequently. The number of batches may increase, but the total number of experiments will be reduced.

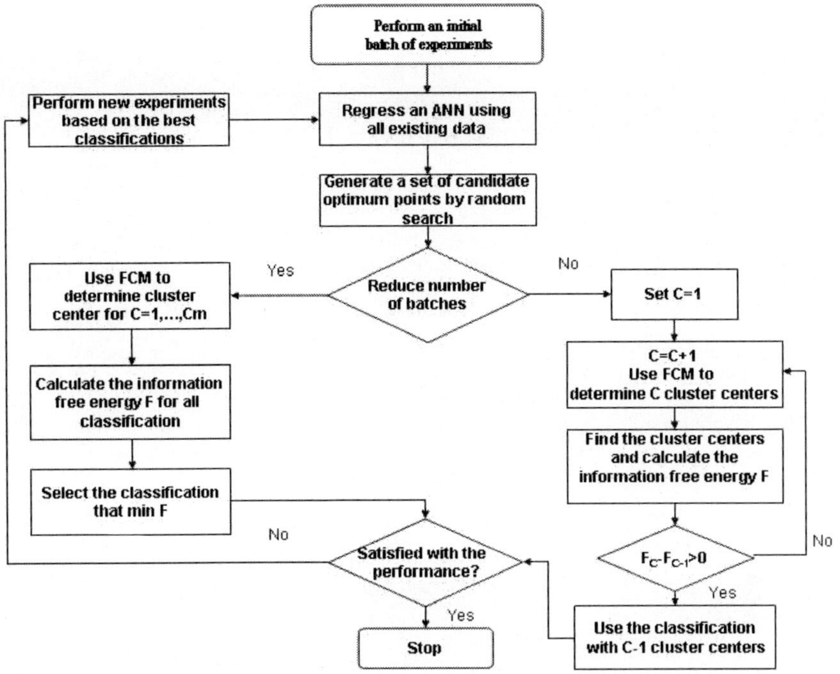

Figure 2: Flow-chart of the intelligent experimental design procedure

2. Case Studies

Four case studies, one a bench mark problem, and three in different industries are presented here to demonstrate the effectiveness of our approach.

2-1 Modified Himmelblau function [16]

The following benchmark problem, a modified Himmelblau function [17] is used to compare the proposed method with the other well known non-convex objective function optimization solvers.

$$H\left(h_1,h_2\right) = (h_1^2 + h_2 - 11)^2 + \left(h_1 + h_2^2 - 7\right)^2 + h_1 + 3h_2 + 57 \qquad (17)$$

defined in the search space $-5 \le h_1 \le 5$ and $-5 \le h_2 \le 5$. The two dimensional nature of the search space allows us to visualize the evolution of sampled points. The real optimum of this problem locates at (-3.8, -3.32), and the optimum value is 43.3. The contour plot of this function is shown in Figures 3a and 3d. The total experimental design approach to discover the real optimum is as the following:

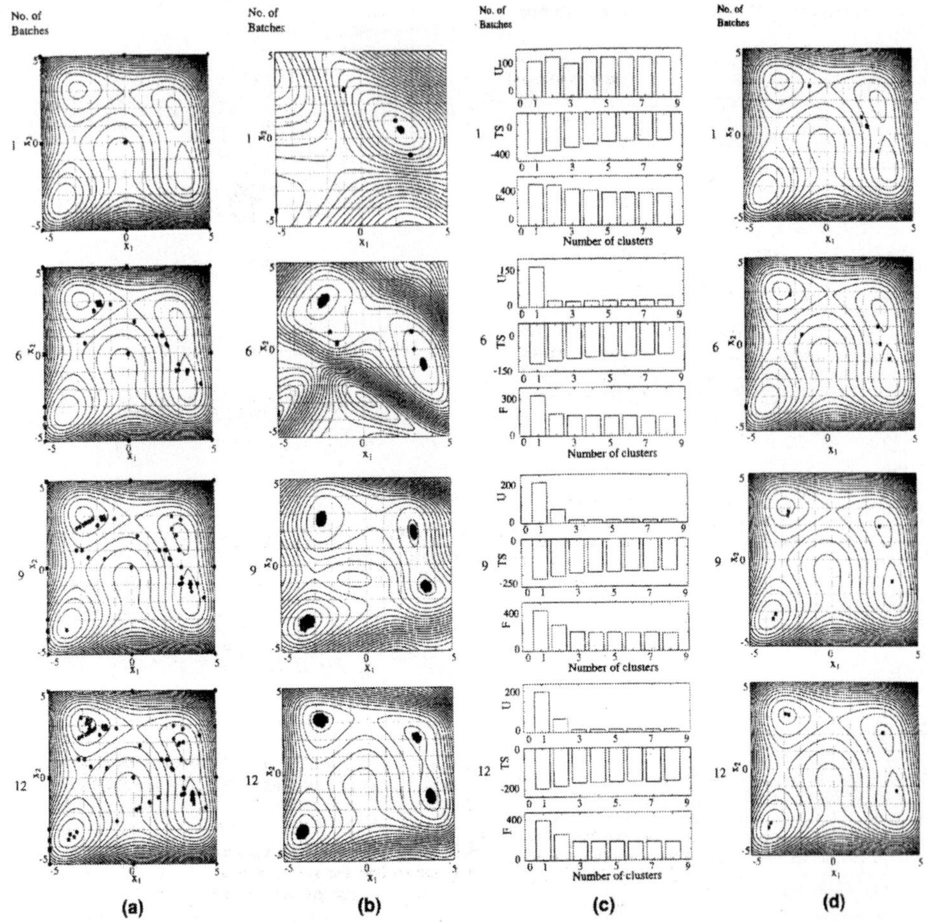

(a) The current and the past experimental points against the contour of Himmelblau function. (b) Corresponding model contour whose solid points represent the found local minimum points. (c) Information analysis plot. (d) The next batch of the new experimental points against the contour of Himmelblau function.

Figure 3: Evolution of sampled points in search space during solution of the modified Himmelblau problem at different batches.

Step 1: A full 3-level factorial design is used to estimated the main effects and interactions of two variables are performed, and function values H(h1,h2) are discovered as shown in column *a* of batch 1 of Figure 3.

Step 2: An ANN model $\hat{y} = ANN(\Xi, h_1, h_2) \approx H(h_1, h_2)$ is built. Column *b* of Figure 3 gives the approximate contour of each iteration batch of Figure 2.

Step 3: Perform regional optimal search described by section 1-2 ad Figure 2 to determine the candidates of optimum points as shown in the column b of Figure 3.

Step 4: Perform the fuzzy classification described in section 1-3 and Figure 2 to determine the cluster centers of the optimum points determined by step 3, and implement the information theory derived by this work to determine the optimal number of clusters as described by section 1-4. Column c of Figure 3 gives the total information free energy as function of number of cluster centers. Column d of Figure 3 depicts the positions of cluster centers that should be evaluated by the experiments.

Step 5: The experiments (or the function evaluations) are performed at the positions determined by the previous step. Column *a* of Figure 3 shows the accumulated experiments (or the function evaluations) performed.

Step 6: The termination condition described in section 1-5 is checked, if the condition is satisfied then the whole approach is ended, otherwise go back to step 2.

Table 1 compares the solution of the above modified Himmelblau function using the proposed approach and several well-known heuristic approaches namely, Nelder-Mead Simplex Method [18], Simulated Annealing [19] and Genetic Algorithm [20]. All results are based on 50 samplings. Gradient based approaches are not included on the comparison the problem because their solution strongly depends on the initial point of the search. Average and standard deviation of optimal values obtained in 10 optimization runs are given so that the results are statistically meaningful. It can be seen that only our proposed approach finds the global optimum in every runs.

Table 1: Comparison of Different Optimization Methods

Optimal Value	Simplex method	Genetic Algorithm	Simulated Annealing	The proposed approach
Average	58.1	66.4	57.0	43.9
Standard deviation	0.573	1.466	0.949	0.173

2-2 Synthesis of Cobalt Blue Color Pigment [16]

Aluminum oxide (Al_2O_3), cobaltous oxide (CoO), zinc oxide(ZnO), magnesium oxide(MgO), potassium nitrate(KNO_3), and potassium chloride(KCl) are the basic ingredients of the cobalt blue color pigment. Al_2O_3 is the bulk material of the pigment and CoO provides the blue color. The color-modifiers, ZnO and MgO, are used during precalcining or premilling to adjust the color of the pigment. The sample can be made greener by adding ZnO; and more red with MgO. Adding mineralizers, KNO_3 and KCl, can reduce the reaction temperature. The sample preparation and color measuring process can be outlined as follows: (1) The six components are weighted and blended. (2) Samples are calcined in a crucible using a preset heating policy. (3) Temperature is ramped to a set point and held constant for a long period. (4) After cooling, the pigments are ground, washed and dried into particles of small size. (5) Body powder and water are added to the pigments. The mixture are powdered in a blender and dried again in an electric fired kiln.

Three color index L, a and b of the sample are measured on a visible spectrophotometer. The color of the final sample is determined by complex interaction between recipe of the sample, and the heating policy in the calcinations process. Only the effect of the recipe is presented here. The objective is to find a recipe that satisfies customer's specification as quickly as possible:

$$\begin{aligned} -1.1 &\le L = L^{exp} - L^{ref} \le -0.9 \\ -0.6 &\le a = a^{exp} - a^{ref} \le -0.4 \\ -2.1 &\le b = b^{exp} - b^{ref} \le -1.9 \end{aligned} \tag{18}$$

where the superscript *exp* and *ref* represent the experimental result and the reference points, respectively. The preceding procedure involves complex chemical reaction. It takes a long time (order of days) to compete the pigment preparation procedure. However a maximum of 10 samples can be processed simultaneously. The results of the suggested experiment are shown in three-dimensional space (L, a and b) in Figure **4**.

Figure **4**(a) shows the results of the all accumulated experiments while

Figure **4**(b) illustrated results of successive batches. The best performance in each batch and the corresponding number of experiments are shown in Figure 5(a) and (b) respectively. The changes in with the number of clusters after first, third and sixth batch of experiments F, U and TS are shown in Figure 6.

The operator's experience provided locations for the first batch of eight experiments, but the results deviate substantially from the desired target. Similarly, in the first few batches, there were suggested experiments that yielded rather unsatisfactory results. However, those experimental points were not wasted. They provided information on the response surface that was incorporated into the neural network model. It should be noticed that a feasible recipe was obtained after the fifth batch, if we allow the procedure to be carried on, a near optimum is always found after the tenth batch Figure 5(a). That means the model become more and more reliable and points around the optimum condition will be sampled (Figure **4**).

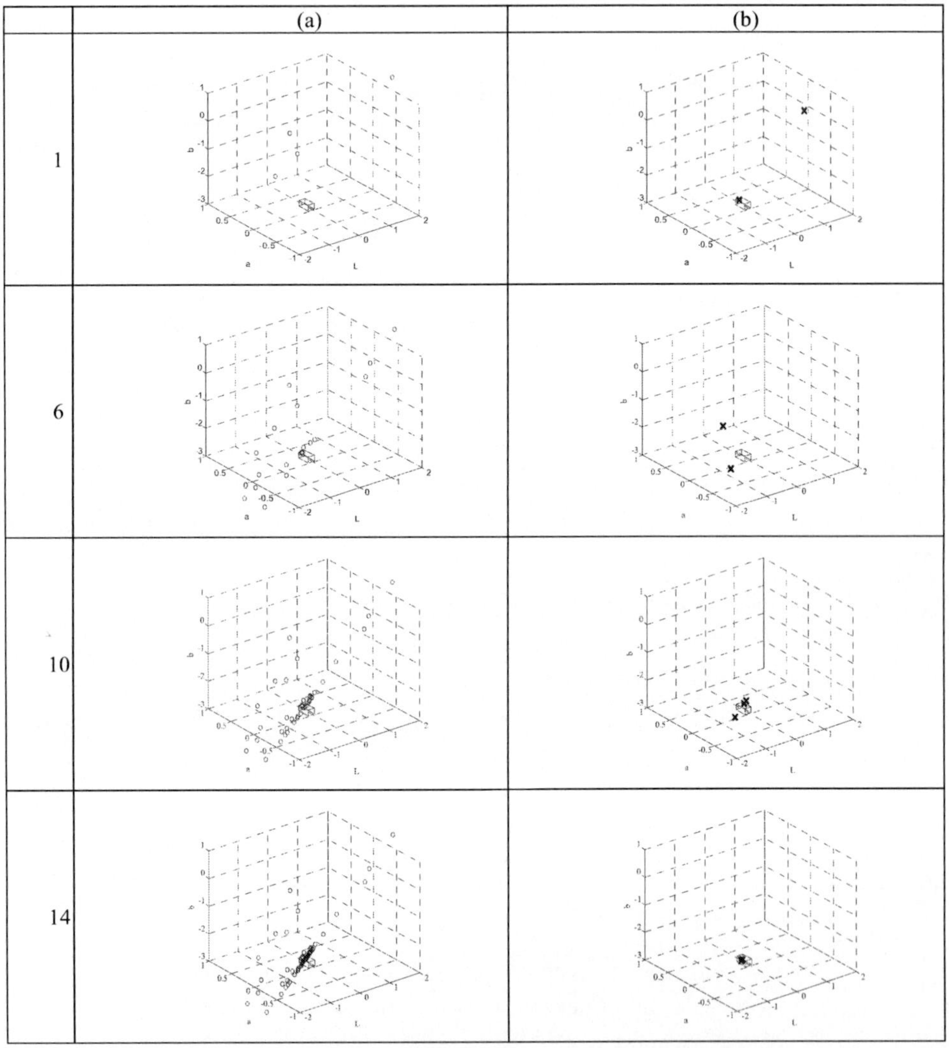

Figure 4: Evolution of performance indices with batches of samples

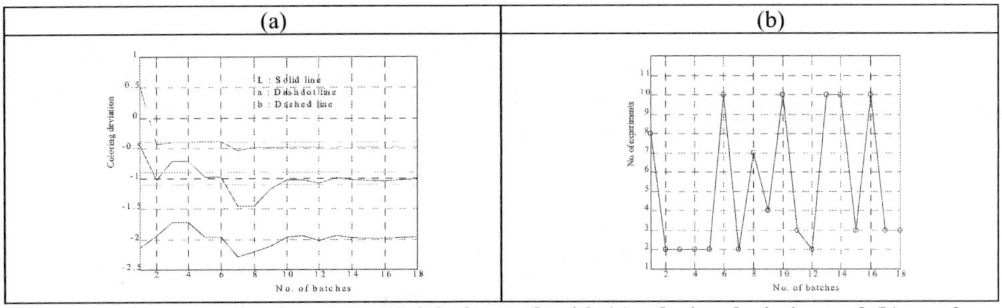

Figure 5: Pigment optimal experimental design path with (a) coloring deviation and (b) number experiments needed.

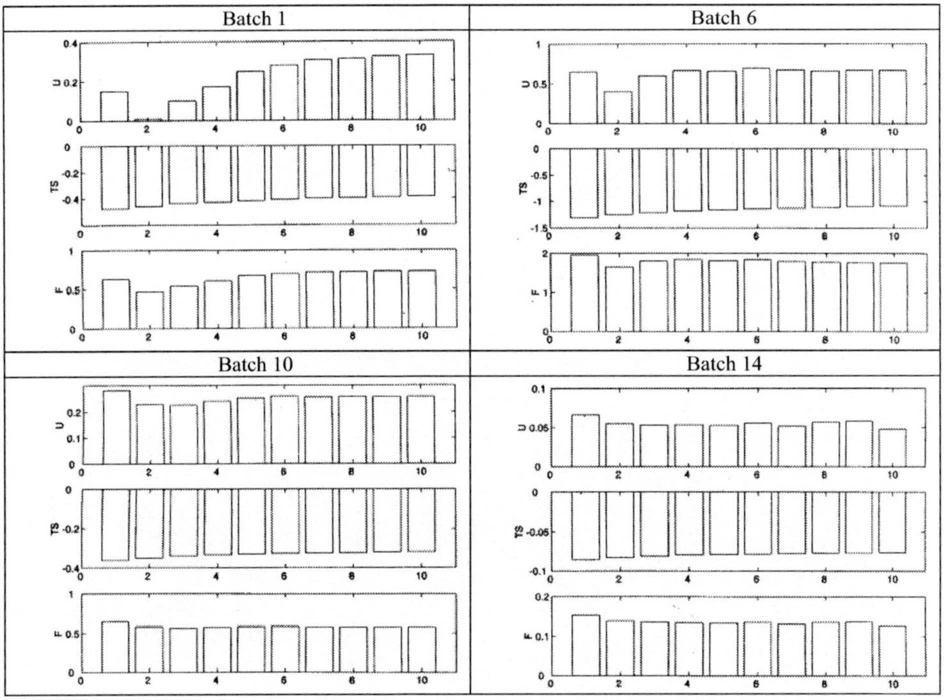

Figure 6: Evolution of information indices at different batches

2-3 Optimal Design of a Plasma Etching Process [21]

The SAS field oxide dry etching process is illustrated in Figure 7. The process is supposed to remove the field oxide effectively and sustain minimum amount of silicon loss. According to the device memory cells' data retention performance data, the silicon loss should be controlled at the level of less than 30nm. In order to maintain a reasonable oxide etching rate (throughput concern) but control silicon loss, the oxide etching recipe's selectivity of oxide to silicon on the patterned wafer should be maintained at the level of greater than 30 to 1. According to the literature data [22], selectivity is primarily determined by the C/F ratio in the etchant chemicals. The increase of the C/F ratio can improve the selectivity of oxide to silicon, so CO gas has been considered as one of the factors which can help improve the oxide etch selectivity to silicon. Besides CO flow, the other critical parameters in this recipe design include RF power, chamber pressure, bottom electrode temperature, and Ar flow. In order to acquire the process data, an etching monitoring system transferring data from the RIE chamber onto a workstation has been designed and implemented. LAM Research RIE mode 4520 single-wafer parallel-plate system operating at 400 kHz is chosen as the plasma etching tool in this study. The patterned wafer with the deposition of SiO2 and polysilicon is etched for 30 seconds and 120 seconds respectively. The remaining thickness of the oxide and polysilicon before and after plasma etching are measured by PROMETRIX FT-750 film thickness probe system on specific pattern opening area. The five input factors and their corresponding range of operation are shown in Table 2.

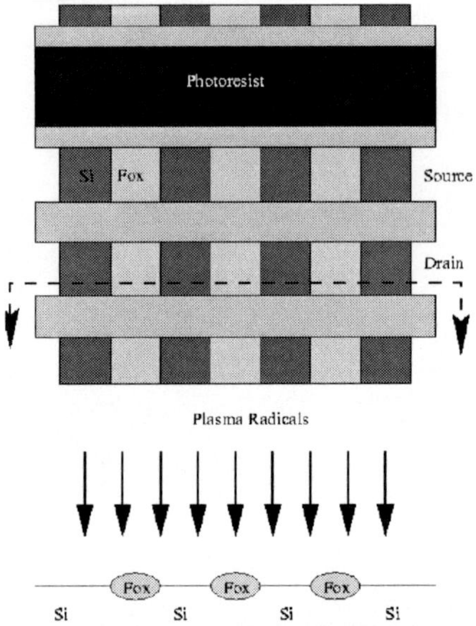

Figure 7: SAS etching process, Fox represent field oxide and Si is silicon.

Table 2: Ranges of Operating Variables of the SAS Etching Process

Variable	Range	Unit
Power	700-900	W
Pressure	200-300	MTorr
Bottom electrode temperature	-20~+10	°C
CO gas flow	100~300	sccm
Ar gas flow	100~300	sccm

The objective is to find a recipe (or an operating condition) that satisfies the following specifications:

1. etching rate of field oxide (OX E/R) > 4000 /min
2. uniformity of field oxide (OX U) < 5%
3. etching rate of polysilicon (Poly E/R) < 100 /min

According to the operator's experience, a total of 14 trials from the past historical data are initially provided. The performance of the first run is shown in two-dimensional space (OX U vs. OX E/R, and OX U vs. Poly E/R) in

(a) view from OX U and Poly E/R directions. (b) view from OX E/R and OX U directions. Target deviation plots for the current (x) and the past (o) experiments.

Figure 8. We can see that none of the experiments satisfy the specifications for OXU and Poly E/R simultaneously. After three additional batches of experiments, a feasible recipe is determined,

Cross-section SEM images obtained by using one of the initial high pressure recipes and the newly developed CO recipe are shown in

(a) High pressure recipe without CO gas at Si loss > 600 A. (b) Recipe with CO gas at Si loss ~300 A.

Figure 9. The results confirm that the newly developed CO recipe could meet the pre-defined process criteria.

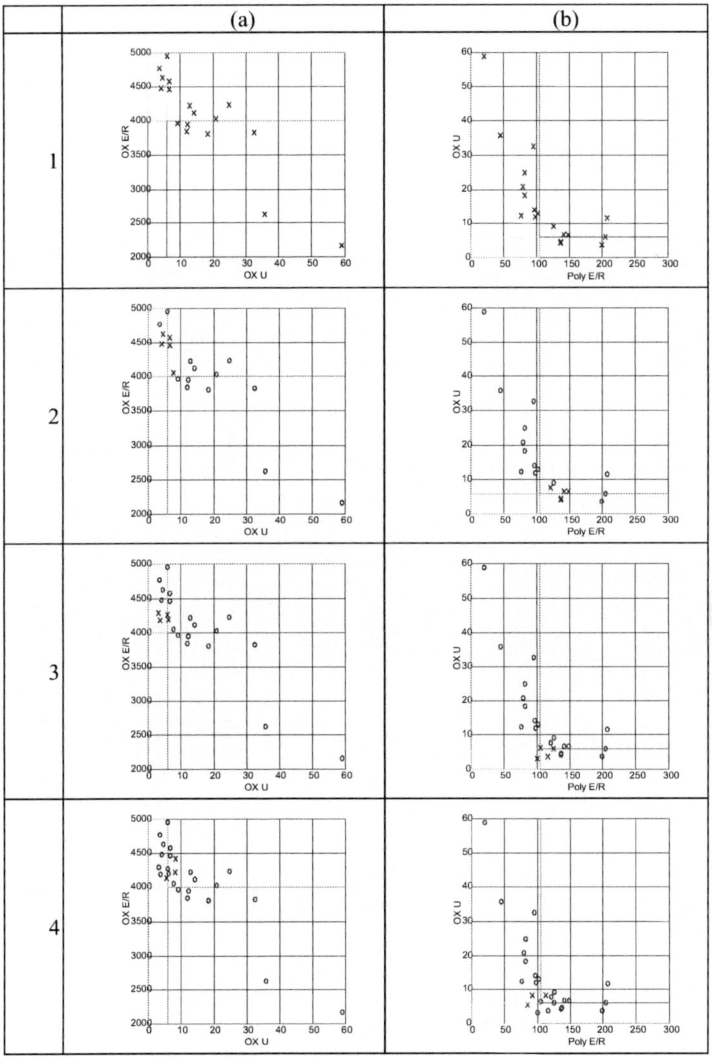

(a) view from OX U and Poly E/R directions. (b) view from OX E/R and OX U directions. Target deviation plots for the current (x) and the past (o) experiments.

Figure 8: Etching process designed experiment results

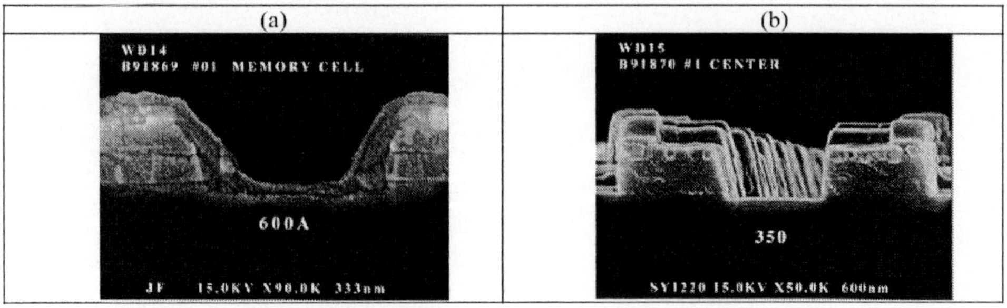

(a) High pressure recipe without CO gas at Si loss > 600 A. (b) Recipe with CO gas at Si loss ~300 A.

Figure 9: Cross-section SEM of SAS etching

2-4 Optimal Design of a Filament Winding [23]

Filament winding is a process which involves the impregnation of fiber strands with resin and the winding of the impregnated strand onto a rotating mandrel in a predetermined fiber path under controlled tension (Figure 10). Due to the advantages of the high production rate, low production cost and ease to make complex part, the filament winding process becomes a widely used technique for the production of high quality composite structures. Although the CNC technique has been developed to control the fiber path more precisely, the optimum parameters to make parts with best properties are not easy to obtain. The traditional method of the design of the filament winding has always been based on time-consuming and costly trail-and-error experiments. Accordingly, the new composite structure cannot be done just in time to meet the needs on the request. If a more efficient method to obtain optimum processing parameters for filament winding could be developed, it will help to accelerate the process design procedure for filament winding.

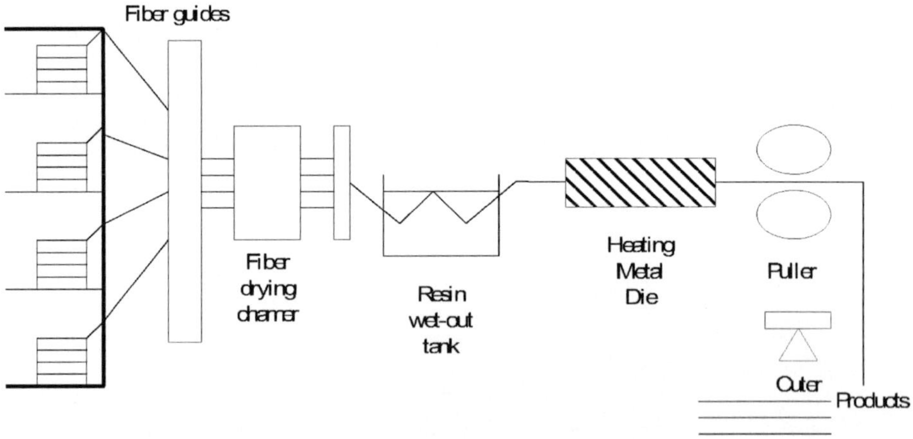

Figure 10: The filament winding process.

In this study, the DGEBA type Epoxy resin, which was used as the matrix in the filament winding process, was supplied by the Nan Ya Plastic Co. Taiwan with Code No. NPEC-124. Glass fibers were used as the reinforcements. Glass fibers (TGFR-P2200) were made by the Taiwan Glass Ind. Co., Taiwan. The objectives of this study are to obtain the best mechanical properties of the composite for the above-mentioned materials by maximum tensile stress σ of the ring test specimen and to increase the compressive load P of tubular test specimen to 50kN via controlling three parameters, including winding angle θ, fiber tension τ, and resin temperature T. The objective function can be defined as:

$$\min_{\theta,T,\tau}\left\{q_1\left(\frac{P-50}{50}\right)-q_2\frac{\sigma}{600}\right\}$$
$$s.t.$$
$$60\le\theta\le 90 \quad (19)$$
$$32\le T\le 45$$
$$10\le\tau\le 40$$

The values, 50 and 600, are normalized for the responses based on the operating region. Factors q_1 and q_2 are used to weigh the relative importance of the two mechanical properties compressive load and tensile stress for different customers.

Figure 11 depicts the evolution of the objective function with each successive batch of experiments. The performance is gradually improved except for the second run. This means that the neural network model can better represent the response surface of the desired design region in the last two runs whose data were getting concentrated toward the desired operating region.

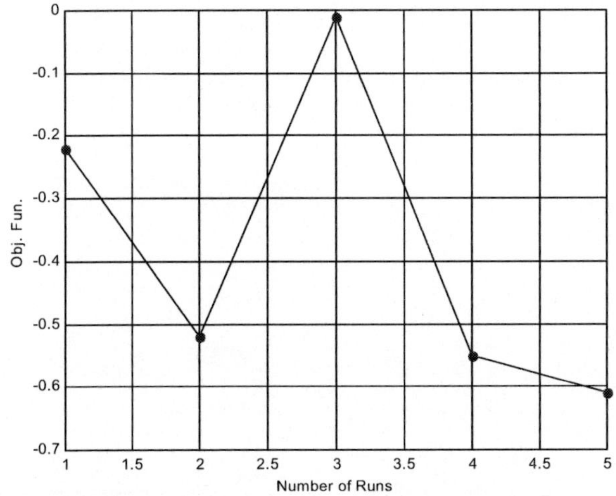

Figure 11: Evolution of objective function winding optimal experimental design path

3. New Trends of the Intelligent Experimental Design

The basic approach of using ANN as a meta-model and information theory to direct evolutionary design has been modified applied to a variety of design problems including: identification of nonlinear dynamic models [24], [25], batch trajectory optimization [26] of a coal fired boiler [27], energy management in petrochemical plant, Wu et al.[28], locating "hot spot" region of pollution [29].

Traditional design usually based upon some understanding of the relevant physical and/or chemical phenomena. Such an approach is "hypothesis-driven". However, in recent years, chemical and biochemical product design and development have attracted increased attention. Combinatorial synthesis and library design become an important optimization technique in drug discovery [30] and functional material development [31], [32]. This approach is data-driven, in which researchers attempt to develop experiments that can evaluate a large number of alternatives using some index of fitness.

There are actually two types optimization problem in combinatorial synthesis. One is a structural search problem, in which components in a library are assembled to form a particular structure. Experiments or calculations are then carried out to see if the synthesized structure has the desired property. Drug discovery is a typical example of structural optimization. Such an optimization is characterized by a highly rugged response surface. A benchmark problem that captures the physics of such optimization is known as the N–K problem [33]. Alternatively, in functional material development, experimental variables are the compositions of the material and processing conditions. The input state-space consists of both discrete and continuous variables and is typically of very high dimensions. The merit function is usually one or a set of specific properties of the material such as superconductivity, luminescence, catalytic activity, tensile strength, etc. Such properties will depend on the particular phase of the product material. Since the change of physical property of the material is discontinuous across a phase boundary, the objective function encountered is only piecewise continuous. Falcioni and Deem [34] proposed a random phase volume (RPV) problem as the benchmark of such optimization.

It was suggested that importance sampling can be used to improve the efficiency of sampling in combinatorial methods. However, without a model function, true importance sampling cannot be performed. Using the same logic as the approach we have presented in this chapter, Yen et al. [35] suggested that a simple prediction model can be constructed using a generalized regression neural network using currently available data. An index of our uncertainty about a point in the search space can also be established using information entropy. An information free energy combined the two indices to direct the search so that importance sampling is performed. Solution of the N-K and RPV model showed that when importance sampling is performed, the combinatorial technique becomes much more effective.

4. Conclusions

An intelligent design is actually a judicious integration of data-driven approach and hypothesis-driven approach. However, we show in this chapter that the organization of knowledge and generation of hypothesis need not be done on a theoretical manner. It can be done by constructing a simple empirical model with sufficient flexibility. Artificial neural net work (ANN) meta-model is used as the tool to summarize all experimental information into a metal model. However features of this meta-model include not only hypothesis of optimal performance, but also model uncertainty, typically measured by information entropy. Hence we define the performance index as information energy. Another index that balances the needs of resolution of model uncertain and validation of optimal performance is defined as the information free energy. New experiment can be designed to minimize information free energy. The applicability of this approach is illustrated using three examples from diverse areas of industrial process design and reference materials of other similar applications are included. Possibility of using such an intelligent design approach to combinatory synthesis and library design is also explored.

References:

[1] Box, G. and N.R. Draper, *Empirical Model-Building and Response Surface*, Wiley, New York, 1987.

[2] Taguchi, G., *Introduction to Quality Engineering*, Asian Productivity Organization, 1986.

[3] Montgomery, *Design and Analysis of Experiments*, 4th ed., John Wiley and Sons, New York, 1997.

[4] Fukunaga, K., *Introduction to Statistical Pattern Recognition*, Academic Press, Boston, 1990.

[5] Saraiva, P.M. and Stephanopoulos, G., "Continuous Process Improvement Through Inductive and Analogical Learning", *AIChE J.*, **38** (1992),2, 161.

[6] Ho, Y. C., R. Sreenivas, and P. Vakili, "Ordinal optimization of discrete event dynamic systems", *Journal of Discrete Event Dynamic Systems*, **2** (1992), 2, 61.

[7] Lau, T. W. E., and Ho, Y. C., "Universal alignment probability and subset selection for ordinal optimization", *Journal of Optimization Theory and Applications*, **93** (1997), 3, 455.

[8] Hornik, K., M. Stinchcombe, and H. White, "Multilayer Feedforward Neural Networks are Universal Approximators," *Neural Networks*, **2**, 359, 1989.

[9] Hornik, K., M. Stinchcombe, and H. White, "Universal Approximation of an Unknown Mapping and Its Derivatives Using Multilayer Feedforward Networks," *Neural Networks*, **3** (1990), 551.

[10] Kalman, B. L., and S. C. Kwasny, "Why Tanh? Choosing a Sigmoidal Function," *Int. Joint Conf. on Neural Networks*, Baltimore, MD, 1992.

[11] Hertz, J., A. Krogh, and R. G. Palmer, *Introduction to the Theory of Neural Computation*, Addison-Wesley, New York, 1991.

[12] Gorodkin, J., L. K. Hansen, A. Krogh, C. Savrer, and O. Winther, "A Quantitative Study Pruning by Optimal Brain Damage," *Int. J. Neural Syst.*, **4** (1993), 159.

[13] Allen, D. M., "The Relationship Between Variable Selection and Data Augmentation and a Method for Prediction," *Technometrics*, **16**(1974), 1, 125.

[14] Bezdek, J., R.J. Hathaway, M.J. Sabin and W.T. Tuker, "Convergence Theory for Fuzzy c-Mean: Counterexamples and repairs", *IEEE Trans. Syst., Man. Cybern*, **17** (1987), 5, 583.

[15] Shannon, C.E., "A Mathematical Theory of Communication," *Bell Syst. Tech. J.*, **27** (1948), 379.

[16] Chen, J.H., Wong, D.S.H., Jang, S.S. and S.L. Yang, "Product and Process Development Using Artificial Neural Network Model and Information Theory", *AICHE J.*, **44** (1998), 4, 876.

[17] Reklaitis, G. V., A. Ravindran, and K. M. Ragsdell, *Engineering Optimization: Methods and Applications*, Wiley, New York, 1983.

[18] J.C. Lagarias, J. A. Reeds, M. H. Wright, and P. E. Wright, "Convergence Properties of the Nelder-Mead Simplex Method in Low Dimensions", *SIAM Journal of Optimization*, vol. 9, no. 1, pp. 112-147, 1998.

[19] Kauffman, S. A. and S. Levin; "Towards A General Theory of Adaptive Walks on Rugged Landscapes," *J. Theor. Biol.*, **11** (1987), 128.

[20] Goldberg, D. E., *Genetic Algorithms in Search, Optimization & Machine Learning*, Addison-Wesley, 1989.

[21] Chen, J.H., S.S. Jang, D.S.H. Wong, and B.T. Chu, "Optimal Design Using Neural Network and Information Analysis In Plasma Etching", *J. Vacuum Sci. & Tech.*, **17** (1999), 145-153.

[22] McAnally, P.S., W. L. Krisa, L. M. Ting, and G. A. Dixit, *International Symposium on VLSI Technonlgy, Systems and Applications*. Proceeding of Technical Papers, (1995), 180.

[23] Chen, J.H., S.S. Jang, S.S., D.S. H. Wong, D.S.H., Ma , C.C.M. and Lin, J.M. , 1999b, "Optimal Design of Filament Winding Using Neural Network Experimental Design Scheme", *J. of Composite Materials*, **33** (1999), 2281.

[24] Lin J.S., S.S. Jang, S.S. Shieh and M. Subramaniam, 1999 "Generalized multivariable dynamic artificial neurol network modeling for chemical processes" *Ind. & Eng. Chem. Res.*, **38** (1999), 12, 4700.

[25] Lin J.S., S.S.Jang SS, S.J. Chien, C.C. Ma, and S.S. Shieh "The enhancement of empirical model capability and optimal/robust design of intractable processes," *Ind. & Eng. Chem. Res.*, **40**(2001), 12, 3951.

[26] Chen J.H., R.G. Sheui "Optimal batch trajectory design based on an intelligent data-driven method" *Ind. & Eng. Chem. Res.*, **42**(2003), 7, 1363.

[27] Chu, J.Z., S.S. Shieh, S.S. Jang, C.I. Chien, H.P. Wan, H.H. Ko, "Constrained optimization of combustion in a simulated coal-fired boiler using artificial neural network model and information analysis" *FUEL* **82**(2003), 6, 693.

[28] Wu T.Y., S.S. Shieh, S.S. Jang, C.C.L. Liu, "Optimal energy management integration for a petrochemical plant under considerations of uncertain power supplies" *IEEE Trans. on Power Syst.*, **20**(3), 1431, 2005

[29] Shieh S.S., J.Z. Chu, S.S. Jang, "An interactive sampling strategy based on information analysis and ordinary kriging for locating hot spot regions" *Math. Geo.*, **37** (2005), 1, 29.

[30] Gordon, E. M., M. A. Gallop and D. V. Patel; "Strategy and Tactics in Combinatorial Organic Synthesis. Applications to Drug Discovery," *Acc. Chem. Res.*, **29** (1996), 144.

[31] Davis, M. E., "Combinatorial Methods: How will They Integrate into Chemical Engineering?" *AIChE J.*, **45** (1999), 2270.

[32] Engstrom, J. R. and W. H. Weinberg; "Combinatorial Materials Science: Paradigm Shift in Materials Discovery and Optimization," *AIChE J.*, **46** (2000), 2.

[33] Kauffman, S. A., *The Origins of Order: Self Organization and Selection in Evolution*, Oxford Univ. Press, New York ,1993.

[34] Falcioni M., M. and W. Deem; "Library Design in Combinatorial Chemistry by Monte Carlo Methods," *Physical Review E.*, **61** (2000), 5948.

[35] Yen, C.H., D.S.H. Wong, S.S. Jang, "Information directed sampling for combinatorial material synthesis and library design" *J. Chem. Eng. Japan*, **36** (2003), 9, 1034.

Integrated Intelligent Systems for Engineering Design
X.F. Zha and R.J. Howlett (Eds.)
IOS Press, 2006

Intelligent Models for Design Conceptualization of Autonomous Vehicle Storage and Retrieval Systems

Miki Fukunari and Charles J. Malmborg[1]

Dept of DSES, RPI, 110 8th Street, Troy, NY 12180, malmbc@rpi.edu

Abstract. Unit load storage systems are pervasive throughout global supply chains. Significant reductions in their automation costs could have significant economic impact. A new alternative to traditional crane-based automation uses flexible autonomous vehicles in storage and retrieval operations. This technology has not significantly penetrated commercial markets for warehouse automation due to a lack of design tools for evaluating its performance. This precludes direct comparisons of autonomous vehicle technology with crane-based technology in the pre-engineering or "design conceptualization" stage of system development where key technology selection decisions are made. To address this problem, this chapter proposes computationally efficient cycle time models for Autonomous Vehicle Storage and Retrieval System that use scalable computational procedures for large-scale design conceptualization. Simulation based validation studies suggest that the models produce high accuracy. The procedure is demonstrated for over 4,000 scenarios corresponding to enumeration of the design spaces for a range of sample problems.

Keywords. Autonomous Vehicle Storage and Retrieval Systems, Random Storage, Storage and Retrieval Cycle Time, Opportunistic Interleaving

Introduction

The prospect of implementing new technology challenges many user communities. This is especially true when a dominant existing technology is perceived as cutting edge, rapidly integrates incremental technological advances to improve performance and cost effectiveness, and technology selection involves large scale and high risk capital expenditures. Systematic assessment of new technologies is particularly unlikely when there is a dearth of accurate and user-accessible design tools that can be applied in the early pre-engineering or "conceptualizing" phase of system development when fundamental technology

[1] Corresponding Author: Professor of Decision Sciences and Engineering Systems, Rensselaer Polytechnic Institute, DSES-CII 5th Floor, 110 Eighth Street, Troy, New York 12180-3590, USA; Email:malmbc@rpi.edu

selections and preliminary design specifications driving the majority of system costs are defined.

Technology selection in unit load storage and retrieval systems (ULS/Rs) provides an illustrative example. Given the pervasiveness of conventional ULS/Rs in industry, a significant increase in the use of automation could dramatically increase the efficiency of supply chains throughout the economy and yield significant cost savings in many service and manufacturing industries. This opportunity has not been fully realized because the dominant automation technologies for ULS/RSs are crane-based automated storage and retrieval systems (AS/RSs) aimed at higher-end systems with respect to throughput performance and storage density. While these objectives apply to a greater or lesser degree to almost all ULS/RSs, the high costs associated with crane-based technologies have raised the adoption threshold of automation to the point where there is a large technologically underserved component of this market. Recent advances in "autonomous vehicle" technologies have created the possibility of drastically reducing the scale of cost effective automation for ULS/RSs, [1].

The primary purpose of this chapter is to introduce a new analytical model for estimating AVS/RS cycle times that will facilitate performance comparisons between AVS/RS and AS/RS technologies in large-scale problems. This model, combined with other analytical tools under development by researchers in this area will eventually yield an integrated analytical tool suite supporting intelligent design tools for accurate and comprehensive head-to-head comparisons of these technologies. The first section of this chapter provides background information on AVS/RS technology including an overview of the major system components and operational features. The second section presents a two-stage cycle time estimation procedure yielding easy and accurate cycle time estimates based on the physical configuration of a storage rack, the number of lifts operating in a system and the autonomous vehicle fleet size. The third section provides a computational illustration of the model including a simulation based validation and application in a conceptualizing type of study. The fourth and fifth sections discuss future trends in AVS/RS conceptualization and offer a summary and conclusions.

1. Background Information

Under normal market conditions, adoption of AVS/RS technology in ULS/Rs would be commensurate with its potential to reduce operating costs. System suppliers would simply demonstrate the performance benefits and cost effectiveness of their products to prospective clients who would adopt the new technology. This has not occurred for three primary reasons. First and foremost, current and potential suppliers of ULS/R automation technologies based on autonomous vehicles face a gap between the advanced state of their hardware and control system technology, and the design conceptualization tools needed to assess the impact of this technology on ULS/R operations. The lack of analytical tools to model system performance for autonomous vehicle technology has been a primary obstacle to the widespread automation of ULS/RSs that operate at low throughput volume to storage capacity ratios. Second, an inherent conflict of interest exists in this industry since current suppliers of crane-based technologies are the most likely future players in the market for systems based on autonomous vehicle technology. This is due to their penetration of the market for conventional and automated ULS/RSs and overlaps among manufacturers of supporting subsystems. The third reason is a business culture that exerts significant time pressure and uncertainty on the process of technology selection. A typical ULS/R design conceptualization study is illustrated in Figure 1.

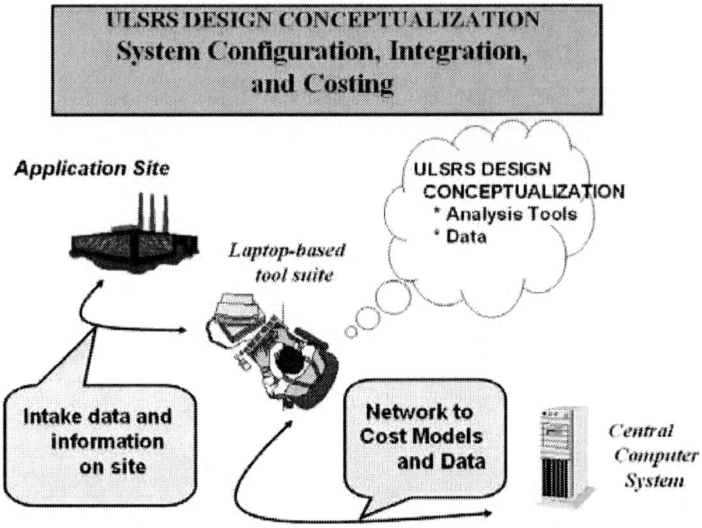

Figure 1. Overview of the ULS/R design conceptualization process.

In a design conceptualization study, a design engineer visits a client location, assesses performance requirements, collects data, accesses networked information on cost and performance parameters, and then prepares a cost proposal. When automation is considered, engineers can access a well-developed suite of analytical and knowledge-based design tools for crane-based technology that are reasonably accurate and computationally efficient. The perceived reliability of these tools has created an environment where early technology selection decisions are made on the strength of limited analysis.

Preliminary design proposals are ultimately subjected to detailed validation but this step is not usually undertaken prior to a significant financial commitment to perform a simulation study or the award of a system installation contract. Although practitioners report a relatively low success rate for preliminary ULS/R automation proposals due to uncertainty in the capital budgeting plans of clients and competition among system suppliers, the risks associated with preliminary proposal development are considerable. Under-promising on system cost for a fixed performance level is likely to result in a non-competitive proposal while over-promising carries serious business risks for the system supplier or consultant as well as the client organization. Therefore, designers have little incentive to propose new technologies.

A fundamental problem in estimating the benefits of autonomous vehicle storage and retrieval systems (AVS/RSs) is the estimation of device cycle times. Unlike crane-based systems, independent vehicles in AVS/RSs operate as the storage/retrieval (S/R) devices. Therefore, a key distinction between AS/RS and AVS/RS technology is the movement patterns of the S/R device. In AS/RSs, aisle-captive storage cranes use simultaneous movement in the horizontal and vertical dimensions to minimize the lengths of material flow paths. In AVS/RSs, vehicles share a fixed number of lifts for vertical movement and follow rectilinear flow patterns in horizontal travel. While the travel patterns in AS/RSs are generally more efficient, AVS/RSs have the potential advantage of adaptability of system throughput capacity to demand by changing the number of vehicles operating in a fixed storage rack. Another advantage of AVS/RS technology is that any S/R device is capable of servicing any storage position in the system. This feature enables AVS/RSs to pool storage and retrieval transactions in a single queue served by all S/R devices as opposed to using parallel queues for individual aisles. These movement and queuing features contribute directly to the expected S/R device cycle time observed in a unit load storage system that, like cost and throughout capacity, represents a key measure of system performance.

To assess the potential impact of this technology, designers must be able to predict storage and retrieval cycles times associated with the use of autonomous vehicle technology in ULS/RSs. However, only a few preliminary studies for estimating autonomous vehicle storage/retrieval (AVS/R) device cycle times are reported in the literature [2,3]. Of particular interest to this chapter is the AVS/RS cycle time model for systems using opportunistic interleaving, [3], which has been shown to yield reasonable accuracy in simulation validation studies but relies on computationally inefficient state equation models that do not scale up easily for large problems. A recently published comparison of the same state equation approach to cycle time estimation in AS/RSs with an alternative approach based on specialized M/G/1 queuing models clearly illustrates the computational difficulties of the state equation modeling strategy despite its accuracy,[4]. As AVS/RS technology matures, it is reasonable to expect that it will challenge the dominance of AS/RS technology in certain applications thereby forcing system designers to confront the technology selection question at a very early stage of system development. Furthermore, the importance of S/R interleaving for maximizing the efficiency of automated unit load storage systems makes this an extremely important special case, [5].

S/R device cycle times are closely related to system throughput capacity. The throughput capacity of a unit load storage system is based on the number of S/R devices and the time required for the most efficient type of cycle with respect to S/R device time per transaction. In unit load systems, this is usually a dual command (DC) cycle where both a storage transaction and a retrieval transaction are completed on the same cycle. DC cycles contrast with single command (SC) cycles where either a storage transaction or a retrieval transaction is completed on a cycle. DC cycles are generally more efficient since they complete two transactions as the S/R device travels loaded to the storage position, empty between the storage and retrieval positions, and loaded from the retrieval position to the load transfer point. Depending on the S/R device dwell point policies, completing a storage and retrieval transaction using two SC cycles generally requires more total S/R device travel time in most types of automated ULS/RSs, [6]. The term "interleaving" refers to the pairing of storage and retrieval transactions on the same cycle to generate DC cycles. The term "opportunistic interleaving" refers to the practice of implementing DC cycles whenever possible and SC cycles otherwise. That is, whenever there are both storage and retrieval transactions pending in the active queue at the start of a cycle, the S/R device implements a DC cycle. If only one type of transaction is pending at the start of the cycle, SC cycles are implemented, i.e., the S/R device does not remain idle to force DC cycles. Opportunistic interleaving is the preferred operating discipline in automated unit load storage systems since it effectively balances the objectives of travel efficiency and device response times. Under an opportunistic interleaving discipline, relatively more DC cycles characterize busy periods while relatively more SC cycles characterize slack periods. Subsequently, the expected cycle time is a weighted average of the DC and SC cycle times and directly influences the key performance measure of S/R device utilization by defining the average amount of travel time per transaction served.

Efficient analytical models to estimate expected cycle times for AS/RSs using opportunistic interleaving based on preliminary design data are available to practitioners, [4]. This not true for AVS/RS technology where the key distinguishing feature is that S/R devices are autonomously operating vehicles as opposed to aisle-captive storage cranes. Figure 2 illustrates the major hardware components of these systems. Unlike an AS/RS, all S/R devices in an AVS/RS can access all storage positions in the system. For horizontal movement, vehicles move in rectilinear flow paths while vertical movement is facilitated by lifts mounted along the periphery of the storage rack. The basic design features of AVS/RS technology are detailed in [2]. The throughput capacity of an AVS/RS for a given storage rack configuration is proportional to the number of vehicles subject to interference effects associated with vehicles accessing shared lifts and, to a lesser degree, vehicle blocking effects within storage aisles. By using aisle captive cranes, AS/RSs avoid these interference effects although it is not difficult to envision situations where the flexibility of decoupling the number of S/R devices from the number of storage aisles is appealing.

Systems with high storage capacity requirements and relatively low throughout requirements, especially those where the rack configuration is restricted to a relatively large number of shallow storage aisles, represent one example where the efficient movement patterns in AS/RS are likely to be offset by the flexibility associated with installing fewer S/R devices than storage aisles. To inform the selection between AS/RS and AVS/RS technology in such cases, it's necessary to model the influences of the unique design features of AVS/RS technology on expected cycle times including differences with respect to interleaving. The relative efficiency of DC versus SC cycles in AVS/RSs is apt to be less than with AS/RSs due to the fact that high throughput, high capital cost storage technologies such as AS/RSs and AVS/RSs generally use random storage to maximize space efficiency. Subsequently, first come first serve

(FCFS) transaction dispatching causes the majority of paired S/R transactions on DC cycles to include storage positions on different tiers. (Interleaving is assumed to preserve FCFS sequencing for storage transactions and retrieval transactions as individual service processes.) The need to access lifts for movement between storage tiers and the rectilinear movement patterns of AVS/RS vehicles will tend to reduce the efficiency advantages of interleaving in AVS/RSs versus AS/RSs. However, this disadvantage would be offset in part by efficiencies gained by pooling storage and retrieval transactions in a single system-wide queue as opposed to the parallel aisle-based, transactions queues in AS/RSs. The pooling effect increases opportunities for DC cycles as well as increasing server utilization during slack periods.

Figure 2. AVS/R system components including vehicle (top), lift (middle) and system view.

Simple travel time models for SC and DC cycle times in AVS/RSs are presented in, [2]. Denoting these values as τ_{SC} and τ_{DC} respectively, the expected S/R device time expended per transaction in AVS/RSs using interleaving can be defined as $\tau = (1\text{-}\alpha)\tau_{SC} + \alpha\tau_{DC}/2$, where α denotes the proportion of cycles that are DC cycles. The state equation procedure to estimate the value of α, [3], defines AVS/RS states based on the number of vehicles in the system and the transactions queue using a vector of the form, $\{v_1, v_2, \ldots, v_V, q_s, q_r\}$, where v_i denotes the state of vehicle i, for i=1…V, q_s denotes the number of pending storage transactions in the active queue, and q_r denotes the number of pending retrieval transactions in the active queue. For each vehicle, the values in the state vector can take on values equal to $v_i = 0, 1, 2$ or 3 indicating that vehicle i is idle, performing a SC storage cycle, performing a SC retrieval cycle or performing a DC cycle, respectively. Values of q_s and q_r can take on the values $q_s = 0, 1 \ldots Q\text{-}q_r$ and $q_r = 0, 1 \ldots Q\text{-}q_s$, where Q denotes a positive integer greater than or equal to the maximum expected number of transactions observed in the active queue during normal system operation.

Using this representation of system states, it is possible to describe random variation in AVS/RS states using equations describing fluctuations associated with state changing events such as service completions and transaction arrivals to the active queue. The value of α is then obtained as a function of the arrival rates of S/R transactions and their corresponding service rates ($1/\tau_{SC}$ and $1/\tau_{DC}$). State equations representing the logic of opportunistic interleaving, (e.g., transitioning to a dual command state when both storage and retrieval transactions are pending following a service completion), are then formulated and used to solve for state probabilities of the form $P_{v1\ v1,\ v2,\ \ldots,\ vV,\ qs,\ qr,}$ (the probability of state $v_1, v_2, \ldots, v_V, q_s, q_r$) where v_i=0,1,2,3, i=1…V, $q_s = 0, 1 \ldots Q\text{-}q_r$ and $q_r = 0, 1 \ldots Q\text{-}q_s$. Once the state probability distribution is obtained, the probability that vehicles are either performing SC cycles or DC cycles at randomly selected intervals can be estimated and the corresponding value of α, i.e., the proportion of transactions served on DC cycles is obtainable as:

$$\alpha = 2P_{DC}/(2P_{DC}+P_{SC}).$$

The implementation difficulty of the state equation approach to estimating α lies in dimensionality since the number of states grows rapidly with the number of vehicles and the maximum queue size. The total number of vehicle states is given by 4^V with the number of queue states given as $\Sigma_{i=0,\ldots,Q}(Q+1-i)$. With opportunistic interleaving, non-empty queue states can only exist for the 3^V vehicle states where there are no idle vehicles. The total number of feasible AVS/RS states is therefore the product of the number of vehicle states with no idle vehicles and the total number queue states, plus the number of vehicle states with at least one idle vehicle (in these states it must be that $q_s=q_r=0$). Thus, the number of system states as a function of the number of vehicles in a system (V) and Q can be written as:

$$S = \Sigma_{i=0,\ldots,Q}(Q+1-i)3^V + (4^V-3^V).$$

Although it is not difficult to program the state equation model since all state equations fall into one of just four generic categories, ($q_s=q_r=0$, $q_s=0$ $q_r\geq1$, $q_s\geq1$ $q_r=0$, and $q_s\geq1$ $q_r\geq1$), the number of equations increases rapidly with Q and V. Although solving the state equations is made less difficult by the fact that the underlying Markov chain is sparse, the dimensionality is still troublesome since Q corresponds to the maximum number of transactions observed in the queue and AVS/RSs use a single pooled transactions queue. Thus, the Q parameter and the corresponding number of state equations could quickly become too large to compute for systems of practical size that have a high rate of vehicle utilization and therefore a long transactions queue. An alternative model designed to address this problem is presented in the following section.

2. An Efficient Cycle Time Model for AVS/RSs

An important aspect of cycle time modeling in AVS/RSs is the lift and vehicle dwell point policies. In the current study, the dwell point policy for lifts is that they remain at the storage tier where the vehicle is discharged until called by a vehicle to perform the next service. The dwell point policy for vehicles is that they return to the load buffer area following service completion, i.e., the location where incoming loads arrive for storage in the system and retrieved loads are deposited for transfer out of the system. In AVS/RSs, the lift queuing system is nested within a separate vehicle queuing system. In the lift queuing system, vehicles are customers and lifts are servers. In the vehicle queuing system, loads are customers and vehicles are servers. The modeling strategy proposed in this chapter is to analyze the two systems iteratively until convergence of their queuing performance measures. In both queuing systems, empty device travel is a component of service time. For the lift system, two transaction types are defined including direct transactions involving service that does not include empty travel, i.e., a vehicle requests service from the same storage tier where the lift completed the last service, and indirect transactions where the lift travels empty to the storage tier where the requesting vehicle is located. Empty travel for the vehicle queuing system is associated with vehicle movement to or from the load buffer area on SC cycles, and travel between the storage and retrieval load positions on DC cycles.

It is also assumed that vehicles release lifts from service following the completion of each vertical movement. That is, lifts do not wait for vehicles to complete the horizontal travel component of SC and DC cycles before becoming available to other vehicles. Under this operating discipline, vehicles request lift travel up to two times on SC cycles and up to three times for DC cycles. To describe the procedure, the following design profile decision variables are defined, {V=number of vehicles, L=number of lifts, A=number of storage aisles, C=number of storage columns and T=number of storage tiers}. In addition, the following parameter values are defined, {λ_S=arrival rate of storage transactions, λ_R=arrival rate of retrieval transactions, t_v=lift travel time for direct (one-way) vertical movements, τ_l=lift travel time for all transactions, t_H=expected horizontal vehicle travel time, δ=load transfer time between vehicles and storage positions, s_v=vertical space allowance per tier, s_h=horizontal space allowance per storage position, v_v=the vertical travel velocity of lifts, v_h=the horizontal travel velocity of vehicles, η=the vehicle charging/discharging time allowance.}

Respectively denoting the proportion of DC cycles in the vehicle queuing system and the proportion of direct transactions in the lift queuing system as α and β, the iterative cycle time estimation procedure is based on the random storage assumption. With random storage, the estimation of β for AVS/RSs is based on the dwell point policies where the ratios λ_R/T and λ_S/T approximate the probability of transactions demand on each storage tier. It follows that the probability that a lift is located on any specific tier excluding the first tier following completion of service can be estimated using $(\lambda_S/T)/(\lambda_S+\lambda_R)$ and the probability that a lift is located on the first tier following completion of service can be estimated using $(\lambda_R+\lambda_S/T)/(\lambda_S+\lambda_R)$ since vehicles always return to the first floor buffer area following service completion. Similarly, the probability of a direct transaction on any specific tier excluding the first tier can be estimated using $(\lambda_R/T)/(\lambda_S+\lambda_R)$ and the probability of a direct transaction on the first tier can be estimated using $(\lambda_S+\lambda_R/T)/(\lambda_S+\lambda_R)$. Subsequently, the probability of direct transactions on lifts is estimated as:

$$\beta = [(T-1)(\lambda_S/T)(\lambda_R/T)/(\lambda_S+\lambda_R)^2] + [(\lambda_R+\lambda_S/T)(\lambda_S+\lambda_R/T)/(\lambda_S+\lambda_R)^2].$$

Under the assumption of random storage, the value τ_l follows a uniform distribution of the form $U(s_v/v_v+2\eta,\ (T-1)\, s_v/v_v+2\eta)$ with point probability values of $p(\tau_l)=1/(T-1)$ for tiers $t=2,\ldots,T$. The mean travel time for the lift system is then given by the expected value:

$$E(t_v) = [(T-1)\, s_v/v_v+2\eta + s_v/v_v+2\eta]/2$$

and

$$\tau_l = (2(1-\beta)+\beta)E(t_v),$$

to reflect the fact that indirect transactions require two lift movements and direct transaction require one lift movement. To obtain the squared coefficient of variation (SCV) of τ_l, i.e., $E[x^2]$, we can apply the general definition of the variance of a random variable, $V(x)=E(x^2)-[E(x)]^2$:

$$E[x^2] = V(x)+ [E(x)]^2$$

or

$$E[t_v^2] = [(T-1)\, s_v/v_v+2\eta - s_v/v_v+2\eta]^2/12 + [E(t_v)]^2$$

for direct transactions, and

$$E[(2t_v)^2] = [(T-1)\, s_v/v_v+2\eta - s_v/v_v+2\eta]^2/3 + 4[E(t_v)]^2$$

for indirect transactions. This yields the SCV for lift transactions as the ratio of the second moment to the square of the mean or:

$$c^2 = \{\beta\, E[t_v^2] + (1-\beta)\, E[(2t_v)^2] - \tau_l^2\}/\, \tau_l^2.$$

These initial estimates of the mean travel time and its SCV for the lift queuing system are applied in estimating vehicle waiting times for lifts based on Whitt's approximation [7] for multi-server queuing systems with generally distributed service times:

$$W_L = W_q(1+ c^2)/2,$$

where W_q denotes the expected waiting time estimate from the corresponding M/M/L/V/∞ queuing model with the arrival rate of $2[(T-1)/T](\lambda_S+\lambda_R)$. This represents the overall transactions arrival rate adjusted to account for the fact that transactions on the first storage tier do not require access to lifts. This initial estimate also assumes 100% SC cycles or $\alpha=0$. The initial cycle time estimate for the lift system is then given by:

$$1/\mu_{lift} = \tau_1 + W_{L.}$$

Given an initial estimate of the lift cycle time, the expected SC cycle time for vehicles can be approximated using:

$$\tau_{SC} = 2\{[(T-1)/T](t_H+1/\mu_{lift}) + t_H/T+\delta\}$$

where t_H is based on the rectilinear travel patterns used in horizontal travel and the random storage assumption (see [2] for illustrations of travel time models based on AVS/RS kinematics). To approximate the DC cycle time for vehicles, it is necessary to consider four possible S/R coupling scenarios and their probabilities. The first is where both transactions are on the first tier with probability:

$$\phi_1=1/(\Phi+2T-1),$$

The second is where both transactions on the same tier exclusive of the first tier with probability:

$$\phi_2= (T-1)/ (\Phi+2T-1),$$

The third is where transactions are on different tiers including one on the first tier with probability:

$$\phi_3= (T-1)/ (\Phi+2T-1),$$

The fourth is where transactions are on different tiers excluding the first tier with probability:

$$\phi_4= \Phi/ (\Phi+2T-1),$$

where Φ is a counter of the form:

$$\Phi=\Sigma_{k=2\ldots T-1}\Sigma_{j=k+1\ldots T}\ (1).$$

To visualize this probability distribution, consider an example with T=5, the number of combinations of tiers that could be included on a DC cycle is 1 + (T-1) + (T-1) + Φ = (Φ+2T-1) = 15. That is, if (t_1,t_2) refers to the (identical) tiers included on a DC cycle, then the possibilities for each combination with T=5 tiers include: case 1: (1,1) – one combination, case 2: (2,2),(3,3),(4,4),(5,5) – four combinations (T-1), case 3: (1,2), (1,3), (1,4), (1,5) – four combinations (T-1) and case 4: (2,3), (2,4), (2,5), (3,4), (3,5), (4,5) – six combinations (Φ). The initial approximation of the DC cycle time is then given by:

$$\tau_{DC} = \phi_1 3t_H + \phi_2(3t_H+2/\mu_{lift}) + \phi_3(4t_H+2/\mu_{lift}) + \phi_4(4t_H+3/\mu_{lift}) + 4\delta.$$

This definition is based on the movement assumptions for each case. Three horizontal movements for cases 1 and 2 correspond to movement from the lift to a randomly selected storage address, movement from that storage address to another randomly selected address on the same tier, and movement back to the lift location. Four horizontal movements for cases 3 and 4 correspond to movement from the lift to a randomly selected storage position and return travel to the lift on each of two storage tiers.

To estimate α_k, the proportion of DC cycles on the current iteration k, the M/M/V/∞/∞ queuing model is applied to approximate the state distribution of the number of transactions waiting and in service ($P_0, P_1, \ldots$), where the arrival rate is ($\lambda_S+\lambda_R$) and the service rate is $\mu_k=(\alpha_{k-1}\tau_{DC}/2+(1-\alpha_{k-1})\tau_{SC})^{-1}$, where α_{k-1} denotes the estimated α value from the previous iteration ($\alpha_0=0$). This state distribution provides estimates of the probability of N transactions in the queue.

For a given value of N, let S denote the number of storage transactions and R denote the number of retrieval transactions where N=S+R. It is possible to approximate the probability of S and R for a given value of N by taking a binomial perspective of the transactions queue. Specifically, a queue of size N corresponds to N trials in a binomial process where each trial yields a storage transaction with probability $\lambda_S/(\lambda_S+\lambda_R)$ and a retrieval transaction with probability $\lambda_R/(\lambda_S+\lambda_R)$. Using this concept, the probability of a given number of storage transactions given N is obtainable using the binomial point probability of the form:

$$P_{S|N} = [N!/(N-S)!S!]\,[\lambda_S/(\lambda_S+\lambda_R)]^S\,[1-(\lambda_S/(\lambda_S+\lambda_R))]^{N-S}$$

Similarly, the probability of a given number of retrieval transactions given N is:

$$P_{R|N}=[N!/(N-R)!R!]\,[\lambda_R/(\lambda_S+\lambda_R)]^R\,[1-(\lambda_R/(\lambda_S+\lambda_R))]^{N-R}$$

Recognizing that SC cycles occur when $N\leq V$, $N_S=0$, or $N_R=0$, the probability that a cycle is of the SC type can be approximated using:

$$P_{SC} = P_0+P_1+..\,P_V + \Sigma_{N=V+1,\ldots\infty}\, P_N\,(P_{S=0|N} + P_{R=0|N}).$$

It follows that $\alpha_k=(1-P_{SC})$ and the vehicle time expended per transaction is given by:

$$1/\mu_{vehicle} = (1-\alpha_k)\tau_{SC} + \alpha_k\tau_{DC}/2.$$

To implement the iterative computational procedure, the initial value of $\alpha=0$ and an error tolerance value is specified (ε). On each iteration k, the change in the estimated value of α from the previous iteration is compared to ε. If $(\alpha_k-\alpha_{k-1})\leq\varepsilon$, the process terminates, otherwise a revised value of $\alpha_{k+1}=(\alpha_k+\alpha_{k-1})/2$ is computed and the process is repeated. The only variation to the lift queuing system procedure following the first iteration is to change the arrival rate in the M/M/L/V/∞ queuing model used to estimate W_L corresponding to new α. Specifically, the arrival rate used on the first iteration:

$$2[(T-1)/T]/(\lambda_S+\lambda_R)$$

is adjusted to reflect the current value of α using:

$$(1-\alpha)2[(T-1)/T](\lambda_S+\lambda_R)+\alpha(\lambda_S+\lambda_R)/2[0\phi_1+2\phi_2+2\phi_3+3\phi_4],$$

to reflect that DC cycles require up to three lift transactions compared to four lift transactions for two equivalent SC cycles. In the next section, the results of simulation based validation studies of the iterative cycle time model are reported and the model is demonstrated in a conceptualizing mode for sample problems.

3. Results from Computational Experimentation

To validate the algorithm, an AutoMod (version 10.0) simulation of an AVS/RS is created for a rack configuration with T=5 tiers, C=10 columns, A=10 aisles and L=2 lifts. Statistics are collected on vehicle and lift utilization and observed α values under opportunistic interleaving. The model is run for two demand scenarios including with $\lambda_S=\lambda_R=50$ (Case I) and $\lambda_S=\lambda_R=100$ (Case II). For the first demand scenario, the model is run for $2\leq V\leq 10$. For the second demand scenario, the model was run with $5\leq V\leq 15$. Key parameter values included in the simulation and analytical model include: $\{\delta=0, \eta=0, s_v=5$ feet, $s_h=5$ feet, $v_v=60$ feet/min, $v_h=250$ feet/min$\}$. To collect statistics on vehicle and lift utilization and α values, a total of five 24-hour simulation runs are implemented with an initial 24-hour warm-up period truncated to eliminate initial biases. Storage and retrieval arrivals are programmed to follow a Poisson distribution with vehicles remaining at or returning to the first tier system I/O point following service completions if there are no interleaving opportunities. To represent a random storage policy, the simulation model is programmed to initiate the timing and location of storage and retrieval demands based on a uniform transactions demand distribution across all storage locations. Although not entirely consistent with storage and retrieval patterns observed in practice under random storage/closest open location dispatching policies [8], the uniform distribution assumption is consistent within the analytical and simulation models. In the simulation, arriving transactions first check for vehicle availability. Interleaving opportunities are exploited when allowable based on the storage and retrieval transaction queue conditions at the start of each cycle and follow FCFS dispatching of transactions within each queue. Acceleration and deceleration effects are ignored in both the simulation and analytical models.

The results from these computational studies show strong consistency between the iterative model and the simulation. With respect to vehicle utilization, the average error is 0.20% for the low demand scenarios and 0.44% for the high demand scenarios. With respect to lift utilization, the average error is 0.66% for the low demand scenarios and 0.59% for the high demand scenarios. Since offsetting errors tend to be confounded in aggregated utilization measures, it is not surprising that the errors observed in results with respect to transaction cycle times are somewhat larger. In percentage deviation terms, the magnitudes of these errors range from 2.28% to 0.25% averaging 0.78% for the low demand scenarios and 1.16% for the high demand scenarios.

With respect to transaction waiting times, the model appears less capable of generating accurate results. In the case of vehicle waiting times for lifts, estimation errors range from 1% to 19% across the two demand scenarios with an average error of 13%. Since waiting times for lifts represent a relatively small component of total transaction cycle times which can also confound offsetting errors from other sources, these errors do not exert much influence on cycle time results. However, they do provide some insight into the accuracy of using Whitt's approximation in estimating waiting times with non-Poisson arrivals and non-exponential service times. A similar observation on the performance of Whitt's approximation with non-exponential service times can be made with respect to transaction waiting times for vehicles where percentage deviations decrease with higher magnitude waiting times. These results are encouraging to the extent that waiting time estimates are reasonably accurate in scenarios where waiting time is likely to be a concern, i.e., significantly larger than zero. The level of accuracy observed in computational results is sufficient for conceptualization purposes where the objective is to define preliminary system profiles that are likely to be cost effective and with adequate operating capacity and slack.

To demonstrate the computational advantage of the iterative procedure relative to state equation based approaches, experimentation with a total of 4,032 sample designs spanning a wide range of the AVS/R design variables is undertaken with $5 \leq T \leq 10$, $2 \leq L \leq 5$, $3 \leq V \leq 10$, and $5 \leq C \leq 25$. For each design scenario, the value of A is the smallest integer satisfying, A=1000/2CT, corresponding to a total of approximately 1000 storage positions. The transactions demand level is set at $\lambda_R = \lambda_S = 50$/hour. Applying the iterative procedure with a tolerance level of, $\varepsilon = 0.0005$, the average number of iterations prior to convergence for the 4,032 design scenarios is 4.92. Furthermore, none of the 4,032 scenarios required more than 10 iterations. The iteration counts prior to convergence for the 4,032 design scenarios are summarized below:

Iterations:	1	2	3	4	5	6	7	8	9	10
Scenarios:	901	256	222	249	287	503	602	994	17	1

Rapid convergence in the estimation of α is a key factor in limiting the computational cost of evaluating AVS/RS performance measures for very large problems. To demonstrate this, the iterative procedure is implemented for a series of problems scaled to practical size, i.e., 10,000 storage positions. In these examples, key parameter values include, $s_h=5$ feet, $s_v=6$ feet, $\eta=0.05$ minutes, $\delta=0.08$ minutes, $v_h=400$ fpm, $v_v=200$ fpm, and $\lambda_R=\lambda_S=50$/hour. In these examples, the range of design variables is T=5, $5 \leq A \leq 10$, L=5, $5 \leq V \leq 10$ with the value of C equal to the smallest integer greater than the ratio, 5000/2AT, (since aisles are two-sided, ACT=5,000 corresponds to 10,000 storage positions). On average, each of these scenarios requires approximately 7 seconds of CPU to evaluate using a 498 Pentium III processor with 128 MB RAM and the Windows 98 operating system. Twelve additional scenarios including three each for values of, ACT= 1000, 10000, 20000 and 30000 are also executed to assess the ability of the procedure to scale up for very large problems. The average CPU times observed over the three runs for each value of ACT, are, 4.53, 7.16, 7.62 and 7.67 seconds, respectively.

The results from computational studies for large scale problems clearly demonstrate the ability of the iterative model to scale efficiently for large problems. This attribute is essential for AVSRS conceptualization since typical application scenarios can involve hundreds of thousands of potential design variable combinations. To facilitate creativity, the system designer must be able to easily undertake broad based exploration of the solution space. This would enable consideration of very different types of solutions with sufficient operating capacity and slack to meet application specifications with respect to transactions demand and total storage capacity requirements. At the same time, design constraints such as site restrictions and building codes can be accommodated implicitly in the iterative design process. Such an approach would usually be more effective that constraint-driven search where the final solution is heavily influenced by starting from an initial set of design constraints.

4. Future Trends

Intelligent design tools for AVS/RS conceptualization will be the means by which this technology can realize its economic potential. Like many new technologies, it is necessary for system suppliers to demonstrate capability before garnering the interest of the user community. It is relatively easy for AVS/RS suppliers to accurately estimate initial system cost based on preliminary design profiles as described in this chapter. There are several reliable ways to do this including direct aggregation of system component costs. As more experience is gained with new installations, it is possible to refine cost estimation capabilities. This knowledge can easily be codified using techniques such as multiple regression models with design attributes as independent variables. It is far more difficult to predict system performance, particularly expected cycle time and S/R device response time, as a function of these same design attributes. The models proposed in this chapter provide a core analytical component for intelligent design tools for AVS/RS conceptualization. The immediate future challenge is to develop a decision support system that optimizes human-computer interaction to facilitate rapid convergence to cost effective AVS/RS design profiles that meet user requirements. Such systems must be developed using designer protocols that reflect the technical and business culture of automated unit load warehousing systems described in the background section of this chapter where high quality design proposals must be developed quickly and with minimum investment of expert time and effort. The availability of such a system could enable direct, head-to-head comparisons of AVS/R and AS/R technology for a given application and potentially lower the cost effective threshold of warehouse automation with respect to transactions throughput. Given the pervasiveness of unit load warehouse systems where the ratio of transactions demand to total storage space requirements to is too low to justify high cost crane-based automation technology, the available of lower cost AVS/R technology has significant economic potential for supply chains worldwide.

5. Summary and Conclusions

A cycle time model for AVS/RSs using opportunistic interleaving is proposed for use in the process of AVS/RS conceptualizing. This is a process where the size of the design solution space can include hundreds of thousands of design profiles but where close to complete enumeration may also be justified given the importance of decisions made early in the design process. However, these design solution spaces may not be small enough to support the use of more accurate but very inefficient modeling tools such as the state equation model described in this chapter. Thus, the model presented in this chapter represents an alternative to previous models having sufficient accuracy but inadequate computational efficiency. The procedure presented in this chapter would be practical for evaluating hundreds of thousands of design profiles in a relatively seamless "spreadsheet like" mode of application. Given that it yields accurate estimates of system performance, it serves the primary goal of the conceptualizing process to identify the most promising set of candidate designs justifying more extensive, simulation-based analysis and validation. To enhance the value of this model to potential users, useful next steps would include the development of design tools optimizing human-computer interactions to facilitate rapid convergence to cost effective AVS/RS design profiles that meet user requirements with respect to transactions demand and total storage space requirements.

References

[1] D.V. Zizzi, Whats New in the Equipment Field, *2000 International Material Handling Research Colloquium*, Material Handling Institute, York, Pennsylvania, 2000.

[2] C.J. Malmborg, Conceptualizing Tools for Autonomous Vehicle Storage and Retrieval Systems, *International Journal of Production Research*, **40**, 8, (2002), 1807-1822.

[3] C.J. Malmborg, Interleaving Dynamics in Autonomous Vehicle Storage and Retrieval Systems, *International Journal of Production Research*, **41**, 5, (2003), 1057-1069.

[4] F. Eldemir, R.J. Graves, C.J. Malmborg, A Comparison of Alternative Conceptualizing Tools for Automated Storage and Retrieval Systems, *International Journal of Production Research*, **41**,18, (2003), 4517-4539.

[5] C.J. Malmborg, Rule of Thumb Heuristics for Configuring Storage Racks in Automated Storage and Retrieval Systems, *International Journal of Production Research*, **39**, 3, (2001), 511-527.

[6] C.J. Malmborg C.J., B. Krishnakumar, Optimal Storage Assignment in Multiaddress Warehousing Systems, *IEEE Transactions on Systems, Man and Cybernetics*, **19**, 2, (1989), 197-204.

[7] W. Whitt, The Queuing Network Analyzer, *The Bell System Technical Journal*, **62**, 9, (1983), 2779-2815.

[8] M.R. Wilhelm, J.L. Shaw, An Empirical Study of the 'Closest Open Location Rule' for AS/RS Storage Assignments, *Progress in Material Handling Research: 1996*, R.J. Graves, L.F. McGinnis, D.J. Medeiros, R.E. Ward, M.R. Wilhelm (eds.), Material Handling Institute, Charlotte, NC, 639-650 1996.

Integrated Intelligent Systems for Engineering Design
X.F. Zha and R.J. Howlett (Eds.)
IOS Press, 2006

Approximate Optimization Using Computational Intelligence and its Application to Reinforcement of Cable-stayed Bridges[1]

Hirotaka Nakayama [a,2], Koichi Inoue [b] and Yukihiro Yoshimori [c]
[a] *Konan University, Dept. of Info. Sci. & Sys. Eng.*
[b] *Mitsubishi Heavy Industries, Ltd., Steel Structural and Civil Eng. Lab.*
[c] *Ryosen Engineers, Co. Ltd., Technical Analysis Center*

Abstract. In many practical engineering design problems, the form of objective function is not given explicitly in terms of design variables. Given the value of design variables, under this circumstance, the value of objective function is obtained by some analysis such as structural analysis, fluidmechanic analysis, thermodynamic analysis, and so on. Usually, these analyses are considerably time consuming to obtain a value of objective function.

In order to make the number of analyses as few as possible, approximate optimization methods using computational intelligence have been developed. In those methods, optimization is performed in parallel with predicting the form of objective function. In this paper, radial basis function networks (RBFN) are employed in predicting the form of objective function, and genetic algorithms (GA) in searching the optimal value of the predicted objective function. One of the most important tasks in this approach is to allocate sample data moderately in order to make the number of experiments as small as possible. The effectiveness of the suggested method will be shown through some numerical examples along with an application to seismic design in reinforcement of cable-stayed bridges.

Keywords. approximate optimization, computational intelligence, black-box objective function, seismic design, reinforcement of cable-stayed bridges

1. Introduction

The aim of this paper is to apply an approximate optimization method for black-box objective functions using computational intelligence to practical problems of reinforcement of cable-stayed bridges. Black-box objective functions are objective functions whose forms are not explicitly known in terms of design variables, but whose values are given

[1]This research was supported by JSPS.KAKENHI13680540.
[2]Correspondence to: Hirotaka Nakayama, Konan University, Dept. of Info. Sci. & Sys. Eng., 8-9-1 Okamoto, HIgashinada, Kobe 658-8501. Tel.: +81 78 435 2534; Fax: +81 78 435 2540; E-mail: nakayama@konan-u.ac.jp

by sampled real/computational experiments. In many engineering design problems, several kinds of analyses such as structural analysis, fluidmechanic analysis, thermodynamic analysis, and so on are applied for this purpose. Usually, these analyses are considerably expensive. Therefore, if these functions are optimized by existing methods, it takes an unrealistic order of time to obtain a solution. For this situation, the number of necessary sampled experiments should be as few as possible.

Response Surface Method (RSM) is probably most widely applied to our aim [6]. Usually, Response Surface Method is a generic name, and it covers a wide range of methods. Above all, methods using experimental design are famous. However, many of them select sample points only on the basis of statisitical analysis of design variable space. They may provide a good approximation of black-box functions with a mild nonlineality. It is clear, however, that in cases in which the black-box function is highly nonlinear, we can obtain better performance by methods taking into account not only the statistical property of design variable space but also that of range space of the black-box function (in other words, the shape of function).

Recently, the authors proposed to apply machine learning techniques such as RBF (Radial Basis Function) networks and Support Vector Machines (SVM) for approximating the black-box function [7], [8], [9]. There, the sample points are given by considering both global and local information of the black-box function. This paper introduce an optimization method for black-box objective functions using computational intelligence, in particular, RBF network and genetic algorithm along with an application to reinforcement of cable-stayed bridges.

2. Incremental Learning in Machine Learning

Since the number of sample points for predicting objective functions should be as few as possible, we adopt some incremental learning techniques which predict black-box functions by adding learning samples step by step. RBF Networks (RBFN) and Support Vector Machines (SVM) are effective to this end. We can update the necessary information for incremental learning very easily by RBFN, while the information of support vector can be utilized in selecting additional samples as the sensitivity in SVM. The details can be seen in [7], [8], [9]. Here, we introduce the incremental learning by RBFN briefly in the following.

The output of an RBFN is given by

$$f(\boldsymbol{x}) = \sum_{j=1}^{m} w_j h_j(\boldsymbol{x}),$$

where $h_j,\ j = 1, \ldots, m$ are radial basis functions, e.g.,

$$h_j(\boldsymbol{x}) = e^{-\|\boldsymbol{x}-\boldsymbol{c}_j\|^2/r_j}.$$

Given the training data $(\boldsymbol{x}_i, \hat{y}_i),\ i = 1, \cdots, p$, the learning of RBFN is usually made by solving

$$\text{Min} \quad E = \sum_{i=1}^{p} (\hat{y}_i - f(\boldsymbol{x}_i))^2 + \sum_{j=1}^{m} \lambda_j w_j^2$$

where the second term is introduced for the purpose of regularization.

In general cases with a large number of training data p, the number of basis functions m is set to be less than p in order to avoid overlearning. However, the number of training data is not so large in this paper, because it is desired to be as small as possible in applications under consideration. The value m is set, therefore, to be equal to p in later sections in this paper. Also, the center of radial basis function $\boldsymbol{c}_i$ is set to be $\boldsymbol{x}_i$. The values of λ_j and r_j are usually determined by cross-validation test. It is observed through our experience that in many problems we have a good performance with $\lambda_j = 0.01$ and a simple estimate for r_j given by

$$r = \frac{d_{max}}{\sqrt[n]{np}} \tag{1}$$

where d_{max} is the maximal distance among the data; n is the dimension of data; p is the number of data.

Letting $A = (H_p^T H_p + \Lambda)$, we have

$$A\boldsymbol{w} = H_p^T \hat{\boldsymbol{y}},$$

as a necessary condition for the above minimization. Here

$$H_p^T = [\boldsymbol{h}_1 \ \cdots \ \boldsymbol{h}_p],$$

where $\boldsymbol{h}_j^T = [h_1(\boldsymbol{x}_j), \ldots, h_m(\boldsymbol{x}_j)]$, and Λ is a diagonal matrix whose diagonal components are $\lambda_1 \ \cdots \ \lambda_m$.
Therefore, the learning in RBFN is reduced to finding

$$A^{-1} = (H_p^T H_p + \Lambda)^{-1}.$$

The incremental learninig in RBFN can be made by adding new samples and/or a basis function, if necesary. Since the learning in RBFN is equivalent to the matrix inversion A^{-1}, the additional learning here is reduced to the incremental calculation of the matrix inversion. The following algorithm can be seen in [10]:

(i) Adding a New Training Sample

Adding a new sample $\boldsymbol{x}_{p+1}$, the incremental learning in RBFN can be made by the following simple update formula: Let

$$H_{p+1} = \begin{bmatrix} H_p \\ \boldsymbol{h}_{p+1}^T \end{bmatrix},$$

where $\boldsymbol{h}_{p+1}^T = [h_1(\boldsymbol{x}_{p+1}), \dots, h_m(\boldsymbol{x}_{p+1})]$.

Then

$$A_{p+1}^{-1} = A_p^{-1} - \frac{A_p^{-1}\boldsymbol{h}_{p+1}\boldsymbol{h}_{p+1}^T A_p^{-1}}{1 + \boldsymbol{h}_{p+1}^T A_p^{-1}\boldsymbol{h}_{p+1}}.$$

(ii) Adding a New Basis Function

In those cases where a new basis function is needed to improve the learning for a new data, we have the following update formula for the matrix inversion: Let

$$H_{m+1} = \begin{bmatrix} H_m & \boldsymbol{h}_{m+1} \end{bmatrix},$$

where $\boldsymbol{h}_{m+1}^T = [h_{m+1}(\boldsymbol{x}_1), \dots, h_{m+1}(\boldsymbol{x}_p)]$.

Then

$$A_{m+1}^{-1} = \begin{bmatrix} A_m^{-1} & \boldsymbol{0} \\ \boldsymbol{0}^T & 0 \end{bmatrix}$$

$$+ \frac{1}{\lambda_{m+1} + \boldsymbol{h}_{m+1}^T (I_p - H_m A_m^{-1} H_m^T)\boldsymbol{h}_{m+1}} \times \begin{bmatrix} A_m^{-1} H_m^T \boldsymbol{h}_{m+1} \\ -1 \end{bmatrix} \begin{bmatrix} A_m^{-1} H_m^T \boldsymbol{h}_{m+1} \\ -1 \end{bmatrix}^T.$$

3. Selection of Additional Samples

If the current solution is not satisactory, namely if our stopping condition is not satisfied, we need some additional samples in order to improve the approximation of the black-box objective function. Now, how to select such additional samples becomes an important issue.

If the current optimal point is taken as such additional sample, the estimated optimal point tends to converge to a local maximum (or minimum) point. This is due to lack of global information in predicting the objective function.

On the other hand, if additional samples are taken far away from the existing data, it is difficult to obtain more detailed information near the optimal point. Therefore, it is hard to obtain a solution with a high precision. This is because of insufficient information near the optimal point.

It is important to get well balanced samples providing both global information and local information on black-box objective functions. The author and his coresearchers suggested a method which gives both global information for predicting the objective function and local information near the optimal point at the same time [8]. Namely, two kinds of additional samples are taken at the same time for relearning the form of the objective function. One of them is selected from a neighborhood of the current optimal point in order to add local information near the (estimated) optimal point. The size of this neighborhood is controlled during the convergence process. The other one is selected far away from the currenct optimal value in order to give a better prediction of the form of the objective function. The former additional sample gives more detailed information near the current optimal point. The latter sample prevents converging to local maximum (or minimum) point.

In the following, in order to avoid scalability effect, the value of k-th component x_{ki} of each pattern $\boldsymbol{x}_i$ is noramlized by

$$\tilde{x}_{ki} = \frac{x_{ki} - \mu_k}{\sigma_k}$$

where μ_k and σ_k are the mean value and the standard deviation of k-th component of the data set $\{\boldsymbol{x}_1, \dots, \boldsymbol{x}_p\}$, respectively.

The neighborhood of the current optimal point is given by a square S, whose center is the current optimal point, with the length of a side l. Let S_0 be a square, whose center is the current optimal point, with the fixed length of a side l_0. The square S is shrinked according to the number C_x of optimal points appeared continuously in S_0 in the past. Namely,

$$l = l_0 \times \frac{1}{C_x + 1}, \tag{2}$$

The first additional sample is selected inside the square S randomly. The second additional sample is selected in an area, in which the existing learning data are sparse, outside the square S as is shown in Fig. 1. An area with sparsely existing data may be found as follows: First, a certain number (N_{rand}) of data (denoted by white stars) are generated randomly outside the square S. Denote d_{ij} the distance from this random data p_i ($i = 1, \dots, N_{rand}$) to the existing learning data q_j ($j = 1, \dots, N$) (denoted by $\bullet$). Select the shortest k distances $\tilde{d}_{ij}$ ($j = 1, \dots, k$) for each p_i, and sum up these k distances, i.e., $D_i = \sum_{j=1}^{k} \tilde{d}_{ij}$. We set $k = 2$ in Fig. 1. Take p_t which maximizes $\{D_i\}_{(i=1,\dots,N_{rand})}$ as an addtional sample outside S. The sample point denoted by black stars $\bigstar$ in Fig. 1 enjoys this property.

The algorithm is sumarized as follows:

Step 1: Predict the form of the objective function by RBFN on the basis of the given training data.
Step 2: Estimate an optimal point for the predicted objective function by GA.
Step 3: Count the number of optimal points appeared continuously in the past in S_0. This number is represented by C_x.
Step 4: Terminate the iteration,

(i) if C_x is larger than or equal to the given C_x^0 a priori,

or

(ii) if the best value of the objective function obtained so far is identical during the last certain number (C_f^0) of iterations.

Otherwise calculate l by (2), and go to the next step.

Step 5: Select an additional sample near the current optimal value, i.e., inside S.
Step 6: Select another additional sample outside S in a place in which the density of the training data is low as stated above.
Step 7: Go to Step.1.

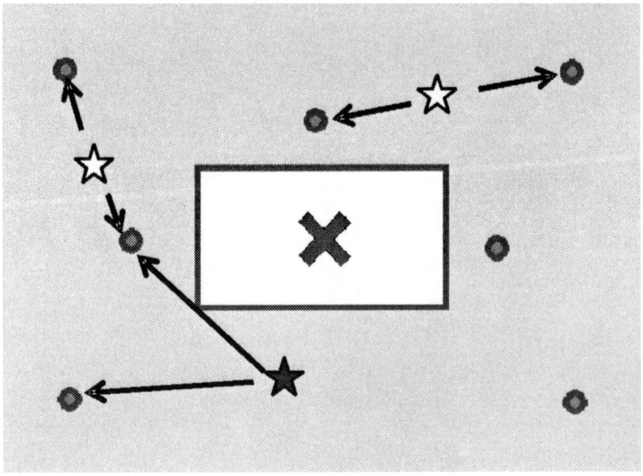

Figure 1. Additional sample point in a sparse area of existing data

4. Optimizing Predicted Objective Functions

We can apply any optimization methods for the predicted objective functions, because they are represented explicitly in terms of design variables. In order to obtain an approximate solution to the global optimum, genetic algorithms (GA) can be effectively applied.

In the following examples, we applied the BLX-α method which can treat in a simple way continuous design variables [11]. The method uses continuous values of design valriables as codes of individuals as they are. In crossover, children are generated randomly inside a hyperbox including their parents. It has been observed that the BLX-α method without mutation is effective in problems with continuous design variables. Several sophisticated methods which can treat continuous variables have been developed (e.g., Arakawa *et al.*[1]).

5. An Illustrative Example

Consider an example given by

$$f(x_1, x_2) = 10\exp(-0.01(x_1 - 10)^2 - 0.01(x_2 - 15)^2)\sin x_1 \quad (3)$$

$$(0 \leq x_1 \leq 15, \quad 0 \leq x_2 \leq 20).$$

This function has a maximum value 9.5585 at $x_1 = 7.8960$ and $x_2 = 15.0000$ as shown in Figure 2.

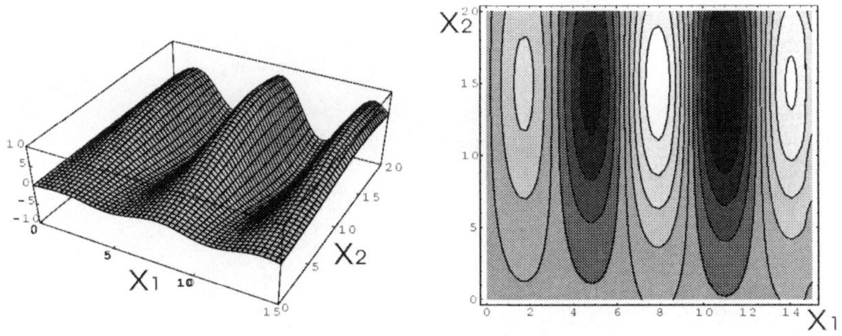

$x_1 = 7.8960, x_2 = 15.0000, f = 9.5585$

Figure 2. Contour of the example

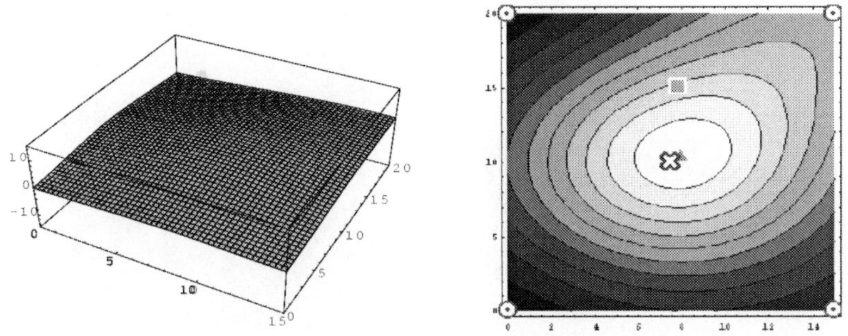

$x_1 = 7.9947, x_2 = 10.5039, f = 6.8400$

Figure 3. Result after the first iteration with 5 training data

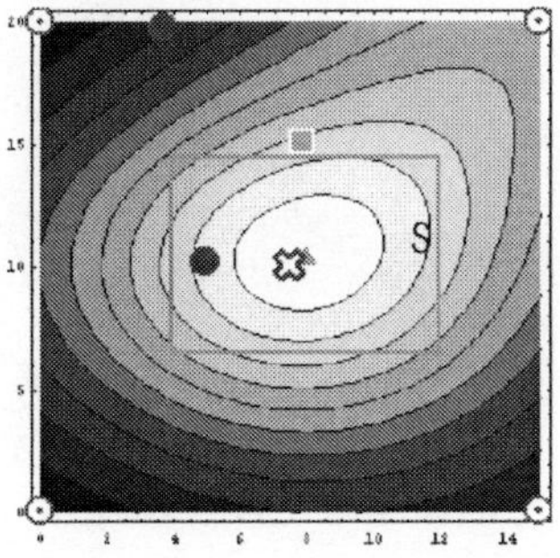

Figure 4. Additional samples denoted by •

Set λ in RBFN: 0.01, population in GA: 10, generation in GA: 50, l_0: 4.0, C_0: 10, N_{rand}: 50, and k: 2. At the begining, five data ($(x_1, x_2) = (0,0), (15,0), (0,20), (7.5,10), (15,20)$) were taken for the initial learning. The contour of the forecasted objective function by RBFN is shown in Figure 3. In Figures 2-4, here after, "$\odot$" represents the training

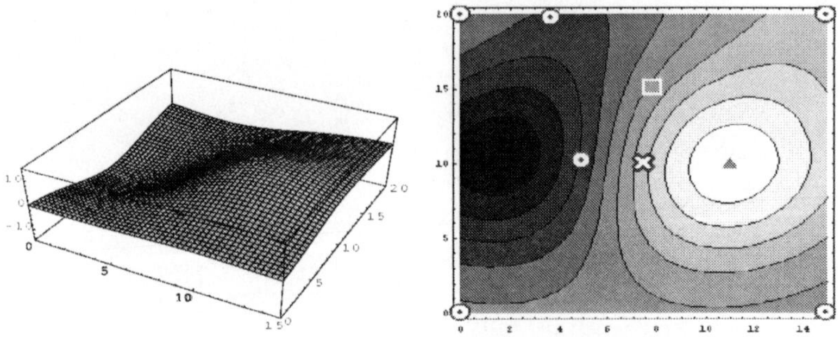

$x_1 = 7.5000, x_2 = 10.0000, f = 6.8730$

Figure 5. The result of our simulation with 7 training data

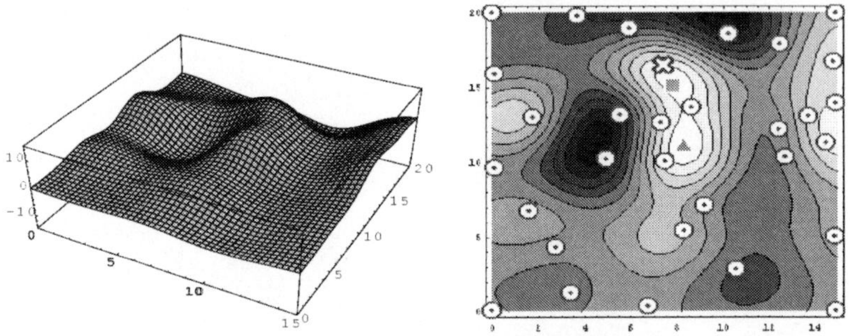

$x_1 = 7.4645, x_2 = 16.4721, f = 8.4892$

Figure 6. The result of our simulation with 31 training data

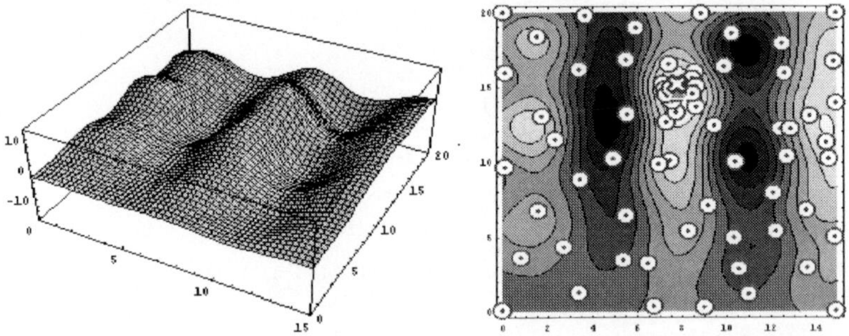

$x_1 = 7.8725, x_2 = 15.2023, f = 9.5519$

Figure 7. The result of our simulation with 69 training data

data, "×" the current optimal point, and "□" the correct optimal point to (2).

Figure 3 shows the result after the first iteration (5 training data). At this stage, two additional samples ("•") are selected according to the method stated in Section 3 (Figure

4). Then, there is no past optimal points in the square S_0. Therefore, the squares S and S_0 are identical at this stage. An additional sample is selected randomly inside the square S, and another additional sample is outside S in an area where the density of training data is low.

Figure 6 shows the result at an intermediate 31 iterations. After 69 iterations, we have a good estimate as shown in Figure 7.

It has been observed in our experiences that our proposed method shows good performance for problems with more variables as well. A comparative test of the proposed method with other existing methods for a bench-mark test problem (pressure vessel design) with four design variables, where two variables are continuous and the other two are discrete, can be seen in [8], [9].

6. Application to Reinforcement of Cable-stayed Bridges

After the big earthquake in Kobe in 1995, many in-service structures are required to improve their anti-seismic property. The criterion in the Specifications for Highway Bridges for resisting huge earthquake called Level 2 Earthquake, which has little possibility to occur during the life of bridges, was revised drastically. According to the revision, relatively small and simple bridges in service, such as viaducts in city area, were verified their anti-seismic property and almost of them have been reinforced. On the other hand, it is very difficult for large and/or complicated bridges, such as suspension bridge, cable-stayed bridge, arch bridge and so on, to be made a reinforcement because of impractical executing method and complicated dynamic response.

Recently many kinds of anti-seismic device have been developed [4]. It is practical in the bridge to be installed a number of small devices taking into account of strength and/or space, and to obtain the most reasonable arrangement and capacity of the devices by using optimization technique. In this problem, the form of objective function is not given explicitly in terms of design variables, but the value of the function is obtained by seismic response analysis. Since this analysis needs much cost and long time, it is strongly desirable to make the number of analysis as few as possible.

To this end, the proposed method in the preceding sections is applied. In this study, radial basis function networks (RBFN) are employed in predicting the form of objective function, and genetic algorithms (GA) in searching the optimal value of the predicted objective function. We made an attempt to apply the proposed method to a problem of anti-seismic improvement of a cable-stayed bridge which typifies the difficulty of reinforcement of in-service structure. In this investigation, we determine an efficient arrangement and amount of additional mass for cables to reduce the seismic response of the tower of a cable-stayed bridge (Fig. 8)..

6.1. Dynamic Characteristics of Cable-stayed Bridge

Though modal coupling among structural elements of cable-stayed bridge, i.e. girder, towers and cables, makes its seismic response complicated, this fact shows that the seismic response can be controlled by dynamic characteristics of each element. Obviously it is the dynamic characteristic of cable that we can change more easily. There are two ways to change characteristics of cables, that is, adding mass to cable and tensing or

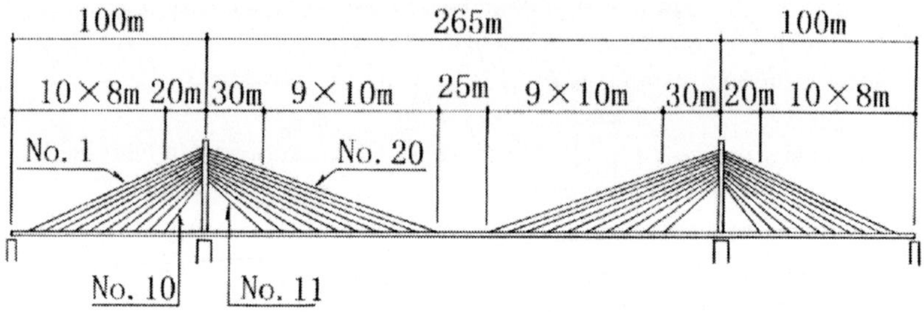

Figure 8. Cable-stayed bridge

loosening cable. The former can be acceptable if the amount of mass is limited appropriately [12], but the latter can not be alternative because it causes bridge an undesirable stress condition. Fig.9 shows an impression of additional mass with cable.

The influence of additional mass on cables was investigated by numerical sensitivity analysis. Analytical model shown in Fig. 8 is 3-span continuous and symmetrical cable-stayed bridge whose 2 × 20 cables are in one plane and the towers stand freely in their transverse direction. The mass must be distributed over cables uniformly to prevent them from concentrating deformation.

Seismic response to be paid attention was the stress at the fixed end of the tower when earthquake occurred in transverse direction. Seismic response analysis was carried out by spectrum method. As there are a lot of modes whose natural frequencies were close to each other, response was evaluated by the complete quadratic combination method. Input spectrum is given in the new Specifications for Highway Bridges in Japan.

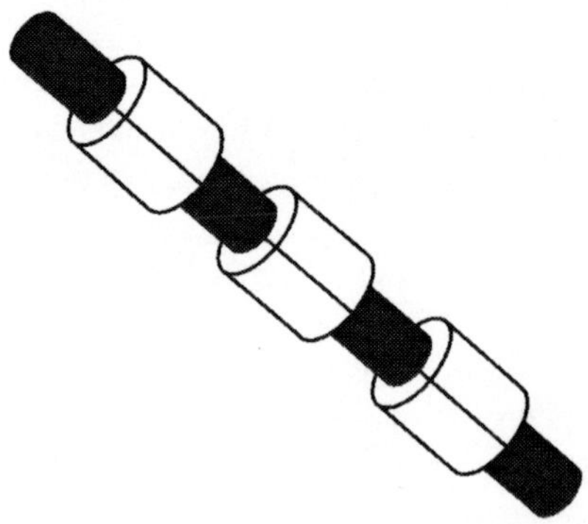

Figure 9. Cable with additional mass

The natural frequencies of modes accompanied with bending of tower (natural frequency of tower alone is 1.4Hz) range from 0.79Hz to 2.31Hz due to coupling with cables. For example, Fig.10 demonstrate half part of symmetrical mode shapes which are coupling among tower, girder and cables.

Change of bending moment at the fixed end of tower with additional mass on one cable distributed uniformly on it is shown in Fig.11. The abscissa means the ratio of additional mass to cable's own mass ranged from 0.0 to 1.0, and the ordinate means response of bending moment at the fixed end of tower. The 2nd upper cable (No.19) in the center span has much more influence on it than any other cables. Then, change of seismic response of bending moment of interest is considered individually with additional mass on each cable when additional mass ratio of No.19 was fixed to 0.5. The result is shown in Fig.12. No.20 cable becomes effective instead of No.1 and No.2. The minimum value of bending moment is 64.6MN·m which is 44% of bending moment without additional mass. As mentioned above, the seismic response of tower can be controlled by additional mass to cables, but each cable influences to one another complexly. Then, the reasonable distribution of additional mass must be decided by optimization technique.

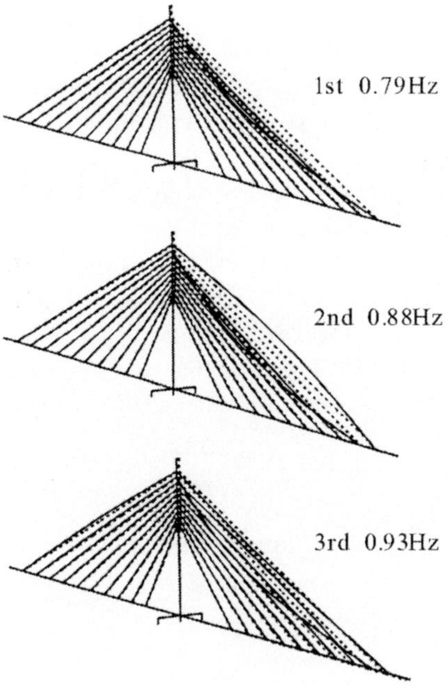

Figure 10. Examples of mode shape

6.2. Case 1

The objective is to minimize the bending moment M at the base of tower. The variables are ratios of additional mass and mass of cables. The number of variables is 20. The lower bound and upper bound of each variable are 0.0, and 1.0, respectively. For comparison, we applied a quasi-Newton method based on approximated differentials as

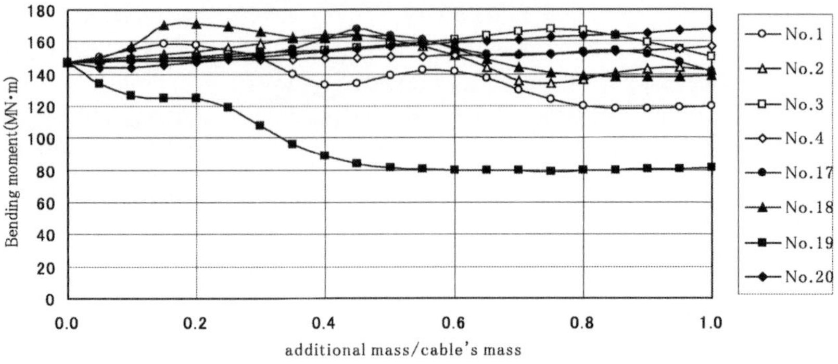

Figure 11. Influence on bending moment by additional mass (No ratio of additional mass is fixed)

an existing method. We made five trials with different initial points in order to obtain a global optimum.

In applying our proposed method, we used BLX-α as a genetic algorithm which is observed to be effective for continuous variables. The population is 10, and the number of generation is 200. We set $\lambda = 0.01$, and decided the value of width r of Gaussian by the simple estimate given by (1).

We started the iteration with 60 sample points. The first 20 sample points are generated randomly with one of variables fixed at the upper bound 1 by turns; the next 20s are generated similarly with one of variables fixed at the lower bound 0 by turns; the last 20s similary with one of variables fixed at the mid-value 0.5 by turns. The parameters for convergence are $C_x^0 = 20$, $C_f^0 = 20$ and $l_0 = 0.1$.

The result is shown in Table 1. It is seen that the proposed method can find out fairly good solutions within 1/10 or less times of analysis than the conventional optimization.

6.3. Case 2

Now, we take the number of cables to be added with masses, N, as another objective function in addition to the bending moment M. Namely, our objective function is

$$F = (M/M_0) + \alpha(N/N_0) \tag{4}$$

where α is a parameter for trade-off between the first term and the second one. M_0 and N_0 are used for normalization of the bending moment and the number of cables, respectively. In this experiment, we set $M_0 = 147.0$MN·m and $N_0 = 20$.

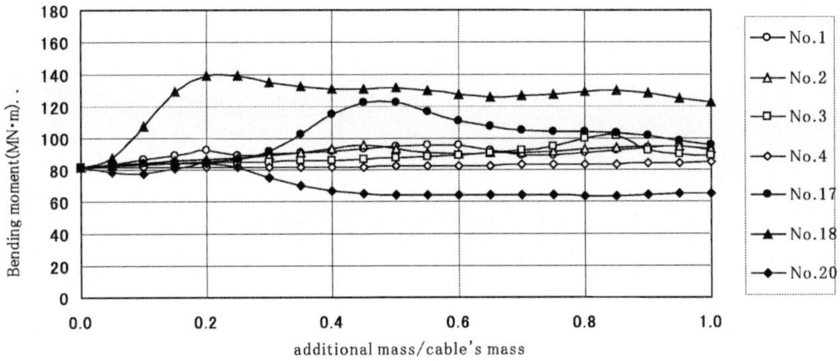

Figure 12. Influence on bending moment by additional mass (No. 19 ratio of additional mass is fixed at 0.5)

In this experiment, we used a simple GA, because some of variables are discrete. The parameters for calculation are the same as in Case 1. The result is given in Table 2. It should be noted that the number of analysis in our proposed method is reduced to about 1/20 of the conventional method. Although the precision of solution in our method is behind the conventional method, it is sufficiently acceptable in practice.

7. Concluding Remarks

In cable-stayed bridge as an example of structure in service, effective countermeasure against earthquake was investigated by a computational intelligence. The conclusions of this paper can be summarized as follows:

(1) Additional mass to cables can sharply reduce bending moment at the fixed end of tower when earthquake occurs in transverse direction.
(2) As there are a lot of conditions of distribution of additional mass, the procedure which reasonably decided distribution by optimization technique must be considered.
(3) Application of conventional optimization technique compels us to analyze seismic response very many times, because objective function is not given explicitly in terms of design variables, but obtained by analysis.
(4) In the problem of cable-stayed bridge, the proposed method can find out fairly good solutions within 1/10 $\sim$ 1/20 or less times of analysis than the conventional method.

Table 1. Result for Case 1

		exisiting method	RBF Network best	RBF Network average
cable No.	1	0.32	0.04	0.40
	2	1.00	0.69	0.84
	3	0.49	0.18	0.51
	4	0.62	0.82	0.80
	5	0.81	0.57	0.64
	6	0.52	0.43	0.56
	7	0.49	1.00	0.39
	8	0.52	0.44	0.66
	9	0.48	0.94	0.50
	10	0.48	0.50	0.56
	11	0.50	0.45	0.47
	12	0.55	1.00	0.74
	13	0.70	0.85	0.71
	14	0.61	0.50	0.30
	15	0.61	1.00	0.58
	16	0.46	0.24	0.37
	17	0.22	0.10	0.13
	18	1.00	0.95	0.91
	19	0.98	1.00	0.94
	20	1.00	1.00	0.91
bending moment (MN·m)		50.3	54.90	63.70
#analysis		1365	150.00	124.80

We introduced a method for optimizing black-box objective functions along with an application to seismic reinforcement of cable-stayed bridge. Not only in the seismic reinforcement of cable-stayed bridge but also in many engineering design, it is much expensive to make precise analyses by computer simulations or real experiments. Therefore, it has a very important implication in practice to decrease the number of analysis up to 1/10 or 1/20.

The well known Response Surface Method (RSM, in short) is competitive with our proposed method. However, our proposed method has been observed to have advantage over RSM especially for highly nonlinear cases. On the other hand, Jones *et al.* [5] suggested a method called EGO (Efficient Global Optimization) for black-box objective functions (see also [13], [14]). They applied a stochastic process model for predictor and the expected improvement as a figure of merit for additional sample points. However, it takes much effort to find a maximum of the expected improvement in EGO, while it is rather simple and easy to add two kinds of additional samples for global information

Table 2. Result for Case 2

		existing method		RBF network	
		best	average	best	average
cable No.	1	0.00	0.00	0.00	0.83
	2	0.00	0.00	0.00	0.09
	3	0.00	0.00	0.00	0.00
	4	0.00	0.00	0.00	0.04
	5	0.00	0.00	0.00	0.00
	6	0.00	0.00	0.00	0.00
	7	0.00	0.00	0.00	0.00
	8	0.00	0.00	1.00	0.99
	9	0.00	0.00	0.00	0.10
	10	0.00	0.00	0.86	0.53
	11	0.00	0.00	0.00	0.00
	12	0.00	0.00	1.00	0.63
	13	0.00	0.00	0.00	0.13
	14	0.00	0.00	0.86	0.53
	15	0.00	0.00	0.00	0.00
	16	0.00	0.00	0.00	0.00
	17	0.00	0.00	0.00	0.00
	18	0.00	0.00	0.00	0.00
	19	0.71	0.74	1.00	1.00
	20	0.86	0.83	0.86	0.79
bending moment (MN·m)		62.6	62.8	67.1	69.7
#cable with additional mass		2	2	6	6.2
objective fn.		0.526	0.527	0.756	0.784
#analysis		4100	3780	199	193.3

and local information for approximation in our proposed method. Further applications to practical problems will be expected to improve the method.

References

[1] M. Arakawa and I. Hagiwara, Nonlinear Integer, Discrete and Continuous Optimization Using Adaptive Range Genetic Algorithms, *Proc. of ASME Design Technical Conferences (in CD-ROM)*, 1997

[2] N. Cristianini, J. Shawe-Tylor, An Introduction Support Vector Machines and other Kernel-based learning method, Cambridge, 2000

[3] S. Haykin, *Neural Networks: A Comprehensive Foundation*, Macmillan College Publishing Company, 1994

[4] M. Honda, K. Morishita, K. Inoue and J. Hirai, Improvement of anti-seismic capacity with damper braces for bridges, *Proc. of the 7-th International Conference on Motion and Vibration Control*, 2004

[5] D.R. Jones, M. Schonlau and W.J. Welch, Efficient Global Optimization of Expensive Black-Box Functions, *J. of Global Optimization*, **13** (1998), 455-92

[6] R.H. Myers, and D.C. Montgomery, *Response Surface Methodology: Process and Product Optimization using Designed Experiments*, Wiley, 1995

[7] H. Nakayama, M. Arakawa and R. Sasaki, A Computational Intelligence Approach to Optimization with Unknown Objective Functions, *Artificial Neural Networks -ICANN2001*, ed. by G. Dorffner, H. Bischof and K. Hornik, pp. 73-80, Springer, 2001

[8] H. Nakayama, M. Arakawa and R. Sasaki, Simulation based Optimization for Unknown Objective Functions, *Optimization and Engineering*, **3** (2002), 201-214

[9] H. Nakayama, M. Arakawa and K. Washino, Optimization for Black-box Objective Functions, *Optimization and Optimal Control*, (eds.) P.M. Pardalos, I. Tseveendorj and R. Enkhbat, World Scientific (2003), 185-210

[10] M.J.L. Orr, Introduction to Radial Basis Function Networks, http://www.cns.ed.ac.uk/people/mark.html, 1996

[11] N.J. Radcliffe, Forma Analysis and Random Respectful Recombination, *Proceedings of the Forth International Conference on Genetic Algorithms*, (1991), 222-229

[12] K. Sasajima, K. Inoue, K. Morishita, J. Hirai and M. Honda, Study on the Optimal Anti-seismic Designing Method of Cable-stayed Bridge, *Proc. of the 3-rd World Conference on Structural Control*, 2002

[13] M.J. Sasena, P.Y. Papalambros, P. Goovaerts : Metamodeling Sample Criteria In a Global Optimization Framework, *8th AIAA/NASA/USAF/ISSMO Symposium on Multidisciplinary Analysis and Optimization*, Long Beach (2000, AIAA-2000-4921

[14] M. Schonlau, *Computer Experiments and Global Optimization*, PhD. thesis, Univ.of Waterloo, Ontario, Canada, 1997

Integrated Intelligent Systems for Engineering Design
X.F. Zha and R.J. Howlett (Eds.)
IOS Press, 2006

Design and Development of Monitoring Agents for Assisting NASA Engineers with Shuttle Ground Processing

Glenn S. SEMMEL [a,1], Steven R. DAVIS [a], Kurt W. LEUCHT [a], Daniel A. ROWE [a], Kevin E. SMITH [a], Ladislau BÖLÖNI [b]
[a] *National Aeronautics and Space Administration (NASA), Kennedy Space Center*
[b]*Dept. of Electrical and Computer Engineering, University of Central Florida (UCF)*

Abstract. The Engineering Development Directorate at NASA Kennedy Space Center has designed, developed, and deployed a rule-based agent to monitor the Space Shuttle's ground processing telemetry stream. The NASA Engineering Shuttle Telemetry Agent increases situational awareness for system and hardware engineers during ground processing of the Shuttle's subsystems. The agent provides autonomous monitoring of the telemetry stream and automatically alerts system engineers when user defined conditions are satisfied. Efficiency and safety are improved through increased automation.

Sandia National Labs' Java Expert System Shell is employed as the agent's rule engine. The shell's predicate logic lends itself well to capturing the heuristics and specifying the engineering rules within this domain. The declarative paradigm of the rule-based agent yields a highly modular and scalable design spanning multiple subsystems of the Shuttle. Several hundred monitoring rules have been written thus far with corresponding notifications sent to Shuttle engineers. This chapter discusses the rule-based telemetry agent used for Space Shuttle ground processing. We present the problem domain along with design and development considerations such as information modeling, knowledge capture, and the deployment of the product. We also present ongoing work with other condition monitoring agents.

Keywords. Agent, monitoring, rule-based expert system

[1] Correspondence to: Glenn S. Semmel, NASA, DX-E1, Kennedy Space Center, FL 32899. Tel.: +1 321 861 2267; E-mail: Glenn.S.Semmel@nasa.gov.

1. Introduction

1.1. Background

NASA Kennedy Space Center (KSC) is responsible for pre-launch ground checkout of the Space Shuttle. The Launch Processing System (LPS) at KSC provides facilities for NASA Shuttle system engineers, contractors, and test conductors to command, control, and monitor space vehicle systems from the start of Shuttle interface testing through various phases including terminal countdown, launch, abort, safing, and scrub turnaround.

LPS continually monitors the Shuttle and its ground equipment including environmental controls and hardware that loads propellants. Consoles with vehicle responsibilities communicate information directly to and from the Shuttle computer systems. Consoles with ground support equipment responsibility communicate information to and from the hardware interface modules which are connected to the numerous ground support systems. See Figure 1. Each module is capable of interfacing to approximately 240 sensors or controls. Overall, some 50,000 temperatures, pressures, flow rates, liquid levels, turbine speeds, voltages, currents, valve positions, switch positions, and many other parameters must be controlled and monitored.

Using LPS, NASA Shuttle engineers and contractors at KSC are responsible for certifying that ground checkout of the Space Shuttle has been performed according to program specifications. The Operations and Maintenance Requirements and Specifications Document[2] lists those procedures. For over 25 years, engineers have used LPS to verify Space Shuttle flight readiness and to control launch countdown. LPS has performed superbly well. Recently, much of the LPS hardware was upgraded assuring its continuance for many more years. However, the system architecture was not changed and software remains basically the same. As a result, the level of situational awareness has not increased proportionally to what would otherwise be possible with more modern software technologies.

After the Shuttle Columbia disaster on February 1, 2003, the Columbia Accident Investigation Board[3] proposed recommendations to improve safety from both an organizational and technical perspective. The Board indicated the need to "[adopt] and maintain a Shuttle flight schedule that is consistent with available resources." Also, both management and engineering support staff must maintain an awareness of anomalies and those must not be lost "as engineering risk analyses [move] through the

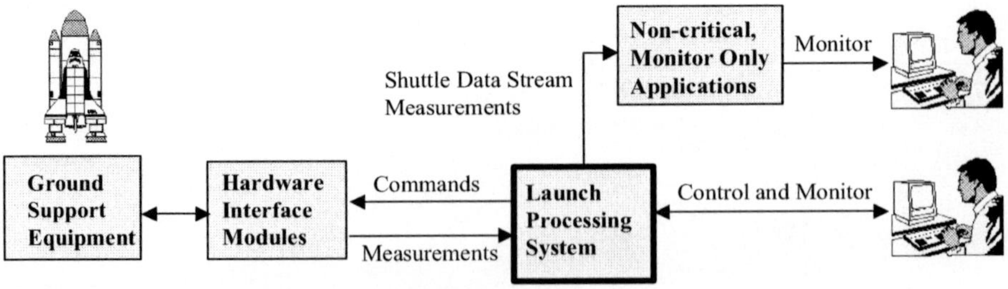

Figure 1. Ground Control and Monitoring at NASA KSC

process." Given two tragic losses of a crew and Shuttle, today NASA engineers have an even greater pressure to be more vigilant in identifying problems. At KSC, ground processing of the Shuttle is performed by thousands of employees, both contractors and civil servants. Anomalies must be detected and reported to prevent problems with Shuttle subsystems, countdown, and launch. The aging LPS hardware has limited resources and precludes the level of automation and notification warranted by this domain.

Contractors at KSC are responsible for the day to day operations, checkout, and maintenance of the Shuttle. They are the primary users of LPS. NASA Shuttle engineers are civil service employees who oversee the contractors. Given the limitations and resource scarcity of LPS, NASA Shuttle engineers needed a tool to provide more insight and situational awareness and oversee the work performed by contractors. An increased insight could help detect anomalies that might otherwise go unnoticed, whether by process error, software or hardware failures in the monitoring equipment, or many other possible causes. A tool was needed to complement LPS that could autonomously and continuously monitor Shuttle telemetry data and automatically alert NASA Shuttle engineers when predefined criteria have been met. In the latter half of 2003, a software tool was proposed to provide better insight into Shuttle ground processing and increase the level of situational awareness. This tool is known as the NASA Engineering Shuttle Telemetry Agent (NESTA).

1.2. Objectives

Data processed by LPS is distributed on a local area network. As shown in Figure 1, the distributed data is known as the Shuttle Data Stream (SDS)[4] and contains real-time vehicle ground processing data. It is used by various human intensive, monitor-only applications. The primary objective of NESTA is to provide full time autonomous monitoring of the SDS and to automatically alert NASA engineers in near real-time when pre-defined criteria have been met. Types of monitoring criteria include expected operational events or milestones (e.g. vehicle power up, start of launch countdown test, etc.) as well as unexpected events or failures (e.g. large difference between redundant sensor values). NESTA allows Shuttle engineers to work on other tasks while minimizing the risk of losing awareness of real-time Shuttle processing data and events.

NESTA acts as a software agent for the NASA engineer. For this discussion, an agent is defined as rule-based, autonomous software that reacts to its environment and communicates results to a human (e.g. NASA engineer). Agents have been extensively researched[5][6]. Agent standards[7] and frameworks[8][9] have also been developed.

The primary objectives for NESTA include:

- Allow a NASA engineer to specify rules to be applied to measurements published in the SDS.
- Generate near real-time notifications and alerts in the form of emails or wireless pages. Notifications may include a text message and measurement values, and may be sent to multiple users when the rule's premises are satisfied.
- Monitor up to four separate SDS sources. This includes four control rooms used for checkout and launch of the Shuttle and its components.

- Process multiple types and subtypes of measurements and read-only commands including discretes (i.e. boolean measurements), analogs (i.e. floating point measurements), and digital patterns (i.e. integer measurements).
- Allow users to create and modify multiple monitoring requests without restarting NESTA.

1.3. Why an AI Solution

NESTA leverages various AI technologies within a rule-based paradigm including forward chaining, fast pattern matching, declarative programming, predicate logic, and more. AI was a natural fit for monitoring the SDS since pattern recognition and analysis are the primary needs. Although pattern identification could be achieved by employing regular expression libraries within various procedural and object oriented languages, those paradigms are not specifically intended for this type of application and have less efficient matching algorithms. The pattern matching algorithms of rule-based expert system shells are highly specialized and tuned. Also, AI, particularly rule-based languages, lends itself better to this domain since pattern recognition wrapped within a premise-action construct closely mirrors the level of abstraction at which the domain experts work.

The type of data signatures sought by Shuttle engineers requires the derivation of rules that are of the same granularity as those typically used in rule-based languages. Fortunately, Shuttle engineers were already accustomed to representing knowledge at a fine grained level. The engineers are adept at either constructing the rules themselves or expressing the knowledge in pseudo code that lends itself well for translation directly into declarative rules. Many of the rules are either standalone or work in conjunction with several other rules. This suggests a highly modular system with a rule being a suitably sized working block.

1.4. Other Attempted Solutions

NESTA is a peripheral advisory tool to the real time control system within LPS. There were three previous projects that attempted to upgrade LPS in the last 15 years. Even though those efforts had significantly greater objectives that spanned well beyond just advisory applications, they were advertised to include many of the capabilities that NESTA provides and much more. Approximately half a billion dollars was spent on those efforts and upwards of 600 people worked on the most recent of those upgrade attempts. There were various technical and political hurdles that initially impeded and then ultimately doomed those full scale replacements of LPS.

NESTA's infusion of state-of-the-art AI technologies and engineering within the legacy launch system, LPS, is particularly notable given the number and size of the preceding attempts to modernize the ground control system at KSC. Those fallen projects, despite having much grander objectives, had little to no spin-offs within the LPS community. In contrast, NESTA is becoming accepted and internalized by members of the launch team and appears to be on its way as a widely used tool. From a business vantage point, NESTA's greatest asset is its development and marketing as a value added product. That is helping pave its path to acceptance.

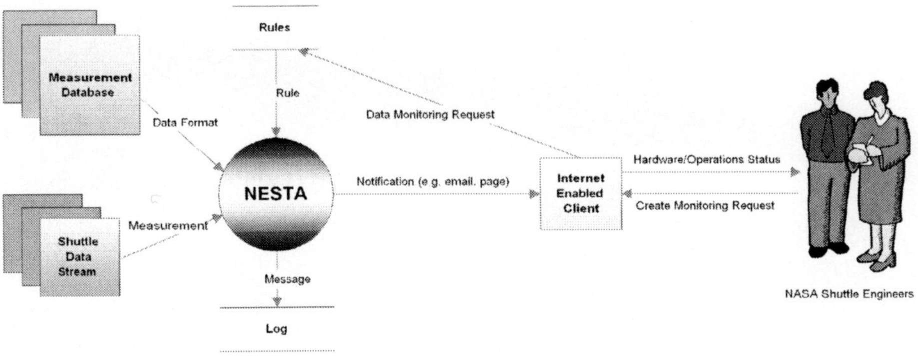

Figure 2. NESTA Context Diagram

2. Application Description

2.1. System Components and How They Interact

Figure 2 shows the context diagram for NESTA. The agent process is represented in the middle circle. It communicates with various sources and data stores. A measurement database is used to decode the SDS into usable measurements. The SDS source broadcasts measurements as data packets over local area networks. NESTA monitors this stream for data patterns specified by the Shuttle engineers. If a pattern is matched, a notification is sent as an email or wireless page. The Rules data store represents the Jess scripts and knowledge base that defines the rules for the monitoring criteria. All messages and relevant agent activities are also locally logged.

2.2. Languages and AI Tools Used

The Java Expert System Shell (Jess)[10] was selected as the rule engine. Jess was developed and supported by another government agency, Sandia National Labs. Jess' forward chaining reasoning system was modeled after production systems such as CLIPS[11] and OPS5[12]. It contains highly efficient and sophisticated pattern matching based on the Rete algorithm[13]. This enables its inference engine to process many rules and data rapidly. The engine repeatedly processes through a match-select-act cycle. As a production system, its consequents can be actions. A conflict resolution strategy determines the precedence of rule firings.

Several hundred monitoring rules have been written thus far for monitoring Shuttle ground telemetry. Jess' predicate logic lends itself to capturing and specifying the heuristics and engineering rules of this spaceport domain. The declarative paradigm of this rule-based agent also makes it highly modular and scalable to span multiple subsystems of the Shuttle. Jess also includes a fourth generation scripting language and interactive command line which are very conducive for prototyping and testing.

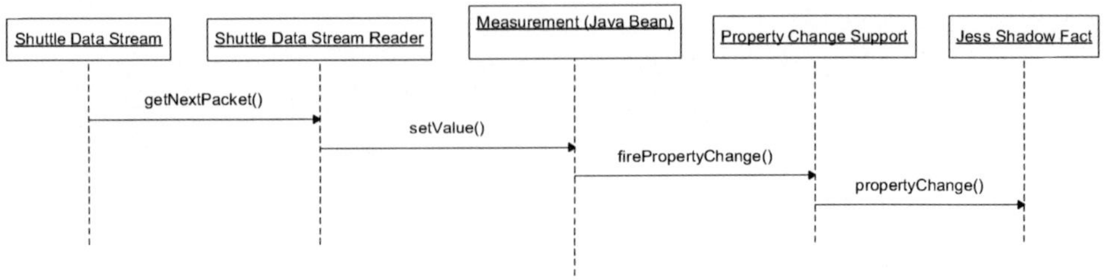

Figure 3. Sequence Diagram Illustrating Update to Jess Working Memory from Shuttle Data Stream

Jess is written entirely in Java and has access to the full Java application programming interface from the scripting language. It provides standard control flow constructs and supports variables, strings, objects, and function calls. Jess automatically converts between its own types and Java types insulating the developer from manually performing the conversions. Its use as a Java library made Jess' selection more appealing since Java supports multiple platforms with its "write once, run anywhere" paradigm. Beyond that, the need for NESTA to support web enabled clients also made Java a natural fit given its origins and strong support for developing Internet based applications.

2.3. Design

Java classes were developed to parse and decode the data stream and represent measurements as facts in Jess' working memory. To interface Jess' rule engine with the SDS, each data measurement is modeled and implemented as a Java bean[14]. Java beans provide a component architecture to enable easier integration of applications. A property change notification mechanism is supported that allows one object to become a registered listener of another object. The listener object will then automatically receive changes from the source object. This is also known as a publish-subscribe or observer pattern[15].

2.3.1. Integrated Object Oriented Rule Based Pattern Matching Architecture

Java beans help enable the integration of the rule based and object oriented architectures. Java beans map to shadow facts that are unordered facts, meaning each fact is composed of attributes with specific properties. The shadow fact is a mirror image of a Java bean, such as a pressure measurement, within Jess' working memory. All shadow facts are registered listeners of their Java bean counterparts. Thus, whenever a measurement changes in the data stream, a property change event is automatically generated for the given measurement and its sibling shadow fact is updated in Jess' working memory. Figure 3 illustrates this path.

After a shadow fact is updated, the Jess pattern matcher will determine if the premises of any rules match the new or modified facts. Rules are compared to working memory to identify premises that are matched by the data in working memory. For NESTA, this data represents measurements from the SDS and rules represent data monitoring criteria submitted by NASA Shuttle and system engineers. Rules with matching premises are activated and placed onto an agenda. Next, the agenda is ordered according to Jess' default conflict resolution strategy. The highest priority rule is then fired and executed. This match-select-act cycle repeats until no more rules are available to fire. An action handler class was developed and is used to build and send the notification message to the Shuttle engineer whenever a rule fires.

2.4. Knowledge Capture and Representation

Figure 4 shows the knowledge acquisition workflow for creating or modifying a rule to monitor specific measurements on the Shuttle data stream. The Shuttle engineer must specify who is responsible for the rule, the contents of the email notifications, the rule's firing conditions (i.e. antecedent, left hand side), and rearming conditions. That is, some rules may need to have a “one shot” behavior and only fire once when activated the first time. Other rules may need to be re-armed after a given time period or when certain types of conditions are met.

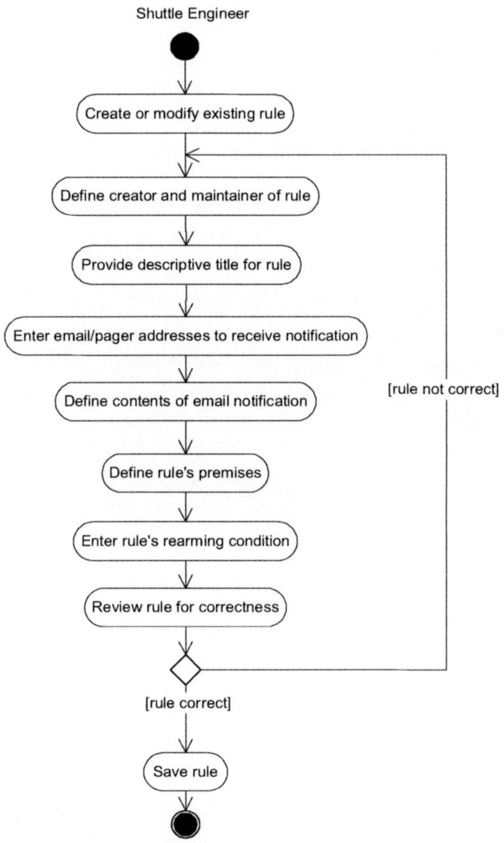

Figure 4. NESTA Knowledge Acquisition Workflow

The current version of NESTA does not have a graphical user interface capturing this workflow, but all of the steps are effectively provided within script files. Those files are editable with a plain text editor by the end users. Hundreds of rules have been produced by the customer.

As the rule database grew, patterns of rules began to emerge. Patterns in software design and modeling have been extensively investigated and reported[15]. Analogous to those design patterns, the development team and customer began recognizing knowledge patterns for this domain and developed rules following these structures. Some patterns include:

- *One shot:* Rule fires once regardless of how many times facts cause the premise to reactivate.
- *Recurring:* Rule fires each time the premise reactivates.
- *Timed:* Rule fires every X minutes as premise remains true.
- *Queued:* Multiple rules will fire but notifications are sent to a queue that gets flushed based on a user configurable amount of time or maximum number of firings. One composite notification is sent when the queue is flushed. That composite notification contains what would have otherwise been multiple emails or wireless pages.

Some sample rules in English prose include:

- *Notify Shuttle Engineer when measurement V79S4126E1 or V79S4132E1 or V79S4138E1 or V79S4143E1 equal ON.* Indicates that Flight Control Power (ASA 1-4) has been activated.
- *Notify Shuttle Engineer when measurement V90Q8001C1 equals 801.* Indicates that a Shuttle is in orbit and is preparing to initiate the on-orbit flight control checkout activity.
- *Notify Shuttle Engineer every 60 minutes with current values of Flight Control launch countdown measurement list when measurement NMAJORTEST equals 7.* Indicates launch countdown test is occurring. While in launch countdown test, send a current value email containing a list of Flight Control measurements every hour.
- *Notify Shuttle Engineer when FD N79IV019D bit masked 0x0001 equals 1.* Indicates that an LPS command and control program has stopped due to a failure and is waiting on the operator for action.

This is an actual NESTA rule written in the Jess scripting language:

```
(defrule vehicle-pwr-on-rule
    "Orbiter electrical power is up."

    (recipient-list (recipient-list-name vehicle-pwr-on-rule))

    ?notPowered <- (vehicle-not-powered)

    (DigitalPatternFd (fdName "NORBTAILNO") )
```

```
    (AnalogFd (fdName "V76V0100A1") (valid TRUE) (value ?val1))
    (AnalogFd (fdName "V76V0200A1") (valid TRUE) (value ?val2))
    (AnalogFd (fdName "V76V0300A1") (valid TRUE) (value ?val3))
    (test
        (and
            (> ?val1 26.0)
            (> ?val2 26.0)
            (> ?val3 26.0)
        )
    )

    =>

    (retract ?notPowered)
    (assert (vehicle-powered))
    (notifyActionHandler nil nil)
)
```

For this rule, if all three analog bus voltage measurements, V76V0100A1, V76V0200A1, and V76V0300A1, concurrently exceed 26 volts, the Shuttle Orbiter is considered to be powered on. Finally, another measurement, NORBTAILNO, is located on the rule's left hand side. In our terminology, we call this an informational measurement as its specific value has no bearing on whether the rule fires, but it is necessary to include it on the rules left hand side so that it becomes part of Jess' activation object and then its value is included in the notification. The action handler parses the fields in the activation object and builds an email with all of the measurements' values that were listed on the left hand side of the rule. The **`notifyActionHandler`** call has two arguments that allow for the notification to be queued. This particular example does not use queuing and simply passes **`nil`** arguments in the call. Queuing is discussed later in the chapter.

Figure 5 shows an email that was generated for the preceding rule. As illustrated, the exact values of all three bus voltages are listed along with the informational

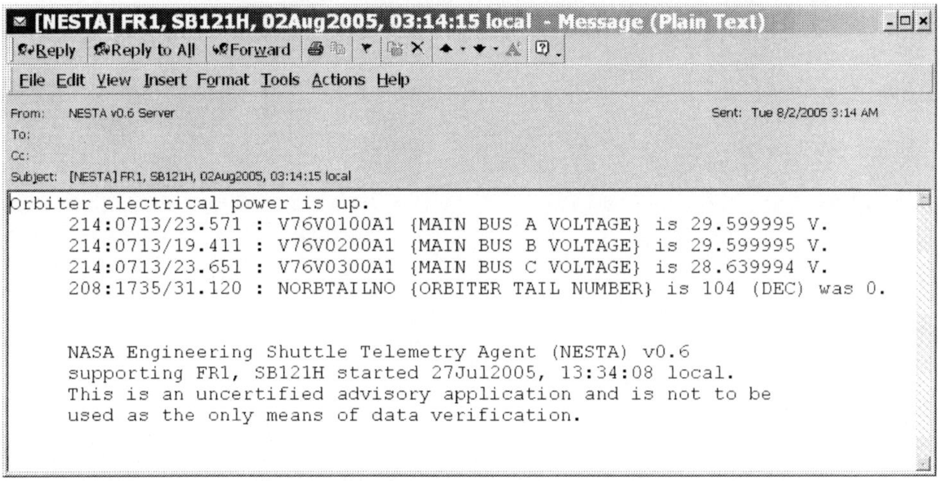

[NESTA] FR1, SB121H, 02Aug2005, 03:14:15 local - Message (Plain Text)

Reply | Reply to All | Forward

File Edit View Insert Format Tools Actions Help

From: NESTA v0.6 Server Sent: Tue 8/2/2005 3:14 AM
To:
Cc:
Subject: [NESTA] FR1, SB121H, 02Aug2005, 03:14:15 local

```
Orbiter electrical power is up.
     214:0713/23.571 : V76V0100A1 {MAIN BUS A VOLTAGE} is 29.599995 V.
     214:0713/19.411 : V76V0200A1 {MAIN BUS B VOLTAGE} is 29.599995 V.
     214:0713/23.651 : V76V0300A1 {MAIN BUS C VOLTAGE} is 28.639994 V.
     208:1735/31.120 : NORBTAILNO {ORBITER TAIL NUMBER} is 104 (DEC) was 0.


     NASA Engineering Shuttle Telemetry Agent (NESTA) v0.6
     supporting FR1, SB121H started 27Jul2005, 13:34:08 local.
     This is an uncertified advisory application and is not to be
     used as the only means of data verification.
```

Figure 5. Email Generated by NESTA

measurement showing which of the three Orbiters was powered up. In this case, 104 refers to Atlantis. The informational measurement proves useful in not only allowing the Orbiter reference to be included in the email, but it does not bind the rule to a particular Orbiter. That is, NASA Shuttle engineers are interested in any Orbiter that may become powered up. The rule's pattern matching provides that level of genericity in a very straight forward representation. Of course, the engineer may be interested in being notified only about a specific Orbiter. This would require a simple modification to the rule. One additional slot would be referenced in the **`DigitalPatternFd`** template narrowing the focus to a particular Orbiter. Thus, minor modifications to the rule demonstrate the rich behavior available to the Shuttle engineer and show the semantic power of pattern matching.

2.5. Applying Knowledge Discovery to Engineering Design

As knowledge about the domain was captured in the growing rule database, patterns emerged that drove partial refactoring of the agent design and architecture. For example, in the initial releases of NESTA, users were allowed to create rules each having unique queue times and queue lengths. Queuing provides a mechanism where multiple messages expected to occur within a short time period are grouped together before being emailed in bulk. For example, four flight control avionics boxes are often powered up in a short time period. Rather than a user receiving four separate flight control emails that may be interrelated, it was necessary to provide a queuing mechanism that allows a user (i.e. email recipient) to tie related emails to the same queue and receive one bulk email that was a compilation of what would otherwise be multiple emails. Both the queue time and queue length are configurable by the end user.

There were complexities associated with the design for queuing. If several emails were all assigned to the same queue, a preemption mechanism was required to allow a newly fired rule to effectively lower the queue's remaining delay to its own maximum delay. This resulted in a design requiring multiple data structures, schedulers, timers, notification mechanisms, and multiple threads. However, as the rule database grew, patterns emerged where it became apparent that the sets of rules assigned to a particular queue all had the same identical queue length and time parameters. Thus, as knowledge discovery spanned many releases, patterns began emerging that enabled many simplifications and refactoring within the design.

A recipient list fact was refactored to become a distribution task fact. The distribution task has a name and is composed of a set of email addresses, a flush time, and a maximum queue length. Rules no longer "owned" the flush time and queue length parameters. Those are now owned by (encapsulated within) the distribution task. A set of rules belong to a distribution task that is then responsible maintaining the remaining flush time and current queue length.

2.6. Hardware and Software Environment

The NESTA application currently resides on a Dell 1.7 GHz Pentium server. The server includes the necessary user and support files such as the facts scripts, rules scripts, measurement database, logs, and more. Currently, the server executes on a Microsoft Windows 2000 operating system. However, since Java was used exclusively along with its virtual machine, the ability to execute software on other types of servers is readily available. Again, this was a primary driver in the selection of Java and Jess so as to not be bound to a particular hardware platform or operating system. Customers receive notification on standard email clients including Windows workstations, wireless pagers, personal digital assistants, cell phones, and more.

2.7. Performance Requirements and Testing

2.7.1. Performance Characteristics of Shuttle Data Stream

At application startup, NESTA connects to a datastream selected by the user. The datastream includes all measurements at their respective change rates. No data changes will be missing from this stream known as FIFO. Another type of stream exists, but for this discussion, only the FIFO stream will be presented as it is the stream of choice.

The datastream averages 5 to 10 packets per second and peaks around 50 packets per second at launch. Each SDS data packet can hold up to 360 measurement changes before rolling over to another packet. This calculates to an average of 1,800 changes per second for the FIFO stream nominally, and 18,000 changes per second peak at launch. During peak data loads, the SDS is throttled at the source and does not maintain true real time updates. It may lag up to 1 minute or so, but all measurement changes are buffered and none is ever dropped from the data stream. Throttling of the data typically begins at T+1 second, that is, just after launch. Even though it is the hypothetical peak limit, 18,000 changes per second is the performance load that NESTA is expected to meet to avoid missing a measurement change. This is referring strictly to updating 18,000 facts per second and not indicating how many rules might fire. In fact, only a small percentage of those facts is expected to result in a small percentage of the total rules to fire at any given time, even during the peak launch data rates.

The measurement data in the stream is refreshed every three minutes regardless as to whether or not it has changed. Since the stream is based on User Datagram Protocol (UDP), this results in an unreliable datagram packet service. When a packet is dropped on the network, all measurements are marked invalid and the measurements change back to valid one by one as refresh data is received until the completion of a three minute refresh cycle.

2.7.2. Performance Testing

Performance testing occurred on an Intel Pentium 4, 1.7 GHz desktop workstation with 768 MB of RAM running Microsoft Windows XP Professional. The SDS reader class in NESTA parses the data stream and updates facts in Jess' working memory. To test the reader class, 12 high speed analog measurements were selected and instantiated as shadow facts. In the range of 18,000 (nominal) to 36,000 (peak at launch) data changes occurred every second in the test-enhanced data stream and were processed by the SDS reader class. This included various types of measurements such as discretes and analogs. 12,000 analog data changes per second were being processed into current values and updated in Jess' working memory by a property change event handler.

Rules were written for 6 of the high speed analog measurements. The other 6 measurements were still relevant to stress the SDS reader class and updating of facts. 5 of the 6 rules fired once every minute. The 6th rule fired once for every single measurement change (1,000 per sec) for two full seconds sustained out of every minute. Thus, a total of 2005 rules fired every minute, with 2000 of them firing within a 2 second period. Analog measurements have considerably more processing overhead than the discrete measurements so it was not possible to sustain thousands of rules containing analogs to fire every second without causing CPU starvation. However, the "fair test" was considered to have only a very small percentage of the measurements that are in the stream actually causing rules to fire. It was considered fair to have short bursts of high rate rule firings but not long term sustained high rate rule firings. NESTA is not intended for users to write rules to notify them via email hundreds or thousands of times each second for a long and sustained period of time.

To summarize, NESTA sustained the above scenario for many cycles on the test-enhanced playback file without CPU starvation and without reporting any packet losses. The CPU utilization on the development workstation was about 90% prior to launch and higher than that after T-0. It was heavily loaded, but NESTA maintained the pace. NESTA performed well considering that the data stream was stuffed with between 1 and 2 times the hypothetical peak load of measurement changes for the performance test. The "long pole" in the process appeared to be the number of rules that actually fired every second sustained. However, even under launch conditions when a heavy data change load exists, there is not expected to be many thousands of rules firing every second. Even several hundred rules firing per minute is considered unrealistically high, but this performance test suggests NESTA could readily handle that load.

3. Development and Deployment

3.1. Application Use and Payoff

At the time of writing of this chapter, the customer had used NESTA almost two years. Hundreds of rules have been written. Along with that, hundreds of NESTA notifications have been generated for multiple NASA engineers. These users have received both emails and wireless pages at KSC and other remote sites. Since the customer is a NASA engineer responsible for oversight of contractors, the notifications act as an extra set of eyes that further assure the quality of government oversight.

To better understand NESTA's payoff, the responsibilities of NASA Shuttle Engineers must be examined. They include:

- Understanding their system and supporting equipment.
- Knowing how their systems are tested and processed.
- Being aware of when their systems are activated, tested, or in use.
- Analyzing performance and data retrievals from any use of a system.
- Being ready to answer questions about their systems such as
  - When was it tested?
  - How did testing proceed?
  - How did the data look?
  - Is it ready to fly?

NESTA has helped Shuttle Engineers meet these responsibilities in varying degrees. Below are two success stories documenting some of the benefits NESTA has provided.

3.1.1. Success Story – Increased Situational Awareness

In one usage, a Shuttle avionics system was powered up over a weekend. The NASA Shuttle Engineer, being responsible for that system, would not have been aware that the system was powered up except for receiving a NESTA notification. In this case, the avionics user was not part of the Shuttle Engineer's immediate organization. Thus, the Shuttle Engineer did not receive any communiqués regarding the system's weekend usage. Due to NESTA, the Shuttle Engineer was better prepared to address questions about his system's usage were they to arise. This has not been an uncommon occurrence. Shuttle Engineers utilizing NESTA began realizing that some of their systems were being utilized much more than previously thought. Situational awareness increased markedly.

3.1.2. Success Story – Increased Efficiency

Some ground operations span 24 hours and include dozens of asynchronous events that are broadcast on the data stream. For example, checkout of flight control hardware in the Orbiter Processing Facility occurred 4 to 6 times within the last year. The checkout included long hydraulic operations, powering up different parts of avionics, pressurizing/depressurizing the Orbiter, and other work. During a recent flow, the NESTA notifications gave exact times of events of interest to the Shuttle Engineer. That allowed the Shuttle Engineer to quickly identify timelines of these lengthy operations. Effectively, a virtual roadmap identifying significant events was automatically generated and that saved labor time. More efficient data retrievals resulted.

3.2. Phased Approach to Implementation and Delivery

Multiple releases of NESTA have been delivered to the customer. The development team has four members each working approximately sixty percent of his time on the project. The team works very closely with the customer. Generally, the team meets with the customer at least once per week and has multiple other correspondences via email and phone.

The initial NESTA release required six months. Thereafter, a release occurred approximately every month. Prior to adopting Java and Jess, some preliminary performance testing was completed to verify that the Java language and Jess rule engine were fast enough to handle the Shuttle data stream rates. Concurrently with that coarse performance testing, the initial set of requirements were being developed.

The software process model employed is a combination of extreme programming and the iterative waterfall model. The team and customer understood the need to anticipate and accommodate changes in the requirements. The customer, as much of the development team, had little experience with rule based systems so there was a learning curve in how best to represent knowledge and interface the data stream with Jess. After about six months, a baseline set of requirements existed but the requirement space is still fluid and undergoes change over time. These changes are seen as a learning process through which we explore the possibilities of the system. As releases are delivered to the customer, new requirements are elicited and old ones may become defunct.

3.3. Development Tools

In addition to Java and Jess, other tools used include:

- Eclipse as an integrated development environment.
- Visio 2000 to develop Unified Modeling Language models.
- CVS for configuration management.
- Ant for automating builds.
- JUnit for automated Java unit testing.
- Emma for Java code coverage including measurements and reporting.
- Optimizeit by Borland for profiling performance and detecting and isolating problems.

3.4. Technical Difficulties

3.4.1. Data Validity

As indicated earlier in the chapter, the data stream is based on User Datagram Protocol (UDP). As such, the connection is not always reliable and packets may get dropped by the network. This poses problems when rules are waiting for data to arrive. Data health and validity become questionable. If the data stream connection is lost entirely or data becomes stale (i.e. not updated), false positives or false negatives may result. That is, notifications of hardware events may never be sent or be sent in error.

To partially address this data validity issue, additional measurements are included in the rules to check for the validity of the stream. Measurements are now marked invalid for a dropped packet(s) or when the source of the measurement becomes bad. There is still a larger problem of false negatives and never receiving an email if the data stream drops packets while a monitored event occurred. Aside from notifying the Shuttle engineer of a data loss when it happens, we have not yet identified a mechanism that guarantees all notifications since the data stream is unreliable.

3.4.2. Measurement Databases Changes

Multiple data streams and control rooms exist. Often, the measurement database, which is used to decode the SDS, dynamically changes on the stream as a result of operations. When that happens, decoding measurements becomes impossible and facts can no longer be updated in Jess' working memory. A short term fix to this problem was to simply notify the NESTA system administrator when the stream changes. A measurement database Java bean was added and is used within a user rule as a fact. When the measurement database changes, the administrator automatically gets an email and may restart NESTA accordingly. Longer term, automatic restarts of the agent will be provided.

3.4.3. Flood of Emails

If an end user incorrectly writes a rule, a possibility existed of flooding the network and servers with hundreds or even thousands of notifications. To prevent that, multiple safeguards, such as user defined limits, were provided to filter emails after a given number have been generated for a particular email account.

3.5. Maintenance

New releases are delivered approximately every month by the development team. Those releases may include bug fixes for problems reported in the former release. However, new releases are generally driven by new functionality as opposed to being driven by software errors.

The design of the NESTA application facilities update by the end user. The application uses a data driven approach for the user files. All of the rules and facts are stored in Jess scripts. When rules have to be created or modified, the user has access to several text based files. A facts file allows a user to add measurements that should be monitored. A rules file allows the entry of new rules. Since these are text-based script files, no compilation is required by the end user. The files are parsed at application startup. This data driven approach is powerful in that it enables the end users to maintain their own files and not be at the mercy of the development team to add new support for new facts and rules.

3.5.1. Web-based Application Maintenance Interface

A Web-based Application Maintenance Interface (WAMI) was developed to aid the users in managing and monitoring the agent. WAMI is based upon JMX[16] and MX4J[17]. Figures 6 and 7 show the Summary and Management Bean Views, respectively. The Summary View shows the current state of the agent, presenting information such as agent starting time, the data stream being monitored, the number of dropped packets, memory usage, and more. The Management Bean page shows a snapshot of the values of a particular set of measurements from the data stream and also allows the customer to query the value of any arbitrary measurement on the data stream. Further information is provided in other pages and views.

4. Launch Commit Criteria Monitoring Agent

Another agent using Jess has also been developed at NASA KSC. The Launch Commit Criteria Monitoring Agent (LCCMA)[18] identifies limit warnings and violations of launch commit criteria. As opposed to being used for day to day operations for which NESTA was developed, LCCMA's scope is targeted for launch countdown activities.

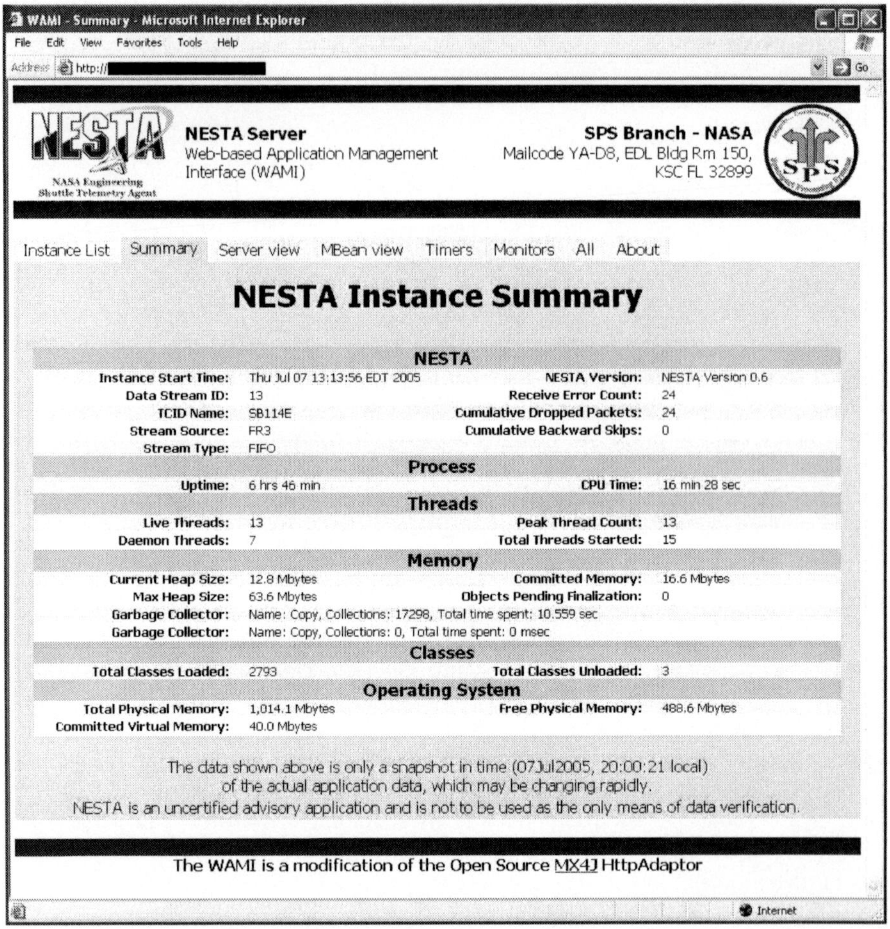

Figure 6. Web-based Application Maintenance Interface Summary Page

During launch countdown, NASA Shuttle engineers are required to monitor shuttle telemetry data for violations of launch commit criteria (LCC) and to verify that the contractors troubleshoot problems correctly. When a violation is recognized by the system engineers it is reported to the NASA Test Director. The problem report, or call, includes a description of the problem, the criticality, whether a hold is requested, and whether a preplanned troubleshooting procedure exists.

The Shuttle is composed of many subsystems (e.g. Main Propulsion, Hydraulics). Each of those subsystems has a team of engineers responsible for troubleshooting problems for that respective system during a launch countdown. Many systems have a large number of measurements with associated LCC limits and a large number of LCC requirements.

Shuttle Engineers must monitor for many types of limit violations ranging from simple high and low limit boundaries to much more complex first order logic expressions. Each team has its own tools for identifying LCC violations. Many of these tools use the LPS software and simply change the color of the displayed data

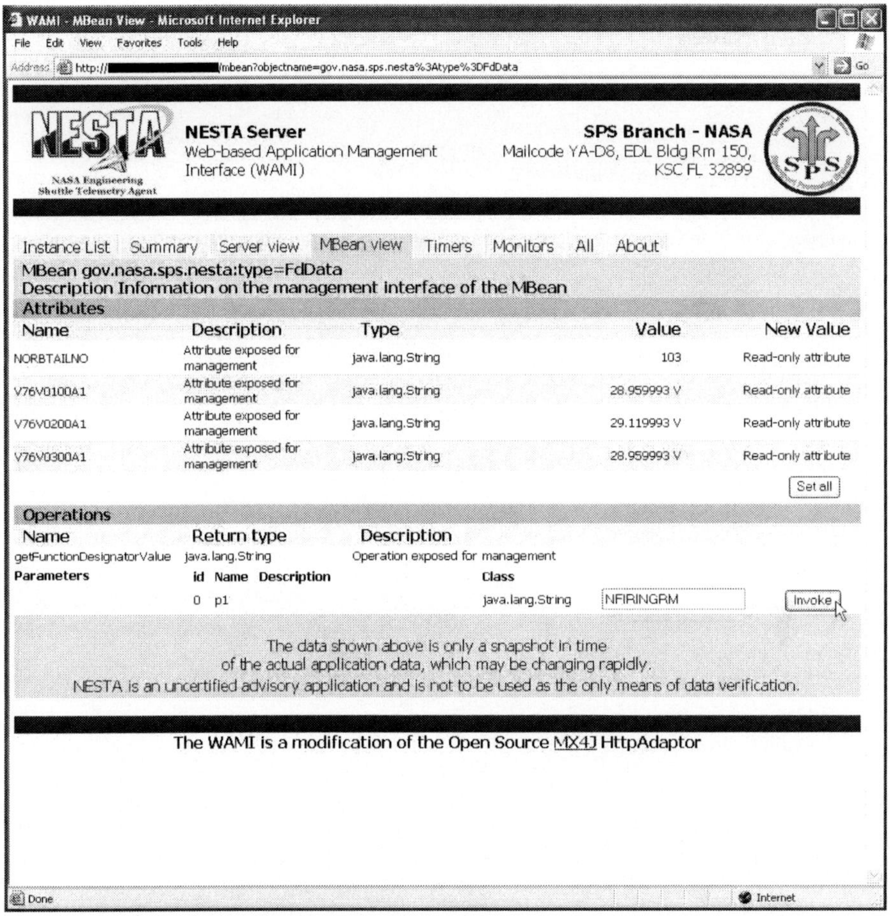

Figure 7. Web Application Maintenance Interface Management Bean View Page

and/or present a text message to the user or set off an audible alarm. Troubleshooting may require other displays such as plots and troubleshooting flowcharts. Valuable time is spent locating these procedures and locating the data that supports them.

With LCCMA, when a launch commit criteria violation is detected, the Shuttle engineer is notified via a Status Board Display on a workstation. Troubleshooting procedures are automatically made available on the Display. This precludes the Shuttle engineer from manually searching for the correct procedure mapped to the given violation.

4.1. Graphical User Interface

A graphical user interface currently exists for the Status Board Display. It is being upgraded and Figure 8 shows a storyboard representative of that future interface. The

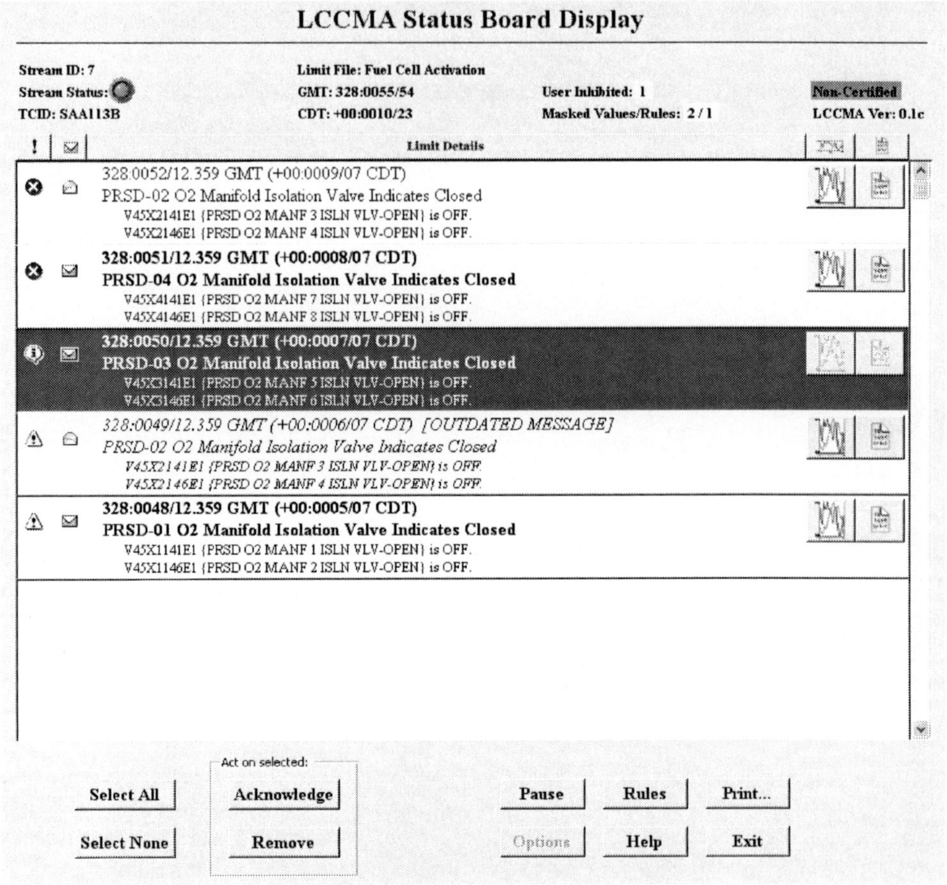

Figure 8. LCCMA Status Board Display

Status Board Display shows the health of the network connection, data stream status, countdown time, and other relevant information.

When LCC limits are violated, the LCC call is displayed in the text box. The user reads the text and, if there is an associated troubleshooting file, clicks the file button next to the text. This brings up a Troubleshooting Display for that particular LCC and limit. The LCC text remains bold until the Acknowledge button is pressed. Message text can be displayed with one of three icons representing a violation, warning, or informational cue. Measurements associated with the LCC may also be plotted.

The text messages can be read over the Operational Intercommunication System as LCC calls during the countdown. Calls will change based on what limit is violated (e.g. warning, LCC, high/low limit), the time criticality of the call, and LCC effectivity. The agent aids the NASA engineer in making a Go/No-Go decision for launch.

5. Conclusion and Future Work

NESTA has increased situational awareness of ground processing at NASA KSC. More and more Shuttle engineers are relying on NESTA each month and are creating additional rules for monitoring the data stream. The infusion of AI technologies, particularly the Jess rule-based library, has proved very fruitful. Interfacing and integrating these modern AI tools within a legacy launch system demonstrates the scalability and applicability of the tools and paradigm.

The knowledge patterns that are evolving within NESTA will make it easier to train new users and also allow faster creation of rules. Many other enhancements are planned such as providing an advanced graphical user interface for creating the rules.

5.1. Future Exploration Agents

As indicated in the national Vision for Space Exploration[19], an increased human and robotic presence will be cultivated in space, on lunar and Martian surfaces, and other destinations. Spaceports will now span from the Earth to the Moon and beyond. A new set of challenges is presented by this Exploration Vision. In particular, the need for autonomy significantly increases as people and payloads are sent greater distances from Earth.

Agents for these future applications will demand much higher degrees of autonomy than today's Shuttle agents. Few or no human experts will reside at remote lunar or Martian sites to correct problems in a timely manner. More automation will be required along with advanced diagnostics and prognostics. This requires higher levels of reasoning.

Today on Earth, system and hardware engineers along with technicians leverage multiple skills when monitoring, diagnosing, and prognosticating problems in Shuttle ground support equipment. For the Exploration Vision, the need for extending these skills to support other vehicles and payloads at remote locations from the Earth to Mars becomes essential. These skills include being rational, collaborative, goal driven, and the ability to reason over time and uncertainty. The agents discussed earlier in the chapter, NESTA and LCCMA, are capable of shallow reasoning of short inference chains within the Shuttle domain. However, these existing agents can be endowed with higher levels of rationality enabling a deeper reasoning. We are investigating how to mature these agents into Spaceport Exploration Agents (SEAs) in support of the Exploration Vision.

SEAs will need to communicate and collaborate along multiple and lengthy logistics chains. This does not simply include agents monitoring pre-flight checkout of vehicles at a terrestrial spaceport (e.g. NESTA monitoring Shuttle ground processing events). Rather, SEAs will reside in multiple locations at great distances. Logistics, scheduling, and planning are just some of the activities that these agents will manage.

Within this virtual collaborative management chain, SEAs will be inundated with massive amounts of data that must be sorted and processed. It becomes necessary for them to revise their sets of beliefs as new data arrives. It is simply not enough to revise singular data points within an agent's working memory and to have an agent blindly react to those changes. Rather, an agent must possess the ability to revise previously concluded assertions based on what may be now stale data. This activity is called truth maintenance[20][21][22], also known as belief revision, and is particularly important when deep reasoning of long inferences is necessary. An assumption based truth maintenance system (ATMS) can reason over many contexts simultaneously. By capturing, maintaining, and deploying spaceport expertise within ATMS-enabled SEAs, the costs and manpower required to meet the Exploration Vision are reduced while safety, reliability, and availability are increased.

References

[1] G.S. Semmel, S.R. Davis, K.W. Leucht, D.A. Rowe, K.E. Smith, and L. Bölöni. NESTA: NASA Engineering Shuttle Telemetry Agent. In *Proceedings of the 20th National Conference on Artificial Intelligence and the 17th Innovative Applications of Artificial Intelligence Conference* (July 2005). AAAI Press. Menlo Park, CA, USA, 1491-1498.
[2] *Operations and Maintenance Requirements and Specifications Document*. NASA. 2005.
[3] H. Gehman, S. Turcotte, J. Barry, K. Hess, J. Hallock, S. Wallace, D. Deal, S. Hubbard, R. Tetrault, S. Widnall, D. Osheroff, S. Ride, and J. Logsdon. *Columbia Accident Investigation Board (CAIB), Volume 1*. NASA. Washington D.C., August 2003.
[4] Lockheed. *PCGOAL Requirements Document*, Technical Report KSCL-1100-0804. Lockheed Space Operations Company, 1991.
[5] M. Wooldridge. *Reasoning about Rational Agents*. Cambridge, Massachusetts. The MIT Press, 2000.
[6] S. Russell and P. Norvig. *Artificial Intelligence: A Modern Approach*. Prentice Hall, 2nd edition, 2003.
[7] FIPA. Foundation for Intelligent Physical Agents Abstract Architecture Specification, 2002.
[8] L. Bölöni and D.C. Marinescu. An Object-Oriented Framework for Building Collaborative Network Agents. In Teodorescu, H.; Mlynek, D.; Kandel, A.; and Zimmerman, H., eds., *Intelligent Systems and Interfaces*, International Series in Intelligent Technologies. Kluwer Publishing House. Chapter 3, 21-64, 2000.
[9] JADE. Java Agent Development Framework. http://jade.tilab.com/, 2004.
[10] E. Friedman-Hill. *Java Expert System Shell*. Greenwich, CT. Manning Publications, 2003.
[11] R.M. Wygant. CLIPS: A Powerful Development and Delivery Expert System. In *Computers and Industrial Engineering, Volume 17*(1989), 546-549.
[12] L. Brownston, R. Farrell, E. Kant, and N. Martin. *Programming Expert Systems in OPS5: An Introduction to Rule-Based Programming*. Reading, MA. Addison-Wesley, 1986.
[13] C.L. Forgy. Rete: A fast algorithm for the many pattern/many object pattern match problem. In *Artificial Intelligence*, volume 19(1)(1982), 17-37.
[14] Sun Microsystems. Java Bean Specification. http://java.sun.com/, 2004.
[15] E. Gamma, R. Helm, E. Johnson, and J. Vlissides. *Design Patterns: Elements of Reusable Object-Oriented Software*. Greenwich, CT. Addison-Wesley, 1995.
[16] Sun. Java Management Extensions (JMX). http://java.sun.com/products/JavaManagement/index.jsp.
[17] MX4J. http://mx4j.sourceforge.net/.
[18] G.S. Semmel, S.R. Davis, K.W. Leucht, D.A. Rowe, A.O. Kelly, and L. Bölöni. Launch Commit Criteria Monitoring Agent. In *The 4th International Joint Conference on Autonomous Agents and Multiagent Systems (AAMAS 2005)*. Association for Computing Machinery. New York, NY, USA, 3 – 10.
[19] NASA. The Vision for Space Exploration. Technical Report, NP-2004-01-334-HQ, 2004.
[20] J. Doyle. A Truth Maintenance System. *Artificial Intelligence*, 12(3): 231-272, November 1979.
[21] J. de Kleer. An Assumption-based Truth Maintenance System. *Artificial Intelligence,* 28(2): 127-162, 1986
[22] K. Forbus and J. de Kleer. *Building Problem Solvers*. MIT Press. Cambridge, MA, 1993.

Integrated Intelligent Systems for Engineering Design
X.F. Zha and R.J. Howlett (Eds.)
IOS Press, 2006

Intelligent Mechanisms for Energy Reduction in Design of Wireless Sensor Networks using Learning Methods

Mitun Bhattacharyya, Ashok Kumar, Magdy Bayoumi
The Center for Advanced Computer Studies, University of Louisiana at Lafayette

ABSTRACT: In this chapter we propose two techniques in two different sub areas of Wireless Sensor Network (WSN) to reduce energy using learning methods. In the first technique we introduce a watchdog/blackboard mechanism to reduce query transmissions, in which an approach is used to learn the query pattern from cluster head. Once the pattern is learnt, data are automatically sent back even without any queries from the cluster head. In the second technique we propose a learning agent method of profiling the residual energies of sensors within a cluster. Here as the agent moves across the different sensor nodes, it profiles the sensors' residual energies. This profile information is provided to sensors to help them make intelligent routing decisions.

Keywords: sensor networks, energy metric, database, learning methods

1. Introduction

In recent years there have been tremendous strides made in small mobile devices. As hardware on-chip area becomes smaller, so do the sizes of these devices. Additionally the features offered by these devices keep increasing. These advances have reintroduced the field of sensors: small, inexpensive sensors that are used in high density to monitor any area of interest. Sensors being small and disposable have restricted resources and energy that they can use. Therefore any action that the sensors take needs to be intelligent and energy-efficient. The following are the common features of wireless sensor networks:

1) *Limited Resources*: Since batteries of sensor networks cannot be replaced, the energy available to a sensor to sense and communicate is limited. Memory size and processing capabilities also are limited.

2) *Harsh Environment*: Sensors are sometimes placed in harsh environments for monitoring purposes. Adequate protection needs to be given to sensors depending on what application they are intended for.

3) *Dynamic Environment*: Sensors could be static, moving due to sudden movement like a strong gust of wind or moving constantly like sensor scattered in a flowing river. Depending on what rate of movement of sensor is expected, connection need to be set up ad-hoc or table based. Also due to limited energy supply, sensors that are overused can die out leaving holes in the sensor network. Sensor network algorithms aimed towards a highly dynamic environment need to manage this phenomenon.

4) *Deployment*: Sensors are densely deployed. This needs to be done, as communication range of sensors needs to be low to limit energy usage. Also having

more sensors helps in optimally operating only a sub group of sensors at a time and let the rest conserve energy. This effectively increases the lifetime of the sensor network. However this also leads to more interference, network congestion and collision of messages if not managed properly.

Wireless sensors can be used in various applications. These applications include high security uses like border monitoring, restricted area monitoring, and health applications and environment monitoring. Each application has its own specification of design. Since sensors are so small they are designed for specific purposes to optimize on their restrictive resources and serve the purpose they are designed for.

As discussed earlier energy is a restricted resource. Energy is optimized at various levels of the communication protocol from a network point of view. Energy is also optimized while designing the hardware for sensors. In network design, algorithms are developed for energy optimization in the area of routing of data, data aggregation, and MAC scheduling and application specification.

Sensors gather raw data while monitoring a process. If this raw data were sent back in its raw form, energy consumption would be very high. For example, consider a situation where the temperature of a given area if being monitored. If all the sensors sent back all the different temperatures that they sense, the network would be flooded with data that may not even make any sense. However, if in-between aggregation is done of the raw data at in-between sensors while routing it to the required destination, more sensible information could be sent back while still optimizing on energy.

Queries that are sent out by the base station need to be intelligently created to target specific sensors if possible. Flooding of queries throughout the network should be avoided as much as possible unless not known where a given data can be sensed.

The organization of the rest of the chapter is as follows. Section 2 deals with related background work. Sections 3 and 4 discuss our work in two different areas for energy reduction. Finally we conclude in the last section.

2. Background

The background covers two different areas of research. The first area is on system energy reduction and optimization in sensor networks. The second area covered is aggregation of data and application level methods of query sent out by base station.

There are a number of papers that deal with different efficient methods of memory aggregation. The work in [1] discusses a distributed geographical information embedded classical index data structure that meets the demands of multidimensional query. The system allows nodes to hash an event to some geographical location nearby. Events whose attributes are “close” are stored at the same or nearby nodes.

The work in [2] discusses a uniformed view of data handling, incorporating long-term storage, multi-resolution data access and spatio-temporal pattern mining. Compressed data as well as details of certain spatio-temporal phase of data is required. The system uses wavelet subband coding to view data at multiple spatio-temporal layers, provide ability to extract abrupt changes of data at all layers, provide easy distributed implementation, and low processing and storage overhead. Long-term storage is achieved at progressively aging the data. Old data is compressed more while still keeping the essence of the data collected to give space to new data. Essentially, in both these methods, data needs to be duplicated at other nodes to make the system fault resistant.

In [3] due to spatial proximity the nodes in the same cluster may sense the same set of data values. Exploiting the locality of values transmitted by a node within a cluster could reduce power consumption.

In [4], a study of clustered sensor network is done. In this work [4] homogeneous, i.e., using one type of sensor in the entire network was considered. The sensor sensed local phenomenon and the data quantified by the sensor were sent to the cluster head. The cluster head aggregated all the data received from all the sensors within that cluster, and transmitted it to the base station. The algorithm used was called LEACH. A distributed algorithm was presented for the cluster head selection. The cluster head expends the most energy, as it needs to make long-range transmissions to the base station. The cluster head function is rotated amongst sensors to achieve load balancing and uniform drainage of energy.

The work in [5] studied a multi-hop clustered wireless sensor network. The cluster head collected data from all the sensors within the cluster, aggregated the data and the sent the aggregated result to a base station located at the center of the region using multi-hop communication. Expressions were provided for the required cluster head densities to minimize the total energy expenditure in the network.

The work in [6] finds the optimal number of cluster heads. The work also does an analysis on single hop versus multi-hop routing to cluster head. Routing within a cluster normally expedites the energy consumption levels of sensors near the cluster head. This work suggests a hybrid scheme where single hop routing is implemented with multiple hop schemes to decrease the above problem.

In [7], two types of nodes are used. Type 0 nodes act like general sensors nodes limited capacity. Type 1 nodes act as cluster heads. Cluster head nodes have higher software and hardware capabilities. In addition they have higher energy capacities. An optimizing problem is formulated to minimize the overall cost of the network and determine an optimal number of cluster heads and battery energies of both types of nodes. Some other routing based papers are [8], [9], [10].

In our work the nodes learn from query patterns sent out from the base station to a created group of nodes. Next, after learning process is done, data is sent back from queried nodes to base station in the absence of queries from base station. We propose two techniques in two different sub areas of Wireless Sensor Network (WSN) to reduce energy using learning methods. In the first technique, we introduce a watchdog/blackboard mechanism to reduce query transmissions. In the second technique, we propose a learning agent method of profiling the residual energies of sensors within a cluster. In this work we practically combine deducing analysis results and presenting it to the sensors. The sensors then can make an intelligent decision with regard to routing.

3. Learning Methods in Database formation

The first proposed technique is to use learning methods in database formation.

System Assumptions: The sensors deployed in our system are considered to be heterogeneous in nature. Each sensor may have more than one type of sensing unit to sense and record more than 1 type of phenomenon. Different sensors may have different types of sensing units. Totally a whole range of different data can be collected by a group of sensors.

Our system setup is as shown in Figure1. As seen from Figure1, the sensors form a group locally to intelligently manage data sent back. For example, if a very low humidity is sensed with increase in temperature, then the probability of having a fire goes up. Detailed sensed data is sent back in this case to more urgently monitor the situation. However if both humidity and temperature are moderate, then incremental values could be sent back with less urgency. This requires intelligent decisions to be taken at the local level.

3.1 Data Monitoring Details

Initially the group of sensors goes through a learning phase. The learning phase involves monitoring the base stations response to various local sensed data sent back. The base station initially takes the decision as to what data needs to be queried in order to get an effective view of the system. The queried response given by base station act like learning steps to the group of sensors set up with a very simple decision making algorithm. As the learning phase progresses, the sensor start taking local decisions themselves that is monitored by the base station. During the final steps, the local set of nodes can make intelligent decisions by themselves as to what data to send back by which routing path.

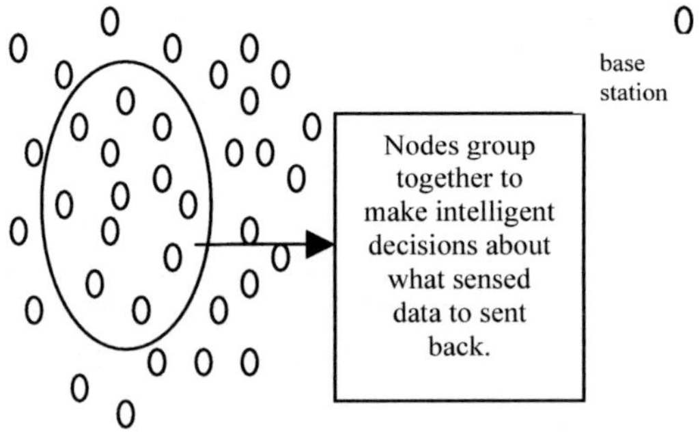

Figure 1: System setup

We have different levels of data, as described next.

1) Critical data (CD) - This data, when sensed, needs to be sent back to the base-station urgently with the other relevant information sensed. The message size for critical data needs to be enough to send all sensed data. For example, if a leak of a dangerous chemical is sensed, that data needs to be sent back along with probably the surrounding temperature and humidity that might make the chemical even more dangerous or its spread faster.

This relationship between a certain chemical being sensed and humidity is learnt by monitoring the queries sent out by the base station previously. Critical data of a region form the set of permanent data that needs to be sent back only when detected.

2) Monitory data (MD) – Data that is sent back at regular intervals for monitoring purposes. Data within this group either can be set up within a sensor before deployment or a sensor could be trained to sense this set of data by repeated queries.

3) Temporary data (TD) – Data that needs to be sensed for a certain period of time. Sometimes when certain phenomenon occurs, they need to be monitored for a certain period of time to get the most significant data. There is limited learning opportunity for this kind of data.

The group of nodes learns what data falls at what level depending on query history of the base station to the sensed data. Once the learning phase is over and data sensed has been graded, the learned results are sent all other local nodes so that they can make intelligent decisions. This approach is much better than fixing up the levels of data sensed in a pre-defined way, as sensor after being deployed, might need to change data levels depending on what type of data it senses. Also not having the base station sent out querying message can help in reducing the number of messages being communicated and helps in conserving energy.

We adopt a black board and watch dog mechanism (Figure 2) for the learning process. The black board is a distributed database kept within the group of nodes. The mechanism of dealing with limited memory size of sensor nodes is explained in sections 3.3 and 3.4. For now let us assume that a black board is present and an appropriate "watch dog" has been set up. The "watch dog" is a group of smart sensors used to monitor the queries and appropriately update the black board.

3.2 Dealings with Memory Constraint Limits

In our system sensors are densely deployed. The sensors are also heterogeneous and collectively sense different kinds of data. We assume that a sensor of each kind is uniformly distributed so that a group formed locally almost can consist of the entire set of sensors. Since the sensors are closely placed, the data set sensed within a region by a group of sensors would be limited.

Data	To be included within a Range	Criticality (critical, monitory, ignored)	Total estimated path delay acceptable
D1	Yes	Monitory	40 time slots
D2	No	Critical	15 time slots
• • • •			

Figure 2: A "watch dog" is a set of messages used to monitor and update the black board as shown above.

Even with a limited set of data, a sensor node will not be able to accommodate storing all the data set due to constraints in memory resources of a sensor. Besides if the node fails or moves away, the entire data set setup is affected. To eliminate the problem we propose storing the data set within a group of sensors. Since the data set within each sensor is small in size a simple scheme of index search would be sufficient to locate a certain data and update it.

3.3 Formation of the Database

Let us assume that there is one data set of each type (CD, MD, TD) that a region is queried with. Let the training data set for the CD category include data (D1, D2, D3, D4), MD (D5, D6, D7, D8) and TD (D9 – D11). A sensor could sense a single data or a group of them. Let a matrix represent the data sensed by sensors within a group. Let the sensors be deployed randomly but close to each other. Let us also assume that the following sets of queries are sent by the base station to the local region before the formation of any group or database (Snap shot of all queries unto time T1).

(CD, D2); (CD, D3); (CD, D4); (CD, D1); (MD, D7, 10).

A seen from the matrix, the sensors S1, S2, S3 and S4 cover the set of data queried by the base station. The problem solving procedure follows on the similar lines of solving the Transportation Problem.

	D1	D2	D3	D4	D5	D6	D7	D8	D9	D10	D11
S1	1		1						1		1
S2				1	1						
S3		1					1			1	1
S4		1							1		
S5					1						1
S6						1		1		1	

Table 1: Shows an example of sensed data, sensor combination

Sensor S1, by monitoring data transmission, discovers that it can form a partial database. It sends out *invitation messages* to S2, S3, S4 sensors. S2, S3, S4 send back the data that they can each sense. S1 and S3 realize that another set of data could be initiated with D7.

Let the next snapshot of queries sent out by the base station unto time T2 be the following:

(MD, D8, 16); (TD, D10, 50); (CD, D4); (CD, D3); (CD, D1); (CD, D2).

Data D8 is sensed by S6. By monitoring the data S1 comes to realize that S6 and S3 can form another database. *Initiating action* is initiated by either S3 or S6. From exchange of *invitation* and *reply* messages, S3, S6 form another database and get to know what sort of data each other sense. D10 is covered by S3. S3 realizes that there is another opportunity for forming a database. The CD series reiterate the fact that D1-D4 need to be grouped together. By monitoring data transmission S6 realizes that D4 is supplied by S2. *Invitation message* is sent out to S2 by S6. After getting the reply back, S6 informs S3 of the finding of the D4 data. The CD group is completed.

Let the snap shot of queries unto time T3 be:

(MD, D7, 10); (MD, D8, 16); (TD, D9, 20); (TD, D11, 40); (MD, D6, 10).

Data D7, D8, D6 is covered S3, S6. The sensors sensing the individual data, note the time interval after which the new data needs to be sent back. D9 can be covered by either S4 or S1. D11 can be covered by S1, S3, S5. S3 starts forming a database with the data sensed before D10 and data found now (D11, D9). *Initiating action* is taken to enter S5 within the group. Three distinct databases have stated forming within this group. It should be noted at this point that even though a sensor within the existing group covers a certain data, a new sensor might be added to the group for redundancy purposes (fault tolerance). For example, the data covered by S5 is all covered by other sensors but even then it is included within the group.

Let the snap shot of queries unto time T4 be:

(MD, D7, 10); (TD, D11, 40); (MD, D5, 5); (MD, D6, 10); (TD, D10, 50); (TD, D9, 30);

The group formed covers all the data. The database is updated within each sensor with the time information sent out by the queries from the base station. Also prioritized connections are set up between sensors that form a part of a database within the group. The final resulting matrix formed is shown in Table 2.

	D1	D2	D3	D4	D5	D6	D7	D8	D9	D10	D11
S1	1		1						1		1
S2				1	1						
S3		1					1			1	1
S4		1							1		
S5					1						1
S6						1		1		1	
	Database 1				Database 2				Database 3		
Time					5	10	10	16	30	50	40

Table 2: Final Resulting Matrix

3.4 Problem Formulation

Optimized number of sensors within group:

Let us consider the following matrix as an example.

	D1	D2	D3	D4	D5	D6
S1	1				1	1
S2			1			
S3	1			1		
S4		1		1		1
S5	1		1	1		1

Table 3: Initial reduction to achieve 2 copies of elements of a matrix

	D1	D2	D3	D4	D5	D6
S1	1				1	1
S2			1			
S3	1			1		
S4		1		1		1
S5	1		1	1		1

Table 4: Matrix state after step 3 is executed (1copy)

To find optimized number of sensors to have one copy of each data within the group

1) Look at the matrix column wise.
2) If there is any data covered by any one single sensor, remove that sensor from the matrix. It forms a primary sensor that needs to be kept within the group.
3) Remove all data covered by primary sensor(s).
4) Next a row reduction is done. That is a sensor that covers most of the remaining data is selected and removed along with the data that the sensor senses.
5) If any data remains again a column reduction is done.
6) Steps 4, 5 are repeated until all data are covered.

	D3
S2	1
S3	
S5	1

Table 5: Final matrix state of Table 3 after reduction

To find optimized number of sensors to have two copies (if possible) of each data within the group

1) Look at the matrix column wise.
2) If there is any data covered by any one single sensor, remove that sensor from the matrix. It forms a primary sensor that needs to be kept within the group.
3) Keep track of data covered by eliminated sensors of step 2 – 1 copy is over (D1, D4, D6)
4) Reduce matrix – remove columns of single copy of data.
5) Check row wise, which sensor/s cover the most set of data (S5)
6) Remove sensor if along with previous selected (primary sensor) two copies of any data are covered. Foe example: S5
7) Keep track of data covered. (2-D1, 2-D4, 2-D6, 1-D3).
8) Reduce the matrix.
9) Check column wise for data still uncovered by sensor/s.
10) Repeat step 4 – 9 to get minimum set of sensors to cover a data set twice.

Distributed algorithm for getting 2 copies of each data:

1) The group is formed as given in Section 1.2. As explained earlier two or more copies of the same data are attempted to be included within the group. After the group is formed the following steps are taken.
2) By small message exchange the sensor that covers the largest number of data within the group is determined.
3) This sensor forms the first primary sensor (PS). The sensor then broadcasts the Id of all the data that it senses, along with the sensor Id and 'awake' time period. Each sensor has a counter associated with it. Any sensor within the group that can duplicate any of the data, sends back that data Id along with its own Id and sleep time interval.
4) The primary sensor then keeps a record of duplicates and also decides which sensor(s) to correspond with. The factors that are taken into account are, the

distance and sleep schedule of the sensor with relation to the PS's. A message is sent back to the appropriate sensor.

5) While this information interchange takes place, the other sensors passively listen and try to get duplicates for their own data (if not covered by first primary sensor).
6) Step 1 is repeated, to find the sensor in the remaining group that has the largest number of unduplicated data.
7) That sensor becomes the primary sensor and Steps 3, 4, 5 are repeated.
8) Steps 6,7 are repeated to find data that are sensed by only 1 sensor within the group. These sensors also form the PS.
9) Sleep schedules of PS are given the first preference. The other adjusts their sleep schedules depending on which PS they are corresponding with.

3.5 Optimized training set for Data Details

The training required to setup database needs to be done with the least amount of extra communications. Most the learning needs to be done with the queries and replies exchanged between base station and the sensors during normal operations of the sensor network.

1) Critical Data: Training of the critical data set is the easiest. The group of data that needs to be sensed together forms the training set. Critical data for a region do not change much. If anything out of the ordinary or extreme (other than the critical data) is sensed: the data is sent back to the base station for further examination.

Block based leaning is used to set the Critical Dataset as shown in Figure 3.

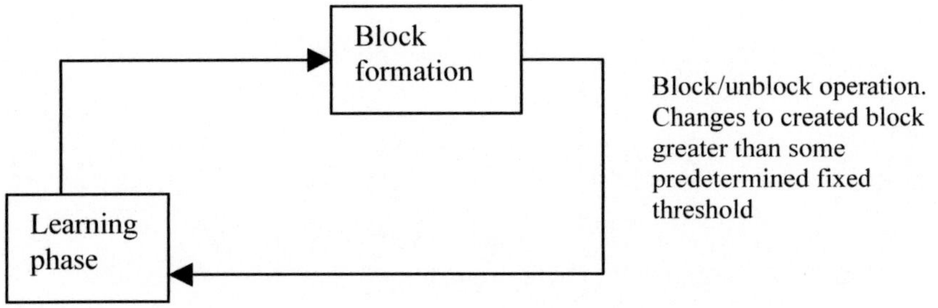

Figure 3: Block based learning

In the case as queries are sent out by the base station, the critical queries are grouped together into essential groups. For example, let us assume that a sensor senses a very high temperature in environment. During normal operation of querying, the base station receives this high temperature reading and asks for the humidity factor for the region. The query is prefixed with association of the code CD. If this is repeated for several instances of querying, the sensors realize that high temperature and high factor of humidity need to be linked together and both are of the Critical Data (CD) group. This forms a block. Different sets of blocks of this sort could be formed to form the critical data set. Once blocks are created several operations can be performed on these blocks of data. For example let us consider the different blocks as shown in Figure 4. Operations that can be performed are: Buildup (Data is added to a block), Remove

(Data is removed from a block), Destroy (If the number of data within a block goes below a predefined threshold number the block is destroyed), Marry (Blocks are sent out together depending on common elements queried).

2) *Monitoring Data*: The data as explained earlier can be representative of a range of values. This is especially true for monitoring data. The time interval of the training set of data is examined for learning. Two methods are employed together as described next.
a) *Learning by Association* – For example let us say that when the temperature is high, the time interval for both humidity and temperature monitoring goes up. In terms of range let us say that when the temperature increases, the range of temperature to be monitored goes up. Also the range of humidity to be monitored goes up. That means if the group of sensors can learn from this, it means that if there is a query for a large range of humidity or a query for a large time interval for humidity, the temperature for a large interval or range could be sensed and sent back even before any query comes.

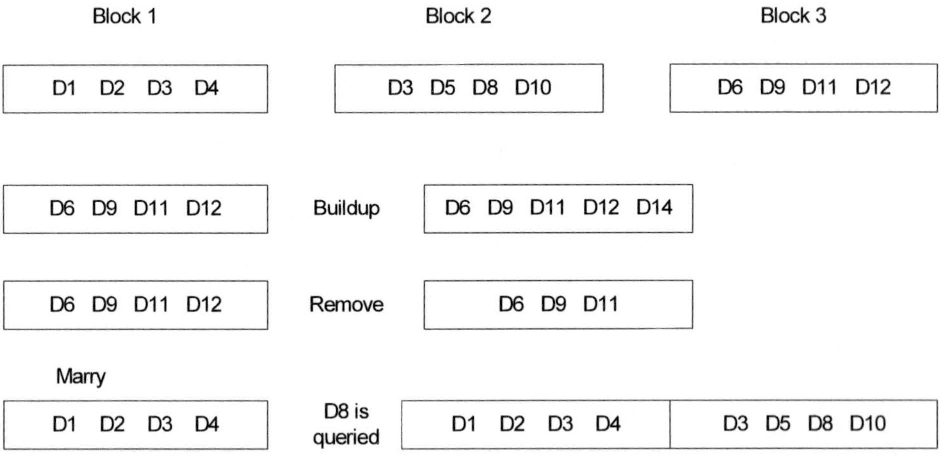

Figure 4: Examples for the buildup, remove and marry operations

b) *Intelligent Base Station* – Initially, the base station sends out queries when required. Analyzing these queries, sensor form databases as explained in 3.3, 3.4. Once the database group is formed, the sensors that form a group send their Ids to the base station. The base station then can set up training patterns knowing which data is sensed in the group and which data will be associated with each other in the sensor group.

In sensor networks normally a range of data is queried, for example, the average or histogram of temperatures within a certain time interval or vice versa. If automatic learning is required of these ranges then incremental learning with weights is employed. Let us assume a certain range 'X' time interval is observed for queries for the temperature of a certain region. Two sets of this observation are kept. The first set is kept as it is and the second set time interval of observation is incremented by certain amount and sent back to base station. The intelligent base station checks the range of queries it has for that particular region and assigns an incremental or decremented weight to the changed time interval. Based on the weighted feedback, the time interval of future queries is learned. The amount of time interval change of the second set depends on the communication costs that can be sustained for learning.

The training set sent out by the base station should be well defined. Initially queries are sent out as required. As database groups are formed depending on matches between locally sensed data and queries, the database sensors Ids are sent back to the base station. The base station identifies the sensors according to their Ids. The base station that has required energy, then sets up training sets for each sensor database.

Let T_{db} be the time required to form a database of sensors. Let the corresponding energy required to form the database be E_{db}. Once the database is formed let E_{sent} be the energy required to inform the base station of the Ids of the database and other information like location. Let the time taken be T_{sent}. The base station then creates the training set for the sensor database and sends it back to the sensors. The energy and time required for this is $E_{training}$, $T_{training}$. Therefore the total energy and time required to initially set up the database is

$$T_{total} = T_{db} + T_{sent} + T_{training} \tag{1}$$

$$E_{total} = E_{db} + E_{sent} + T_{training} \tag{2}$$

Let E_{query} and T_{query} be the energy and time required to send out a query and get a response back from a database. For the trained database to be efficient the number of query messages that need to be eliminated is 'k' such that

$$E_{query} * k \geq E_{total} \tag{3}$$

$$T_{query} * k \geq T_{total} \tag{4}$$

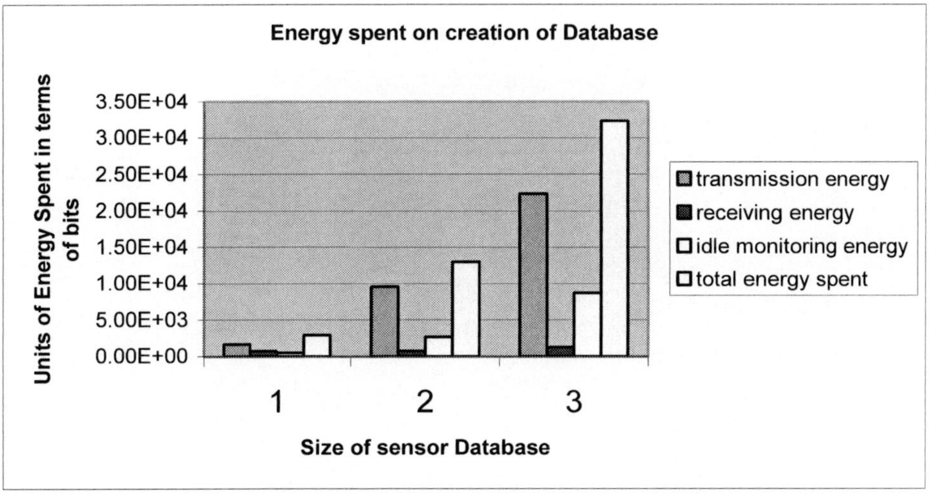

Figure 5: Shows the energy spent in creating different sizes of databases
1 = 6 sensors database, 2 = 10 sensors database, 3 = 15sensors database

Simulations were carried out in MATLAB to estimate the amount of energy spent on the creation of the database. The energy spent for transmission and receiving of data is given by equation 5,6. Energy spent depends on number of bits transmitted and received. The graph, Figure 5, shows the number of bit transmitted and received for the creation of the database. Idle monitoring is like receiving of data bits with minimal amount of processing within sensor. It is seen that as the data base size increases the number of elements that can be covered in a local region increase but the amount of energy required also increases significantly. From simulations it is seen that having smaller database are easier to manage with reasonable amount of energy expenditure. Having too small a database may not cover all elements of a localized region. Therefore a tradeoff exists between the size of the database and energy expenditure limits. Also, having low changing datasets helps in regaining the energy lost while creating the database. That is the reason this algorithm is most suitable for slowly changing datasets.

3.6 Routing Details

As seen from the above data set we have groups of data with different priority levels. Critical data has the highest priority level and the time required for travel to base station should be the shortest. Monitoring and Temporary data are assigned the same levels of priority unless specified otherwise by the base station. Those data requiring special attention in the Temporary or Monitoring group of data are tagged. Care should be taken that the number of priority levels should be small in number due to memory restrictions. Having different priorities of data set, routing method/algorithm needs to take care of different routing requirements.

3.6.1 Sensor Observations

Each node keeps a record of the time required for a message to reach the base station when sent through a particular neighbor. Each node then decides independently at each stage decide which neighbor to pass a particular message through depending on priority level. Each node tries to balance out the load by only passing critical data through neighboring nodes with the shortest time and sending all other through other neighbors. There is a learning process involved here.

3.6.2 System Observation

At regular intervals one of the sensed data from the Monitoring or temporary group of data is sent to the base station, it is tagged with a special bit, starting time and sent through one of its neighbors. Each node that receives this data message with the special bit sends it through a neighbor and notes the time stamp. Finally when the base station receives this tagged message it sends a small time (received time) message back as soon as it receives the message tagged with the special bit. A number of such messages are sent through different neighbors.

4. Intelligent Profiling Techniques

The second proposed technique is intelligent profiling used to get energy scans of sensors. Details are discussed below. Sensor nodes within a cluster need to make intelligent decision while routing data from one point to another. One of the most important parameters that need to be considered is the residual energy. Residual energy refers to the battery energy left within a sensor. It is seen that in order for batteries to have an optimal lifetime, they need to be operated in a pulsed format. That means that the sensor need to work and then rest for a period of time for the battery to be used optimally. Sensors need to make an intelligent decision regarding routing a data and conserving its battery energy levels. To do this the sensors need to be presented with a profile view of the residual energies. Our work addresses the issue of learning to provide the profile view to the sensor node.

In our setup, the cluster heads have superior capabilities then that of the other sensors in the cluster. We propose a control group (explained later) type of control. As in any biological or chemical experiments, control groups are set up to notice the change in the group under experiment. To optimize on energy a local group of nodes are set up as the experimental group. Next a control group is created either from randomly selecting energy levels from various nodes or the set of energy vectors from surrounding nodes by an "energy packet" as shown in Figure 6. The energy aggregate information (can be the average) is stored in the Energy Agent message. In this work Energy Agent and Energy message can be used interchangeable.

To increase node lifetime, energy needs to be optimized both globally and locally. Each node can have different initial energy levels. The general energy model for radio communications is as given below

$$E_{Tx}(k,d) = E_{elec} * k + \varepsilon_{amp} * k * d^2 \quad (5)$$

$$E_{Rx} = P_{rec} * T_{on} \quad (6)$$

$E_{Tx}(k,d)$ = Energy required to transmit k bits of data through distance 'd';
E_{elec} = 50nJ/bit;
ε_{amp} =100pJ/bit/m^2;
E_{Rx} = energy consumed while receiving data during T_{on} seconds.

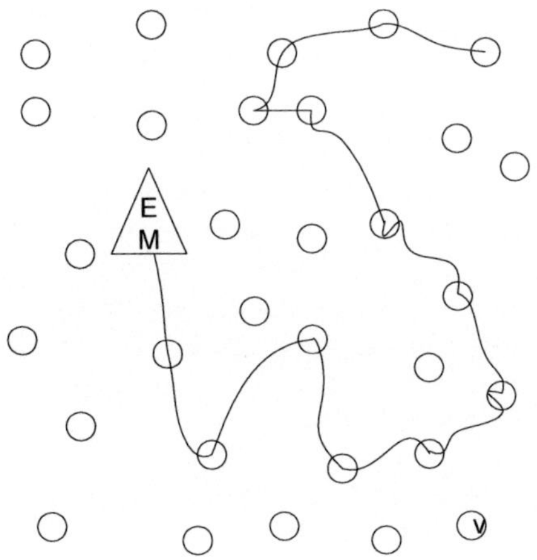

Figure 6: EM: Energy Agent moves randomly across the network, selecting random node to get global energy aggregate.

```
Pseudocode of the Algorithm for the Energy Message
Let S = {s1, s2, s3, .......sk}
Let Msg_sent (aggregate, initial node, destination node) be
the energy message that is sent from one node to another.

At the Cluster Head (CH)
Msg_sent ( aggregate, CH, sx)   sx – random node 1 ≤ x ≤ k

Among sensor nodes
For sy = sx to upper_time_threshold
     Msg_sent ( aggregate, sy, sy+1)
End

At each sensor si
   Sensor receives {Msg_sent (aggregate, initiating node, si)
    /* Aggregate comparison */
   Compare {(si[residual energy]), (Msg_sent[aggregate])}
    /*Update aggregate to get new aggregate*/
   Msg_sent (new_aggregate, si, sj)
   Where sj = another random node not visited by Msg_sent
   previously.
```

Figure 7: Pseudocode for Energy message

The Energy Agent is initiated at the cluster head. The message (Msg_sent) is sent from node to node and at every node a comparison and update of aggregate is done. The pseudocode is given in Figure 7. The next section explains how the control and experimental group is created. It also specifies the energy message format, the different time stamp thresholds and an illustrative example of the aggregation process.

4.1 Group Formation Details

As seen from the psuedocode in Figure 7, the Energy Agent is created by the cluster head and then passed from node to node. It is seen from Equations 5,6 that transmitting and receiving any kind of messages requires energy expenditure by the sensor nodes. Therefore the number of hops to get the aggregate in the experimental group and control group needs to be limited. We define the lower time threshold (hop count) as the threshold above which the control group is initiated. The upper time threshold (hop count) is the threshold above which the Energy message is discarded. The pseudocode for the creation of the experimental and control group is given in Figure 8. Here time refers to the hop count. At every hop of the Energy message from node to node the time is incremented by one.

This aggregate is calculated and tagged to the energy message along with an incremented time stamp. It should be noted at this point that any sensor that is at the point of dying out, would not contribute to the aggregate. This is due to the fact that routing algorithms avoid sensors with low residual energy. The format of the Energy message is given in Equation 7.

Energy message = {aggregated value (for example average), Indicative bits, time stamp, initiating time, lower time threshold (T_{lower}), upper time threshold (T_{upper})} 7

```
Pseudocode for Experimental group
Among sensor nodes (Experimental Group)
    For s_y = s_x to lower_time threshold
        /* Formation of the experimental group*/
        Sensor receives {Msg_sent (aggregate, initiating
         node, s_y)
        /* Aggregate comparison */
         Compare{(s_y[residual energy]),
                                (Msg_sent[aggregate])}
        /*Update aggregate to get new aggregate*/
         Msg_sent (new_aggregate, s_y, s_y+1)
         Where s_y+1 = another random node not visited
         by Msg_sent previously.
    End
Pseudocode for Control group
Among sensor nodes (Control Group)
    For s_z = lower_time threshold to upper_time threshold
        /* Formation of the experimental group*/
        Sensor receives {Msg_sent (aggregate, initiating
         node, s_z)
        /* Aggregate comparison */
         Compare{(s_z[residual energy]),
                                (Msg_sent[aggregate])}
        If Msg_sent [time stamp] > Msg_sent
                                 [upper_time threshold]
             Discard Msg_sent
         Else{
            /*Update aggregate to get new aggregate*/
            Broadcast  Msg_sent (new_aggregate, s_z, s_t)
            Where s_t = all nodes within receiving range
            of s_z
            }
    End
```

Figure 8: Gives the Pseudocode for formation of the Experimental and Control group.

Next, we address the validation period of the energy message. The aggregate is valid between two pre-determined threshold time stamp limits. The lower level of the time stamp threshold is required, as we do not want to get wrong estimation of energy levels by having an aggregation of too little data. The upper limit of the threshold time stamp is required since we want to limit the energy expended on an Energy message. Any node that receives an Energy message with a time stamp greater then the upper threshold drops the Energy message.

The cluster head initiates certain number of Energy messages at predefined internals of time as shown in Figure 9. The Experimental group sensors have an updated aggregate approximation of the sensors within the system of sensors. These (Experimental group) then can estimate as to how each sensor fares, energy wise, with respect to other sensors and can make a conscious decision for routing or other decisions by the protocol stack layers of the sensor. Intuitively, better decisions are made as aggregate shows status of energy beyond the scope of localized regions.

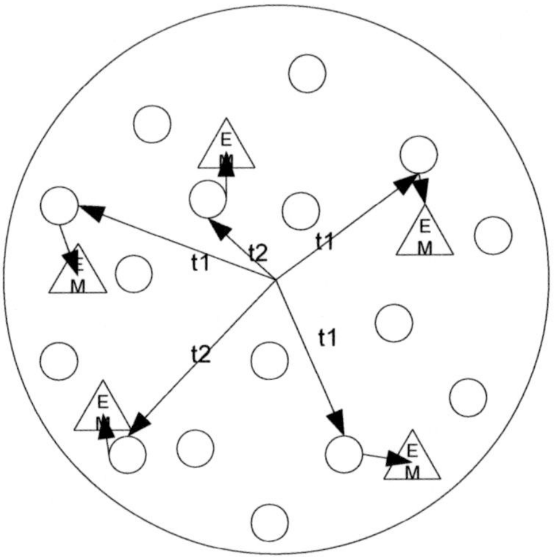

Figure 9: Initiation of Energy Agents

There could be a situation where a node receives more than 1 *energy* message. The information that the sensor gets from the different *energy* messages would be a very informative one. The aggregation of data takes place between *energy* messages that have the same initiating time. There are two cases when a sensor receives two energy messages as follows:

1st case: Both have the same initiating time → energy level data is aggregated between the two messages along with the sensors' energy information.

2nd case: Both have two different initiating times → energy data is aggregated individually for each message with the sensors' energy data.

4.2 Aggregate Computations

We consider a network of N x N nodes that wish to compute a scalar function of their energy states. We consider 2 types of aggregates, range based (RB) and location based energy approximation (LBEA). The results are shown in Figures 10 and 11. The results show that at a much lesser hop count (maximum being Energy Agent hopping through all sensors), a pattern matching aggregate is obtained that matches the final aggregate pattern.

4.2.1 Range Based Aggregate

The sensor energy levels are broken down to ranges. For example if the battery value varies from 0-5 Joules then the 0-1J, 1-2J, 2-3J, 3-4J, 4-5J would form the different ranges. These ranges are fixed. As the Energy message travels through the cluster sensors, the energy levels of each sensor are compared to the fixed ranges. If a sensor's energy level falls within an energy range, then the counter for that range is incremented. The indicative bits display the counters. Each position of the indicative bits indicates which range it represents. For example the first 4 bits could be the counter for the number of sensors having energy levels between the 0-1J ranges. The next 4 bits could indicate the number of sensors in the 1-2J range and so on.

4.2.2 Location based energy approximation (LBEA)

We first break up the cluster into sectors. In this work we assume that each node knows by some means which sector it belongs to within a cluster. We logically break up the cluster into 8 sectors. Each sector is identified by its unique position in the energy message. The value the bits of these unique positions indicate some information about that sector – indicative bits.

As explained earlier each node the energy message visits calculates an average. As each energy message moves from one sector to another, the energy levels of the sector are indicated by the unique position bits. The unique position bits are explained as given below.

00 01 10 11 01 01 01 01

Each two bit indicates the positions of each individual sector. The first 2 bits is for the 1st sector, the next two bits are for the second sector and so on. The values are indicative as given below.

00 – Not enough information.
01 – Average calculated is below the median value for the sector.
10 – Average calculated is above the median value for the sector.
11 – Average calculated is same as the median value for the sector.

We assume that each sensor has a specific battery capacity due to same type of batteries being used in each sensor. If the maximum capacity of a battery is X Joules and the minimum 0, then the median value of the energy level would be (X-0)/2. If the average calculated within each sector sensors are compared with this median value we can find out how a sector fares with relation to the given medium. These values are set after the energy message reaches the Lower Threshold. These values would help in making routing decisions. Nodes that have energy levels below median would be avoided as much as possible to recover their energy levels. Another scenario would be that the median is calculated with X Joules as the maximum and B (battery threshold as the minimum). The median would be (X-B)/2.

4.3 Energy profiling

Let Energy-ratio be the ratio of the residual energy of the concerned sensor to the profile of energy aggregates gathered by the energy message from the other sensors. The profile of Energy-ratio obtained gives an indication of which category of sensors does the sensor fall under. If the Energy-ratio of the high-end aggregate is small, that means that there are a lot of energy rich sensors compared to the sensor under observation. However if the ratio almost equals 1 or higher, that means the sensor under observation is an energy rich sensor compared to the others.

Let sy[residual energy] be the residual energy for the sensor that gets the energy message and is the sensor under observation. Let the different profiles obtained from the energy message be H_1, H_2, H_3 … where H_1 be the profile for the highest energy range group and the H_n the lowest. This is for the 1st aggregate (Range Based Aggregate). Let Energy-ratio$_{H1}$ be the ratio as described earlier.

Energy-ratio$_{H1}$ = sy[residual energy]/H_1

If Energy-ratio$_{H1}$ < 1

sy - energy low sensor compared to sensors in this category.

If Energy-ratio$_{H1}$ >= 1

sy – energy rich sensor compared to sensors in this category.

Similarly Energy-ratio$_{H2}$, Energy-ratio$_{H2}$ … are calculated and weighted decisions are made by the sensor under observation.

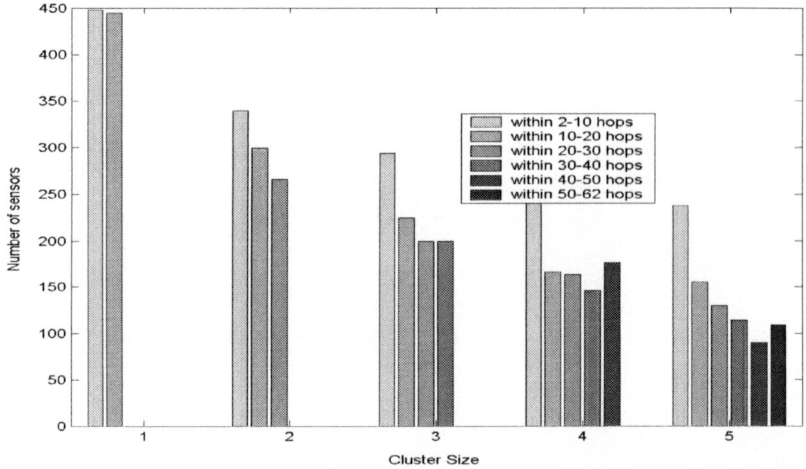

Figure 10: First time pattern-matching distribution for Range Based Aggregate

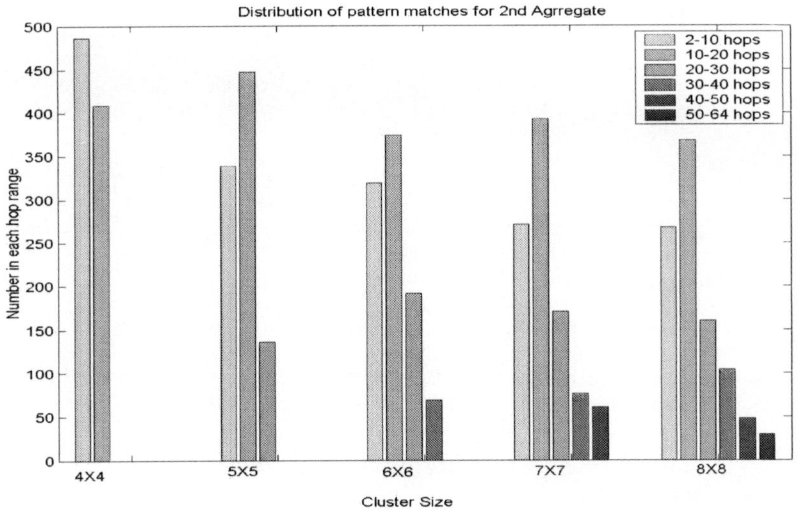

Figure 11: First time pattern-matching distribution for Location based Energy Aggregate

5. Future Trends

Sensor Network is fast growing field. The challenge of applying any artificial intelligence algorithms to sensor network is due to the constraint in resources of sensors. AI normally required a lot of computational and memory resources, which are limited in the case of wireless sensors. Developing a distributed database type of system that can efficiently communicate and share resources would help in elevating the problem. However more algorithms are required as the training set needs to be very small. Also the data that each sensor can hold and temporarily analyze to get a pattern for future prediction is small. This makes the problem even more challenging.

6. Conclusion

In this chapter we have discussed two aspects of sensor networks and how learning methods could reduce the energy expenditure of the sensors. In the first technique energy was saved for a slow data rate system by leaning and automatically sending back data without having any queries coming from the base station. In the second technique we described a system where intelligent profiling was done to the energy values of the sensors. This information was then used by each sensor to make intelligent comparative decisions about its energy expenditure. It is seen from experimentation that more energy rich sensors where chosen for routing when aggregate information is present than without.

References

[1] X Li, Y Kim, R Govindan, W Hong; "Multi-dimensional Range Queries in Sensor Networks", *Proceedings of the 1st international conference on Embedded networked sensor systems,2003* , pp: 63-75

[2] D Ganesan, B Greenstein, D Perelyubshiy, D Estrin, J Heidermann; "An Evaluation of Multi-resolution Storage for Sensor Networks" *Proceedings of the 1st international conference on Embedded networked sensor systems, 2003*, pp: 89-102.

[3] Saputra H, Vijaykrishnan N, Kandemir M, Brooks R, Irwin J," Exploiting value locality for secure-energy aware communication", *Signal Processing Systems, 2003. SIPS 2003,* 27-29 Aug. 2003, pp: 16-121.

[4] W.Heinzelman, A. Chandrakasan, H. Balakrishnan, "An application-specific protocol architecture for wireless micro-sensor networks", *IEEE Transactions on Wireless Communication* (04) (2002) pp: 660-670

[5] S. Bandyopadhyay, E. Coyle, "An energy efficient hierarchical clustering algorithm for wireless sensor networks", *IEEE Infocom*, San Francisco, CA, 2003.

[6] V.Mhatre, C.Rosenberg. "Design guidelines for wireless sensor networks: communication, clustering and aggregation". *Elsevier, Ad hoc networks 2* (2004) pp: 45-63.

[7] V. Mhatre, C. Rosenberg, D. Kofman, R. Mazumdar, N. Shroff. "A minimum cost surveillance sensor network with a lifetime constraint" *IEEE Transaction on Mobile Computing,* January 2004.

[8] C. Intanagonwiwat, R. Govindan, and D. Estrin, "Directed diffusion: a scalable and robust communication paradigm for sensor networks," *Proceedings of ACM MobiCom* '00, Boston, MA, 2000, pp: 56-67.

[9] W. Heinzelman, J. Kulik, and H. Balakrishnan, "Adaptive Protocols for Information Dissemination in Wireless Sensor Networks," *Proc. 5th ACM/IEEE Mobicom Conference* (MobiCom '99), Seattle, WA, August, 1999. pp: 174-85.

[10] J. Kulik, W. R. Heinzelman, and H. Balakrishnan, "Negotiation-based protocols for disseminating information in wireless sensor networks," *Wireless Networks*, Volume: 8, pp: 169-185, 2002.

Integrated Intelligent Systems for Engineering Design
X.F. Zha and R.J. Howlett (Eds.)
IOS Press, 2006

Integrated Knowledge-based System for Product Design in Furniture Estimate

Juan C. VIDAL[1], Manuel LAMA, and Alberto BUGARÍN
Department of Electronics and Computer Science
University of Santiago de Compostela, Spain

Abstract. This chapter describes a knowledge-based system approach that combines problem-solving methods, workflow and machine learning technologies for dealing with the furniture estimate task. The system integrates product design in a workflow-oriented solution, and is built over a workflow management system that delegates activities execution to a problem-solving layer. An accurate estimation of the manufacturing cost of a custom furniture client order allows competitive prices, better profits adjustment, and increments the client portfolio too. Nevertheless, task scope is even broader. On one hand, it fixes future material and storage capacity requirements. On the other hand, it defines the manufacturing plan and logistic requirements to fulfil the client order in time. However, these objectives cannot be achieved without an adequate product design, which relates client order requirements with a manufacturing and assembly-oriented design.

Keywords. Workflow modelling, Problem-Solving Methods, Product design.

Introduction

Custom furniture industry is facing the challenge of globalization and unprecedented levels of competitiveness. Market demands force organizations to find solutions in order to increment its productivity and to reduce its costs. A way to achieve these objectives is the improvement of the processes related to the furniture estimate task, that predetermines the cost of manufacturing a client order (usually composed of hundreds, and even thousands, of units) and evaluates how the product development will modify the manufacturing plan in case the order is accepted. A key stage for being competitive is the product design stage. Through design improvement time to manufacturing a product is also reduced, and hence its cost. For this purpose, the design phase must take the factors associated with the life cycle of the product into account: manufacturing, assembly, testing, maintenance, reliability, cost and quality [1].

Accurate estimation of a large scale manufacturing of a custom furniture is a difficult task. On one hand, furniture composition plays an important role in price estimation. Custom furniture consists of different kinds of materials such as metal, wood and wood-based products, plastic, melamine foils, laminate, PVC, and so on. This variety

[1] Corresponding Author: Department of Electronics and Computer Science, University of Santiago de Compostela, 15782 Santiago de Compostela, A Coruña, Spain; Email: jvidal@dec.usc.es.

of materials makes manufacturing evaluation hard and defines specific manufacturing and assembly rules related to the kind of material the furniture is made of.

On the other hand, the lack of previous manufacturing experiences for most of client orders provokes that new products are developed. The manufacturing process of furniture is complicated and one aspect of the design may affect its fabrication and assembly costs. In this sense, the importance of a *product design* strategy in the furniture industry is even greater, and methodologies like Design for Assembly (DFA) [2,3] are one of the trends used to reduce the cost and improve the quality of products. DFA accomplishes this issue by providing design advice on how the product can be more efficiently and economically manufactured, and considers furniture design simulations, manufacturing, assembly and cost related knowledge. For example, *joinery* is the method by which different pieces of wood are attached, and it is often an indication of the quality of a piece of furniture. One of the cost saving practices introduced by means of DFA guidelines is the reduction and standardization of the number of types of joints that can be used in manufacturing (Figure 1) and also the introduction of rules in order to use the most suitable type of joint per furniture considering the strength required, cost and final appearance of the product.

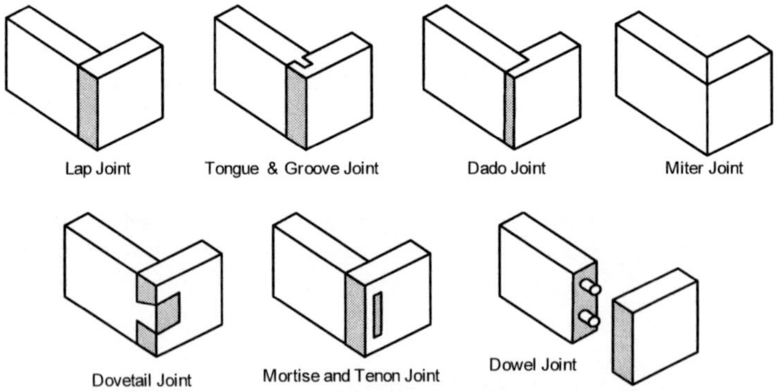

Figure 1. Types of wood joints.

A new product development is more than a creative activity and designers need time to hatch their ideas and to take into account both how to more closely watch the customer request and also the cost effects of their decisions. However, custom furniture industry has time limit to develop their products and so product designs are done in haste. For this reason, organizations need to formalize and structure its reviewing procedures. Moreover, organizations need to find a way to monitor how these procedures are fulfilled and to facilitate collaborative analysis of a product design. In this sense, the monitoring process must cover all the steps of the product design life cycle, guiding the expert knowledge through the whole process and coordinating both tasks and people involved. For example, the assessment of the furniture Computer Aided Design-based (CAD) designs cannot be driven by the same kind of knowledge than the furniture manufacturing cost evaluation. The first evaluation must compare the CAD product design with its conceptual design constraints, while the second evaluates the product design based on cost and quality criteria. In order to deal with both the collaborative analysis and the description process related to the product design, *business process management* (BPM) can be used.

BPM is an emerging technology that supports the whole business process development and management [4,5]. In this context, a business process consists of an ordered sequence of activities and task executors that define the work and knowledge flows. BPM allows the explicit representation of the business process logic in a process-oriented view, and is increasingly used as a solution to integrate engineering/manufacturing processes. However, BPM modelling techniques are not well suited for dealing with expert knowledge.

Organizations need to focus on process effectiveness and this feature must be achieved through *knowledge management* [6]. The increasing importance of an adequate knowledge management within process automation is a key factor for organizations future competitiveness. Design, manufacturing and assembly knowledge need to be fixed, reused and even enhanced through some learning process. By collecting, sharing and enhancing knowledge, organizations will better implement their business processes thus improving its knowledge flow. As regards, knowledge management does not only deal with static knowledge, such as ontologies [7,8] but also handles the dynamic knowledge of organizations [9,10]. For example, methods used in the solution of some task are handled as a piece of knowledge. This feature lets dynamic knowledge to be reused independently from the domain knowledge and also enables to change the behaviour of the system on run time.

In this chapter, we propose a knowledge-based BPM system that deals with the product design in the context of the furniture estimate task. The business process related to the furniture estimate has been modelled as a workflow [11] composed of a set of activities that use domain knowledge. As regards, our workflow is defined around the product design life cycle. This structure is mostly composed by a number of knowledge-based tasks that evaluate the furniture designs against the main factors involved in its manufacture and assembly. In order to incorporate the task knowledge, we have developed a framework that integrates workflow-modelling techniques with knowledge management technology. For this purpose, we decided to use the Unified Problem-solving method Modelling Language (UPML) framework [12, 13] in order to define and reuse both static and dynamic knowledge. The result of this integration is that the workflow structure represents the operational description of composite problem-solving methods (PSM) [10], while the leaf activities of these workflows represent tasks solved by non-composite PSM. Automation of the furniture estimate task through a BPM system that uses knowledge management techniques allows us to deal with a great amount of data and knowledge, and to coordinate the activities carried out by software programs and users.

The paper is structured as follows: section 1 describes a proposal for workflow modelling that combines workflow and knowledge perspectives. Based on this proposal, section 2 describes a workflow that models the furniture estimate task and a system where this model has been implemented. Finally, sections 3 and 4 present the future trends and the conclusions of this work.

1. Workflow Modelling

In this chapter we present a solution for the furniture estimate process that has been developed in a custom furniture manufacturing company. Several reasons exist for undertaking this effort. Manually-based processes have traditionally increased the number of mistakes in the production process cycle and promoted the lack of communication between sections of the company. Each designer or manufacturing expert has its own know-how and does not take advantage of the knowledge of the other members in the organization. In this sense, design and cost results are questionable since they are possibly based on inconsistent information and are done in a non-deterministic way. Collaboration and information sharing across functional areas must deepen in order to increase the product quality and reduce costs.

The automation of furniture estimate process is an important advance for improving organizations productivity. This process is a step forward to reducing product costs, getting new customers and defining a better manufacturing. Furthermore, the quality of its solutions will define the profits of the company. However, the automation of manually-based processes is not enough to get the desired improvements. The system needs: *(i)* to support a high level of collaboration, synchronize people, departments or resources involved in the process; and *(ii)* monitor its behaviour and detect events in a defined way. Workflows [11] and DFA [2, 3] are the starting points to get these objectives. Workflows will define the framework and provide the technology while the introduction of DFA will integrate the domain, manufacturing and assembly knowledge in the product design stages.

Approaching business process modelling by means of workflows entails organizations to develop and adopt new working models and infrastructures [11]. Moreover, the underlying philosophy of this approach requires supporting process management, reengineering and monitoring capabilities [14]. Most of the current enterprise information systems do not provide these features and their integration capabilities are very limited [15]. Nevertheless, some systems have shown a notable skill for dealing with workflows, such as Workflow Management Systems (WfMS), Business Process Management Systems or Enterprise Resource Planning Systems (ERP) with an extra module for workflow support. All these systems focus on the business process automation, data transfer and information sharing across the organization but differ from each other in their applicability and the technological approach they are based on [15].

Workflows provide a way to understand the business process in order to improve it. The fact of designing a workflow forces itself to examine the whole business process. This is a time-consuming process that makes it possible to define the business process structure, people, and resources involved in it. In this sense, the workflow design helps to get the right information to the right person. As regards product development, workflow design forces us to find a way to improve the furniture estimate task behaviour, looking for *concurrent engineering* [1] and deepen the DFA practices. As a result of this stage, we proposed a workflow where furniture designs are evaluated continuously as they progress.

Workflow is increasingly being used to model engineering or manufacturing processes [16,17,18]. However, the complexity of furniture estimate process, which integrates product design, manufacturing and assembly, makes necessary to add strategies for efficient knowledge handling [25]. Knowledge management [6] is a field specially designed for this purpose. Knowledge management deals with how to leverage knowledge as a key asset in organizations. On one hand, knowledge management helps us to manage the domain knowledge of custom furniture manufacturing. Manufacturing and assembly rules, guidelines, or furniture features represent this static knowledge, which is expressed by means of ontologies [7,8]. On the other hand, the dynamic knowledge is handled like another piece of knowledge. Methods are the greatest exponent of such knowledge and are defined independently from the static knowledge they manage in order to be reused. For example, it is possible to use the same assessment method to assess the client furniture order or to assess the final estimate. This is done through the definition of knowledge adapters that maps the static knowledge managed by tasks with the one of methods.

1.1. Framework for Workflow Modelling

In order to define workflows that use knowledge, we created a new framework (Figure 2) based on both workflow [4] and UPML frameworks [13]. This new framework extends the workflow framework into the new knowledge dimension that deals with knowledge modelling [19]. More specifically, it deals with the definition and modelling of the static knowledge by means of ontologies and the knowledge-intensive tasks by means of PSM. PSM describe explicitly how a task can be performed. PSM are *reusable* components applicable to different, but similar, domains and tasks. For this purpose, this dimension follows the UPML framework to define the knowledge components. UPML is an architectural description language that provides knowledge components (tasks, methods, domain models and ontologies), adapters, and a configuration of how these components should be connected using adapters [12].

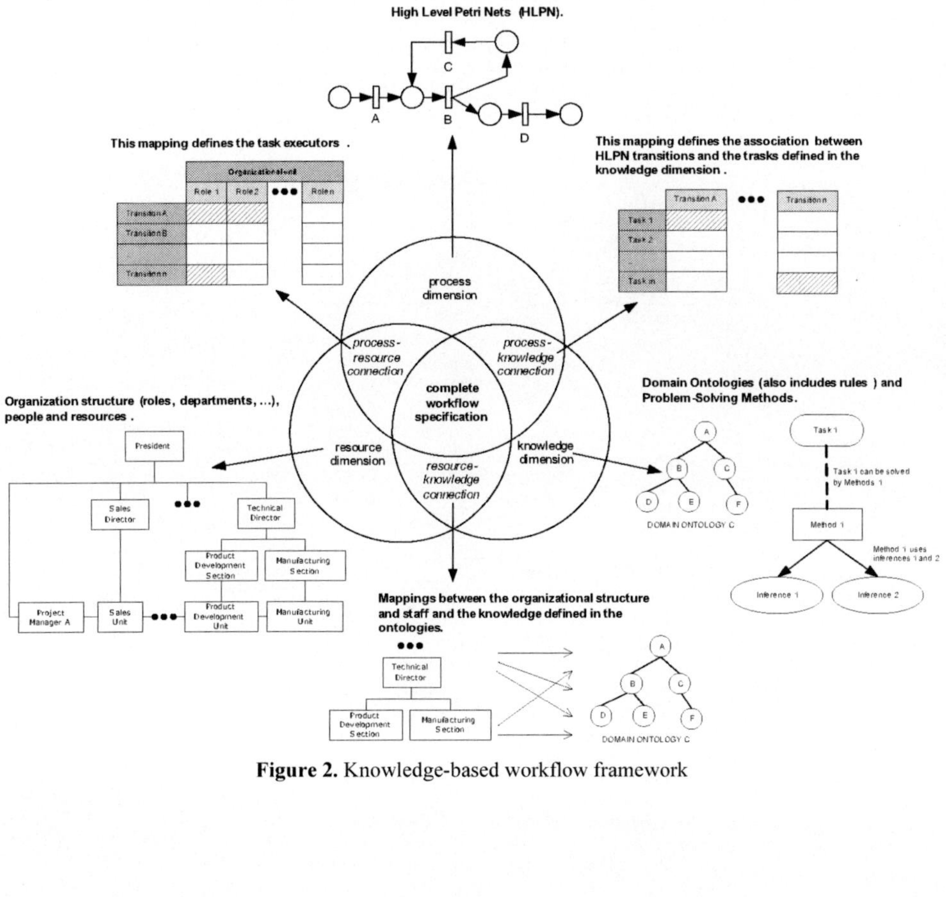

Figure 2. Knowledge-based workflow framework

In this framework, the task describes the operations to be solved during the execution of a method, specifying the required input and output and pre- and post-conditions. This description is independent of the method used to solve the task. The method details the control of the reasoning process to achieve a task. It also describes both the decomposition of the general tasks into subtasks and the coordination of those subtasks to achieve the required result. This control flow is carried out by means of High Level Petri Nets (HLPN) [20].

Workflow framework has two other dimensions. The first one, the process dimension, uses HLPN to deal with the definition of the processes structure. At present there is no standard language for workflow specification, but Petri Nets [21,22] with its solid mathematical foundation have proved to be a good choice for workflow representation [4,23]. Through this formalism, the workflow specification models tasks by means of transitions, conditions by means of places and cases by means of coloured tokens. The second dimension is the resource dimension. This dimension deals with the organization model definition, and particularly with the definition of organizational elements that take part in the workflow definition, such as organizational units and roles. From these elements, both human and non-human resources that participate in the workflow execution are classified. The intersection between these three dimensions defines the relationships between resources, processes and knowledge.

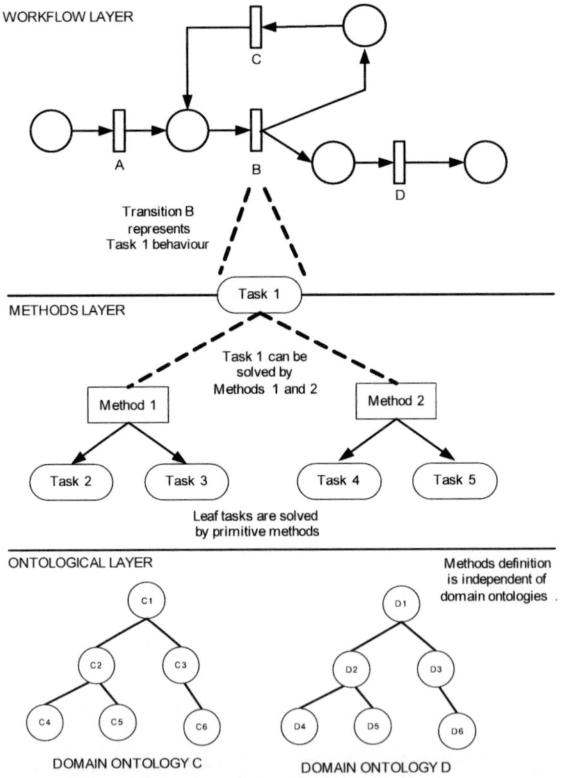

Figure 3. Workflow and PSM integration.

The intersection between process and resource dimensions defines the privileges that a user must be endowed with in order to carry a process out. Specifically, this intersection defines the organizational units and roles allowed to execute a HLPN transition. The intersection between process and knowledge dimensions relates *(i)* HLPN transitions with the tasks they represent and *(ii)* HLPN pages [20] with the control flow of methods they solve. In this way, transitions and methods are related by means of knowledge tasks. The last intersection, between resources and knowledge defines the owners of knowledge assets.

From the process structure perspective, the result of the integration between process and knowledge dimensions is a two layer model that delegates the execution of non-composite PSM to the UPML framework implementation (Figure 3). Complex tasks will be solved by means of composite methods. These tasks are represented by means of substitution transitions [20,24] and they can be solved by one or more methods. In execution mode, a selection process based on the problem assumptions relates these tasks to the PSM. We must remark that the task specification is the union point between the workflow activities and the PSM that solve them. Tasks only describe its inputs/outputs and its behaviour, but they do not introduce any execution details. These details are defined by means of PSM at the knowledge-level [9], enabling that a task could be carried out by several PSM.

2. Furniture Estimate Task Automation

We remark that the solution described in this chapter is attached to the characteristics of a specific company. However experience tells us that most of companies face similar troubles when promoting the automation of this task. Therefore, they could take advantages of the solution herein described. This solution is based on the framework proposed on section 1.1 and defines workflows in order to accelerate the introduction of new, revised and better product designs. In the interest of achieving this objective, workflows will automate the product development stages, notifying the meetings, analysis, evaluations or revisions involved in the process. In this sense, a broad range of departments will be affected by such a change. Inventory, purchasing, production, sales and cost accounting would all be able to more quickly determine the impact of the change on their areas.

The workflow structure has been thought in order to support the design for manufacturing and assembly (DFA) rules and guidelines. These guidelines have been adapted to the furniture-manufacturing domain by the company experts, and are evaluated in several ways in the system: some guidelines are introduced as check-list constraints (evaluated by users) or as text, while others are directly introduced as knowledge in the system.

In this section we propose the use of knowledge-based workflows for modelling the way the designers and reviewers interact with the environment and to coordinate the different PSM in the achievement of DFA goals. The objective is to incorporate the knowledge of conceptual design, CAD designs, manufacturing and assembly in the product design. This strategy combined with artificial intelligence planning and learning is used to obtain cost savings.

2.1. Furniture Estimate Workflow

The workflow for furniture estimate is represented by means of the Petri net described in Figure 4. For the sake of simplicity, we use a place-transition view to present the workflow structure, although the final workflow model is described with HLPN. The behaviour of the workflow is modelled through a PSM known as *Propose, Revise and Modify*, which is a generic constraint-satisfaction method [6,26]. This method obtains a viable and cost-optimised furniture design, through the evaluation and modification of the requirements and constraints of the conceptual design.

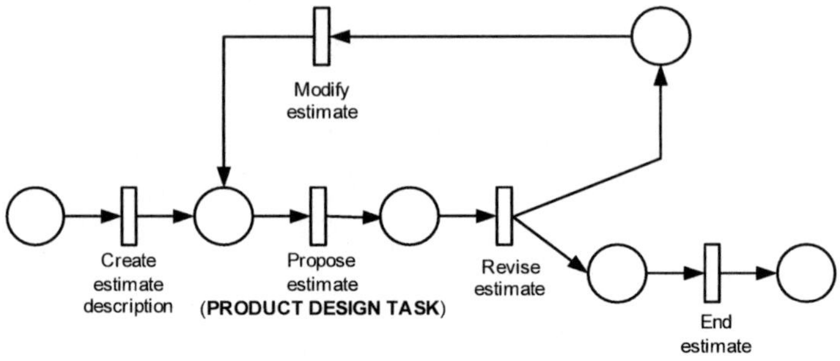

Figure 4. Furniture estimate workflow.

Let us suppose, as an example, that the marketing department of our company has received a furniture estimate request which includes the manufacture of one hundred hotel rooms. On a first step, the activity *Create estimate description* defines the main requirements of the furniture and its conceptual model (Figure 5). On this stage, the project manager introduces a description of the furniture, *time* and *cost restrictions*, and so on, extracted from the client specifications. This description includes design characteristics, expected wood type and qualities, finishes, and so on. Following with the example, marketing staff must get the rooms architectural design; that is,-they must define the furniture (a double bed, two bedside tables, a desk and a wardrobe) and the quality criterions that define the manufacturing. Next, taking the broad description into account, the technical staff decides the main features and restrictions. For example, they define the wood finishes based on the quality demanded by the client, the durability, the moisture, environmental concerns or safety regulations. This is a brainstorming stage where technical staff provides a basis for making design decisions. As regards the hotel materials, one aspect to be defined is that wood or wood-based materials must be treated with flame retardant chemicals. Once the client order has been assessed (by using workload and cost criteria), a first conceptual design is built with some manufacturing and assembly constraints.

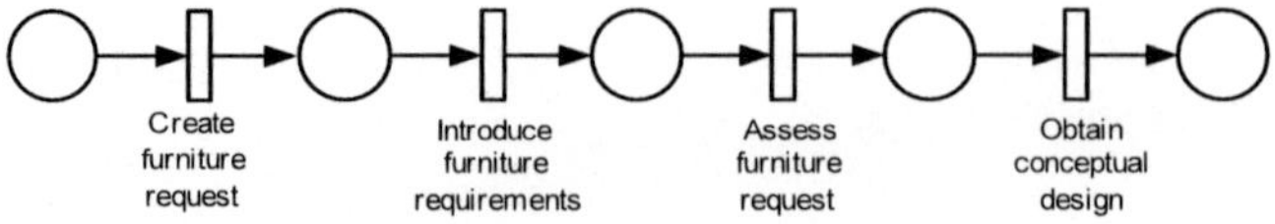

Figure 5. Create estimate description.

However, the main activity of the furniture estimate task is to propose a design that should be consistent with the conceptual design. The activity *Propose estimate* is in charge of giving such functionality. Figure 6 shows the task decomposition. It is structured to execute several evaluations for the furniture CAD designs. First, designers create the CAD designs. These designs combine a creative activity with the fulfilment of the conceptual designs and the guidelines of DFA. Designers are supposed to use standard components and materials (an ontology of materials and components has been created to support this rule), to develop a modular design, to reduce the cost of handling with large pieces of furniture, and to test the robustness of designs. For example, the product cost may vary depending on the type of joint used to assemble the furniture (Figure 1). Some types of joints require to assemble the furniture before the finishing, thus increasing the manufacturing, packaging and/or shipping costs. As regards, the strength required and the quality requirements are the main factors for joint type selection rule.

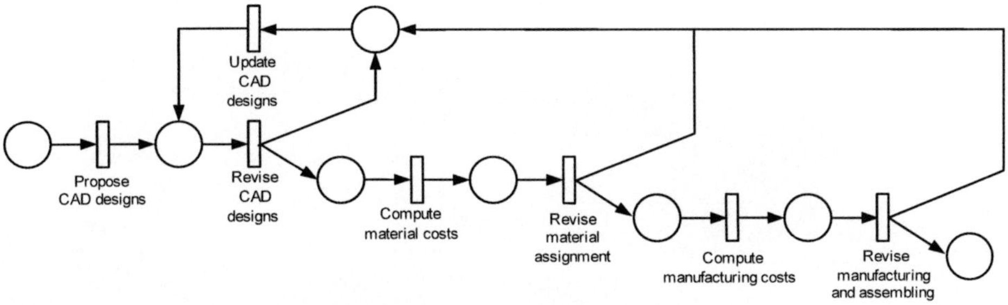

Figure 6. Propose estimate (product design) task.

Once the decomposition of the furniture in its parts, part dimensions, parameters and materials have been defined, the *Propose CAD designs* is in charge of defining manufacturing processes to be applied to the pieces of furniture. If possible, assembly parts are linked each other by means of its description and expert knowledge related to furniture components catalogue. Pieces of furniture are assigned to the manufacturing processes based on a precedence diagram developed to check the sequence of manufacture and assembly. As regards, the parameters of pieces of furniture define most of manufacturing processes to be applied. For example, the number of holes to be drilled for a dowel joint may be standardized and they are based on the length of the piece. Another example is that panels are always cut in order to obtain the pieces of the furniture. This operation arranges the pieces of furniture in the panel in order to minimize the waste.

The *Revise CAD designs* is the first review process of product designs. This stage ensures that each component or piece of furniture is necessary, since components are often included because of invalid or historical reasons. The same happens with manufacturing processes. If some mistake is detected in the review process, CAD designs are taken to the *Update CAD designs* task.

The furniture estimate cost is basically composed by both material and manufacturing costs (including assembly and packing). The correct estimation of these costs is the key to obtain an accurate estimate. This cost is the sum of the parts cost (material and manufacturing) of the furniture.

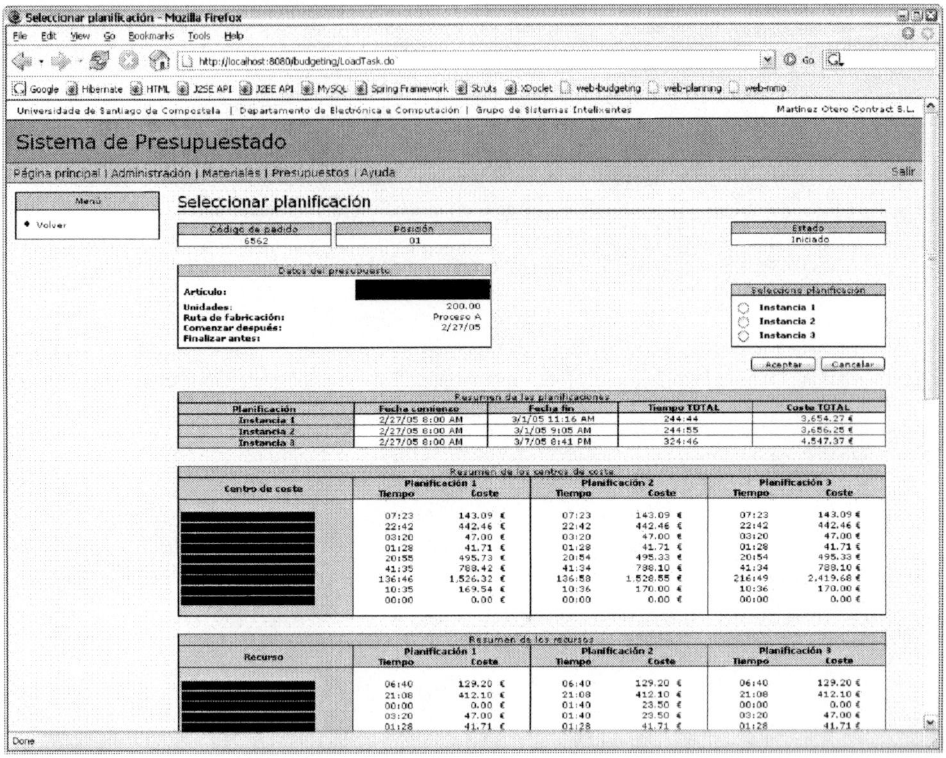

Figure 7. Screenshot of the manufacturing plan selection.

Compute material costs defines the cost of materials. This cost is computed adding to direct material costs (raw material and logistic), the cost of additive material (veneer, varnish, glue, and so on), and the cost of material part rejects. The assessment of this cost (*Revise material assignment*) is based on the information provided by the suppliers about its sale prices. The selection of these suppliers depends on its reliability to facilitate the material in manufacturing time. Material qualities and tolerances are also checked. Moreover, furniture parts are also checked in order to minimize part numbers, assembly surfaces or simplify assembly sequences. This evaluation looks for semi-finished components (such as doors, drawers, or semi-finished panels) that should simplify the manufacturing and assembly process and reduce the workload (perhaps at the expense of the cost). For example, the doors of the hotel wardrobe can be bought semi-manufactured if the workload cannot assume its manufacturing.

The *Compute manufacturing costs* task is also solved by the adaptation of a composed PSM called *Scheduling* [6,26]. This task determines the time needed to manufacture the furniture order. In order to obtain this time, it is necessary to decide which human and machine resources will execute the manufacturing operations taking the foreseen workload into account. Operations are determined by means of knowledge rules over the furniture design specification. This is a supervised procedure where the system defines the operations to be carried out for each part. The result of the scheduling task is the availability of several manufacturing plans with different operation sequences and resources (Figure 7). These plans propose different solutions for furniture planning with different assembly methods (for example, manual assembly or robot assembly). In order to obtain the manufacturing cost, the costs associated to the different manufacturing processes and the machine costs are added. Machines costs are defined as a function of the machine amortization time and the tooling cost. Packaging costs such as shipping cartons, bags and blister packages are also included in the computation.

Manufacturing plans are the input for the *Revise manufacturing and assembly* task. This task assesses the different plans in order to minimize the cost of manufacturing and assembly, and to verify that the manufacturing plan fulfils the constraints imposed in the conceptual design. It is another review process and basically checks for problems and bottlenecks in the manufacturing. The evaluation of these assessments could provoke the redesign of some aspects of the final product, which would suppose the execution of a new workflow cycle. If some minor problem is detected, some solutions could be proposed. For example, the authorization of extra hours, a third shift, etc.

Time computation is difficult to obtain. The manufacturing processes time is directly related to the resources or group of resources involved in it (both humans and machines). This time also depends on both the material and the operation descriptions. We used an estimation method based on influence parameters in order to determine a polynomial approach for estimating the time of resource operations. Once these parameters were identified by means of regression equations (over a set of well balanced examples), we approximate the coefficients of these polynomials. For example, the polynomial that obtains the time for drying kiln operations should take in consideration parameters such as the intended use of the product (inside or outside use), the product volume, or the moisture content among other aspects. As regards, in order to improve the time estimations, we built a learning system that polishes these coefficients (see section 2.2).

Finally, shipping costs, general administration costs and profit costs are added to the estimate (Figure 8).

2.2. System Implementation

The final step of furniture estimate task automation is to translate the previously described specification (see section 2.1) into a real system implementation. However, organizations are dubious of assuming the change of its business processes. The lack of flexibility of most of the information systems implies infrastructure changes in order to acquire workflow management capabilities. Moreover, this change may also affect business process reliability, and workflow technology is recent and is not still enough mature.

We must remark that some furniture manufacturing software already contributes with a furniture estimate capability. Nevertheless, most of these solutions cannot be applied to our problem. Manufacturing simulations and price estimations are one of the existing trends to perform furniture estimations. Both trends give a CAD solution that estimates the price of a furniture from design and manufacturing plant characteristics. However, this solution has limitations: it does neither contemplate materials, machines and human resources complexity nor manufacturing workload.

Many organizations support their business processes by means of some kind of enterprise information systems. These systems are mainly in charge of performing some business process but they have not capabilities for changing its structure or handling its behaviour. Moreover, most of these systems have the business process hard-coded in their applications, so no modification is applicable. In addition, most of the existing enterprise information systems are not endowed with this ability so it is convenient provide an additional technological support. Information technologies focussing on process management and improvement are a good option to assume this challenge, and WfMS and ERP are their most representative solutions.

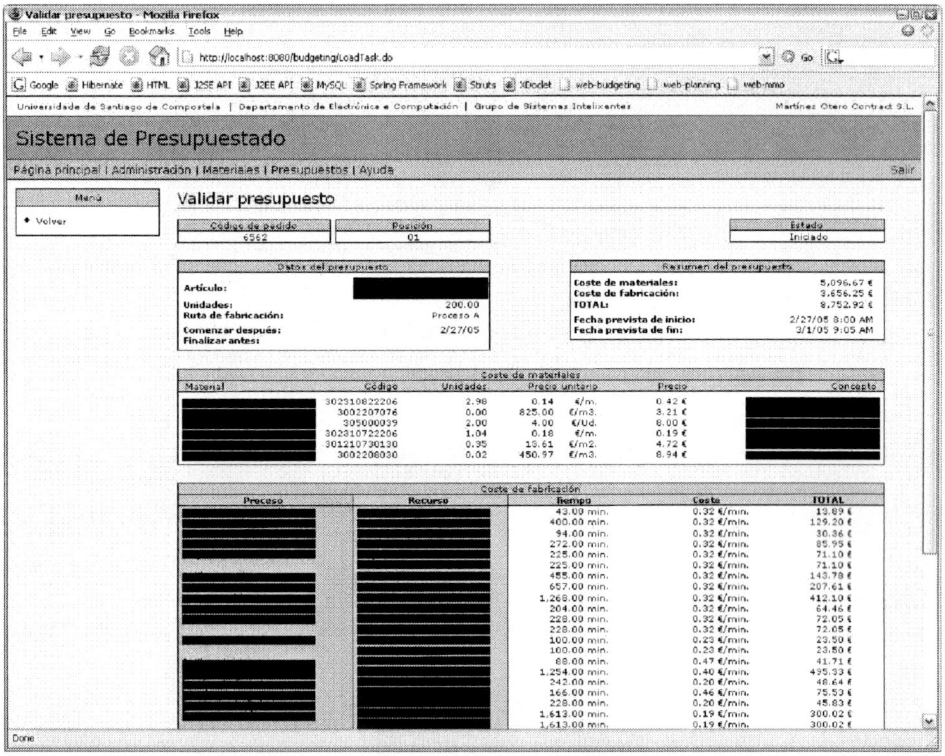

Figure 8. Furniture estimate.

Despite these systems solve the same problem, their differentiation lies in their technological approach and applicability. On one hand, WfMS are a set of applications and tools for workflow definition, creation, and management. These systems are able to interpret the process definition (business process logic), interact with workflow participants and, where required, invoke the external use of applications and tools [14]. On the other hand, ERP systems are prefabricated applications developed for specific domain applications. Unlike the traditional approach, companies must learn to adapt to the ERP software and, in some cases, modify their processes and structures to meet the software specifications.

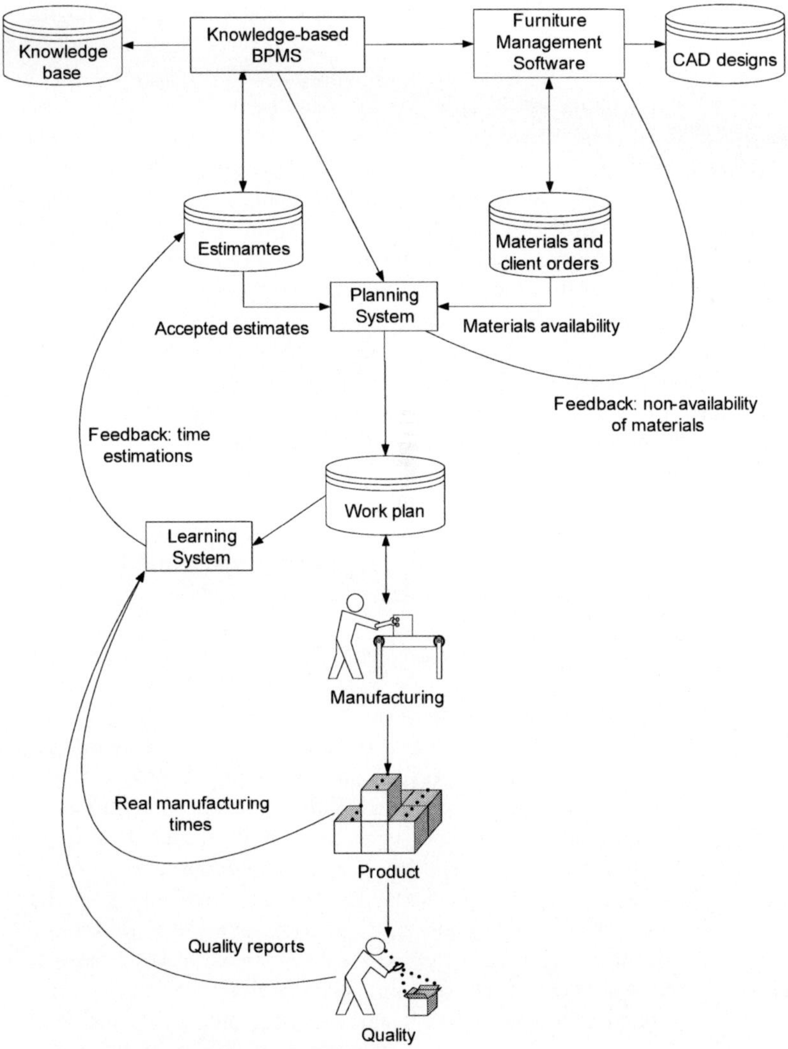

Figure 9. System description.

Our solution (Figure 9) is a mixture between workflow and knowledge-oriented approaches. The core of the system infrastructure is a WfMS enriched with knowledge-based capabilities. We decided to use WfMS instead of ERP because of WfMS are more suitable to model workflows involving humans and software systems, especially if the systems are autonomous and heterogeneous [15]. With its BPM components, the system executes the workflows and performs the coordination of users and software; while with its knowledge components, it defines the knowledge-enriched workflows and gives the desired behaviour to them. Our BPMS is based on the Workflow Reference Model architecture [14] proposed by the Workflow Management Coalition (WfMC), but introduces some significant differences to enable the knowledge management.

Workflow technology combines several paradigms like enterprise application integration, cooperative information systems, computer aided work, groupware and so on. This feature makes WfMS a good solution for companies with several independent software applications because of coupling with Enterprise Application Integration technologies [27]. These systems have the ability to integrate heterogeneous applications into a coherent environment through a hub architecture. This architecture uses adaptors for each legacy application reducing the integration problem to a centralized integration. As shown in Figure 9, the furniture estimate process interacts with several applications and legacy systems. The centralized architecture facilitates the integration with legacy applications within organization Intranet. This architecture also avoids information duplication, and facilitates a transparent information access.

A significant part of the information used in furniture estimate, like furniture designs, material catalogue prices or material stocks, are managed by external applications. For example, the furniture management software module showed in Figure 9. However, the core of the knowledge used in the system is directly stored in its knowledge base or in its estimate database, which includes the furniture manufacturing domain ontology and the set of rules and constraints for product design evaluation. The system also manages the ontology instances defined in order to classify accessibility and visibility of each user. Other kind of knowledge such as mappings between ontologies and other data accessed by means of external applications are also stored in the database.

Other components developed to give support to the whole furniture estimate problem are the *Planning system* and the *Learning system*. Both systems were included in order to support the manufacturing planning and a feedback to improve time estimations. Planning and learning systems are based on the paradigm of genetic algorithms [28]. The first system uses a genetic algorithm to carry out a search to allocate the manufacture of the product in the supply chain workload. The chromosome of the algorithm associates the operations executed for each part of the design to the resources. The resource assignment is based on a first phase that selects the method of assembly and on a second phase where the resources are selected. The state space of this search is broad and genetic algorithms are a good solution to reduce the search computation time. The learning system is a standalone system whose objective is the adjustment of the coefficients of the regression equations that defines the times needed by the resources to do a manufacturing process (each operation carried out by a different machine can have a different equation). In our case, the learning system runs a genetic algorithm for each equation. The encoded chromosome represents the coefficients of the equation that are crossed and mutated in order to get a better behaviour.

3. Future Trends

The system described in section 2.2 is the first step of an R&D project. Next steps currently undergoing include the development of new workflows in order to support other business processes of the organization. As regards, we plan to apply some of the results of the workflow described in section 2.1 to the pre-manufacturing process. Usually, the client accepts the furniture estimate but requests changes in the design or in the materials of the furniture. The motivation is to use the product design workflow specified for the estimate process but adding additional restrictions to meet the new specifications of the client.

Although the results of the system developed have been good enough, and the time and costs of the furniture estimations have reached an acceptable level of accuracy, some aspects of the product design should be improved:

1. To enrich the CAD design with more information that is helpful for better describing the manufacturing and assembly processes. In this sense, we plan to enrich our design ontology for improving the descriptions of the components of a product: we will add ontology axioms that constraints the possible combinations of furniture pieces in order to obtain a product.
2. To use the design history in order to reduce the product development time. Processed by some kind of case-based reasoning, its knowledge should be useful to define the manufacturing route and to improve the manufacturing time estimations. In order to do this, the surrounding knowledge of the product design process must also be taken into account, that is, it is necessary to justify the reasons of a decision.

4. Conclusion

This chapter describes a way to approach the product design of custom furniture by means of knowledge-enriched workflows. As regards, a framework for knowledge-based workflows has been developed. This framework combines the advantages of workflow and knowledge management technologies. It provides a workflow-oriented view of the business process that people can easily understand and encourages them to take part in the business process definition and optimization. However, the main contribution is that this framework integrates the UPML framework in order to manage the knowledge of the workflows by means of ontologies and PSM.

A knowledge-based workflow based on this framework has been created in order to support the behaviour of the furniture estimate task. This task has been structured taking in consideration the factors that influence the product design. As regards DFA rules and guidelines are introduced in the different stages of the workflow. In this sense, DFA can take advantage of the knowledge and process organization of the proposed framework since only agents with a specific permission and knowledge can execute a task.

As a real application of our approach, the chapter also describes the implementation of the furniture estimate task. This implementation has been developed for a custom furniture company. Workflow automation is based on WfMS. WfMS and ERP could be thought as the best solutions for workflow implementation. The selection of WfMS was based on the company software previously existing infrastructure but also on the fact that WfMS are more suited to model workflows involving humans and software. Some of the benefits of this solution are related to the process behaviour:

1. Resource productivity (both human and software) has been improved. Tasks are assigned to the most suitable resource taking the availability, restrictions and knowledge factors into consideration.
2. Work delegation has also been improved. The competence of each resource is clearly defined.
3. Process monitoring has been improved. This improvement affects the time to perform the product design since it reduces the reaction time.

However, the main benefits of this solution are directly related to the product design improvement:

1. Cost savings on product manufacturing and assembly. Product materials, components, manufacturing processes and so on have been standardized and regulated.
2. Cost savings in storage. Material needs can be long-term planned allowing a constant flow of materials and avoiding storage taxes.
3. Time estimations errors have been reduced and are continuously being improved.

Acknowledgments

This work is been carried out in the framework of a R+D contract with Martínez Otero Contract, S.A., supported by the Dirección Xeral de I+D of the Xunta de Galicia through grant PGIDIT04DPI096E. Also financial support from grant PGIDIT04SIN206003PR is acknowledged.

References

[1] P. O'Grady, R. E. Young, Issues in Concurrent Engineering Systems, *Journal of Design and Manufacturing: The Research Journal of Concurrent Engineering* 1:1-9, 1991.
[2] G. Boothroyd and P. Dewhurst, *Design for Assembly – A Designer's Handbook*, Boothroyd Dewhurst Inc., Wakerfield, Rhode Island, 1983.
[3] G. Boothroyd, P. Dewhurst and K. A. Knight, *Product Design for Manufacture and Assembly*, 1994.
[4] W. M. P. van der Aalst, The Application of Petri Nets to Workflow Management, *The Journal of Circuits, Systems and Computers,* 8(1):21-66, 1998.
[5] W. M. P. van der Aalst, A. H. M. ter Hofstede and M. Weske, Business Process Management: A Survey, In W.M.P. van der Aalst, A.H.M. ter Hofstede, and M. Weske, editors, *International Conference on Business Process Management (BPM 2003)*, volume 2678 of *Lecture Notes in Computer Science*, pages 1-12. Springer-Verlag, Berlin, 2003..
[6] G. Schreiber, H. Akkermans, A. Anjewierden, et al., *Knowledge Engineering and Management: The CommonKADS Methodology*, The MIT Press, Cambridge, Massachusetts, London, England, 1999.
[7] T. R. Gruber, Toward Principles for the Design of Ontologies Used for Knowledge Sharing, *Int. Journal of Human-Computer Studies*, 43:907-928, 1995.
[8] M. Uschold and M. Gruninger. Ontologies: Principles, Methods, and Applications, *Knowledge Engineering Review*, 11(2), 93-155, 1996.
[9] A. Newell, The Knowledge Level, *Artificial Intelligence*, 18(1):87-127, 1982.
[10] V. R. Benjamins and D. Fensel, Special Issue on Problem-Solving Methods, *International Journal of Human-Computer Studies (IJHCS)*, 49(4): 305-313, 1998.
[11] D. Georgakopoulos, M. F. Hornick, and A. P. Sheth, An Overview of Workflow Management: From Process Modeling to Workflow Automation Infrastructure, *Distributed and Parallel Databases*, 3(2):119-153, 1995.
[12] D. Fensel, E. Motta, Frank van Harmelen, et al., The Unified Problem-solving Method Development Language UPML, *Knowledge and Information Systems* 5(1):83-131, 2003.
[13] B. Omelayenko, M. Crubézy, D. Fensel, et al., *UPML version 2.0*, Free University of Amsterdam, IBROW Deliverable D5, 2000.
[14] D. Hollinsworth, *The Workflow Reference Model*, Technical Report TC00-1003, Workflow Management Coalition, 1994.
[15] J. Cardoso, R. P. Bostrom and A. Sheth, Workflow Management Systems vs. ERP Systems: Differences, Commonalities, and Applications, *Information Technology and Management* 5(3):319-338, 2004.
[16] I. Choi, C. Park and C. Lee, A transactional workflow model for engineering/manufacturing processes, *Int. J. Computer Integrated Manufacturing*, 15(2):178-192, 2002.
[17] Q. Xu, R. Qiu, and F. Xu, Integration of Workflow and Multi-agents for Supply Chain Coordination, *Computer Engineering*, 29(15):19-21, 2003.

[18] S. Huang, Y. Hu, C. Li, A TCPN based approach to model the coordination in virtual manufacturing organizations, *Computers and Industrial Engineering*, 47(1):61-76, 2004.

[19] J.C. Vidal, M. Lama, A. Bugarín and S. Barro, *Problem-Solving Analysis for the Budgeting Task in Furniture Industry*, Proceedings of the Seventh International Conference on Knowledge-Based Intelligent Information & Engineering Systems, 2:1307-1313, 2003.

[20] K. Jensen, Coloured Petri Nets. Basic Concepts, Analysis Methods and Practical Use, *EATCS monographs on Theoretical Computer Science*, Springer-Verlag, Berlin, 1992.

[21] C.A. Petri, *Kommunikation mit Automaten*, Institutes für Instrumentelle Mathematik, Germany, 1962.

[22] T. Murata, Petri Nets: Properties, Analysis and Applications, Proceedings of the IEEE 77 (1989), 541-580.

[23] W. M. P. van der Aalst, Workflow Verification: Finding Control-Flow Errors Using Petri-Net-Based Techniques, *Business Process Management: Models, Techniques, and Empirical Studies*, 1806:161-183, 2000.

[24] L. Gomes and J.P. Barros, Structuring and Composability Issues in Petri Nets Modeling, *IEEE Transactions on Industrial Informatics*, 1(2):112-123, 2005.

[25] X. F. Zha, S. Y. E. Lim and S. C. Fok, Integrated knowledge-based approach and system for product design for assembly, *Int. J. Computer Integrated Manufacturing*, 12(3):211-237, 1999.

[26] J. Breuer and W.V. de Velde, *The CommonKADS Library: Reusable Components for Artificial Problem Solving*, IOS Press, The Netherlands, 1994.

[27] M. Marin, Business Process Technology: From EAI and Workflow to BPM, *Workflow Handbook 2002*, Lighthouse Point, FL, 2002.

[28] D. E. Goldberg, *Genetic Algorithms in Search, Optimization and Machine Learning*, Addison-Wesley, Boston, MA, 1989.

Integrated Intelligent Systems for Engineering Design
X.F. Zha and R.J. Howlett (Eds.)
IOS Press, 2006

Dynamic hardware-based optimization for adaptive array antennas

Martin Böhner [a] and Hans Holm Frühauf [b] and Gabriella Kókai [a,1]
[a] *Department of Computer Science, Programming Languages*
Friedrich-Alexander University of Erlangen-Nürnberg
[b] *Department of RF- and Microwave Design*
Fraunhofer Institute for Integrated Circuits

Abstract. The following chapter describes and discusses the suitability of ant colony optimization (ACO) to an employment with blind adaptation of the directional characteristic of antenna array systems. Due to the special advantage of the ACOs their robustness and high adaptation rate - this very young heuristically optimization strategy allows its application in the field of high radio frequency systems. In order to fulfil the hard real time constraints for beam forming in ranges of few milliseconds a very efficient hardware implementation for a highly parallel distributed logic is proposed in this chapter. The application requirements are given because of the high mobility of wireless subscribers in modern telecommunication networks. Such a dynamic alignment of the directional characteristic of a base-station antenna can be achieved with the help of a hardware based Ant Colony Optimization, by controlling the steerable antenna array system parameters as digital phase shifts and amplitude adjustment. By means of extensive simulations it was confirmed that the suggested ACO fulfils the requirements regarding the highly dynamic changes of the environment. Based on these results a concept is presented to integrate the optimizing procedure as high-parallel digital circuit structure in a customized integrated circuit of a reconfigurable gate array.

Keywords. Ant Colony Optimization, Adaptive Antennas, Hardware Implementation.

1. Introduction

Within the scope of the current research project[2] adaptive antennas were developed in the last years as a universal demonstrator system in the 2.45*GHz* ISM volume. In addition to classical applications such as localization by direction estimation or capacity extension with *MIMO* procedures [5] a further research is the adaptive

[1] Correspondence to: Gabriella Kókai, Department of Computer Science, Programming Languages Friedrich-Alexander University of Erlangen-Nürnberg Martensstr. 3, D-91058 Erlangen, Germany. Tel.:+49 9131 85-28996 ; Fax: +49 9131 85-28809; E-mail: kokai@informatik.uni-erlangen.de

[2] This project is a co-operation Fraunhofer Institut for Integrated Circuits (IIS), Department of High-Frequency Engineering with the Technical Faculty of the University of University Erlangen, Department of Programming Languages and the Department of Information Technology, Communication Electronics

adjusting of the directional characteristic of the used array antenna system to achieve optimal efficiency and maximum transfer quality while simultaneously eliminating the influences of noise. This procedure is called *beam forming*.

Advantage of adaptive array antennas is the possibility to change their directional radio pattern (also called antenna pattern) with the help of few adjustable parameters in a semi deterministic way. Talking about smart antennas implies, that not the antennas themselves but the antenna systems are smart. Generally, being collocate with a base station, a smart antennae system combines an antennae array with a digital signal-processing capability to transmit and receive power in an adaptive, spatially sensitive manner ([21]). In other words, such a system can automatically change its patterns in response to the signal environment. This can dramatically improve the performance characteristics (for example, increase the capacity) of a wireless system. To process directionally sensitive information, an array of antennas (typically 4 to 12 - but not limited to) is required, whose inputs are combined to adaptively control signal transmission. The antennae amplitudes and phases are adjusted (weighted) electronically to generate the desired radiation patterns. A radiation pattern for far-field conditions is usually represented graphically in either the horizontal or the vertical plane.

In a significant number of real world scenarios both the desired transmitters and additional disturbing transmitters are mobile. *Disturbing transmitters* are transmitters which arise at the same time in the same frequency and local area of the base station and communicate (with other systems), as for example foreign WLAN-participants.

Therefore the controlling of the directional characteristics has strict real time conditions. If these conditions are not met, the communication connection gets lost and the adaptation goal cannot be achieved. Due to this, the efficiency of the system is substantially reduced. For the real time adaptation of patterns the following three fundamental solutions are worked out: The **classical method** [18]. With the help of direction estimation algorithms, as for example *ESPRIT* or *MUSIC*, the position of all participants and disturbers is determined [17,18]. Subsequently the appropriate parameter settings are derived from the prepared tables. Because of the high expenses these tables must be pre-computed and can be validated only for a limited number of scenarios.

The second approach comes up to the **current state of the art**. By means of modern learning procedures, preferably *neuronal nets*, the transmitter positions are linked to favourable parameter settings by training the net. The efficiency of this procedure however depends on the quality and on the number of trained scenarios as well as on the complexity of the assigned learning method [27]. However both approaches have serious disadvantages. On the one hand they are bound to a static environment. In case of a change in the environment either the tables must be computed again or scenarios have to be trained again. In addition it is always necessary to locate desired and undesired transmitters within a short time period in a complex operation.

A possibility to avoid these problems is offered by the use of **blind adaptive beam forming** ([24]). Under blind adaptive beam forming we understand the adjustment and continuous adaptation of the directional radio pattern of an electrically and/or electronically controllable array antenna without the need of measuring, simulating or elsewhere knowing the directional characteristics. Likewise it is not necessary to know the position of the mobile station. Instead a so called fitness value is determined, which is computed from characteristic parameters of the electrical field of the array antenna at the position of the mobile station. This can be derived for example from the field strength, the signal to noise ratio or the bit error rate of the transferred signal. A promising way is the employment of heuristic optimization procedures for the determination of a suitable parameter configuration for the antennae ([16]).

In contrast to conventional approaches like MUSIC or ESPRIT algorithms ([12,19]) the basic idea of our solution is that an adaptive antennae receiver is combined with an adaptive control system and this system is applied in an unknown environment. Our work substantially increased the receiving quality of desired mobile transmitters and allows a maximum elimination of disturbing mobile stations.

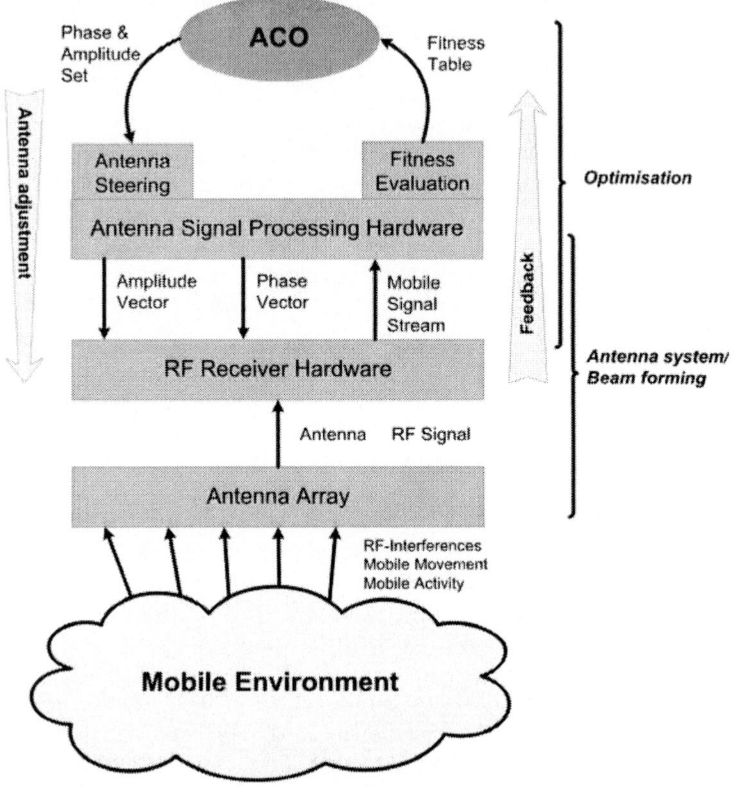

Figure 1. System Overview

In this book chapter the applicability of Ant Colony Optimisation (ACO) is discussed for solving the problem definition of blind adaptive beam forming. The system overview is presented in figure 1 where one can see the three main parts – the *mobile environment*, an *antenna system* and an *optimisation* unit. The mobile environment comprises mobile (moving) transmitters as well as environmental interferences. The antenna system consisting of an antenna array, receiver hardware and signal processing hardware is responsible to send and receive, but also to electrically and/or electronically adjust the antenna configuration. The optimisation unit on the one hand generates new settings for the antenna system and on the other hand evaluates these settings on the basis of feedback it gets back from the antenna system to influence its following search for better settings.

Chapter 2 continues with an introduction to the basic principles of adaptive antennas and the principles of beam forming methods followed by a description of the structure of the demonstrator and the associated simulation environment. The chapter closes with a description of the state of the art in the topic of ACO regarding dynamic parallelism and hardware implementation.

In chapter 3 two new hardware ACO variants are discussed. The suggested ACO algorithms are examined in the simulation environment on the basis of different scenarios and the results are presented. Chapter 4 discusses future and emerging trends and finishes with a conclusion.

2. Background

This chapter provides basic background information on beam forming, our demonstrator for adaptive antenna systems and a software simulation environment for qualified testing of new optimization algorithms. Furthermore several attempts for adjacent problems in literature are introduced and discussed.

2.1. Introduction to beam forming

The directional radio pattern of an antenna in the radiation zone is presented graphically in form of a radiation pattern and/or an antenna pattern. The left part of figure 2 shows such *antenna patterns*. If several individual antenna elements are arranged to each other, one speaks of an *antenna array*. The patterns of the different individual antennas overlay. Additionally an electromagnetic interconnection effect between the antennas occurs which substantially modifies the pattern of each individual antenna element.

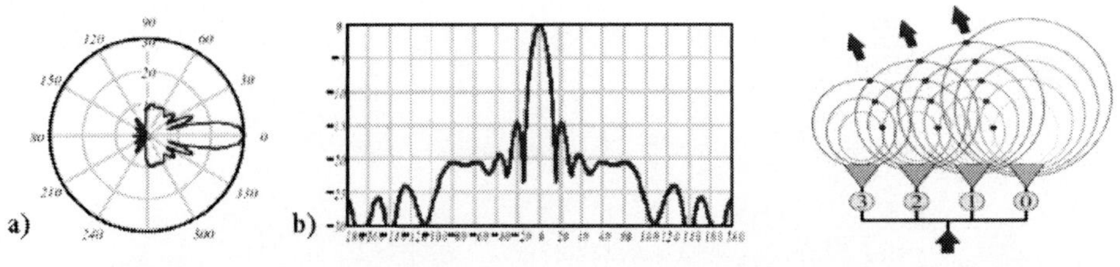

Figure 2. Pattern of one antenna elements and the basic principle of array antennas [13]

Therefore the pattern of the whole antenna array is defined by its geometrical parameters and the pattern of the elementary antennas. If the antenna is activated only in *electromagnetic mode* [15], there is no non-mechanical control possibility to manipulate the patterns.

An *intelligent antenna system* consists of three main components: an antenna array, the corresponding receivers, as well as the digital signal processing unit. If overlapping effects between the elements of an array antenna system are considered it is possible to get different interference patterns and therefore non mechanical *beam forming*. This can be achieved through specific changes of phase and amplitude of the individual antenna signals.

This basic principle is illustrated in figure 2. The arrows point towards maximum sensitivity. By shifting the phase of an individual antenna signal the position of the wave front changes (represented as circles) and thus the directional effect of the entire antenna array is modified. The directional antenna characteristic can be modified by a continuous adaption of the amplitude and phase-shifting units to a dynamic environment.

Beside the array antenna and its elements, also the channel between the mobile transmitter and the output of the base station has crucial influence. Related to this channel the input power *P* at the receiver is estimated as follows [14,26]:

$$P = P_{tx} + G_{tx} + G_{rx} + P_{rx} - L_{fs} - F$$

P_{tx} Transmitting power of the transmitter

G_{rx} Antenna gain of the transmitter

G_{tx} Antenna gain of the array antenna (1)

P_{rx} Sensitivity of the baes station

L_{fs} Transmission loss in the free area

F Fade Margin, further losses

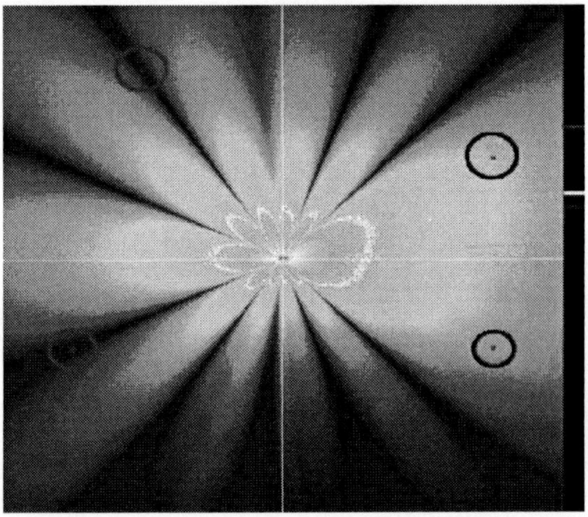

Figure 3. Antenna pattern of an array antenna. The circles mark positions of transmitter

It must be considered that the antenna gain of the array antenna depends on the direction of the transmitter to the antenna. Furthermore receiver sensitivity can be controlled by an adjustable amplifier in the HF receiver path of the appropriate antenna element. Additionally further effects which are difficult to emulate, like channel noise, phase and amplitude errors of the hardware, caused by electromagnetic disturbances, construction unit tolerances, variations in temperature, nonlinearity, quantization effects as well as limited word lengths, play a large role.

Beam forming procedures, which are based on model based mathematical algorithms, are computationally complex, relatively inaccurate and their applicability is therefore restricted. Figure 3 illustrates an adapted antenna pattern. The bright regions (green) mean a high, dark (blue) one a low power density.

2.2. Demonstrator system and simulation environment

To establish a connection between research work and a real world system Fraunhofer IIS developed the demonstrator as a basic system. In parallel a software simulation environment was implemented which extensively incorporates the measured characteristics of the real system. A goal was to afford the possibility to transfer gained perception in the simulation environment directly to the demonstrator.

2.2.1. Demonstrator

The real time demonstrator system for 2.45*GHz* ISM volume is shown in figure 4 including a multiplicity of identifiable mobile transmitters realized with 100*mW* 2.45*GHz* GMSK transmitters. The antenna array, consisting of four individual patch antennas - receives the signals of these mobile transmitters. The signals are filtered and amplified in the corresponding *super heterodyn* receivers as well as converted to a lower intermediate frequency. Subsequently, the four individual signal paths are merged and, with the help of a digital analogue decoder, passed on to digital signal processing as digital data stream with a data rate between 1.12*GBit*/s and 1.47*GBit*/s. The signal processing is implemented on a *Field Programmable Gate Array* (FPGA) [9].

Figure 4. Adaptive antenna demonstrator: upper left: mobile transmitter (without antenna), upper right: array antenna, lower left 4-chanel-HF-receiver, lower right Virtex II FPGA board and ADC

A central component of this system is the Virtex-II-FPGA from Xilinx. FPGAs actually belong to the class of programmable logic devices. They possess a universal circuit structure, which can be reprogrammed by the user via appropriate configuration. Decisive for the application as technology for the optimization procedures was the simple configurability and the potentiality to achieve maximum computing speeds by high-parallel processing units.

Crucial for the adaptation in the demonstrator system are on the one hand two controllable attenuators per channel in the high frequency receiver. On the other hand four digital actuators for amplitude correction and four electronically controllable phase shifters in the FPGA are available. Altogether sixteen controllable parameters influence the directional characteristic of the antenna. The first two attenuators use 12 Bit configuration words each, the amplitude correction needs 18 Bit and a configuration word for a phase shifter is 8 *Bit* long. All in all the antenna system has a number of 2^{200} possible parameter settings.

Real time constraints usually arise from the communication protocol used (e.g. 2ms for 802.11 WLAN or 625µs for Bluetooth). Within these time slots a new best configuration has to be found and the antenna system must be configured with these new settings. If one considers about 1000 fitness evaluation to find a new best setting and on top that only 10% of the available time can be used to find such a setting (rest is used for real data transfer) then in our scenario we come to an estimated time period for generating one new candidate solution within 18µs. These hard real time constraints make a hardware implementation mandatory.

2.2.2. Simulation environment

The demonstrator described in the last section was metrologically characterized and forms the basis for a software simulation environment written in C++. In addition this simulation environment covers the topics fitness evaluation, environment definition and simulation of movement of the mobile transmitters.

The virtual environment is defined by specifying maximum coordinates of possible positions with a standardized spatial resolution of one meter. Afterwards a fixed number of array antenna systems can be placed within this environment. For our fundamental tests we always used just one antenna system placed in the centre of the environment. Further a number of transmitters can be initialized. These are assigned to one of four classes: as an unwanted transmitter which should be masked out or as desired participants with one of the following priorities: *high*, *medium* or *low*. The simulation environment can distinguish between *active* and *inactive* states of the transmitters. Each transmitter takes an active state with a configurable probability and stays in that state for a stochastically controlled time interval, until it becomes inactive again. The mobile transmitters *move* only along predefined horizontal and vertical paths, which can be specified by the user. The definition of different *speeds* is possible by varying the frequency and size of the movement steps.

The available antenna power at the transmitters can be continuously determined. However this measurement is always afflicted with stochastic (e.g. noise) and biased errors (e.g. offset) in real world scenarios. The fitness calculation of the simulation environment incorporates these possibilities only in a limited way, as an additional stochastic error is added to the power ratings.

2.2.3. Simulation scenarios

To compare the efficiency of algorithms with each other in differently complex environments in an objective manner, both for static and in dynamic simulations different complexity classes were defined. In each scenario the performance of each algorithm is evaluated and can be compared to the results of other methods. The environment has always a quantity of 4000 x 4000 units with the adaptive antenna located in its centre.

In static scenarios the mobile transmitters are set on a random position in each test run. By doing so methodical errors can be avoided. Per scenario altogether 500 independent test runs each with 20000 fitness evaluations the following three scenarios are examined:

- **Scenario 1 (static-basic):** Two transmitters, one disturbing and one with high priority (1).
- **Scenario 2 (static-standard):** Four transmitters, two disturbing and two with priority 1 und 2.
- **Scenario 3 (static-complex):** Nine transmitters, three disturbing and respectively two with priority 1, 2 and 3.

In the dynamic case the mobile transmitters move on defined movement paths (roads). The initial positions of the transmitters are identical for each test run. Every 1000 fitness calculations they are moved by 2 or 3 units. In order to be able to collect significant data during each run, 10000 best values must be simulated. The dynamic scenarios are specified as follows:

- **Scenario 4 (dynamic-basic 1):** Two transmitters, one disturbing and one desired. The movement is done stochastically on a square with edge length of 3000 units.
- **Scenario 5 (dynamic-basic 2):** Two transmitters, one disturbing and one desired. These move on a straight line, which passes the basis station very closely.
- **Scenario 6 (dynamic-standard):** Four transmitters, two disturbing and two desired. The mobile transmitters move on a small rectangle, whose sides have the distances 200, 400, 800 and 1600 to the antenna.
- **Scenario 7 (dynamic-complex):** Nine transmitters, three disturbing and six desired. The movement paths consist of five horizontal and five vertical roads, which cut themselves in 25 crossings.

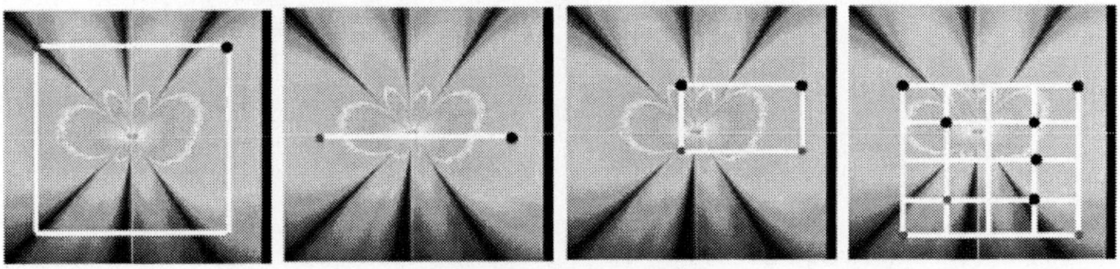

Figure 5. Dynamic scenarios

Figure 5 illustrates the movement paths of the transmitters in the dynamic scenarios. The grey (red) transmitters are the disturbers, the black marked ones are the desired mobiles.

2.3. Ant Colony Optimization - State of the art

Nowadays metaheuristics are applied usually for static problems. A lot of optimization problems are however dynamic and require methods, which continuously adapt to the changing conditions. Branke summarizes in [3] different possibilities to cope with these challenges.

Ant colony optimizing algorithm concerns a multi-agent system, which was developed between 1991 and 1999 by Dorigo et al. ([2,7]) particularly for the Travelling Salesman Problem (TSP). An agent corresponds thereby to an ant. Its characteristic consists of the fact that the natural models of the agents ([10]), the ants do not communicate directly with each another, but indirectly via changes of their environment. Therefore ants are placing pheromone traces on their ways. Following ants are affected by these pheromone traces in their choice of route.

In the following an overview about ACO for application to dynamic problems, as well as existing hardware implementations published in literature are presented. After that, an estimate is given, how these methods can be applied or transferred to our problem or why this is not so easily possible.

2.3.1. ACO for dynamic systems in literature

Dynamic environment means that during an optimization process the environment and therewith the evaluation criteria changes. For the search it is crucial whether these changes are made continuously or in discrete time intervals. In the former the extrem case might happen that each evaluated solution in an optimization run will be evaluated in a different environment than the other candidates. These changes which might even be relatively moderate are however given and influence fitness evaluation. The optimizing procedure must therefore exhibit a certain tolerance in relation to these unequal conditions.

If discrete time periods are used, then the environment and thus the evaluation criteria remain the same during an optimization run. Changes are made if a time period ran off. The next search is done in the changed environment.

Since with the environment, also the evaluation criteria change, very good solutions at time t_1 lose their quality after short time. All optimization procedures and thus also any ACO variant compare the fitness value of new solutions S_{new} with the separately stored best solution so far S_{best} and replace S_{best} if a new solution yields a better fitness value. By the change of the evaluation criteria also the fitness of the stored solution S_{best} is changed. Therefore it is necessary to re-evaluate S_{best} after some iterations.

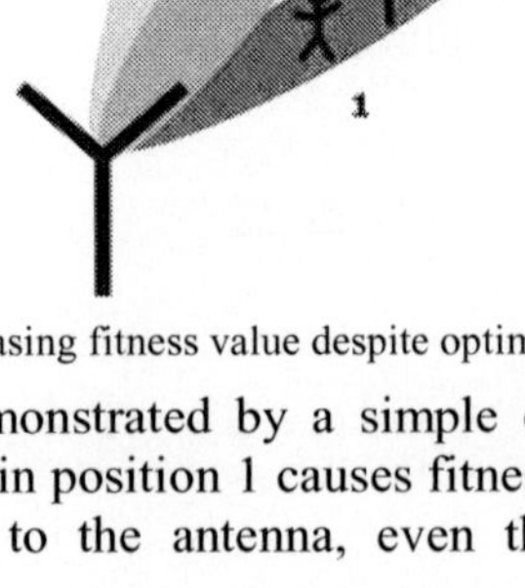

Figure 6. Decreasing fitness value despite optimal solution

This connection can be demonstrated by a simple example shown in figure 6. While a very good configuration in position 1 causes fitness with a high value, because the transmitter is placed close to the antenna, even the optimal solution for the transmitter at position 3 will produce a worse fitness value, because of the larger distance to the antenna. Even though it is necessary to adapt the antenna configuration in the selected example, because the angle is changed, newer configurations compulsorily will yield a lower maximal fitness value than the originally best solution. Therefore the regular re-evaluation of S_{best} is necessary.

Increasing pheromone values on the single ways through the problem graph conduce to guide the search for successful solutions toward the current best solution. If the values of the pheromones arise monotonously with the progression of the optimization run, then a change of the configuration and thus adaptation to new conditions are strongly limited.

Angus and Hendtlass examined in [1] the adaptation ability of ant algorithms. For that purpose a travelling salesman problem (TSP) is regarded. Dynamics of the system are obtained by changing the number of cities after a certain starting time. After the algorithm found a very good solution at the beginning, a city of the TSP is for example removed. Angus and Hendtlass used a modified ACO algorithm that normalizes the pheromone values if a change in the TSP is detected. That avoids the already addressed problem of poor adaptation ability due to high pheromone values on specific way in the problem graph. After different examples presented in [1] the two authors come to the conclusion that ant algorithms are able to adapt faster to a slightly modified problem definition than to determine a completely new solution for the changed scenario. Since the quality of the adapted solution increases over time, the described procedure is very well suited for problems, where an as good as possible solution has to be found within a defined time barrier.

Despite the interesting results the procedure of Angus and Hendtlass cannot be applied directly to the problem examined here. On the one hand the algorithm requires an explicit notification, as soon as the underlying problem is changed. Using adaptive antennas it can be assumed that the optimization problem changes continuously and it is the task of the ACO to pursue these changes as close as possible. The changes must either occur constantly, or they impact the adaptation ability of the antennas. On the other hand the normalization operations of the pheromone values require divisions and real number notations. These operations are extremely inefficient in the limited hardware available.

So far the probably most successful applications of ACO meta heuristic to dynamic systems are algorithms for adaptive routing in communications networks. The most well-known representative of these algorithms is *AntNet* [4] from Di Caro and Dorigo. This kind of problem concerns a stochastically distributed multi-goal optimization (performance and delay in the network). This means that the choice of route can only take place on the basis of local and approximated information about the current and future status of the network. In order to be able to adequately react to changing load conditions in the network, ants are constantly send through the net. These ants update the routing tables in the particular nodes based on their experiences on the way from a starting to an end node. The updating is done in a way to meet the two concurrent goals high data rates and small delay at the same time as good as possible.

AntNet uses two kinds of ants to solve this problem - so called *forward* (F) and *backward ants* (B). In regular intervals forward ants $F_{s \rightarrow d}$ go from each node s of the network to a randomly selected destination node d and store their way on the basis of the node identities k. To choose the next node to take ants use local information stored in a table in each node. If cycles arise in their way while searching, they delete the cycle path from their memory. As soon as an ant F arrives in the destination node d it gives its chosen path to a second ant $B_{d \leftarrow s}$. This ant goes back on the same way in reverse direction to node s and updates the pheromone traces in the individual nodes. For more exact details, particularly on the update of the local information in the nodes it is referred here to [4].

Di Caro and Dorigo compare AntNet with a set of other partly common, partly experimental route algorithms and come to the conclusion that AntNet supplies the best result in this simulation. In particular it works under high net load better than any other algorithm.

Against a direct adoption of the principles of AntNet for the controlling of adaptive antennas speaks the complexity of the two main procedures. As mentioned before a hardware implementation of the algorithms is compulsory to meet the strict time constraints of our problem, but already the representation of the pheromone values as floating point numbers makes a hardware implementation very complex. In addition to that mathematical computations, like multiplications, exponentiations or normalizations of pheromone values are very complex to realize in hardware.

Even if none of the two presented algorithms for dynamic systems can be transferred directly to our application purpose, the two examples show clearly that ant algorithms are particularly suitable to find good solutions in dynamic systems.

2.3.2. Parallelism

In the literature parallelism of ACO was given only little attention so far, despite the natural parallelism of the procedure. Stützle examines in [25] the simplest parallelism procedure, the simultaneous parallel and independent execution of multiple ACO instances. Such an arrangement makes sense due to the random decisions taken during an optimization run. As final result simply the best solution of all parallel determined best solutions is selected. Advantages of the method are the fact that the procedure introduces no additional communication overhead and an appropriate implementation requires only smallest auxiliary effort. For our purposes the principle is however no suitable, since on commercial FPGAs the place for several parallel implementations of ACO is missing.

Randall categorizes in [22] different parallelism strategies for ACO. Beside the very rough granular parallelism of Stützles independent ACOs the next fine-granular structure represents parallel interacting systems. At certain times, for example between single iterations, information between parallel working ACOs is exchanged. The island system ([20]) for example works according to this principle. Parallel ants form the next finer step. The calibration of respective pheromone values between individual ants however clearly increases communication costs. Further more fine-granular variants presented in [22] cannot be considered for us due to their problem specifics. The previous suggestions are mainly intended for software implementations on so called *Multiple Input Multiple Data (MIMD)* computers. Delisle and colleagues propagate in [6] the use of a *Shared Memory System* that can reduce the communication costs, but at the same time however large effort for synchronisation, especially for the simultaneous treatment of the pheromone matrix must be made. That yields to speed losses again.

Unfortunately none of the presented solutions is suited for our project. In chapters 3 a new parallelism concept is therefore presented, which is very well suited for hardware implementation.

2.3.3. Hardware concepts for ACO

Scheuermann et al [11,23] for the first time made the successful attempt to implement a variant of the ACO metaheuristic, the so called *Population Based ACO (P-ACO)* on a commercial FPGA in hardware. Compared to standard ACO, P-ACO is a quite strongly modified algorithm. Instead of passing a complete pheromone matrix from one generation to the next, in P-ACO only a relatively small population of the *k* best solutions will be transferred between generations. Thus the communication effort is reduced substantially. The population is stored in a *n* x *k* matrix *Q*, where *n* stands for the number of cities of the TSPs and *k* defines the size of the population. After the ants generated new solutions P-ACO updates the population matrix based on a standard elite strategy, i.e. only the best solution of every iteration is taken up in *Q*. Additionally the previous best solution is hold. If one solution taken up into the population is better, than the best solution so far, the best solution is replaced with the new solution.

If already *k* solutions are in *Q*, then the new solution replaces the oldest one in *Q*, it means a FIFO[3] strategy is applied. To make hardware implementation possible P-ACO gives up the well-known pheromone matrix of an standard ACO in favour of the population matrix *Q*. At the beginning each pheromone value is initialized with an initial value τ_{init}. As soon as a new solution enters the population the corresponding pheromone values are increased. If a solution leaves the population, then the appropriate values are decreased again. Thus the evaporation of the pheromones in the standard ACO is unnecessary, since an upper limit $\tau_{init}+\xi_{ij}\Delta$ for the pheromone exists, whereby ξ_{ij} describes the number of solutions in *Q* which contain a transition from *i* to *j*. The parameter $\Delta>0$ defines the value by which the pheromone concentration is changed in each case.

With these modifications Scheuermann and colleagues implemented a variant of ACO in hardware on a commercial FPGA. The speed increases obtained in relation to a comparable software variant are factor two to four depending on configuration. For our real time constraints in order to be able to keep the strict barriers this gain is definitely too small.

As shown, applications for ant algorithms to dynamic problems are already defined and also a first attempt of a hardware implementation is already made, but a combination of both is not known at this time to the author of this work. Additionally our problem has strong real time requirements for the supply of new solutions.

2.4. Related work

We already published adaptive beam forming with the help of genetic algorithms ([16]), for which also a complex hardware implementation is suggested. A problem of this solution was the low adaptation rate; therefore this solution can be used so far only in static, and/or semi static environments. Furthermore as we know, there is no other work for an implementation in hardware for this task.

[3] FIFO: First In First Out

3. Main Thrust of the Chapter

This chapter provides two new ACO methods for blind adaptive beam forming. These new ACO variants clearly aim at a very fast implementation in hardware. Several design issues are based on these requirements. Further on in this section several simulation results are introduced and discussed.

3.1. Continuous ACO method for blind adaptive beam forming

We want to optimize parameter settings for our adaptive antenna system. The system consisting of 4 identical antenna modules can be controlled by 16 parameters with a total of 200Bit.

There are different possibilities to map the problem definition to a graph representation. One might think of a representation as a bit vector where the optimization procedure has 200 "switches" each with the possibilities 0 and 1. The opposite extreme would be a representation consisting of 16 switches or in analogy to a TSP called *cities*, with four times (one for each single antenna) 2^{12}, 2^{12}, 2^{16} and 2^{8} possibilities. Of course each step between these extreme suggestions is supposable. A comparison of those two representations and a third one in between them (4^{6}, 4^{6}, 4^{9} and 4^{4}) shows clearly, that the bit variant is best suited for use with ACO.

For an optimization with ACO a problem representation as graph is required, whose edges represent parts of a possible problem solution. Despite a usual TSP our problem described here is characterized by the fact that the order of a possible solution is important. Each permutation of a solution leads to a totally different parameter setting and therewith to a different fitness value whereas in a TSP all permutations of a solution yield the same result.

Figure 7 shows the bit variant of the parameter settings transferred to a graph. One can clearly see the similarities with a typical TSP except for the strong order of the graph. A huge advantage over a normal TSP representation is the reduced memory usage of the ordered variant. In addition Dorigo and Stützle report in [8] increased searching results in large TSPs if the number of possible successor cities is reduced by using a so called candidate list. In our example shown in figure 7 only two possible successor cities exist. Metaphorically speaking an ant has to start on the left side of the graph and walk through the cities to the right. In each city it has to decide whether to take way 0 or way 1. At the end a concatenation of all its decisions forms a candidate solution.

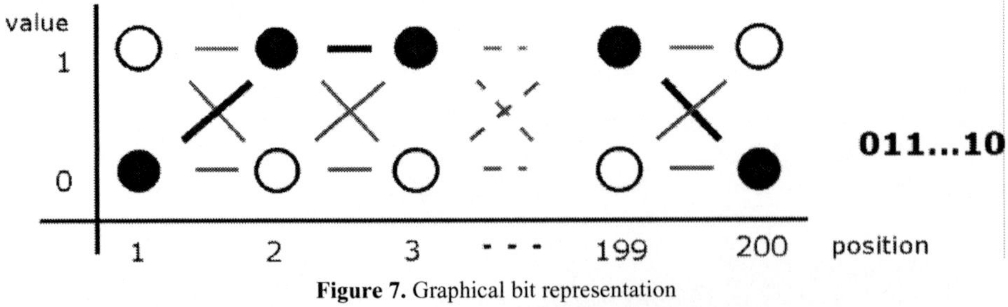

Figure 7. Graphical bit representation

In the following two ACO variants are presented that differ in their pheromone manipulation strategies but both abandon expensive mathematical operations in favour of a subsequent hardware implementation.

3.1.1. *ACO with memory* `acoBrain`

To avoid the problem of reduced adaptability due to high pheromone amounts on favoured paths in the graph this ACO variant called *acoBrain* introduces a lower and upper bound $[\tau_{min},\tau_{max}]$ for pheromone values. The boundaries are realized with a kind of a memory - the brain - that keeps the last *k* best solutions in mind. Only these *k* solutions influence the decisions of following ants. All pheromone values are initialised with $\tau_{init}=\tau_{min}$. If already *k* solutions are memorized they are replaced with new ones by a simple FIFO strategy. This procedure guaranties the compliance with the boundaries defined before. On the first sight `acoBrain` looks very similar to P-ACO defined by Scheuermann et al. The main differences are outlined in chapter 3.1.2.

To keep up the pheromone concentration in `acoLU` as a whole a modified pheromone update rule is used. Cities keep track of the amount of pheromone that is removed on their branches since the last update. If a global update follows, the amount of reduced pheromone is allocated to the branches according to the best solution found during that period. That means after two global updates each city has always the same initial amount of pheromone. The only thing that changes is the allocation of that amount to the branches of a city. Different to most other ACO variants this strategy allows a very fast redistribution of the pheromones within a city and so very good adaptation capabilities.

Algorithm 1.. ACO with local update in pseudocode

```
Initializing all pheromone values in the cities with
τ_i = τ_init, i ∈ [1, number of cities];
LocalUpdateCounter^i = 0;
repeat
    Position ants in starting node of the graph
    Each ant produces a solution S_m, m ∈ [0, number of ants-1]
    Solutions S_m = NULL;
    repeat
        /* Each ant produces a solution */
        for all Ant do
            for all City do
                Ant selects branch j in city i, with j ∈ [0, number of branches per city−1];
                /* Ant applies local updating rule */
                τ^i_j = τ^i_j − Δτ;
                LocalUpdateCounter^i + +;
            end for
        end for
    until Solutions completed
    for all Solution S_m do
        Determine fitness fit(S_m) of the solution
        /* Find the best of all m solutions */
        if fit(S_m) > fit(S_curBest) then
            fit(S_curBest) = fit(S_m); /* Renew globally best solution */
            if fit(S_curBest) > fit(S_best) then
                fit(S_best) = fit(S_curBest);
            end if
```

```
        end if
    end for
    /* Apply global updating rule */
    for all City do
        τ^i_best = τ^i_best + Δτ · LocalUpdateCounter^i
        LocalUpdateCounter^i = 0
    end for
until Termination criterion fulfilled
```

Algorithm 1 once again summarizes the steps of `acoLU`. After initializing all cities at the beginning, each of the m ants starts to produce a solution S_m. For this purpose it successively moves from city to city. In each city i it makes a decision for one branch, considering local pheromone traces τ_i. This decision causes a decrementing of the corresponding pheromone value on branch j. The decrementing is recorded by incrementing the $LocalUpdateCounter^i$. If all ants constructed a solution m new candidate solutions exists. Each of these solutions is evaluated by determining its fitness value $fit(S_m)$ and checking if it is better than the best solution of the current run $S_{actBest}$ and secondly if it is better than the current globally best solution S_{best}. To complete the whole process the global updating rule is applied which means that the already mentioned redistribution of the decreased pheromones in each city is done. Therefore the pheromone value on a city's branch which is part of $S_{curBest}$ is incremented by the amount of pheromone that was decreased before on all branches of that city. If that is done all *LocalUpdateCounters* are reset.

Both presented ACO variants restrict the maximum pheromone values. `acoBrain` defines an upper bound on the pheromone levels while `acoLU` just redistributes an initial pheromone amount. This and the avoidance of complex mathematical operations allow an optimization on the one hand of dynamic systems and on the other hand an implementation in hardware.

3.1.2. Hardware concepts

Main goal of the hardware concept for our new ACO variants is, along with a good feasibility, the attainable speed up over software implementations. Scheuermann and colleagues have already presented a hardware implementation of ACO in [11,23]. Their main goal was to demonstrate the feasibility of ACO in hardware in principle. Our main focus is to achieve an adequate speed to meet the hard real time constraints for adaptive antenna controlling.

Figure 8. Slightly modified graphical bit representation

Both `acoBrain` and `acoLU` work with the same problem representation and therefore use the same structures. They only differ in their pheromone update strategies. To transform the problem representation of chapter 3.1 into an adequate hardware construction the following steps were made. Figure 8 shows a slightly modified version of the graph. If we state that an ant knows the branch it took in the predecessor city we can decide which two of the four possible transitions (0→0, 0→1, 1→0 or 1→1) are available in the current city.

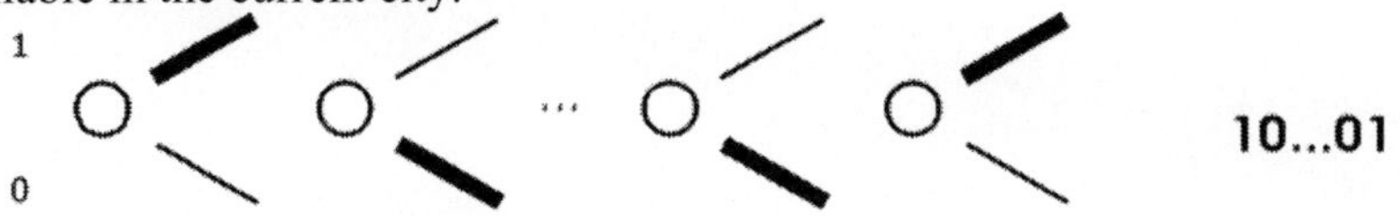

Figure 9. Further simplified bit representation

Simulations in our software environment have shown that the information from the predecessor city is negligible. Ants that used just the local information available in their current city reached even slightly better average results than ants with predecessor knowledge. Abandoning that knowledge allows us to further simplify the problem graph as shown in figure 9.

Now the decision which branch to take in a specified city depends only on information available in that city. It is negligible on which way the city was entered. With this simplification ACO of course loses some learning capabilities, but for the here analysed optimization in the highly dynamic environments this does not matter.

To speed up solution generation in hardware two typical methods can be used *parallelization* and *pipelining*. With P-ACO Scheuermann et al. divided their process of generating new solutions into different modules and arranged these modules in a pipeline. We in contrast do not only divide the generating process into few modules but even split a single solution into many different partial solutions. To achieve that ants are discarded. Instead the cities have to do the decision process from here on. That is possible because all decisions are done with local information in each city and do not depend on predecessor cities. To further parallelize the attempt and to optimize accessibility of the pheromone information in each city also the centralized pheromone matrix is discarded. Instead each city saves its own part of the pheromone information. That allows all cities to access their pheromone values in parallel. In one run each city produces one part of a solution. All those parts together form a candidate solution to the optimisation problem. This approach allows working in parallel on all the different partial solutions.

To evaluate a candidate solution the constructed parts are read out and used as a setting for the phase samples of the array antenna. The result word is divided into the different configuration parameters and these are transferred to the setting devices of the array antenna, for example the analogue and/or digital phase shifters, as well as the amplitude control units. The reconfiguration yields to a measurable change of the directional characteristic of the antenna, whose quality at the position of the mobile station is measurable. The quality value is made available again to the ACO as fitness value. A feedback about the quality of the found solution to the cities is done by changing the artificial pheromone values. Those in turn affect the cities decisions in following searching attempts.

Synchronisation between cities is only necessary if the pheromone values are updated due to a new best solution entering the memory in `acoBrain` or a global update is done when using `acoLU`. Even the updating process of all cities can be done in parallel.

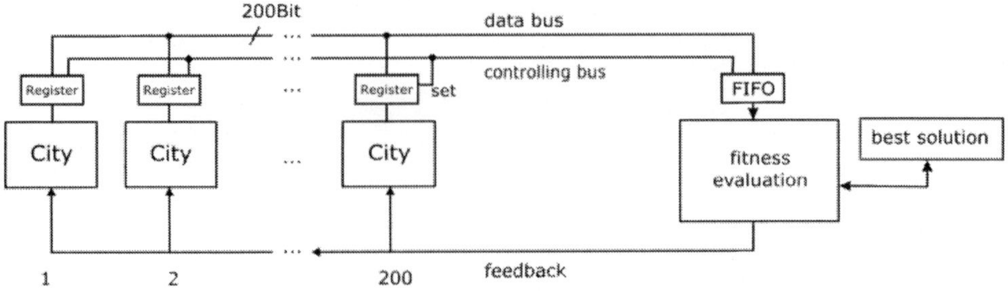

Figure 10. Hardware structure for a highly parallel ACO optimizer

Figure 10 gives an overview of the concept of interconnected cities. One can see cities, a fitness computation module with a FIFO queue and a storage facility for the best solution found. Each single city has its own register and all modules are interconnected via different busses.

The partial solutions generated by cities are written into module-own registers. From these registers they are passed on over a data bus to the FIFO queue of the fitness calculation. The integration of the fitness evaluation in the whole process can be seen in the figure just as well as the feedback mechanism from the fitness evaluation module to the cities for updating pheromone values. Candidate solutions are stored in a FIFO queue to wait for their fitness evaluation. As reference for the evaluation of new solutions the system additionally keeps the best found solution in a separate memory. At the end of a search this best solution forms the configuration of the adaptive antenna.

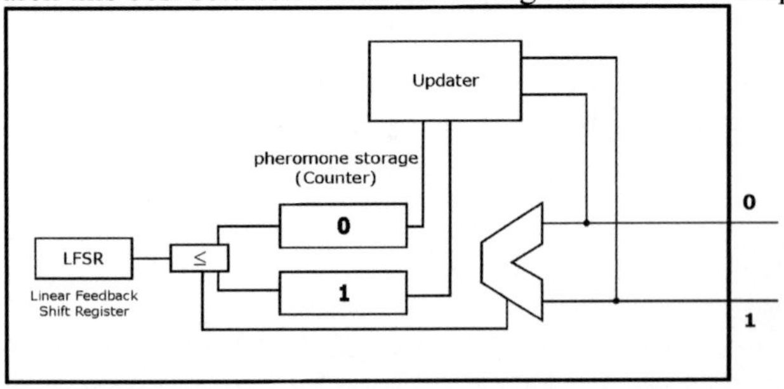

Figure 11. Structure of a single city module for the highly parallel ACO procedure

The hardware scheme for a single city is shown in figure 11. One can see a *Linear Feedback Shift Register (LFSR)* for random number generation to decide – in conjunction with the pheromone values – which branch to take. The *pheromone storages* for each branch are realized as counters. The actualization unit (*updater*) is responsible to update the values in the pheromone memory depending on the selected pheromone actualization strategy. Let's assume the displayed city belongs to an `acoLU` then the updater decrements the corresponding pheromone values if a decision is made. If that is done, an internal counter in the updater is increased to count the amount of decreased pheromone. When a global update is necessary the value of the counter in the updater unit is added to the appropriate counter of one of the two pheromone storages and the internal counter is reset to zero. `acoBrain` uses the same city model in principle but the updating unit works according to the strategy described in section 3.1.1.

This section first introduced two new ACO variants which were derived from the most promising concepts found in literature to meet our needs. For both alternatives a hardware concept has been presented that allows a highly parallel processing approach in solution generation. This was made possible by developing a new problem representation and the consequent abandonment of complex mathematical operations. In the end only additions and subtractions are used instead of multiplications and divisions.

3.2. Simulation results

In this section some simulation results of ACO with memory (`acoBrain`) and ACO with local actualization (`acoLU`) are presented. To compare the performance of different optimization methods in an objective manner defined scenarios are used. For our simulations we defined seven different scenarios, three in a static environment and the other four in a dynamic one. In addition complexity was divided into three different categories - basic, standard and complex. A more detailed scenario description is given in section 2.2.3. Primary goal of the static tests is to demonstrate the general optimization ability of the methods used. The tests in dynamic environments are to proof the applicability to our presented problem, the real time optimization in dynamic environments.

3.2.1. Simulation effort

To determine representative average values in static environments altogether 500 independent runs with 20000 fitness evaluations were done per method and per scenario. Thus a total of 500 x 20000 fitness evaluations were completed by each algorithm. Only the environmental condition in form of the transmitter positions differed randomly from run to run. The values of an optimization run can be compared however directly with each another, since all algorithms were tested in each randomly generated environment.

For simulations in dynamic environments altogether 200 x 10000 x 1000 fitness evaluations per scenario and method were executed - 200 independent runs with 10000 displayed values, which represent the best value after 1000 fitness evaluations respectively.

3.2.2. Static environment

The simulations in static environment have to prove to one that the several methods are in principle suitable for the optimization of the problem and document on the other hand the convergence behaviour of the different algorithms.

Basic

Left part of the figure 12 represents the average of the determined values of the three methods for the basic scenario in a static environment. ACO with local actualization delivered the best results. ACO with memory supplies still good values and achieves clearly better results than the comparison algorithm GA.

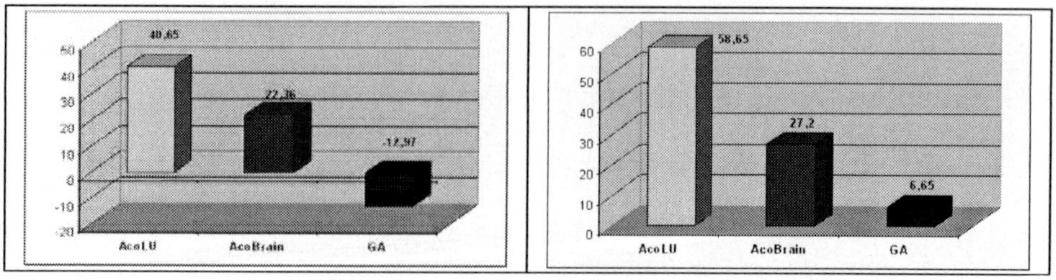

Figure 12. Average fitness value in static environment - basic and standard

Standard

A very similar result as in the basic scenario is obtained in standard scenario, however with clearly higher average values for the standard scenario. The average values on the right side of figure 12 show that ACO with local actualization delivers best result.

The direct comparison in figure 13 proofs that the already presented average numerical values occupy good explanatory power about the actual behaviour of the individual methods. In this diagram only a very small section with 20 values is represented, since otherwise clarity suffers. The values represent in each case the maximally reached fitness after an optimization run, thus after 20000 fitness evaluations. In the diagram the results of 20 of the altogether 500 repetitions of all optimizations runs are shown. Each of these runs took place in another environment which also means with different evaluation criteria. Figure 13 proves that these averaged values are nevertheless significant. It can be seen that general performance levels correspond to the average results in almost any case. The fitness values in figure 13 are directly comparable with one another although the transmitters are positioned randomly in the environment, because subsequent to a generation all three methods start their optimization runs successively in that generated environment.

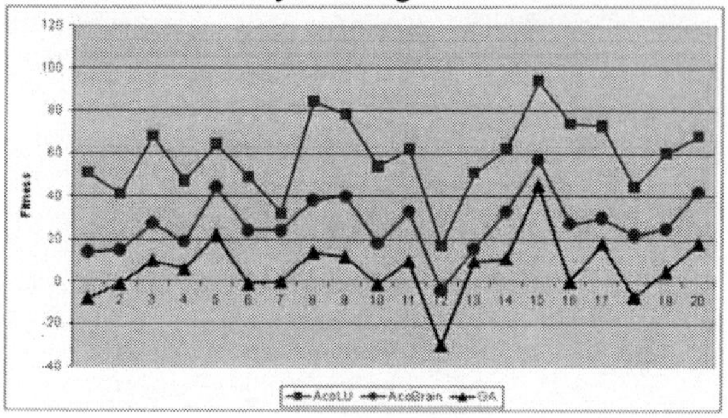

Figure 13. Comparison of the curve progression of all three procedures

Complex

In complex scenario the antenna configuration for an environment with nine transmitters is optimized. Figure 14 shows fitness values averaged over 500 independent runs. One recognizes very clearly that the levels of the values decreased compared to standard scenario. The order of the algorithms however remains unchanged. Over all static scenarios a very constant picture is shown. Algorithm `acoLU` delivers best results while `acoBrain` achieves medium values in all comparisons and the (not far optimized) comparison algorithm GA drops back clearly.

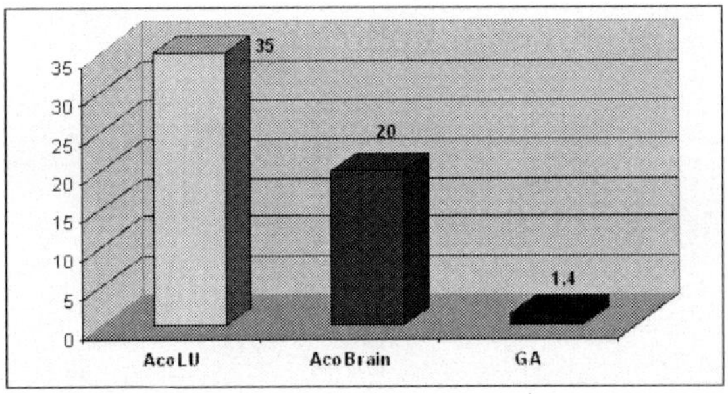

Figure 14. Average fitness value in static environment - complex

3.2.3. Dynamic environment

The simulations in dynamic environments are to proof the applicability for real time optimization in dynamic environments.

For clarity reasons also in dynamic scenarios over all average results are displayed, which are determined from regular intermediate results during an optimization run. These intermediate values are taken before a `move` operation, which arises in the environment every 1000 fitness evaluations. Thus intermediate optima are received, which would be the basis for an antenna setting for data communication in a real world application.

Diagrams 15 - 18 show progression of the intermediate optima for the search process of algorithms `acoLU` and `acoBrain` in different dynamic environments. The diagrams intend to give a good overview of the adaptability of the algorithms in the different scenarios. A good optimisation procedure keeps these values continuously on a relatively high level (fitness value > -90), without producing many intermediate values in the strongly negative range (fitness value < -200).

Each diagram represents only a small cut-out of all produced values. This is due to the limited representation capability of usual spread-sheet programs and for clarity reasons. The cut-outs reflect however in each case a representative picture of the optimization process with altogether 200 x 10000 x 1000 fitness evaluations. Please keep in mind that the diagrams use different scales.

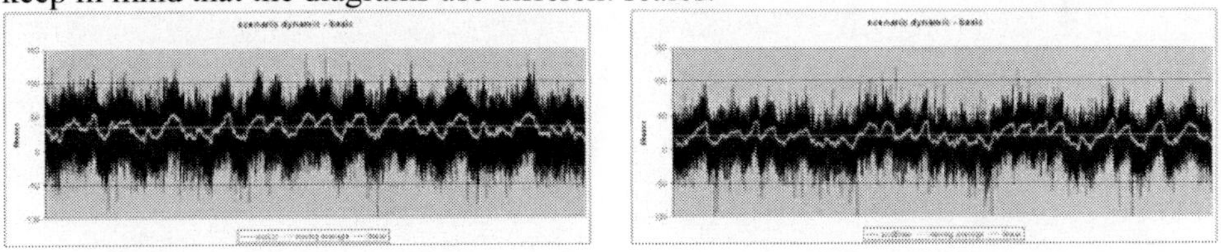

Figure 15. Progression of the dynamic basic scenarios (1) - `acoLU` and `acoBrain`

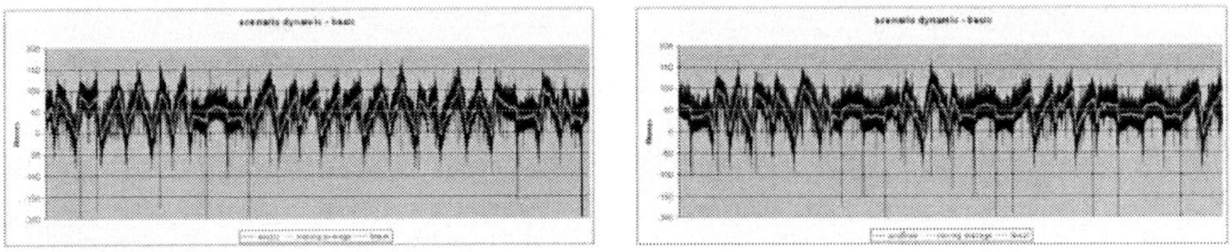

Figure 16. Progression of the dynamic basic scenarios (2) - `acoLU` and `acoBrain`

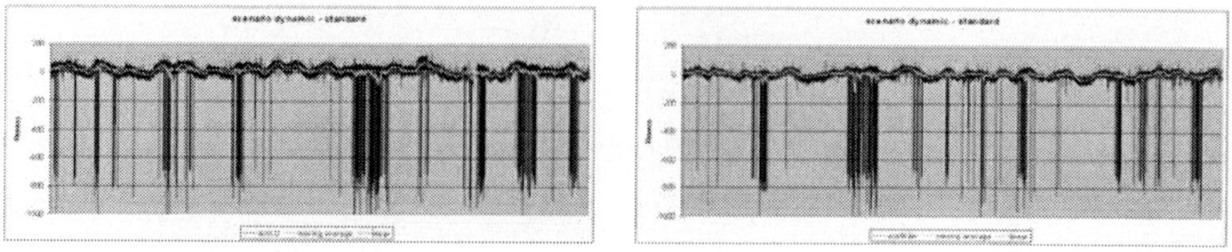

Figure 17. Progression of the dynamic standard scenarios - `acoLU` and `acoBrain`

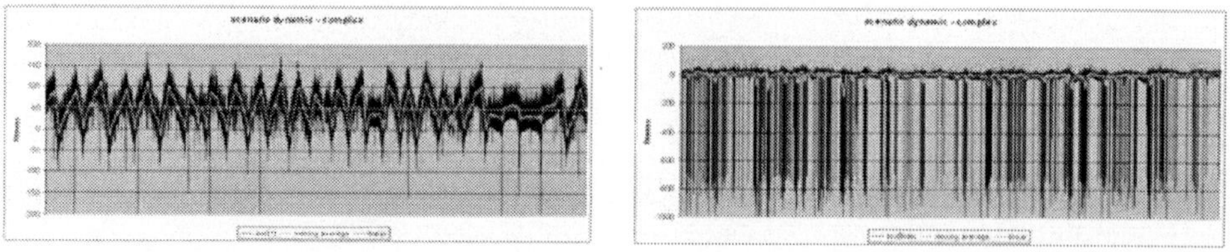

Figure 18. Progression of the dynamic complex scenarios - `acuLU` and `acoBrain`

Basic

We sketched two simple scenarios for dynamic environments to examine fundamentally different antenna behaviour. The left diagram in figure 19 shows the determined average values for the simple square scenario with two mobile transmitters. The main task here is to adjust the antenna system to the changing angels between base station and mobiles while distances remain more or less the same.

The right side of figure 19 presents the average values in case of two meeting mobiles which move on a single straight road that passes the antenna very closely. Each transmitter starts at one end of the road and moves to the opposite one, turns around and moves back and so on. While moving towards the antenna fitness values rise due to physical laws and depreciate again when the mobiles depart from the antenna. Nevertheless antenna configuration must be adapted to the slightly changing angels and especially to the heavily changing distances. The oscillating behaviour of rising and decreasing fitness values shown in that scenario can be seen in figure 20.

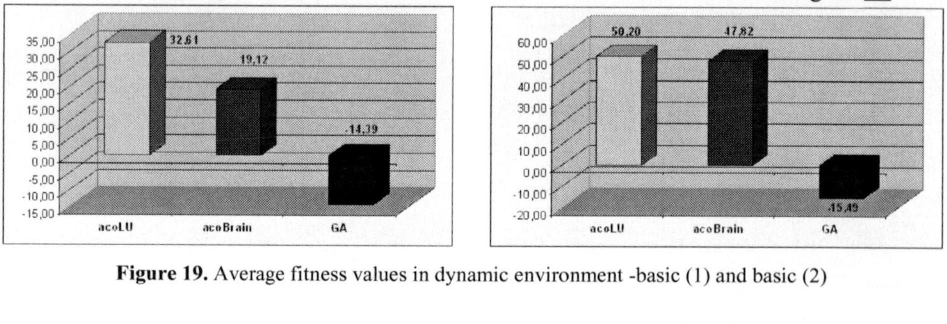

Figure 19. Average fitness values in dynamic environment -basic (1) and basic (2)

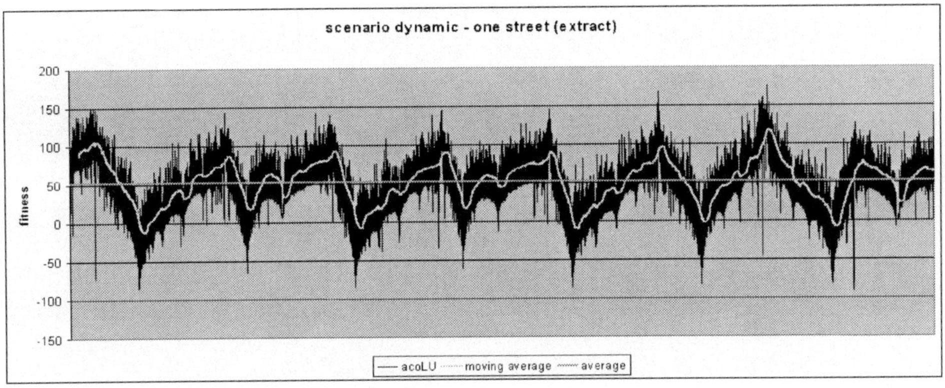

Figure 20. Reciprocating fitness values

Over all it can be stated that with dynamic conditions the picture remains the same. The variant `acoLU` still works best with the conditions while both ACO variants outperform the reference algorithm by all means.

Standard

The challenge of the standard scenario in dynamic environment consists on the one hand of the fact that both radial and tangential movements towards the antenna can occur. On the other hand two crossings were placed at relatively large distance to the antenna, while one crossing lies in direct neighbourhood of the antenna. Thus it is a hard task for the optimizer to find a suitable balance of transmitting power. With very high transmitting power mobile parts far away are well covered, but other ones placed close at the antenna run however the risk of being irradiated with too much power. Similar applies in the reverse case.

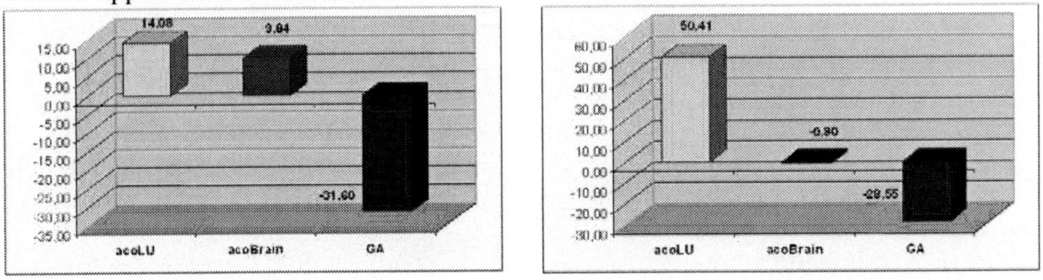

Figure 21. Average fitness value in the dynamic scenario - standard and complex

Figure 21 left shows the average obtained fitness value for standard complexity. The increased difficulties are reflected in clearly lower average results. Again the two ACO variants show the best results, while GA reaches relatively bad total results in comparison.

Complex

As illustrated in the diagram on the right side of figure 21 the repeated increase of the complexity to nine mobile transmitters, which can move between 25 crossings yields the expected behaviour of decreasing fitness values for `acoBrain`. `acoLU` and `GA` in contrast can show even increased average results compared to standard scenario.

3.2.4. Conclusion from the Simulations

For the presented algorithms here one can state that the results obtained in static environments were approved under dynamic circumstances. The variant that copes clearly best with all tasks is `acoLU`. For `acoBrain` it is remarkable to mention that the algorithm converges especially in the static scenario very early in search phase against a suboptimal solution. After that it usually cannot reach better regions of the search space. Figure 22 illustrates such behaviour during a typical search process. Compared to the ACO variant with memory, the procedure with local actualization seems to profit on the one hand by the wider spread search and on the other hand by the ability to faster redistribute its pheromone traces.

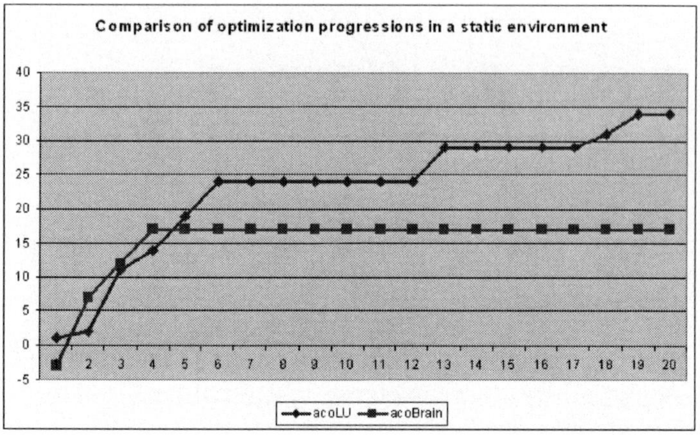

Figure 22. Comparison of the value patterns of both ACO variants

While at the beginning `acoBrain` obtains even better values, as you can see in figure 22, the search stagnates very fast on a medium fitness level. In the dynamic environment this circumstance is balanced by the fact that all algorithms have substantially fewer evaluations per configuration - 20000 in static scenarios to 1000 in dynamic ones - to find an as good as possible solution. The GA serving as rough guideline supplies a very balanced picture over all scenarios but can never reach the solution qualities of the two new ant algorithms.

Hence an important criterion for an optimization in dynamic scenarios seems to be the ability to transfer a certain amount of knowledge about the preceding search process into the changed environment. At the same time the method must be however able to adapt to the new conditions. `acoLU` profits from the ability to reallocate its pheromone values relatively fast and to cover a broader search space by local updating.

4. Future Trends

The future of the beam forming algorithms based on machine learning techniques is dominated by the growing demand for mobile communication systems. In order to illustrate this, two application examples are discussed this chapter.

One upcoming application is the reduction of power consumption for communication networks. The advantage of the reduced power consumption towards the - possibly moving - communication partner was already mentioned in the introduction. In this publication just the motion of several communication partners in relation to a fixed antenna array was discussed. The upcoming challenge is to adopt the beam forming process to communication networks where all objects of interest are equipped with adaptive antennas and extended beam forming algorithms. A real world scenario for this is given by modern vehicle to vehicle communication in the automotive industry as illustrated in left part of figure 23.

Figure 23. a.) Beam forming in case of communication networks and b.) Satellite communication through a very small beam

The necessary beam forming algorithm must be

1. real time adaptive,
2. realizable in VLSI Technology (For this requirement the basics are given by the described ant colony algorithms. But additional an extended ant colony beam forming algorithm has to be considered.) and
3. adaptive to the behaviour of other independent beam forming algorithms of the communication partners.

These new requirements can be reformulated into a research task as: How can hardware ACO interact in one real-time environment. A comparable approach for a similar problem is already known for particle swarm optimization algorithms in the field of multi-swarm algorithms.

It appears clever to build a multi ant-colony optimization and to enable the independent colonies (each in one vehicle) to exchange some promising optimization results. The second application example deals with tap-proof satellite communication. It is well known that under application of a high number of antenna elements an antenna array can be configured so that just a very small beam is produced. This beam links directly to the satellite without having detectable side lobes which could be misused for unauthorized interception or localization. Right part of figure 23 illustrates this.

Such a communication antenna array for the KU-Band (10*GHz*) exists of about 500 to 1300 antenna elements. These antenna elements are connected to the same number of phase shifters and amplitude controllers. This means that the beam forming algorithms have to deal with an immensely increased number of variables than in the current demonstrator setup and the according simulator. Again the application of multi ant-colony optimization is a feasible strategy. Basic idea is to partition the antenna array into separated segments and to apply an own colony optimization algorithm to each segment.

Figure 24 shows the principle of the clustering.

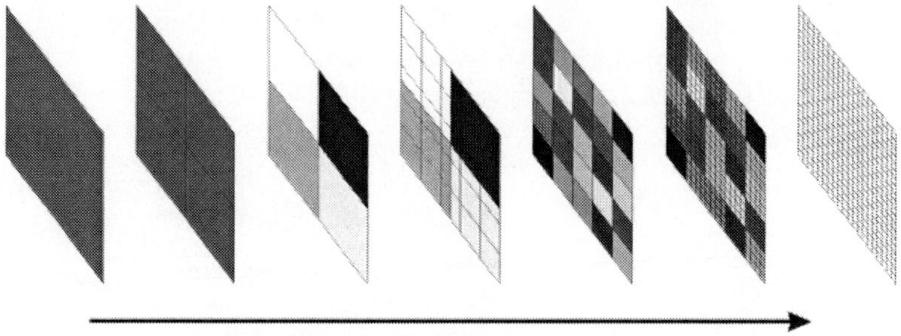

Figure 24. Principle of search with massively increased number of antennae

If a colony optimization algorithm is in saturation and the fitness value can not be improved then the neighbouring partitions are combined to a larger portion. In addition the high number of configurable variables by the optimization algorithm a second constraint is given by real word applications. Due to the fact that for such large antenna array CMOS based VLSI Technology is used and the power consumption is directly related to the toggle rate, it is required that only a limited number of variables can be changed at one time.

5. Conclusion

Our work presents two new ACO variants that are capable of optimizing configuration parameters for smart adaptive antenna systems in dynamic environments. The idea of *smart antennae* is to use base station antennae patterns that are not fixed, but adapt to the current radio conditions. This can be visualized when the antennae directs a beam only toward a communication partner. Smart antennae uses the power much more efficiently and increases the useful received power as well as they can reduce interference. In order to adapt the directional characteristics of an array antenna to variable conditions, a set of hardware parameters has to be adjusted during operation. These adjustments have substantial influence on the received quality of the signals of static and mobile stations within the range of the array antenna. Both algorithms show significant increases in obtainable results over existing methods. To measure the performance of the new algorithms in an objective manner extensive testing and analysing was done and results are presented. Furthermore we developed highly parallel hardware concepts that allow extremely increased computing performance. Thus it makes possible to use adaptive antennas not only in dynamic environments but also to optimize the antennae configuration in real time.

Both the higher efficiency of the new algorithms as well as the extremely increased computing performance due to our hardware concepts open many new application fields for these methods. To conclude several ideas are presented for such future fields of application.

5.1. Bibliography

[1] D. Angus and T. Hendtlass. Ant Colony Optimisation Applied to a Dynamically Changing Problem. In *Proceedings of the 15th International Conference on Industrial and Engineering, Applications of Artificial Intelligence and Expert Systems*, volume 2358, pages 618-627. Springer-Verlag, 2002.

[2] E. Bonabeau, M. Dorigo, and G. Theraulaz. *Swarm intelligence: from natural to artificial systems.* Oxford University Press, Inc., 1999.

[3] J. Branke. Evolutionary Approaches to Dynamic Optimization Problems - Introduction and recent Trends. In J. Branke, editor, *GECCO Workshop on Evolutionary Algorithms for Dynamic Optimization Problems*, pages 2-4, JUL 2003.

[4] G. D. Caro and M. Dorigo. AntNet: A Mobile Agents Approach to Adaptive Routing. Technical Report IRIDIA/97-12, IRIDIA, Université Libre de Bruxelles, 50, av. F. Roosevelt, CP 194/6, 1050 - Brussels, Belgium, 1997.

[5] C.-N. Chuah, D. N. C. Tse, J. M. Kahn, and R. A. Valenzuela. Capacity scaling in mimo wireless systems under correlated fading. *IEEE Transactions on Information Theory*, 48(3):637-650, März 2002.

[6] P. Delisle, M. Krajecki, M. Gravel, and C. Gagné. Parallel Implementation of an Ant Colony Optimization Metaheuristic with OpenMP. In *International conference of parallel architectures and complication techniques (PACT), Proceedings of the third European workshop on OpenMP (EWOMP 2001)*, pages 8-12, Barcelona, Spain, September 8-9 2001.

[7] M. Dorigo, V. Maniezzo, and A. Colorni. Positive feedback as a search strategy. Technical Report 91-016, Dipartimento di Elettronica e Informatica, Politecnico di Milano, Italien, 1991.

[8] M. Dorigo and T. Stützle. The Ant Colony Optimization Metaheuristic: Algorithms, Applications, and Advances. In F. Glover and G. Kochenberger, editors, *Handbook of Metaheuristics*, volume 57 of *International Series in Operations Research and Management Science*. Kluwer Academic Publishers, 2003.

[9] H. H. Frühauf, J. Heubeck, S. Lindebner, R. Wansch, and M. Schühler. Evaluation of direction of arrival location with a 2.45*ghz* smart antenna system. In *Proceedings of Signals, Sensors and Devices 2005*, Sousse, 2005.

[10] S. Goss, S. Aron, J. Deneubourg, and J. Pasteels. Self-organized shortcuts in the Argentine ant. *Naturwissenschaften*, 76:579-581, 1989.

[11] M. Guntsch, B. Scheuermann, H. Schmeck, M. Middendorf, O. Diessel, H. ElGindy, and K. So. Population based Ant Colony Optimization on FPGA. In *Proceedings of the IEEE International Conference on Field-Programmable Technology (FPT), Hong Kong, 2002*, pages 125-133, 2002.

[12] M. Haardt and J. Nossek. Unitary esprit: How to obtain increased estimation accuracy with a reduced computational burden. *IEEE Transaction on signal Processing*, 43(5), May 1995.

[13] D. Hyman. Rugged rf-mems packaging strategies. In *Microelectronics Reliability and Qualification Workshop*, Manhattan Beach Marriott, Manhattan Beach, CA, December 6-7, 2005.

[14] S. Inc. White paper on rf propagation basics. San Francisco, April 2004.

[15] R. W. H. Jr. and D. J. Love. Dual-mode antenna selection for spatial multiplexing systems with linear receivers. Technical report, Dept. of Electrical and Computer Engineering The University of Texas at Austin Austin, TX 78712, 2004.

[16] G. Kókai, H. H. Frühauf, and X. Feng. Development of a hardware based genetic optimizer to adjust smart antenna receiver. *International Journal of Embedded Systems Special Issue on Hardware-Software Codesign for Systems-on-Chip*, pages -, 2005.

[17] J. C. Liberti and T. S. Rappaport. *Smart Antennas for Wireless Communications.* Prentice Hall PTR, 1999.

[18] J. Litva and T. K.-Y. Lo. *Digital Beam forming in Wireless Applicatons.* Artech House Norwood, 1996.

[19] K. Liu, D. O'Leary, G. Stewart, and Y. Wu. An esprit algorithm for tracking time-varying signals. Technical report, University of Maryland, Havard, 1992.

[20] R. Michel and M. Middendorf. An island model based ant system with lookahead for the shortest supersequence problem. In *Proceedings of the 5th International Conference on Parallel Problem Solving from Nature*, pages 692-701. Springer-Verlag, 1998.

[21] F. Miller. Antennen für den Mobilfunk der Zukunft. *Fraunhofer Magazin*, 1:52-53, 2002.

[22] M. Randall and A. Lewis. A parallel implementation of ant colony optimization. *J. Parallel Distrib. Comput.*, 62(9):1421-1432, 2002.

[23] B. Scheuermann, K. So, M. Guntsch, M. Middendorf, O. Diessel, H. ElGindy, and H. Schmeck. FPGA Implementation of Population-based Ant Colony Optimization. *Applied Soft Computing*, 4:303-322, APR 2004.

[24] K. K. Shetty. A novel algorithm for uplink interference suppression using smart antennas in mobile communications. Master's thesis, Electrical and Computer Engineering, Department of Florida State University, 2004.

[25] T. Stützle. Parallelization Strategies for Ant Colony Optimization. *Lecture Notes in Computer Science*, 1498:722-731, 1998.

[26] F. Vilbig. *Lehrbuch der Hochfrequenztechnik*. Akademische Verlagsgeselschaft Geest und Portig K.-G., Leipzig, 5., 1953.

[27] A. E. Zooghby, C. Christodoulou, and M. Georgiopoulos. Neural network-based adaptive beam forming for one and two dimensional antenna arrays. *IEEE Transactions on Antennas and Propagation*, 46(12):1891-1893, Dec. 1998.

Integrated Intelligent Systems for Engineering Design
X.F. Zha and R.J. Howlett (Eds.)
IOS Press, 2006

Embedding Intelligence into EDA Tools

Ankur Agarwal, Ravi Shankar, A. S. Pandya
Florida Atlantic University, USA

Abstract: Multiprocessor system on chip (MpSoC) platform has set a new innovative trend for the system-on-chip (SoC) design. Demanding Quality of Service (QOS) and performance metrics are leading to the adoption of a new design methodology for MpSoC. These will have to be built around highly scalable and reusable architectures that yield high speed at low cost and high energy efficiency for a variety of demanding applications. Designing such a system, in the presence of such aggressive QOS and Performance requirements, is an NP-complete problem. In this paper, we present the application of genetic algorithms to system level design flow to provide best effort solutions for two specific tasks, viz.., performance tradeoff and task partitioning.

Keywords: Electronic Design Automation (EDA), Intelligent Algorithm, Data Mining, Genetic Algorithms, Non-Recurring Expense (NRE), System Level Design, Network On Chip

Introduction

In today's frenzied electronics field, design engineers face major challenges in keeping up with increasing silicon complexity and shrinking time-to-market windows [1]. Creativity and market driven motivation are adding ever more demanding applications. Other complications include new technology development, low end-product cost, high quality of service (QoS), training, software testing and optimization, and ambiguous and changing user preferences [2] [3]. The ability to reduce cost per function by an average 25-30% each year represents one of the unique features of the semiconductor industry and is a direct consequence of the market pressure to deliver twice the functionality on chip every two to three years at the same or lower price [4]. Pressure has been further accentuated by the customer demands for high QoS in the shipped products and the end-users' need for prompt responses to quality issues during the product usage. This has led systems companies to seek ways to get the most out of their design resources. Electronic Design Automation (EDA) tools are key to this effort and are likely to play a very important role in optimizing the entire product development life cycle [5] [6].

The evolution of integrated circuit (IC) implementation tools has brought a major revolution in the way IC design is performed. High quality designs, developed from powerful EDA tools and interface software, ensure efficient use of resources and lead to high quality designs. EDA software product development teams constantly strive to increase the intelligence and functionality of the EDA tools. One effective way of achieving this is by incorporating more and more sophisticated algorithms in the tools to improve the system architecture and meet the rapidly changing requirements [7] [8]. This should also increase the level of automation and reduce cost - customers in today's competitive market environment are resistant to even "moderate" increase in cost and the Moore's law of doubling functionality per chip is taking longer than the earlier norm of 2 years or less [9]. The semiconductor manufacturers and systems companies must seek a new model to deliver the same cost-per function reduction that has fueled the industry growth.

Figure 1 compares pre-production development cost for two different design methodologies and consequent levels of support for design, verification, and test [4]. This cost of the chip, also referred to as manufacturing non-recurring engineering (NRE) cost, would have been $1.5 billion in year 2005 with Register Transfer Level (RTL) design methodology alone; but further improvements in the engineering design flow with the addition of new and innovative EDA tools have kept the NRE cost of the chip to only $20.5 million. This translates to a productivity improvement of about 70 fold during the past 12 years (1993 to 2005). The productivity improvement due to EDA and fabrication tools, during 1975 to 2005, has probably been 1000 fold. Figure 1 lists the factors that have brought about these productivity improvements. In house P&R in general applies to the various levels of design automation brought by the extensive use of the EDA tools especially in the areas of synthesis, placement and routing [4] [10]. Small block reuse implies the use of the pre-designed and characterized component libraries [12] [13]. Large block reuse, incorporated around 1999, came from the formation of new companies that specialized in the design of large intellectual property (IP) blocks, such as CPUs and I/O interfaces [14]. These blocks were designed to particular specifications; design engineers were then able to simply drop these blocks into their design and compile the complete design. This is very similar to the current use of the component based design methodology in software development.

Typically, about 50% of the product design cycle time was spent on testing the product [12] [14]. Thus, more automation in the verification and testing phases of circuit design became inevitable. This reduced the product design time and effort. Several test methodologies evolved during this and earlier phases. This included assertions, RTL test benches, verification languages (such as Vera), fault coverage, and automatic test generation. It also saw the introduction of formal verification methods with languages such as Sugar [15] [16]. However, these more abstract and early-stage verification tools are yet to be adopted widely.

As per the shifts in technology trends, ITRS (International Technology Roadmap for Semiconductor) predicts that "The dimensional scaling of CMOS (Complementary Metal Oxide Semiconductor) device and process technology, as it is known today, will become much more difficult as the industry approaches 16 nm (6 nm physical channel length) around the year 2019 and will eventually approach an asymptotic end. Beyond this period of traditional CMOS scaling, it may be possible to continue functional scaling by integrating alternative electronic devices on to a silicon platform. These alternative electronic devices include, one dimensional structures (such as CNTs (Carbon Nanotubes) and compound semiconductor nanowires), RTDs (resonant tunneling diodes), SETs, molecular and spin devices. Most likely, these options will unfold their full potential only in combination with new and appropriate nanoarchitectures." [17].

Some EDA vendor product development teams believe that "in order to succeed in the technologically-evolving EDA market, they need to respond to the dynamic technology and time-to-market needs of their customers by releasing new technology and successive revisions very quickly" [18]. Other EDA vendor product development teams believe that "the key element in moving the EDA industry ahead is to continuously expand on the feature sets" [5]. This will ensure that they satisfy the needs of all EDA business segments (semiconductors, wireless communication, memory, etc.) and the whole range of customers (experts, main stream users, and novices). Still others believe that the key element in gaining leadership and market share lies in providing good quality of service, maximizing the level of support and becoming a strategic partner with their customer company (the systems company). Thus, market pressures and the chosen path to meet those market pressures would influence the short-term response of the EDA industry. However, in the longer run, the EDA industry as a whole has met and exceeded the needs of the practicing engineer. However, as shown below, the new challenges encompass domains such as specification, architectural trade-offs, and software-hardware co-design, and integration there of, which culturally are alien fields for typical EDA vendors. While some leading EDA vendors seem to have made this transition, others are lagging behind.

Figure 1 indicates that by 2010 it would cost $100 million to design a state-of-the-art chip, even with all the current and expected EDA tools. Assuming that half of this is the engineering cost, the engineering cost for the chip would be $50 million in year 2010. With the total estimated cost of an engineer (salary + fringe benefits) around $200K per year, it would require 250 engineer-years to complete a chip design from specification to the final fabricated end product. Thus, if we devote a team of 25 engineers to work on a chip design, it would take them 250/25, or 10 years, to complete the design process, that is, by the year 2020. Groups of 7 to 10 members tend to be the most productive, on a per-person basis, due to communication and organizational inefficiencies. Thus, merely increasing the number of engineers to 100 will not decrease the chip development cycle to 2.5 years, the current upper limit.

Today, companies have to bet on consumer preferences two years hence and start developing a product to meet that demand which may not come to pass. A wrong bet may adversely the fortunes of a company. Therefore, the product design cycle needs to be further shortened, not just maintained at 24 to 30 months. Hence, more innovation is needed, especially at the earlier phases of the product development since errors introduced there get magnified disproportionately at later stages. As an example, NASA estimates that an error at the specification stage will require 138 times the effort to fix it at the prototyping stage and 536 times the effort to fix it in the field, as it would at the specification stage [19]. It is expected that future systems will have an increasing role for design automation, reusability, and componentization, thus increasing the market share for the EDA industry.

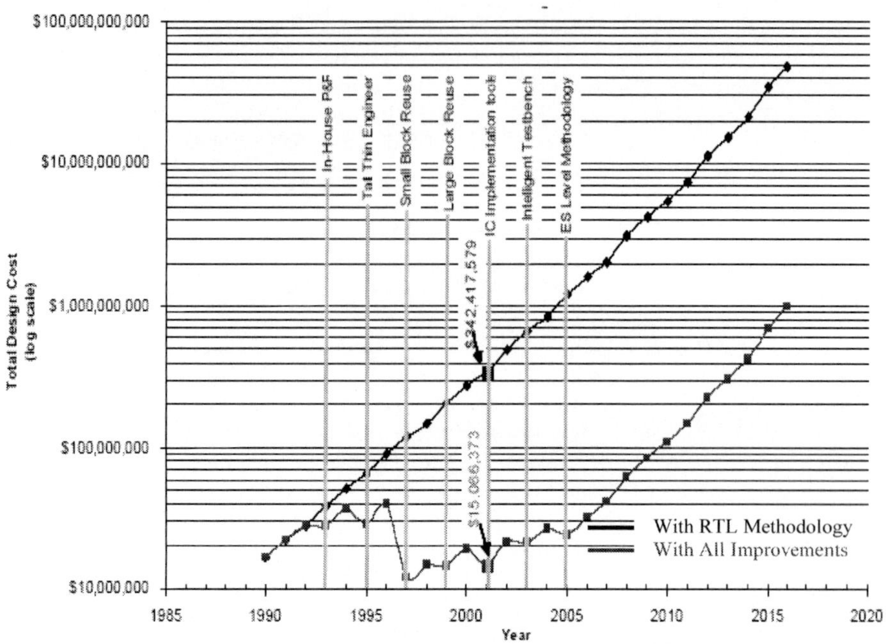

Figure 1. Impact of Design Technology on System Implementation Cost [4]

With the continuing exponential growth of circuit complexities, early planning and understanding of the different trade-offs earlier in the design cycle becomes very critical, especially given the huge cost of correcting an early stage mistake later on. EDA companies attempt varying strategies for balancing the goal of satisfying their immediate customer (the engineering design community at a given company), while developing and introducing their next generation products which their customer company will inevitably need to enhance their design productivity in the longer run. Examples of this balancing act may be seen in the formal verification and system level design tools, such as Sugar and Virtual Component Co-design (VCC), respectively. Both were introduced at the appropriate time, but without enough customer and library support. Both would have fared better via closer dialog between the EDA and systems companies.

The demand for the next decade will be in developing sophisticated higher level system level tools, with abstraction (top-down decomposition) and annotation (bottom-up approximation of realities). This paper discusses various ways by which companies are trying to get the most out of their design resources, where EDA tools are able to play a very vital role in optimizing the entire product development life cycle. We discuss some intelligent algorithms and their applications in traditional IC design flow in the background section. Specifically, the concept of the genetic algorithm has been expanded upon. We then use the concepts of intelligent algorithms and apply them to the present-day technology trends. Finally, we discuss the future technology trends for system design and the use of intelligent algorithms to address these future trends.

1. Economics of EDA Tools

It may come as a surprise that EDA software is very expensive. A single annual license for a commercial grade tool, for any one of the chip design stages, may cost $100,000 to $1M. Thus the support of 25 seats of fully integrated EDA tools, from various EDA vendors, both large and small, may exceed $10 to $20 Million. The reason for such expensive tools is self evident: They enhance the engineering design productivity of a company by 10 or more fold, and are technical marvels. It is one of the few high-tech groups of stocks that show consistently increased valuation. Each software tool incorporates and integrates so many engineering advances and algorithms that they are in a class by themselves.

To quote some numbers, the EDA Consortium's Market Statistics Service (MSS) announced is April 2005 that the electronic design automation (EDA) industry revenue for Q4 (forth quarter) of 2004 was $1,078 million, a 3% increase over Q4 2003. For the full-year in 2004, revenue totaled a record $4,019 million, which was 3% more than the $3911 million reported in 2003. As per the statement by Walden C. Rhines, Chairman of the EDA Consortium and Chairman and CEO of Mentor Graphics Corporation, "In 2004, The EDA industry crossed the $4 billion mark,". The revenue figures in this section are taken from [20] [21] [22]. This compares well with the revenue figures for Motorola, a hot company in a hot market: In 2005, it had an income of $4.58 billion from revenue of $ 36.84 billion [23].

1.1 Revenue by Product Category

EDA's largest tool category, Computer-Aided-Engineering (CAE), that mainly includes the domain of back-end tools for placement and routing, generated revenue of $523 million in Q4 2004, 9% more than the same period in 2003. CAE revenue for all of 2004 totaled $1,919 million, a 5% increase over 2003 [20].

IC Physical Design & Verification decreased 6% to $326 million in Q4 2004 over the same quarter in 2003. For the full year 2004, IC Physical Design & Verification revenue totaled $1,165 million, a 4% decrease over 2003 [20].

Revenue for *Printed Circuit Board* (PCB) and *Multi-Chip Module (MCM) Layout* totaled $91 million in Q4 2004, 9% greater than in Q4 2003. PCB and MCM Layout revenue totaled $341 million for all of 2004, 3% above 2003[20].

The EDA industry's *Semiconductor Intellectual Property* (SIP) revenue totaled $71 million in Q4 2004, 2% less than Q4 2003. For 2004, SIP revenue increased to $314 million (vs. $281 million in 2003) due in part to new company participation [20] [21].

The new domain of Electronic System Level (ESL), long predicted to be the main engine of growth this decade (2005 on), seems slow in taking off. There are several reasons for that: ambiguity (mapping requirements to specifications); concurrency (software-hardware co-design and co-verification); poor communication (for example, between software and hardware designers); and automation of the design flow to much higher levels, relative to the current status. ESL also needs longer path for back annotation (to account for practical limitations of ICs). Further, technology constraints have launched newer architectures, such as MpSoC and NoC, so Moore's law can hold good much longer.

EDA Services revenue was $67 million in Q4 2004, up 5% from Q4 2003. Services revenue totaled $280 million in 2004, a 12% increase over 2003[22].

The fluctuations in the revenue figures for different categories do not come as a surprise if we relate these figures to technology trends. It can be seen that the market share for EDA tools involving higher levels of abstractions (up to a point) and responsible for integration of components, are increasing their revenue; Share of the tools involving lower level of abstraction is declining. However, as pointed out earlier, ESL tools, absolutely needed to gain further multiple fold increase in productivity, need to get better, in terms of user friendliness, component libraries, tool integration, automation, and synthesis.

2. Background

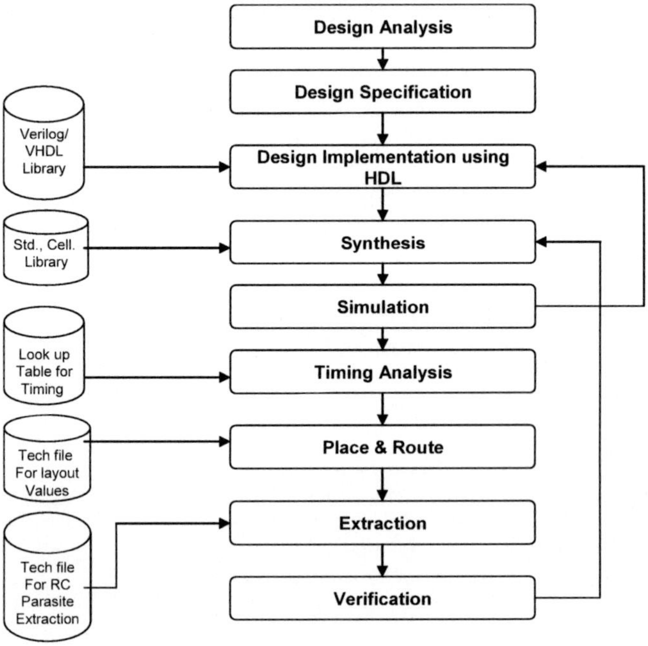

Figure 2. Traditional Digital ASIC Design Flow

Traditionally, EDA tools have been divided into more or less independent realms, namely, synthesis level, logic-level and layout-level. This traditional design flow for digital application specific integrated circuit (ASIC) is as demonstrated in Figure 2. Each of these levels have their own set of synthesis, verification, and analysis toolsets, which were more or less unaware of the underlying overall perspective linking them to each other and to the final design. In the early 1990s, typical design process would consist of design specification, followed by manual or automated synthesis/analysis/verification, and so on, until the layout of the design was completed. If the design did not meet its timing/power/area/noise constraints, there would be a re-iteration and the whole process or part of it would be redone. This resulted in a major waste of design effort and increased time-to-market, especially for analog and mixed mode designs. This was less of an issue for digital chip designs. However, since the early 2000s, as the feature size has shrunk below the 100 nm mark the analog nature of the transistor can no longer be ignored by the digital designers. Going forward, this will become even more of a challenge.

EDA industry has accommodated many intelligent algorithms at different levels in the design flow (see Figure 2) to achieve impressive productivity gains [24]. Genetic Algorithm (GA) is one such example that has been deployed for various optimization problem encountered in VLSI (very large scale integration), or state-of-the-art IC chip, design. This includes partitioning [25], automatic placement and routing [26], technology mapping for field programmable gate arrays (FPGA) [27], automatic test generation (ATG) [28], and power estimation [29]. In this section we will discuss some of the issues at each step in the traditional digital ASIC design flow and show an application of GA to get a best effort solution.

2.1 Genetic Algorithm (GA)

Genetic algorithms are motivated by the natural process of evolution (competition or survival of the fittest) and inheritance (a child's inheritance of the parent's genetic makeup). Algorithms were first introduced by researchers at the University of Michigan [31]. GA has been applied to optimization of various NP-complete problems. Many of the EDA algorithms are NP-complete. GA includes the evolution of population over a number of generations, with a specific individual in the population assigned a specific fitness value. New generation is then produced by reproduction, selecting individuals from the population and then crossing them for producing the next generation. The new individuals are mutated with some low mutation probability. At this point GAs offers two solutions to different problems depending upon the nature of the application. One is complete replacement of the old population with the new population generated and other is a mixture of the old population with the new one. One can try to keep the size of the population constant by keeping the best fit individual while discarding others. GAs may use one of the two basic processes for evolution: inheritance and competition. The latter is also known as the survival of the fittest which results in removal of the unwanted features from the product.

2.1.1 Steady State Algorithm for GA

Steady state algorithms are for applications where incremental change in an existing design rather than a complete re-design, is the focus. In the process of generation, two individuals with adequate fitness levels are crossed with a high probability Pc to form the offspring. The resultant product of the crossing can be mutated with probability Pm and inverted with the probability Pi. The new generated population can be introduced if it is better than the older one; otherwise it is rejected.

This algorithm is represented in Figure 3 [31]. At the evaluation stage in the algorithm, fitness level of each individual object is determined. The individuals are selected based on the levels of fitness, which can be determined by various schemes, such as stochastic universal selection, tournament selection, and roulette-wheel selection. The selection process determines the rate of the convergence of the solution. 'Roulette-wheel' refers to a somewhat proportionate process similar to the process of natural selection; 'Stochastic Selection' is less noisy and is able to perform better in a noisy environment where the behavior is expected to deviate much from their assigned/inherited properties.

Crossover is the main operator of the reproduction process in which the individual is able to reproduce with probability P_C. Crossover operation is used for generation of the new population. One-point and two-point crossovers are employed for this purpose. This works in the same way as the biological crossover that takes place among human beings. Another form of crossover, the uniform crossover, is performed with a given probability. The amount of the crossover can be controlled with the crossover probability. This results in a pair of solutions. From the solutions, chose the ones which better meet the specific properties and cross them again to eliminate the unwanted properties from the offspring. Such an incremental solution is performed until an expected result is obtained.

Mutation refers to an incremental change brought about in the offspring. This enables one to bring new changes in the offspring, with a probability P_m. This can be done by flipping a bit at any specific position to bring randomness in the string. If the resulting string performs better than the previous one, we keep the string, otherwise we discard the changes. Such a step is found to be highly applicable to performance evaluation of circuits where the changes in the performance matrix can be weighted against each other to provide an optimum solution.

Fitness scaling is used to realize a substantial difference, viz., between the best case and the worst case values. Therefore, the main task of fitness scaling is to set the selection pressure. There are various ways of using this scaling factor. Either raw scaling can be used or the scaled and normalized values can be used for choosing the fitness levels of the chromosomes. Use of the mean and the standard deviation are typical for choosing or discarding the offspring. The chromosomes are chosen such that there is a wide range of options available for optimization. Inversion refers to no change being brought into the solution represented by the chromosomes, but replacing the chromosomes themselves. The generational gap would refer to the fraction of the population replaced from the previous population. The explained GA algorithm has found its application in many domains which have NP-complete complexity. For such a problem the best solution is not guaranteed. Thus often the best effort solution is used. GA as explained above can be applied to any such NP-complete problem for identifying the best effort solution.

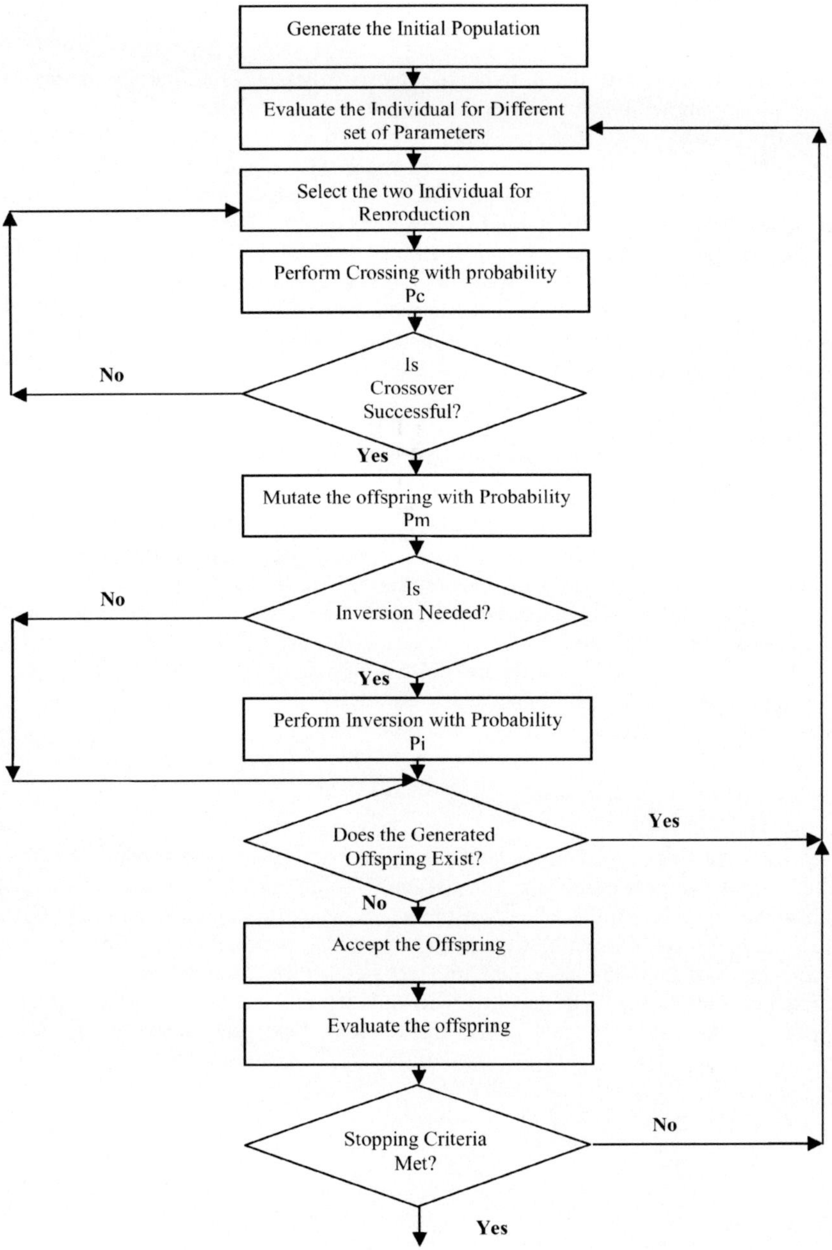

Figure 3. Flowchart for the Steady State Algorithm

2.2 Partitioning

In the general form, the partitioning problem consists of dividing the nodes of the graph into two or more disjoint subsets such that the sum of the weights of the edges connecting the nodes in different subsets is minimized; at the same time, the sum of the nodes in each subset does not exceed a given capacity. For such a partitioning problem different forms of cut sets have been proposed which include the min-cut, max-cut and ratio-cut algorithms that can be applied to different graphs. Partitioning a circuit with n number of components into k number of partitions with each block containing p number of partitions would have $[(n!) / (k! * (p!)^k)]$ unique ways of doing this [32]. This explicitly defines the problem as having an exponential time complexity. As the number of components, i.e. n, increases, the computing complexity grows exponentially. Such a problem is NP-complete.

2.3 Placement

The objective of the placement algorithm is to minimize total chip area and wire length in order to increase functionality in a chip. This reduces capacitive delays in the circuit (which achieves higher operational frequency), cost of the silicon used, and the power consumed. To determine the best plausible solution it is required to evaluate every possible placement. This process would take a computation time proportional to the factorial of the number of cells. Since the chip contains millions of gates, such an approach would be highly inefficient in terms of the time and computing resources needed. Thus, in order to search through a large number of configurable solutions, a best-effort algorithmic approach is needed. One such genetic algorithm is shown in Figure 4 [33].

2.4 Automatic Test Generation (ATG)

Testing and verification often account for about 50% of the total product design time. Sequential circuit test generation using a deterministic algorithm is time consuming since the number of test vectors to be generated to test the design completely will be large. Due to the complexity of the problem this area has been widely researched and innovative ideas such as the stuck-at-fault model, automatic test pattern generation, and assertion based verification languages such as Sugar have been conceptualized. A GA algorithm has also been applied to the automatic test generation domain. One such algorithm is shown in Figure 5.

1. Input the gate-level netlist from the cell libraries with abstracted values of the delay, power, size and speed parameters;
2. Read the GA parameter values such as CrossoverRate, PopulationSize, InversionRate, MutationRate;
3. NumberOfOffspring = CrossoverRate * InitialPopulationSize;
4. NumberOfgenerations = NumberOfGenerations/CossoverRate;
5. Generate initial Population randomly;
    a. For j=1 To PopulationSize Do
        i. Evaluate (Population j);
6. For i=1 To NumberOfGenerations Do
    a. For j=1 To PopulationSize Do
        i. invert (j, InversionRate);
7. NewPopulation = NIL
8. For j=1 To NumberOffSpring Do
    a. Select two parent from population;
    b. Align slot ID numbers of parents 1 & 2;
    c. mutate (offspring, MutationRate);
    d. Add offspring to NewPopulation;
    e. evaluate (NewPopulation, j);
9. Population = reduce (Population, NewPopulation);
10. Solution = individual with the highest fitness in final Population;

Figure 4. GA Algorithm for Placement

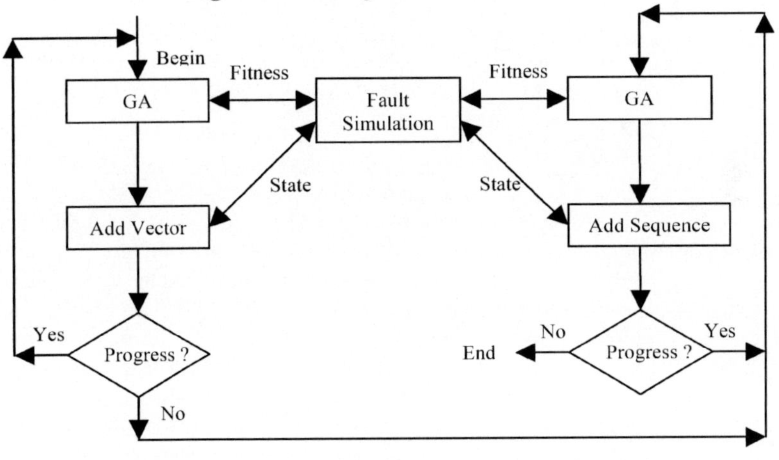

Figure 5. Flowchart for Application of GA Algorithm in ATG

3. Incorporating Intelligence into System Level Design and Issues

Design productivity at the system level may be enhanced by allowing design tradeoffs to be made at higher levels of abstraction. Figure 6 plots the number of design options against various levels of abstraction. It is seen that there are larger number of possible implementations of the system at a higher level of abstraction. As we move down the abstraction levels, down to the transistor level design, the number of the design alternatives becomes increasingly limited. It can also be seen that there are more design decisions to be taken at a higher level of abstraction. So far GAs have been used only for lower level issues such as routing, synthesis, and technology mapping among others; however there is a larger number of solution vectors at higher levels of abstraction; thus GAs may also be usable at the system level design to provide best effort solution.

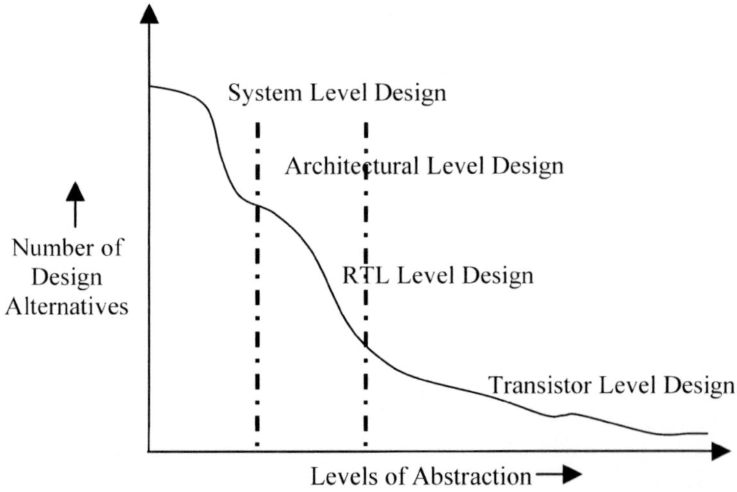

Figure 6. Number of Design Alternative at Various Levels of Abstraction

As an example, Figure 7 shows a simple block level design of a digital camera [34]. Each of these components can be implemented in either software (an application running on a general purpose processor) or as a hardware element (a dedicated hardware element, such as an ASIC (application specific integrated circuit), or a FPGA. (field programmable gate array) Hardware implementation enhances performance and throughput and reduces latency, but might cost more and increase power consumption. These consumer products are always market driven thus tradeoffs will have to be considered. For such system profiling there many be so many alternative implementations that the problem can be presumed to be NP-complete for a system with a large number of components. Thus GA can be used to recommend a best effort solution for such a case.

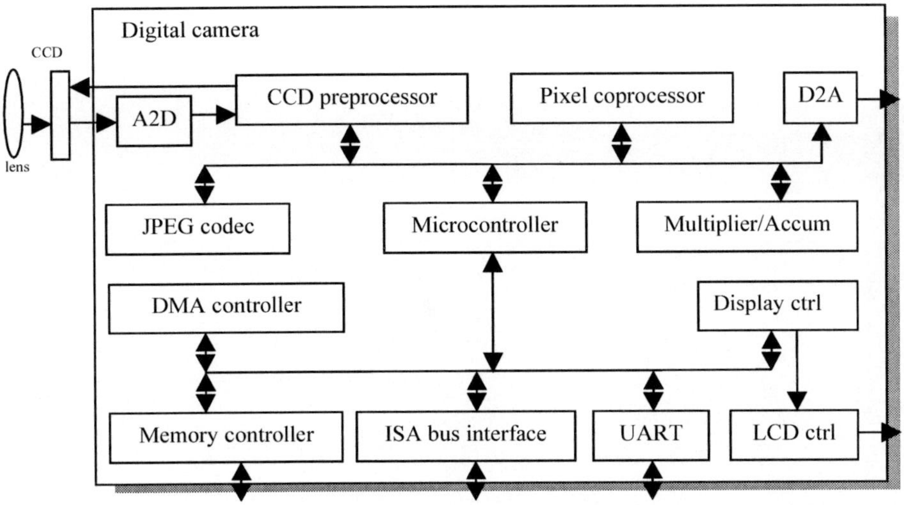

Figure 7. Component based block Diagram for Digital Camera

3.1 Application of GA into System Level Design

Figure 8 represents typical system level design flow [35]. System level modeling (the stage after the specification stage), is used for evaluating performance tradeoffs for various combinations of software and hardware components. The performance metrics considered could include parameters such as performance (cost, latency and throughput), power dissipation, noise (or interference), security, real-time constraints, and QoS (quality of service) metrics. These metrics are derived from the requirement and specification documents. The best effort solution(s) may recommend one or more possible combinations that meet these matrices. The next stage of software-hardware co-design involves replacing the abstract view of the software and hardware components with appropriate software (e.g. C++) and hardware (e.g. SystemC and Verilog) code, along with process scheduling (in general purpose processors), to perform software-hardware co-design.

The hardware and software design may then be co-simulated. If the result of the co-simulation meets with the system functional specifications the design is sent for final testing and prototyping. If the design does not meet the requirements, a new architecture is selected with a different portfolio of hardware and software components and the cycle is repeated again. This is continued until we meet the metrics and the functional specifications.

Under such a scenario of multiple iterations for finding a feasible solution, taking appropriate decisions based on expert's knowledge and other sources of reliable information becomes extremely important. A decision process, which consists of a simple "Yes" or "No", becomes extremely critical for the system level design so as to decide the right step to iterate from. If a wrong decision is taken at this point, it would result in a delay

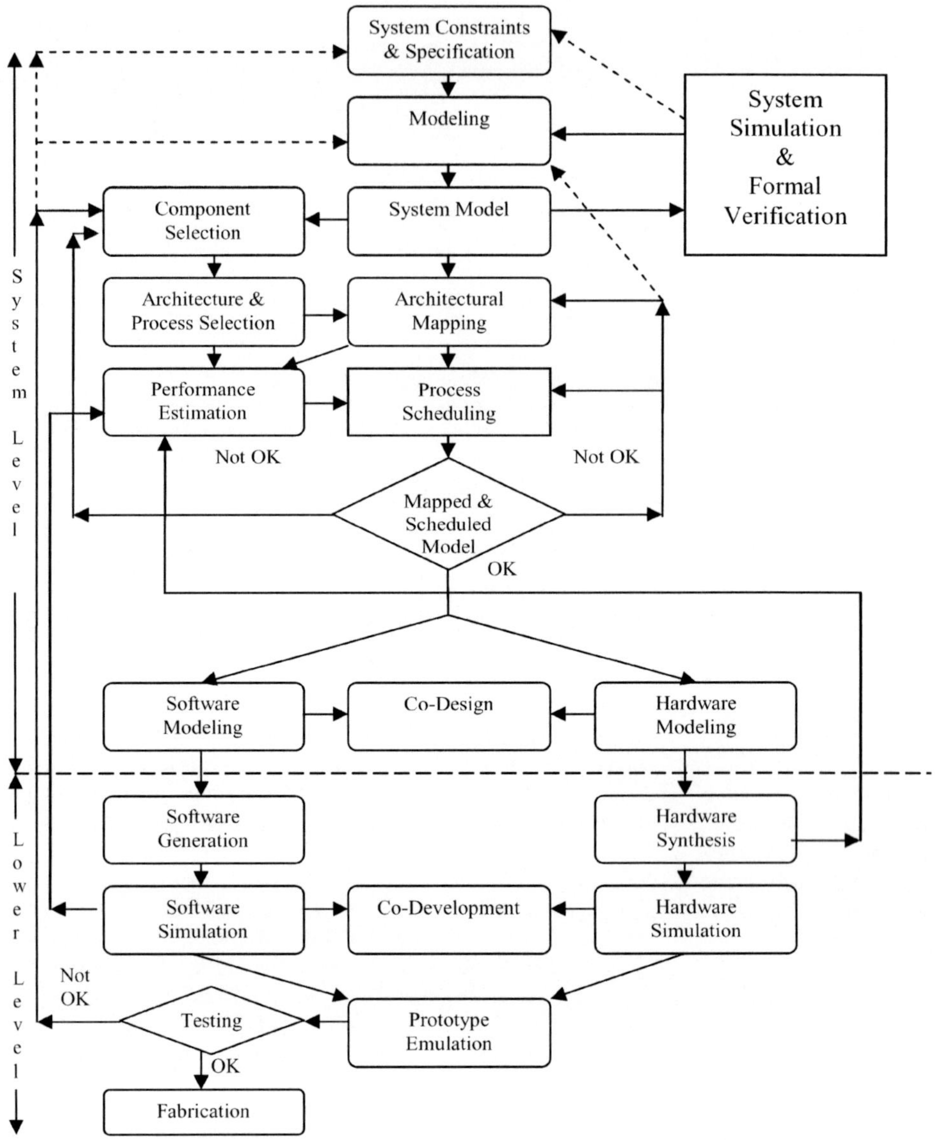

Figure 8. System Level Hardware/Software design Flow

in the design cycle which ultimately may hurt the overall revenue generated from the product. Thus intelligent decision theory could be utilized during the modeling, simulation, and design phases in order to guide the process in the right decision. Experts system and data mining can lead to significant enhancements.

We propose an algorithm for choosing a vector of design parameters to yield an optimum solution. See Figure 9. This algorithm is based on the greedy approach and includes all plausible solutions in the solution vector space. Then, by a randomized adaptive search, it narrows down its solution vector space. It stores the solution vector in the form of a list. The adaptive nature of the algorithm derives from the fact that the new parameter value selected is based on the value of the previous iteration.

```
1. Body ( )
      a. Get all the components of the design
      b. Get the QoS parameters for the components with different
         implementation
      c. do
              i. Parameter Selection Phase (greedy, random, adaptive)
             ii. Required Solution (Initial Solution)
            iii. Get the Best Effort solution (Previous Solution)
2. Parameter Selection Phase ()
      a. While Parameter Selection Incomplete
              i. Greedy: Create a list of all possible solution
             ii. Random: Select a particular matrix from the list prepared
                 above
            iii. Adaptive: Include the other parameter to the randomly
                 selected matrix
             iv. Feasibility: Check is the values are feasible
      b. End While
3. Required Solution ()
      a. Get the initially selected value
      b. While optimum solution is not found
              i. get the new solution from modifying the older one
      c. End while
4. Return the best solution()
```

Figure 9. Algorithm for Greedy Approach for System Level Hardware/Software profiling

Once the hardware-software profiling is completed, we perform the process of task allocation or scheduling. We decompose the task based on the cost of inter-process communication. This is a mapping problem. We try to map the tightly coupled components next to each other (if hardware), or to the same processor (if software). This will reduce routing traffic, thereby reducing the latency in communication on the system. Once the task decomposition and evaluation of inter-process communication is achieved we can apply the

GA for mapping of the tasks on to the cores. We propose one such algorithm for task mapping as shown in Figure 10. After task scheduling, the task synthesis can be performed using GA. These algorithms are well established. Thus, we will be able to optimize the system level design flow by incorporating genetic algorithms at two different levels: hardware/software profiling and task mapping/scheduling.

Initialize( )

1. Generate the initial population with the information of task dependency and inter-process communication stored in the variable t_d, and ipc. These variables are the genetic mutators.

Body( )

2. Calculate the fitness value for each individual based on t_d & ipc
3. Crossover would consist of controlling the fitness level in the population while discarding the unfit tasks
   a. Select the individual with the most fitness value.
   b. Assign this individual (most fit) to a central region.
   c. From the remaining population find another task using the selection operator
   d. Allocate the task next to the previous task.
   e. Iterate to find the complete placement.
   f. Store the placement result in a variable
4. Mutate the tasks:
   a. Find out the task with the similar fitness level.
   b. Exchange their placement
   c. Iterate until done
   d. Store the result in a variable
5. Compare all the result
6. Keep the best placement

Figure 10. GA for Task Scheduling for System Level Design Flow

4. Future Trends & Challenges

Future system applications will require significantly increased computing power along with scalability and reusability to reduce product development time. One main reason that has contributed to this fact is the exponential decrease in the transistor size, enabling faster transistor switching times and more densely integrated circuits. This trend is shown in Figure 11 [36]. The gate length of the transistor in year 2010 would be around 9 nm. This will result in substantial increase in the integration density of transistors onto a chip. Such increase in computation power will cause communication bottleneck which is elegantly resolved in the NOC architecture, with packet-based communication channels, similar to the Internet [3] [4]. NOC has a multi-core globally asynchronous and locally synchronous system level architecture [37] [38] [39] [40].

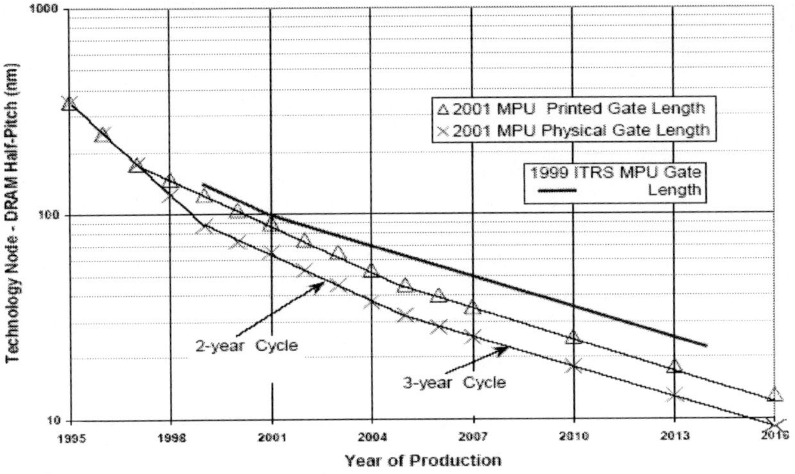

Figure 11. Gate Length Trends

Future innovations would be required to maintain and/or enhance design productivity under such constraints. The NOC architecture is represented in Figure 12. In Figure 12 the larger "R" block represents a resource, which in turn may consist of memory, processor, cache memory resource interface network, etc., connected by a local bus-based connection. The smaller "S" block in the figure represents a NOC router and switch. In NOC different local regions (synchronous regions) would communicate with other synchronous regions by switches and routers. As a whole, the system would be asynchronous in nature.

Design of such a system would need a systematic approach. Such a design solution is tractable based on the best effort approaches, such as GA. A key insight in this context is that several of the optimization problems in this design flow are NP-complete problems. This is due to the flexibility NOC offers in its design platform. A new system level design flow for NOC will have to be evolved. Genetic and Heuristic algorithms are expected to be increasingly applied to this new domain.

Conclusion

This paper presented application of genetic algorithms and intelligent decision theory for system level design. Incorporating such algorithms in EDA tools would increase design productivity by providing for optimum solutions at higher levels of abstractions. Such early phase optimization may reduce the number of design iterations, thus enhancing design productivity further.

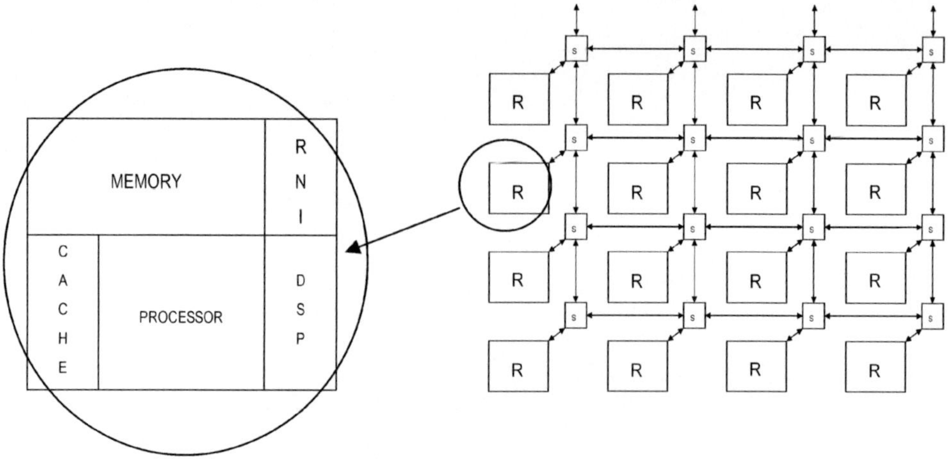

Figure 12. Network on Chip Architecture

References

[1] L. Benini and G. De Micheli. Networks on chip: a new SOC paradigm, *IEEE Computer*, Volume 35, No. 1, January, 2002, 70-78.

[2] A. M. Amory, É. Cota, M. Lubaszewski, and F. G. Moraes, Reducing test time with processor reuse in network-on-chip based systems, *Proceedings of the 17th ACM symposium on Integrated circuits and system design*, 2004, Pages: 111 - 116

[3] Y. Xiong and E. A. Lee, "An extensible type system for component-based design", *6th International Conference on Tools and Algorithms for the Construction and Analysis of Systems*, Berlin, Germany, April 2000.

[4] Semiconductor Industry Association, The international Technology Roadmap for Semiconductors. 2001. http://public.itrs.net/Files/2001ITRS/Home.htm

[5] A. Prabhu, Facets of Growth in EDA Market, Design and Test of Computer IEEE, Vol. 13, Issue 2, Summer 1996, 5-6

[6] T. Hiwatashi, EDA Roadmap in Japan, Proceeding of Design Automation Conference, ACP-DAC, January 1999, 5.

[7] G. B. Yaccob, E. P. Stone, and R. Goldman, Applying Moore's Technology Adoption Life Cycle Model to Quality of EDA Software, International Symposium of Quality Electronic Design, March 2001, 76-80

[8] W. Roethig, IEEE journal of Design and test of Computer, Vol 20 Issue 6, Nov 2003, 98-99

[9] Semiconductor Industry Association, The international Technology Roadmap for Semiconductors. 2003. http://public.itrs.net/Files/2003ITRS/Home2003.htm

[10] T. Das, C. Washburn, P. R. Mukund, S. Howard, K. Paradis, J-G. Jang, and J. Kolnik, Effects of technology and dimensional scaling on input loss prediction of RF MOSFETs, *International Conference on VLSI Design held jointly with 4th International Conference on Embedded Systems Design*, 2005, pp. 295-300

[11] E. A. Lee, and Y. Xiong, System level types for component-based design, *Workshop on Embedded Software*, California, October 2001

[12] M. Keating, and P. Bricuad, Reuse methodology manual for System on Chip Design, 2nd Edition, Kluwer Academic Publication, Norwell 1999

[13] C. Jones, System on Chip and need for Reusable Design Methodology, Electronic product design, IML group plc, Tonbridge, England February 1999

[14] K. Clarke, Multisource IP Integration challenges and design flow for SoC, Proceeding of Intellectual property System on Chip conference, Miller Freeman, Santa Clara, CA USA, March 1999

[15] D. Dill and J. Rushby, "Acceptance of Formal Methods: Lessons from Hardware Design," Computer, vol. 29, pp. 23-24, 1996

[16] E.M. Clarke and J.M. Wing, "Formal Methods: State of the Art and Future Directions," ACM Computing Surveys, Dec. 1996.

[17] Semiconductor Industry Association, The international Technology Roadmap for Semiconductors. 2003. http://www.itrs.net/Common/2005ITRS/ERD2005.pdf

[18] EDA Today, http://www.edat.com/NEA17.htm

[19] T. L. Bennett, P. W. Wennberg, "Eliminating Embedded Software Defects Prior to Integration Test", CrossTalk- Journal of Defense Software engineering, December 2005

[20] EDA Consortium, "EDA Industry Reports 3% Revenue Growth in 4th Quarter and for Full Year", http://www.edac.org

[21] EETimes, "Ever resilient, EDA is growing", http://www.eet.com

[22] Chip Design, "Dot.org—The Metrics of Success", http://www.chipdesignmag.com.

[23] S. Silver, "Motorola's Profit Rises 86%, Driven by Handset Sales", The Wall Street Journal, January20,2006

[24] K. A. De Jong, "Genetic Algorithms are Not Function Optimizers", Foundations of gentic Algorithms 2, San Mateo, CA: Morgan Kaumann, 1993.

[25] J. I. Hidalgo, J. Lanchares, R. Hermida, "Partitioning and Placement for multi-FPGA using Genetic Algorithms", 26th IEEE Proceeding of Euromicro Conference, Sept 2000, Vol. 1, 204-211.

[26] M. Handan, M. E. El-Hawary, "Multicast Routing with Delay and Delay Variation Constraints Using Genetic Algorithms", IEEE Canadian Conference, May 2004, Vol. 4, 2263-2366.

[27] R. Pandey, S. Chattopadhyay, "Low Power Technology Mapping for LUT based FPGA – A Genetic Algorithm Approach", IEEE Proceeding on 16th International Conference on VLSI Design, January 2003, 79-84

[28] S. Khor, P. Grogono, "Using a Genetic Algorithm and Formal Concept Analysis to Generate Branch Coverage Test Data Automatically", IEEE Proceeding on 19th International Conference on Automated Software Engineering, 2004, 346-349

[29] M. S. Hsiao, E. M. Rudnick, J. H. Patel, Peak Power Estimation of VLSI Circuit: New peak Power Measures", IEEE Transaction on VLSI Systems, Aug 2000, Vol 8, Issue 4, 435-439

[30] R. A. Bergamaschi, S. Bhattacharya, R. Wagner, C. Fellenz, M. Muhlada, F. White, W. R. Lee, and J.-M. Daveau. Automating the design of SOCs using cores. *IEEE Design and Test of Computers*, 18(5), September 2001 32-44

[31] P. Mazumder, E. M. Rudnick, "Introduction to GA Terminology", Genetic Algorithm for VLSI Design, layout, & Test Automation, Prentice Hall PTR, Upper Saddle River, NJ, USA, 1999.

[32] T. N. Bui, B-Ro, Moon, "GRCA: A Hybrid Genetic Algorithm for Circuit Ratio-Cut Partitioning", IEEE Transactions on Computer Aided Design of Integrated Circuits and Systems, March 1998, Vol. 17, Issue 3, 193-204

[33] K. Shahookar, H. Esbensen, P. Mazumder, "Standard Cell and Macro Cell Placement", Genetic Algorithm for VLSI Design, layout, & Test Automation, Prentice Hall PTR, Upper Saddle River, NJ, USA, 1999, Page 73

[34] F. Vahid, T. Givargis, Embedded System Design A Unified Hardware/Software Introduction, John Wiley & Sons Inc., New York, 2002

[35] Lecture Notes, Hardware Software Codesign, Dept of Computer and information Science, Linkoping University, http://www.ida.liu.se/~patel/codesign/

[36] Semiconductor Industry Association, The international Technology Roadmap for Semiconductors. 2001. http://public.itrs.net/Files/2003ITRS/Home.htm

[37] J. Xu, , W. Wolf, J. Hankel, S. Charkdhar, A Methodology for design, modeling and analysis for networks-on-Chip, *IEEE International Symposium on Circuits and Systems*, May 2005, 1778-1781

[38] S. Leibson, J. Kim, Configurable Processor: A new Era in Chip Design, IEEE Computer Society, Vol. 28 Issue 7, July 2005, 51-59

[39] L. Benini and G. De Micheli. Networks on chip: a new SOC paradigm, *IEEE Computer*, Volume 35, No. 1, January, 2002, 70-78.

[40] P. Pande, C. Grecu, M. Jones, A. Ivanov, R. Saleh, Performance evaluation and design tradeoffs for network on chip interconnect architecture, *IEEE Transaction on Computers*, vol. 54, Issue 8, August 2005, 1025-1040

Author Index